Plutonium and Highly Enriched Uranium 1996
World Inventories, Capabilities and Policies

Stockholm International Peace Research Institute

SIPRI is an independent international institute for research into problems of peace and conflict, especially those of arms control and disarmament. It was established in 1966 to commemorate Sweden's 150 years of unbroken peace.

The Institute is financed mainly by the Swedish Parliament. The staff and the Governing Board are international. The Institute also has an Advisory Committee as an international consultative body.

The Governing Board is not responsible for the views expressed in the publications of the Institute.

Governing Board

Professor Daniel Tarschys, Chairman (Sweden)
Dr Oscar Arias Sánchez (Costa Rica)
Dr Ryukichi Imai (Japan)
Professor Catherine Kelleher (United States)
Dr Marjatta Rautio (Finland)
Dr Lothar Rühl (Germany)
Dr Abdullah Toukan (Jordan)
The Director

Director

Dr Adam Daniel Rotfeld (Poland)

sipri

Stockholm International Peace Research Institute
Frösunda, S-171 53 Solna Sweden
Cable: SIPRI
Telephone: 46 8/655 97 00
Telefax: 46 8/655 97 33
Email: sipri@sipri.se
Internet URL: http://www.sipri.se

Plutonium and Highly Enriched Uranium 1996
World Inventories, Capabilities and Policies

David Albright, Frans Berkhout and William Walker

OXFORD UNIVERSITY PRESS
1997

Oxford University Press, Walton Street, Oxford OX2 6DP
Oxford New York
Athens Auckland Bangkok Bogotá Bombay
Buenos Aires Calcutta Cape Town Dar es Salaam
Delhi Florence Hong Kong Istanbul Karachi
Kuala Lumpur Madras Madrid Melbourne
Mexico City Nairobi Paris Singapore
Taipei Tokyo Toronto
and associated companies in
Berlin Ibadan

Oxford is a trade mark of Oxford University Press

Published in the United States
by Oxford University Press Inc., New York

© SIPRI 1997

All rights reserved. No part of this publication may be reproduced,
stored in a retrieval system, or transmitted, in any form or by any means,
without the prior permission in writing of Oxford University Press.
Within the UK, exceptions are allowed in respect of any fair dealing for the
purpose of research or private study, or criticism or review, as permitted
under the Copyright, Designs and Patents Act, 1988, or in the case of
reprographic reproduction in accordance with the terms of the licences
issued by the Copyright Licensing Agency. Enquiries concerning
reproduction outside these terms should be sent to the Rights Department,
Oxford University Press, at the address above.
Enquiries concerning reproduction in other countries should be sent to SIPRI.

British Library Cataloguing in Publication Data
Data available

Library of Congress Cataloging in Publication Data

Albright, David.
Plutonium and highly enriched uranium 1996: world inventories,
capabilities, and policies / David Albright, Frans Berkhout, and William Walker
— (SIPRI monographs)
Rev. ed. of: World inventory of plutonium and highly enriched uranium, 1992. 1993.
"Stockholm International Peace Research Institute"
Includes index.
1. Plutonium. 2. Uranium. 3. Reactor fuel reprocessing—Safety measures.
I. Berkhout, F. (Frans) II. Walker, William.
III. Albright, David. World inventory of plutonium and highly enriched uranium, 1992.
IV. Stockholm International Peace Research Institute. V. Title. VI. Series.
TK9360.A39 1996 338.4'762148335—dc20 95–50906
ISBN 0–19–828009–2

Typeset and originated by Stockholm International Peace Research Institute
Printed and bound in Great Britain on acid-free paper by
Biddles Ltd., Guildford and King's Lynn

Contents

Preface   xv
Acknowledgements   xvi
Glossary   xvii
Abbreviations, acronyms and conventions   xxviii

Part I. Introduction

1. Reasons, aims and sources   3
    I. Introduction   3
    II. Four security contexts   4
    III. The need for greater transparency   6
    IV. The limits to accuracy   8
    V. The scope of the book   9
    VI. Sources   10

2. Characteristics of highly enriched uranium and plutonium and their production processes   12
    I. Introduction   12
    II. Highly enriched uranium   12
    III. Plutonium   18
Table 2.1. Plutonium half-lives, and weapon-grade and reactor-grade isotopic concentrations, at given fuel discharges   20
Table 2.2. Neutron cross-sections   23

Part II. Military inventories in the nuclear weapon states

3. Inventories of military plutonium in the nuclear weapon states   29
    I. Introduction   29
    II. The production process   30
    III. Methods of estimating military plutonium inventories   33
    IV. The United States   37
    V. The former Soviet Union   50
    VI. The United Kingdom   59
    VII. France   66
    VIII. China   76
Table 3.1. Historical sources of weapon-grade plutonium   32
Table 3.2. US production reactors   38
Table 3.3. US Department of Energy total production of weapon- and fuel-grade plutonium (book inventory), 1947–89   40

Table 3.4.	US Department of Energy inventory of plutonium, by grade, February 1996	42
Table 3.5.	US Department of Defense (DOD) and Department of Energy inventories of plutonium, by location	45
Table 3.6.	Total plutonium inventory differences at US production sites, cumulative to February 1996	46
Table 3.7.	US inventory of weapon-grade plutonium declared to be excess to weapon requirements	47
Table 3.8.	Plutonium in the DOE waste inventory	48
Table 3.9.	The US plutonium inventory held by the Departments of Energy and Defense, February 1996	49
Table 3.10.	Estimated plutonium production by military reactors in the former Soviet Union, 31 December 1993	54
Table 3.11.	Estimated Soviet military plutonium output, calculated from estimated krypton releases and plutonium arisings, 31 December 1983	57
Table 3.12.	Total inventory of military plutonium of the FSU, 31 December 1993	58
Table 3.13.	Estimated British inventory of military plutonium, 31 December 1995	65
Table 3.14.	Estimated French inventory of military plutonium, 31 December 1995	68
Figure 3.1.	Plutonium in warhead production	30
Figure 3.2.	US Department of Energy weapon-grade plutonium production, 1947–88	41
Figure 3.3.	Plutonium inventories of the US Department of Energy, 31 December 1993	44

4. Inventories of highly enriched uranium in the nuclear weapon states 79

    I. Introduction 79
    II. Overview of enrichment programmes 79
    III. The United States 81
    IV. The former Soviet Union 94
    V. The United Kingdom 116
    VI. France 121
    VII. China 126

Table 4.1.	HEU stocks dedicated to nuclear weapons, reserves and associated categories, including excess, as of 31 December 1995	80
Table 4.2.	Estimates of US consumption of HEU, to 31 December 1994	85
Table 4.3.	Estimated allocation of US highly enriched uranium, 31 December 1993	91
Table 4.4.	US highly enriched uranium inventories, declared excess, 6 February 1996	92
Table 4.5.	Russian enrichment capacity in the early 1990s	103

Table 4.6.	Estimated increase in the separative capacity of Soviet centrifuges	106
Table 4.7.	Estimated Soviet consumption of separative work units, to 31 December 1987	112
Table 4.8.	Estimated uranium-235 content of the HEU at the Ulba Plant	116
Table 4.9.	Estimated British inventory of HEU, 31 December 1995	120
Figure 4.1.	Highly enriched uranium inventories of the US Department of Energy, 31 December 1993	82
Figure 4.2.	Maximum and minimum estimates of annual Soviet gas-centrifuge capacity, 1957–92	108
Figure 4.3.	Maximum and minimum estimates of annual Soviet uranium enrichment capacity, 1950–92	109
Figure 4.4.	Maximum and minimum estimates of cumulative Soviet separative output, 1950–88	110

Part III. Principal civil inventories

5. Plutonium produced in power reactors — 133
    I. Introduction — 133
    II. The fuel cycle in civil reactor systems — 135
    III. Fuelling strategy and fuel burnup — 136
    IV. A sketch of methods — 138
    V. Discharges of spent fuel and plutonium from civil reactors — 141
    VI. Conclusions — 147

Table 5.1.	Fuel characteristics of power reactors	136
Table 5.2.	Past discharges of spent fuel from nuclear power reactors, to 31 December 1993	140
Table 5.3.	Past discharges of plutonium from nuclear power reactors, to 31 December 1993	142
Table 5.4.	Estimated discharges of spent fuel and plutonium from nuclear power reactors, 1994–2000 and 2001–10	143
Figure 5.1.	The nuclear fuel cycle including reprocessing	134
Figure 5.2.	The once-through nuclear fuel cycle	135
Figure 5.3.	Specific plutonium production as a function of fuel burnup, natural uranium fuel	137
Figure 5.4.	Specific plutonium production as a function of fuel burnup, enriched uranium fuel	138
Figure 5.5.	World spent-fuel and plutonium discharges from power reactors by decade, 1961–2010	144
Figure 5.6.	Spent-fuel discharge profiles by decade for Canada, France, Japan, Russia, the UK and the USA, 1961–2010	145
Figure 5.7.	Plutonium discharges from power reactors in six regions: 1961–2010	146

6. Reprocessing programmes and plutonium arisings   148
    I. Reprocessing in the nuclear fuel cycle   148
    II. The evolution of fuel-cycle strategies   150
    III. A sketch of methods   155
    IV. Overview of power-reactor fuel reprocessing   155
    V. Commercial reprocessing programmes   157
    VI. Summary of power- and fast-reactor fuel reprocessing, 1960–2000   183
    VII. Projections of plutonium separation to 2010   190
    VIII. Conclusions   192
Table 6.1. National spent-fuel management policies, 1960–2000 and beyond   154
Table 6.2. World industrial-scale reprocessing plants   156
Table 6.3. Cumulative past plutonium separation at Sellafield and Dounreay, at the end of 1970, 1980, 1990 and 1993   160
Table 6.4. Existing contracts for fuel reprocessing at THORP, 1995   162
Table 6.5. Existing contracts for oxide-fuel reprocessing at La Hague, 1995   168
Table 6.6. Cumulative past plutonium separation at La Hague and Marcoule, at the end of 1970, 1980, 1990 and 1993   172
Table 6.7. VVER-440 fuel dispatched to RT-1, 1976–93   174
Table 6.8. Cumulative separation of plutonium from power-reactor and research-reactor fuel, at the end of 1970, 1980, 1990, 1993 and projected to 2000   184
Table 6.9. Distribution of cumulative plutonium separation from power-reactor fuel, at the end of 1970, 1980, 1990, 1993 and projected to 2000   188
Table 6.10. Cumulative plutonium separated in the UK, France and Russia from fuel from non-nuclear weapon states, at the end of 1980, 1990, 1993 and projected to 2000   189
Table 6.11. Cumulative discharged power-reactor plutonium which has been separated, at the end of 1970, 1980, 1990, 1993 and projected to 2000   190
Table 6.12. Projected annual spent-fuel throughputs and cumulative plutonium separation at industrial-scale reprocessing plants, three scenarios for 2001–10   191
Figure 6.1. Past and projected plutonium separation from power-reactor magnox fuel at the British B205 reprocessing plant, 1960–2010   158
Figure 6.2. Projected rates of plutonium separation from oxide fuel at the British Sellafield THORP reprocessing plant, 1994–2010   163
Figure 6.3. Past and projected plutonium separation from gas-graphite power-reactor fuel at the French Marcoule UP1 and La Hague UP2 reprocessing plants, 1960–2000   166

Figure 6.4.	Total plutonium separation from gas-graphite power-reactor fuel in France, 1960–2000	167
Figure 6.5.	Plutonium separation from oxide-fuel reprocessing at the La Hague UP2 and UP3 reprocessing plants in France, 1975–2000	169
Figure 6.6.	Past and projected quantities of plutonium separated at the German WAK and Japanese Tokai-mura reprocessing plants, 1970–2000	178
Figure 6.7.	World annual separation of civil plutonium, 1960–2000	186
Figure 6.8.	Rate of civil plutonium separation at industrial-scale reprocessing facilities in the UK, France and Russia, 1960–2000	187

7. Commercial and research and development uses of plutonium — 193

I.	Introduction	193
II.	Fast-reactor fuel cycles	194
III.	Plutonium use in fast reactors	195
IV.	Past and projected plutonium use in fast reactors	197
V.	Plutonium use in thermal reactors	209
VI.	National programmes for thermal plutonium recycling	212
VII.	Summary of plutonium use in thermal reactors: past and projected	223
VIII.	Commercial and R&D plutonium use compared with quantities separated	229
IX.	Conclusions	237
Table 7.1.	Fast reactors: retired and operating in 1995	196
Table 7.2.	Plutonium fuel fabrication facilities	197
Table 7.3.	Estimated plutonium consumption in nuclear R&D reactors in Russia and Kazakhstan, 31 December 1993	205
Table 7.4.	Plutonium consumed in fast and experimental reactor fuel, 31 December 1993, and high and low scenarios for 1994–2000 and 2001–10	209
Table 7.5.	Plutonium consumption in LWR–MOX fuel fabrication, up to 31 December 1993	223
Table 7.6.	Scenario for plutonium consumption in LWR–MOX fuel, 1994–2000	224
Table 7.7.	LWR–MOX fabrication capacity, 1993–99	227
Table 7.8.	Projected MOX fuel fabrication capacity, high and low scenarios for 2001–10	229
Table 7.9.	Power-reactor plutonium separation and use, to 31 December 1993	230
Table 7.10.	Civil plutonium balances as declared by states, as of 31 December 1995	233
Table 7.11.	National plutonium balances assuming utility MOX policies are implemented, 1994–2000	235

Figure 7.1. LWR–MOX production at the German Hanau fuel fabrication plant, 1972–92 — 213
Figure 7.2. LWR–MOX production at the Belgian Dessel P0 fuel fabrication plant, 1986–95 — 216
Figure 7.3. MOX fuel reloads at Électricité de France reactors, 1987–95 — 218
Figure 7.4. Projected world LWR–MOX fuel fabrication capacity: committed facilities, 1990–2005 — 226
Figure 7.5. Estimated national plutonium balances by ownership and location, 31 December 1993 — 232
Figure 7.6. Plutonium consumption forecasts: committed MOX fabrication capacity and utility MOX plans, 1994–2010 — 234
Figure 7.7. Projected world and European/Japanese plutonium surpluses, 1990–2010 — 236

8. Civil highly enriched uranium inventories — 238
    I. Introduction — 238
    II. Civil suppliers of highly enriched uranium — 239
    III. Civil reactors using HEU fuels, 1995 — 241
    IV. Converting to low-enriched uranium fuels — 242
    V. Spent fuel take-back and reprocessing of HEU fuels — 245
    VI. US exports of highly enriched uranium — 248
    VII. Civil inventories of highly enriched uranium — 253
Table 8.1. Summary of the distribution of US HEU exports — 248
Table 8.2. United States HEU exports and retransfers for non-Euratom countries — 249
Table 8.3. United States HEU exports and external retransfers for Euratom countries — 250
Table 8.4. Amount of US-origin HEU projected to remain overseas — 251

Part IV. Material inventories and production capabilities in the threshold states

9. De facto nuclear weapon states: Israel, India and Pakistan — 257
    I. Israel — 257
    II. India — 264
    III. Pakistan — 271
Table 9.1. Estimated plutonium production in the Israeli Dimona reactor, 31 December 1994 — 262
Table 9.2. Estimated inventories of Israeli weapon-grade plutonium, at the end of 1994, 1995 and 1999 — 263
Table 9.3. Estimated inventories of Indian weapon-grade plutonium, at the end of 1994 and 1995 — 269

Table 9.4.	Estimated production of weapon-grade uranium at the Pakistani Kahuta centrifuge enrichment plant, 1986–91	277
Figure 9.1.	SPOT image showing a probable plutonium reactor complex near Khushab, Pakistan	280

10. North Korea — 282
    I. Introduction — 282
    II. An unsafeguarded reactor emerges — 282
    III. Initial safeguards declaration — 283
    IV. Reactor defuelling — 288
    V. Plutonium production reactors — 295
    VI. Plutonium separation — 303
    VII. How much plutonium does North Korea have? — 306
    VIII. What about enrichment activities? — 307
    IX. Has North Korea built nuclear weapons? — 307

11. A special case: Iraq — 309
    I. Introduction — 309
    II. The Iraqi nuclear weapon programme — 311
    III. Summary of Iraqi fissile material production plans before 1991 — 313
    IV. The Iraqi enrichment programme — 317
    V. Projected indigenously produced weapon-grade uranium inventory for Iraq — 341
    VI. The Iraqi plutonium programme — 342
    VII. The crash programme — 344
    VIII. Post-war activities — 349

Table 11.1.	Selected EMIS separator design specifications	320
Table 11.2.	An R120 separator deployment schedule for 70 machines, declared by Iraq (but not achieved)	322
Table 11.3.	Actual R120 separator deployment schedule at Al Tarmiya	323
Table 11.4.	Maximal estimated centrifuge production	338
Table 11.5.	Projected Iraqi weapon-grade uranium inventories	341
Table 11.6.	Iraq's safeguarded fuel	345
Figure 11.1.	Map of Iraq showing the approximate locations of the main inspection sites	316

12. Countries of concern: Iran, Algeria, South Korea and Taiwan — 351
    I. Introduction — 351
    II. Iran — 352
    III. Algeria — 363
    IV. South Korea — 365
    V. Taiwan — 366

13. Countries backing away from nuclear weapons: Argentina, Brazil and South Africa — 369
    I. Introduction — 369
    II. Argentina and Brazil — 369
    III. South Africa — 377
    Table 13.1. Illustrative output of the Y-Plant — 387

Part V. Conclusions

14. Overview of present and future stocks of plutonium and highly enriched uranium — 395
    I. Introduction — 395
    II. Inventories at the end of 1994 — 395
    III. Types of inventory — 397
    IV. Military inventories in nuclear weapon states — 398
    V. Weapon-related inventories and capabilities in countries other than the acknowledged nuclear weapon states — 401
    VI. Civil inventories of plutonium and HEU — 404
    VII. Material under international safeguards — 406
    VIII. Possible future trends in plutonium and HEU inventories — 411
    Table 14.1. Central estimates for civil and military inventories of plutonium and HEU, 31 December 1994 — 397
    Table 14.2. Central estimates for inventories of plutonium and HEU by type, 31 December 1994 — 398
    Table 14.3. NWS inventories of highly enriched uranium, after losses and draw-downs, 31 December 1994 — 399
    Table 14.4. NWS inventories of military plutonium, after losses, 31 December 1994 — 400
    Table 14.5. Central estimates for current and former de facto nuclear weapon states' inventories of plutonium and HEU, produced for nuclear weapon purposes — 402
    Table 14.6. Central estimates for de facto nuclear weapon states' inventories of weapon-grade plutonium and WGU, produced for nuclear weapon purposes — 402
    Table 14.7. Civil stocks of plutonium by NPT status, 31 December 1994 (and 1993) — 404
    Table 14.8. Civil plutonium separation and use, to 31 December 1993 — 405
    Table 14.9. Approximate quantities of plutonium under IAEA safeguards, 31 December 1993 — 408
    Table 14.10. Projection of cumulative spent-fuel discharges, plutonium separation and unrecycled stocks, 31 December 1993, 2000 and 2010 — 412
    Table 14.11. Illustrative inventories of weapon-grade plutonium and HEU inside and outside operational nuclear weapons — 414

CONTENTS xiii

15. The control and disposition of fissile materials: the new policy agenda — 416

    I. Introduction — 416
    II. Two industrial and regulatory systems — 418
    III. Contemporary pressures to achieve universality and transparency — 421
    IV. Measures against undeclared activities in NNWS parties to the NPT — 424
    V. Bilateral initiatives to strengthen controls in the FSU — 427
    VI. Extending the multilateral framework for material controls — 430
    VII. The disposition of excess plutonium and HEU — 437
    VIII. HEU disposition — 444
    IX. Plutonium disposition — 446
    X. Obstacles to an international disposition strategy — 453
    XI. Conclusions: moving towards the framework of control required by complete nuclear disarmament — 454

Table 15.1. Illustrative inventories of plutonium available for disposition — 443
Table 15.2. HEU disposition scenarios — 445
Table 15.3. Capacities for disposition through plutonium recycling in power reactors — 447

Appendices

Appendix A. Weapon-grade plutonium and highly enriched uranium production — 461

    I. Weapon-grade plutonium production — 461
    II. Highly enriched uranium — 464

Table A.1. Representative conversion factors for reactors producing weapon-grade plutonium — 462
Table A.2. Conversion factors for gas-graphite reactor with natural uranium fuel — 463
Table A.3. Gas-graphite, air-cooled reactor with natural uranium fuel — 463
Table A.4. Overview of principal enrichment technologies — 468
Table A.5. Common examples of enriched uranium output — 469
Table A.6. Weapon-grade uranium production (93% enriched) — 469
Table A.7. Current activities on uranium separation and the level of such activities, by country — 470
Table A.8. Commercial enrichment plants — 471
Figure A.1. The basic elements of a gaseous-diffusion plant — 465
Figure A.2. A gas centrifuge and a centrifuge rotor — 466
Figure A.3. EMIS configuration — 467

Appendix B. Calculation of plutonium production in power reactors 472
Table B.1. Plutonium discharge rate by reactor type 473

Appendix C. Separation of plutonium from power-reactor fuel at reprocessing plants 479
Table C.1. Plutonium separation from British Magnox power-reactor fuel at B205 479
Table C.2. Plutonium separated from foreign magnox fuel at B205 480
Table C.3. Plutonium separated from Magnox power-reactor fuel at UP1 (France) 481
Table C.4. Plutonium separated from Magnox power-reactor fuel at UP2 (France) 482
Table C.5. Plutonium separated from oxide fuel at UP2 (France) 483
Table C.6. Plutonium separated from oxide fuel at UP3 (France) 483
Table C.7. Plutonium separated from oxide fuel at WAK (Germany) 484
Table C.8. Plutonium separated from oxide fuel at Tokai-mura (Japan) 484
Table C.9. World annual separation of civil plutonium, 1965–2000 485

Appendix D. Research reactors (>1 MWth) using HEU fuel 486
Table D.1. US operating research and test reactors with power >1 MWth using HEU fuel (as of mid-1995) 486
Table D.2. US-supplied operating research and test reactors with power >1 MWth using HEU (>90%) fuel (as of mid-1995) 487
Table D.3. Russian operating research and test reactors with power >1 MWth using HEU fuel (as of mid-1995) 488
Table D.4. Russian-supplied operating research and test reactors with power >1 MWth using HEU fuel (as of mid-1995) 488
Table D.5. Chinese and Chinese-supplied operating research and test reactors using HEU fuel (as of mid-1995) 489

Index 491

Preface

Extending greater control over plutonium and highly enriched uranium, the fissile materials used in nuclear weapons, has always been a fundamental objective of nuclear non-proliferation policy. Its importance has been increased by the arms reductions undertaken by the nuclear weapon states in recent years and will increase still further if progress is to be made towards the ultimate goal of ridding the world of nuclear weapons.

Control must begin with understanding and with transparency. In 1993, SIPRI published *World Inventory of Plutonium and Highly Enriched Uranium 1992*, by David Albright, Frans Berkhout and William Walker, with the intention of contributing in a modest way to the realization of these objectives. The book provided the first authoritative survey of the quantities of fissile material produced for nuclear weapons and in the civilian fuel cycle.

Almost as soon as the book was published, we realized that there would be benefit in preparing a new edition. Fresh information was becoming available and various developments—such as the progress being made by the USA and Russia in nuclear arms reduction and weapon dismantlement, the denuclearization of Belarus, Kazakhstan, South Africa and Ukraine, the destruction of the Iraqi nuclear weapon programme and the expansion of civil reprocessing—had implications for assessments of and policies towards plutonium and highly enriched uranium. When David Albright, Frans Berkhout and William Walker began the task of preparing the new edition, they believed that it would involve a few months' work and be a relatively simple undertaking. They were mistaken. Nearly every part of the book ended up being completely rewritten, so that what lies before us is a totally new work. Readers will appreciate that it has required a formidable effort by the authors.

I am pleased that this book also contains a more extensive assessment of the policy agenda than was attempted in the previous edition. It is recognized that the control of fissile materials and their associated technologies must be the core of any nuclear non-proliferation and disarmament regime. Although the authors acknowledge that many difficulties—technical, political and economic—will have to be overcome if a fully effective material control regime is to be established, they have demonstrated its feasibility in identifying the main steps that need to be taken for its realization. I was particularly struck by the observation in the concluding chapter that today's need for stringent control of nuclear materials and technologies is little different from that which would be required for total nuclear disarmament, implying that policies should be fashioned as if this were about to occur, irrespective of its feasibility or desirability. For once, realism and idealism lead to the same recommendations. My hope is that this book will therefore be regarded as a significant contribution to policy debates as well as a compendium of information on nuclear material inventories and capabilities.

Adam Daniel Rotfeld
Director of SIPRI
December 1996

Acknowledgements

This book is the result of work carried out over several years at the Institute for Science and International Security, Washington DC; the Science Policy Research Unit, University of Sussex, UK; and the Center for Energy and Environmental Studies, Princeton University. It has been made possible by the generous financial support of the Rockefeller Brothers Fund, the Joseph Rowntree Charitable Trust, the Ploughshares Fund, an anonymous donor of the Rockefeller family, the Carnegie Corporation of New York, Francis Close, the John Merck Fund, Prospect Hill Foundation, the Scherman Foundation, the W. Alton Jones Foundation and the Stockholm International Peace Research Institute.

Many people have helped us compile the book. Our thanks go to Oleg Bukharin, George Bunn, Matthew Bunn, Thomas B. Cochran, Tom Zamora Collina, Mary Davis, Anatoli Diakov, Harold Feiveson, Steve Fetter, Jim Finucane, David Fischer, Corey Gay, Lisbeth Gronlund, Richard Guthrie, Wolfgang Heni, Roger Heusser, the late William Higginbotham, Frank von Hippel, Naru Jasani, Robert Kelley, David Kyd, Paul Leventhal, Hans Meyer, Harald Müller, Robert S. Norris, Kevin O'Neill, Frank Pabian, Annette Schaper, Mycle Schneider, Tom Shea, John Simpson, Sandy Spector, Waldo Stumpf, Tatsujiro Suzuki, Jinzaburo Takagi, Theodore Taylor, Steve Thomas, Johann Viljoen, David Wright and Gernot Zippe. There is a much longer list of people in government, industry and international organizations, and of retired officials, who have greatly assisted us but who prefer to remain anonymous.

Billie Bielckus at SIPRI has been as patient and meticulous as ever in editing the manuscript and preparing the illustrations, and has been a tremendous support throughout the project. We are also very grateful to Adam Daniel Rotfeld and Connie Wall for having paved the way for the book's publication. Finally, we owe a great debt to Hilary Palmer, formerly of the Rockefeller Brothers Fund. This project, and so many others besides, would not have taken place without her personal interest and commitment.

Preparing this book has involved long hours of work, day and night, weekdays and weekends. Our thanks and apologies to Ulrike Weinrich, Diane Moody and Carolyn Walker.

David Albright, Frans Berkhout and William Walker
December 1996

Glossary

Aerodynamic enrichment method
: A process of uranium enrichment that is based on centrifugal effects of a fast-moving uranium hexafluoride gas in very small curved-wall chambers.

Alpha particle
: A charged particle emitted from the nucleus of an atom, having a mass and charge equal in magnitude to a helium nucleus.

Americium (Am)
: Transuranic element with atomic number 95. Americium-241, an alpha and gamma emitter, is a decay product of plutonium-241.

Atomic bomb
: A bomb whose energy comes from the fission of uranium or plutonium.

Atomic number
: The number of protons in the atomic nucleus of an element.

Beryllium metal
: A highly toxic steel-grey metal, possessing a low neutron absorption cross-section and high melting point, which can be used in nuclear reactors as a moderator or reflector. In nuclear weapons, beryllium surrounds the fissile material and reflects neutrons back into the nuclear reaction, considerably reducing the amount of fissile material required. Beryllium is also used in guidance systems and other parts for aircraft, missiles or space vehicles.

Beta decay
: Radioactive decay involving the emission of a beta particle. This is a charged particle with the mass and charge equal to that of an electron or positron.

Blanket
: A layer of fertile material, such as uranium-238 or thorium-232, placed around the core of a reactor. During operation of the reactor, additional fissile material is produced in the blanket.

Blend-down of enriched uranium
: The process of reducing the fraction of uranium-235 in enriched uranium by diluting it with lower enriched uranium.

Boiling water reactor
: A light-water nuclear reactor in which steam is produced in the reactor and passed directly to the turbogenerator.

Burnup
: The percentage of heavy metal atoms fissioned or the thermal energy produced per mass of fuel (usually measured in megawatt days per tonne, MWd/t).

Calutron
: (CAlifornia University CycloTRON). An electromagnetic uranium enrichment machine that uses the electromagnetic isotope separation technique. It was used in the Manhattan Project to produce HEU for the Hiroshima bomb and further

	developed in the Iraqi bomb programme. Alpha machines are the first stage, producing LEU from natural uranium; beta machines are the second stage, producing HEU from the output of the alpha machines.
CANDU	(Canadian deuterium–uranium reactor.) The most widely used type of heavy water power reactor. The CANDU reactor uses natural uranium as a fuel and heavy water as a moderator and a coolant.
Cascade	A connected series of enrichment machines, material from one being passed to another for further enrichment.
Centrifuge process	An enrichment method that separates gaseous isotopes by rotating them rapidly in a spinning tube, thereby subjecting them to a centrifugal force. To increase the amount of separation in a centrifuge, various techniques are used to induce a vertical 'countercurrent' flow in the gas. Centrifuges are either 'subcritical' or 'supercritical'. A subcritical centrifuge rotor has a length to diameter ratio such that it runs optimally at an angular velocity below the first fundamental flexural critical frequency. At these critical frequencies, the rotational energy of the spinning rigid body is transferred into large displacements from the axis of rotation, breaking the rotor unless mechanical actions are taken to reduce the displacement amplitudes. A supercritical centrifuge operates above the first critical frequency, and avoids the damaging effects associated with resonances by mechanical methods such as damping mechanisms and bellows (flexible joints connecting rotor tubes together that act like a spring).
Chain reaction	The continuing process of nuclear fissioning in which the neutrons released from a fission trigger at least one other nuclear fission. In a nuclear weapon an extremely rapid, multiplying chain reaction causes the explosive release of energy. In a reactor, the pace of the chain reaction is controlled and sustained.
Chemical enrichment	This method of uranium isotope separation depends on a slight tendency of uranium-235 and uranium-238 to concentrate in different molecules when uranium compounds are continuously brought into contact. Catalysts are used to speed up the chemical exchange.
Chemical processing	Chemical treatment of materials to separate specific usable constituents.
Cladding	The material which encases the nuclear fuel, reducing the risk of radioactive materials leaking from the fuel.

Coolant	A substance circulated through a nuclear reactor to remove or transfer heat. The most common coolants are carbon dioxide, water and heavy water.
Core	The central portion of a nuclear reactor containing the fuel elements and usually the moderator. Also the central portion of a nuclear weapon containing HEU or plutonium.
Critical mass	The minimum mass required to sustain a chain reaction. The exact mass varies with many factors such as the particular isotope present, its concentration and chemical form, the geometrical arrangement of the material and its density. When fissile materials are compressed by high explosives in implosion-type atomic weapons, the critical mass needed for a nuclear explosion is reduced.
Depleted uranium	Uranium with a smaller percentage of uranium-235 than the 0.7 per cent found in natural uranium. It is a by-product of the uranium enrichment process, during which uranium-235 is culled from one batch of uranium, thereby depleting it, and added to another batch to increase its concentration of uranium-235.
Disposition	The disposal of plutonium or enriched uranium, especially stocks arising from dismantled nuclear weapons, using reactor and non-reactor options such as vitrification.
Diversion	The deliberate removal of fissionable material in civil fuel cycles for other uses.
Draw-down	A policy of consuming stocks of nuclear material.
Dry storage	Storage of irradiated nuclear fuel in a gas (either air or an inert gas) environment.
Electromagnetic isotope separation	Separation of isotopes by the use of electromagnetic fields.
Enrichment	The process of increasing the concentration of one isotope of a given element (in the case of uranium, increasing the concentration of uranium-235).
Facilities list	The list of nuclear facilities declared by states parties to the Treaty on the Non-Proliferation of Nuclear Weapons to the IAEA that may be subject to safeguards. In non-nuclear weapon states this includes all nuclear facilities, in nuclear weapon states it includes only facilities designated by the state.
Facility attachment	The detailed plan for applying safeguards at a particular plant. This usually defines the material balance areas and indicates the strategic points to which the IAEA inspector may have access during inspections and at which safeguards instruments may be installed.

Fast breeder reactor	A nuclear reactor in which fuel is irradiated with high-energy neutrons and which produces more fissile material than it consumes, a process known as breeding. Fissile material is produced both in the reactor's core and through neutron capture in fertile material placed around the core (blanket).
Feedstock	Material introduced into a facility at the start of the process, such as uranium hexafluoride in an enrichment plant.
Fertile material	Material composed of atoms which readily absorb neutrons to produce fissionable materials. One such element is uranium-238, which becomes plutonium-239 after it absorbs a neutron. Fertile material alone cannot sustain a chain reaction.
Fission	The process by which a neutron strikes a nucleus and splits it into fragments or 'fission products'. During the process of nuclear fission, several neutrons are emitted at high speed and radiation is released.
Fissionable material	Material, whose nuclei can be induced to fission by a neutron.
Fissile material	Material composed of atoms which fission when irradiated by slow or 'thermal' neutrons. The most common examples of fissile materials are uranium-235 and plutonium-239. The term is often used to describe plutonium and HEU, e.g., a cut-off in the production of fissile materials for weapons.
Fuel element	Engineered bundle of nuclear fuel pins.
Fuel pin	Single rod of basic chain-reacting material, including both fissile and fertile materials.
Fusion	The formation of a heavier nucleus from two lighter ones (usually hydrogen isotopes), with the attendant release of energy (as in a hydrogen bomb).
Gamma radiation	High-energy electromagnetic radiation emitted from nuclei as a result of nuclear reactions and decay.
Gas-centrifuge process	See centrifuge process.
Gas-cooled reactor	A nuclear reactor employing a gas (usually CO_2) as a coolant, rather than water or liquid metal.
Gaseous diffusion	A method of isotope separation based on the fact that gas atoms or molecules with different masses will diffuse through a porous barrier (or membrane) at different rates. The method is used to separate uranium-235 from uranium-238. It requires large plants and significant amounts of electric power.
Gas-graphite reactor	A nuclear reactor in which a gas is the coolant and graphite is the moderator.

GLOSSARY

Graphite	One of the two elemental forms of carbon, used as a moderator in some thermal reactor types (Magnox, RBMK).
Heavy water	Water containing significantly more than the natural proportion (1 in 6500) of heavy hydrogen (deuterium) atoms to ordinary hydrogen atoms. (Hydrogen atoms have one proton, deuterium atoms have one proton and one neutron.) Heavy water is used as a moderator in some reactors as it slows down neutrons more effectively and absorbs them less (than light, or normal, water) making it possible to fission natural uranium and sustain a chain reaction.
Heavy water reactor	A reactor that uses heavy water as its moderator and natural uranium as fuel. See CANDU.
Highly enriched uranium (HEU)	Uranium in which the percentage of uranium-235 nuclei has been increased from the natural level of 0.7 per cent to some level greater than 20 per cent, usually around 90 per cent. All HEU can be used to make nuclear explosives, although very large quantities are needed for HEU enriched to 20 per cent.
Hot cells	Heavily shielded rooms with remote handling equipment for examining and processing radioactive materials. In particular, hot cells are used for examining spent reactor fuel.
Hydrogen bomb	A nuclear weapon that derives its energy largely from fusion. Also known as a thermonuclear bomb.
Irradiation	Exposure to a radioactive source; usually in the case of materials being placed in an operating nuclear reactor.
Isotope	Atoms having the same number of protons, but a different number of neutrons. Two isotopes of the same atom are chemically similar and are therefore difficult to separate by ordinary chemical means. Isotopes can have very different nuclear properties, however. For example, one isotope may spontaneously fission readily, while another isotope of the same atom may not fission at all. An isotope is specified by its atomic mass number (the number of protons plus neutrons) following the symbol denoting the chemical element (e.g., uranium-235 is an isotope of uranium).
Kilogram	A metric weight equivalent to 2.2 pounds.
Kiloton	The energy of a nuclear explosion that is equivalent to an explosion of 1000 tons of TNT.
Laser enrichment method	A still experimental process of uranium enrichment in which lasers are used to separate uranium isotopes.
Light water	Ordinary water (H_2O), as distinguished from heavy water (D_2O).
Light water reactor	A reactor that uses ordinary water as moderator and coolant and low-enriched uranium as fuel.

Light water-cooled, graphite-moderated reactor	A reactor cooled by water and moderated with graphite.
Low-enriched uranium (LEU)	Uranium in which the percentage of uranium-235 nuclei has been increased from the natural level of 0.7 per cent up to 20 per cent, usually 3 to 5 per cent. With the increased level of fissile material, LEU can sustain a chain reaction when immersed in light water and is used as fuel in light water reactors. It cannot be used in nuclear explosives, however. The uranium-238 at the concentrations found in LEU 'denatures' the material, effectively preventing such use.
Magnox fuel	Uranium metal fuel clad with a magnesium alloy (magnox).
Magnox reactor	Gas-cooled, graphite-moderated reactor built principally in the UK and France.
Maraging steel	Special hardened steel used in the fabrication of centrifuge rotors and rocket motors.
Mass number	The number of protons and neutrons in the atomic nucleus. Elements may occur in forms displaying a range of mass numbers—i.e., plutonium-238, -239, -240, -241, -242.
Medium-enriched uranium	Uranium in which the percentage of uranium-235 nuclei has been increased from the natural level of 0.7 per cent to between 20 and 50 per cent. A sub-category of HEU; thus, potentially usable for nuclear weapons, but very large quantities are needed.
Megawatt	One million watts-electric (MWe): used in reference to a nuclear power plant, one million watts of electricity. One million watts-thermal (MWth): one million watts of heat.
Metric tonne	1000 kilograms. A metric weight equivalent to 2200 pounds or 1.1 tons.
Milling	A process in the uranium fuel cycle by which ore containing only a very small percentage of uranium oxide (U_3O_8) is converted into material containing a high percentage (80 per cent) of U_3O_8, often referred to as yellowcake.
Mixed-oxide fuel	Nuclear fuel containing both uranium and plutonium. Most fissions in the fuel will be of plutonium nuclei.
Moderator	A component (usually water, heavy water, or graphite) of some nuclear reactor types that slows neutrons, thereby increasing their chances of fissioning fertile material.
Natural uranium	Uranium as found in nature, containing 0.71 per cent of uranium-235, 99.27 per cent of uranium-238 and a trace of uranium-234.

GLOSSARY xxiii

Neutron	An uncharged elementary particle, with a mass slightly greater than that of a proton, found in the nucleus of every atom heavier than hydrogen. Nuclear fission is caused when a nucleus is irradiated with neutrons. Fissions may be caused by relatively low energy (thermal) neutrons, or by high energy (fast) neutrons. Fission reactors are therefore classed as either 'fast reactors' or 'thermal reactors'.
Nuclear energy	The energy liberated by a nuclear reaction (fission or fusion) or by spontaneous radioactivity.
Nuclear fuel	Basic chain-reacting material, including both fissile and fertile materials. Commonly used nuclear fuels are natural uranium and low-enriched uranium; highly enriched uranium and plutonium are used in some reactors.
Nuclear fuel cycle	The set of chemical and physical operations needed to prepare nuclear material for use in reactors and to dispose of or recycle the material after its removal from the reactor. Existing fuel cycles begin with uranium as the natural resource and create plutonium as a by-product. Some future fuel cycles may rely on thorium and produce the fissionable isotope uranium-233.
Nuclear fuel element	A rod, tube, plate or other mechanical shape or form into which nuclear fuel is fabricated for use in a reactor.
Nuclear fuel fabrication plant	A facility where the nuclear material (e.g., enriched or natural uranium) is fabricated into fuel elements to be inserted into a reactor.
Nuclear power plant	Any device that converts nuclear energy into useful power. In a nuclear electric power plant, heat produced by a reactor is used to produce steam to drive a turbine that in turn drives an electricity generator.
Nuclear reactor	A heat engine configured to sustain a controlled nuclear chain reaction when fuelled with fissionable materials. Reactors are of three general types: electric power reactors, plutonium production reactors and research reactors.
Nuclear waste	The radioactive by-products formed by fission and other nuclear processes in a reactor. Most nuclear waste is initially contained in spent fuel. If this material is reprocessed, new categories of waste result.
Nuclear weapons	A collective term for atomic bombs and hydrogen bombs. Weapons based on a nuclear explosion. The term is generally used throughout the text to mean atomic bombs only, unless used with reference to the nuclear weapon states (all five of which have both atomic and hydrogen weapons).
Nucleus	The part of an atom containing protons and neutrons.

On-load refuelling	Re-fuelling of nuclear reactors under power (i.e., Magnox and CANDU reactors).
Pit	The shaped core of a nuclear weapon containing fissile material, a tamper and a reflector.
Plutonium-239 (^{239}Pu)	A fissile isotope generated artificially when uranium-238, through irradiation, captures an extra neutron. It is one of the two fissile materials that have been almost exclusively used for the core of nuclear weapons, the other being uranium-235. (A small amount of nuclear explosives have been made with uranium-233.)
Plutonium-240 (^{240}Pu)	An isotope produced in reactors when a uranium-239 atom absorbs a neutron instead of fissioning. Its presence complicates the construction of nuclear explosives because of its high neutron emission and its high heat output. Unlike uranium-238, plutonium-240 does not 'denature' plutonium, enabling plutonium with high concentrations of plutonium-240 to be usable in nuclear explosives.
Plutonium recycle	The re-use of separated plutonium as fuel in nuclear reactors.
Pond storage	Storage of irradiated fuel under water.
Power reactor	A reactor designed to produce electricity as distinguished from reactors used primarily for research or for producing radiation or fissionable materials.
Primary	The fission explosive detonated first in a thermonuclear warhead containing two or more stages (secondaries).
Production reactor	A reactor designed primarily for large-scale production of plutonium-239 by neutron irradiation of uranium-238.
Proton	A positively-charged nuclear particle, one of the two principal components of nuclei, with a mass similar to a neutron.
Radioactivity	The spontaneous disintegration of an unstable atomic nucleus resulting in the emission of sub-atomic particles.
Radioisotope	A radioactive isotope.
Reprocessing	Chemical treatment of spent nuclear fuel to separate the plutonium and uranium from unwanted radioactive waste by-products and (under present plans) from each other. Spent fuel is often handled in batches known as 'campaigns'.
Research reactor	A reactor primarily designed to supply neutrons for experimental purposes. It may also be used for training, materials testing and production of radioisotopes.
Safeguards	Technical and inspection measures for verifying that nuclear materials are not being diverted from civil to other uses.
Secondary	See primary.

	GLOSSARY xxv
Separative work	A measure of the effort required in an enrichment facility to separate uranium of a given uranium-235 content into two fractions, one with a higher percentage and one with a lower percentage of uranium-235. The unit of separative work is the kilogram separative work unit (kg SWU), or separative work unit (SWU) for short. The initial material is called the 'feed'. The fraction with a higher proportion of uranium-235 is called the 'product', the other is called the 'tails'.
Significant quantity	The approximate amount of nuclear material (not just fissile material) which the IAEA considers a state would need to manufacture its first nuclear explosive. Eight kilograms of plutonium are considered significant and 25 kilograms of weapon-grade uranium are significant.
Solvent extraction	Technique for separating plutonium, uranium and fission products in a reprocessing plant using solvents.
Spent fuel	Fuel elements that have been removed from the reactor after use because they contain too little fissile and fertile material and too high a concentration of unwanted radioactive by-products to sustain reactor operation. Spent fuel is both thermally and radioactively hot.
Tails	The waste stream of an enrichment facility that contains depleted uranium. It is expressed as a percentage of the uranium-235 content and called the 'tails assay.'
Thermal reactor	See neutron.
Thermal recycle	See plutonium recycle.
Thermonuclear bomb	A hydrogen bomb.
Thorium-232	A fertile material.
Tritium	The heaviest hydrogen isotope, containing one proton and two neutrons in the nucleus, produced typically by bombarding lithium-6 with neutrons. In a fission weapon, tritium is used with deuterium in a fusion process known as 'boosting' to produce excess neutrons, which set off additional fissions in the core. In this way, tritium can either reduce the amount of fissile material required, or multiply (i.e., boost) the weapon's destructive power many times. In fusion reactions, tritium and deuterium bond at very high temperatures, releasing a neutron with 14 million electron-volts of energy.
Uranium (U)	A radioactive element with the atomic number 92 and, as found in natural ores, an average atomic weight of 238. The two principal natural isotopes are uranium-235 (0.71 per cent of natural uranium), which is fissionable, and uranium-238 (99.27 per cent of natural uranium), which is fertile.

Uranium-233 (^{233}U)	A fissile isotope bred in fertile thorium-232. Like plutonium-239 it is theoretically an excellent material for nuclear weapons, but is not known to have been used for this purpose except in research programmes. Can be used as reactor fuel.
Uranium-235 (^{235}U)	The only naturally occurring fissile isotope. Natural uranium contains 0.71 per cent ^{235}U; light water reactors use about 3 to 5 per cent and weapon-grade uranium has more than 90 per cent ^{235}U.
Uranium-238 (^{238}U)	A fertile material. Natural uranium is composed of approximately 99.3 per cent ^{238}U.
Uranium dioxide (UO_2)	Purified uranium. The form of natural uranium used in heavy water reactors. Also the form of uranium used to fabricate enriched uranium fuel elements.
Uranium oxide (U_3O_8)	The most common oxide of uranium found in typical ores. U_3O_8 is extracted from the ore during the milling process. The ore typically contains only 0.1 per cent U_3O_8; yellowcake, the product of the milling process, contains about 80 per cent U_3O_8.
Uranium hexafluoride (UF_6)	A volatile compound of uranium and fluorine. UF_6 is a solid at atmospheric pressure and room temperature, but can be transformed into gas by heating. UF_6 gas (alone, or in combination with hydrogen or helium) is the feedstock in most uranium enrichment processes and is sometimes produced as an intermediate product in the process of purifying yellowcake to produce uranium oxide.
Urenco centrifuges	Urenco is a commercial consortium involving Britain, Germany and the Netherlands that has developed the gas centrifuge to make LEU for nuclear power reactors. It has built several generations of centrifuges. A first-generation machine is a G1 machine. It is a subcritical machine with one maraging-steel rotor tube that spins at about 450 m/s and has a separative capacity of about 3 SWU per year. The G2 supercritical centrifuge has a rotor consisting of two roughly 50-cm maraging-steel rotor tubes connected by a flexible maraging-steel joint (called a 'bellows') that acts as a spring. The bellows prevents the rotor from acting like a rigid body and thus allows the rotor to pass through flexural resonances that would otherwise destroy the rotor. The separative capacity of the G2 centrifuge is about 5–6 SWU per year. A mid-1980s design is a 3-metre long machine, composed of 6–7 carbon-fibre rotor tubes connected by a bellows. It spins at over 600 m/s and has an estimated separative capacity of about 20–30 SWU per year.
Vessel	The part of a reactor that contains the nuclear fuel.

Weapon-grade material	Nuclear material of the type most suitable for nuclear weapons, i.e., uranium enriched to over 90 per cent ^{235}U or plutonium that is primarily ^{239}Pu.
Weapon-usable material	Usually separated plutonium or HEU. Because of the use of this term in UN Security Resolution 687, the IAEA Action Team formally defined it as uranium enriched to 20 per cent or more in uranium isotopes 233, 235 or both; plutonium containing less than 80 per cent ^{238}Pu; any of the foregoing in the form of metal, alloy, chemical compound or concentrate; and any other goods containing one or more of the foregoing, other than irradiated fuel.
Yellowcake	A concentrate produced during the milling process that contains about 80 per cent uranium oxide (U_3O_8). In preparation for uranium enrichment, the yellowcake is converted to uranium hexafluoride gas (UF_6). In the preparation of natural uranium reactor fuel, yellowcake is processed into purified uranium dioxide. Sometimes uranium hexafluoride is produced as an intermediate step in the purification process.
Yield	The total energy released in a nuclear explosion. It is usually expressed in equivalent tons of TNT (the quantity of TNT required to produce a corresponding amount of energy).
Zirconium	A greyish-white lustrous metal which is commonly used in an alloy form (i.e., zircalloy) to encase fuel rods in nuclear reactors.

Sources: Congressional Research Service, *Nuclear Proliferation Factbook* (US Government Printing Office: Washington, DC, 1977); Energy Research & Development Administration, *U.S. Nuclear Power Export Activities* (National Technical Information Service: Springfield, Va., 1976); Fischer, D. and Szasz, P., ed. J. Goldblat, SIPRI, *Safeguarding the Atom: A Critical Appraisal* (Taylor and Francis: London, 1985); Nero, A. V., *A Guidebook to Nuclear Reactors* (University of California Press: Berkeley, Calif., 1979); Nuclear Energy Policy Study Group, *Nuclear Power: Issues and Choices* (Ballinger: Cambridge, Mass, 1977); Office of Technology Assessment, *Nuclear Proliferation and Safeguards* (Office of Technology Assessment: Washington, DC, 1977); *Nuclear Power in an Age of Uncertainty* (Office of Technology Assessment: Washington, DC, 1984); Spector, L. S., *Nuclear Ambitions* (Westview Press: Boulder, Colo., 1990); Wohlstetter, A., *Swords from Plowshares: The Military Potential of Civilian Nuclear the USA, Nuclear Proliferation: A Citizen's Guide to Policy Choices* (UNA-USA: New York, 1983); and the authors.

Acronyms and conventions

ABACC	Brazilian–Argentine Agency for Accounting and Control of Nuclear Materials
AEB	Atomic Energy Board (South Africa)
AEC	Atomic Energy Commission (USA)
AEC	Atomic Energy Corporation (South Africa)
AEOI	Atomic Energy Organization of Iran
AFR	Away from the reactor
AGNS	Allied-General Nuclear Services (USA)
AGR	Advanced gas-cooled reactor
APM	Atelier Pilote Marcoule (France)
ATR	Advanced thermal reactor
AVLIS	Atomic Vapor Laser Isotope Separation Project (USA)
BARC	Bhabha Atomic Research Centre (India)
BNFL	British Nuclear Fuels Ltd (now plc)
BWR	Boiling water reactor
CANDU	Canadian deuterium–uranium reactor
CD	Conference on Disarmament
CEA	Commissariat à l'Énergie Atomique (France)
CETEX	Army Technological Centre (Brazil)
CFDT	Confédération Française Democratique du Travail
CIA	US Central Intelligence Agency
CIS	Commonwealth of Independent States
CNEA	National Atomic Energy Commission (Argentina)
CNEN	National Nuclear Energy Commission (Brazil)
Cogema	Compagnie Générale des Matières Nucléaires (France)
Comecon	Council for Mutual Economic Assistance (CMEA)
CTA	Aerospace Technical Centre (Brazil)
CTBT	Comprehensive Nuclear Test Ban Treaty
DAE	Department of Atomic Energy (India)
DATR	Demonstration ATR (Japan)
DFBR	Demonstration FBR (Japan)
DFR	Demonstration Fast Reactor (UK)
DMTR	Demonstration Materials Test Reactor
DNPDE	Dounreay Nuclear Power Development Establishment
DOE	Department of Energy (USA)
DTI	Department of Trade and Industry (UK)
DWK	Deutsche Gesellschaft für Wiederaufarbeitung von Kernbrennstoffe
EC	European Community

EdF	Électricité de France
EIS	Environmental impact statement
EMIS	Electromagnetic isotope separation
ENEL	Ente Nazionale per l'Energia Elettrica (Italy)
EU	European Union
Euratom	European Atomic Energy Community
FBR	Fast breeder reactor
FBTR	Fast Breeder Test Reactor
FFTF	Fast Flux Test Facility (Hanford, USA)
FMCT	Fissile Material Cut-Off Treaty
FRG	Federal Republic of Germany
FSU	Former Soviet Union
GCR	Gas-cooled, graphite-moderated reactor
GE	General Electric (USA)
HAO	Haute Activité Oxyde (France)
HEP	Head-End Plant (Windscale, UK)
HEU	Highly enriched uranium
HTR	High-temperature reactor
HWR	Heavy-water-cooled and -moderated reactor
IAEA	International Atomic Energy Agency
ICBM	Intercontinental ballistic missile
ICM	Isolate, contain, monitor
INEL	Idaho National Engineering Laboratory
INF	Intermediate-range nuclear forces
IPEN	Institute of Energy and Nuclear Research (Brazil)
IPS	International Plutonium Storage
ISIS	Institute for Science and International Security
JAEC	Japan Atomic Energy Commission
JNFS	Japan Nuclear Fuel Services Company
KAPS	Kakrapar Atomic Power Station (India)
KARP	Kalpakkam Reprocessing Plant (India)
KEDO	Korean Peninsula Energy Development Organization
KfK	Kernforschungszentrum Karlsruhe
KNK	Kompakte Natriumgekühlte Kernreaktoranlage
LEI	Isotope Enrichment Laboratory (Brazil)
LEU	Low-enriched uranium
LIS	Laser isotope separation
LWGR	Light-water-cooled, graphite-moderated reactor
LWR	Light water reactor
Magnox	Magnesium oxide
MAPI	Ministry of Atomic Power and Industry (former USSR)
MAPS	Madras Atomic Power Station (India)
MPC&A	Material protection control and accountancy
MDF	MOX Demonstration Facility
Minatom	Ministry of Atomic Energy (Russia)

MITI	Ministry of International Trade and Industry (Japan)
MLIS	Molecular laser isotope separation
MOX	Mixed-oxide fuel
MTR	Materials Test Reactor
MUF	Material unaccounted for
MZFR	Mehrzweckforschungsreaktor
NAPS	Narora Atomic Power Station (India)
NAS	National Academy of Science (USA)
NE	Nuclear Electric (UK)
NEA	Nuclear Energy Agency, OECD
NERSA	Groupement Central Nucléaire Européen à Neutrons Rapides
NFS	Nuclear Fuel Services (USA)
NIIAR	Scientific Research Institute for Nuclear Reactors, Dimitrovgrad
NMMSS	Nuclear Materials Management and Safeguards System
NNWS	Non-nuclear weapon state
NOL	Normal operating losses
NPT	Treaty on the Non-Proliferation of Nuclear Weapons
NRC	Nuclear Regulatory Commission (USA)
NRDC	Natural Resources Defense Council (USA)
NWS	Nuclear weapon state
OECD	Organisation for Economic Co-operation and Development
OMV	Ongoing monitoring and verification
PCM	Plutonium contaminated materials
PFBR	Prototype fast breeder reactor (India)
PFDF	Plutonium Fuel Development Facility
PFFF	Plutonium Fuel Fabrication Facility (Japan)
PFPF	Plutonium Fuel Production Facility (Tokai, Japan)
PFR	Prototype Fast Reactor (UK)
PIE	Post-irradiation examination (Daeduk, South Korea)
PNC	Power Reactor and Nuclear Fuel Development Corporation (Japan)
PREFRE	Power Reactor Fuel Reprocessing (India)
Pu	Plutonium
Purex	Plutonium–uranium extraction
PWR	Pressurized water reactor
R&D	Research and development
RAPS	Rajasthan Atomic Power Station (India)
RBMK	Reaktor Bolshoy Moshchnosti Kanalniy/Kipyashchiy [High-power, channel-type reactor], (former USSR)
RepU	Reprocessed uranium
RERTR	Reduced Enrichment Research and Test Reactors (USA)
RMP	Rare Materials Plant (India)
rpm	revolutions per minute
rps	revolutions per second

RWE	Rheinische-Westfälische Elektizitätswerk AG
SAGSI	Standing Advisory Group on Safeguards Implementation (IAEA)
SAP	Service de l'Atelier Pilote
SBH	Siemens Brennelementwerke Hanau
SBK	Schnell-Brüter Kernkraftwerkgesellschaft mbH
SMP	Sellafield MOX plant
SNL	Scottish Nuclear
SPD	Social Democratic Party (Germany)
SPIN	Separation–incineration
SPRU	Science Policy Research Unit (UK)
SQ	Significant quantity
START	Strategic Arms Reduction Talks/Treaty
SWU	Separative work units
TAPS	Tarapur Atomic Power Station (India)
THORP	Thermal Oxide Reprocessing Plant (UK)
TOP	Traitement d'Oxydes Pilote (France)
TOR	Traitement d'Oxydes Rapides (France)
TRR	Taiwan Research Reactor
U	Uranium
UCOR	Uranium Enrichment Corporation (South Africa)
UCS	Union of Concerned Scientists
UEK	Ural Electrochemistry Kombinat
UKAEA	United Kingdom Atomic Energy Authority
UNGG	Uranium Naturel Graphite Gaz (France/Spain)
UNSCOM	UN Special Commission
Urenco	Uranium Enrichment Company
USEC	US Enrichment Corporation
VAK	Versuchsatomkraftwerk Kahl
VDEW	Vereinigung Deutscher Elektrizitätswerke
VVER	Vodo-Vodyanoy Energeticheskiy Reaktor [Water–water power reactor]
WAK	Wiederaufarbeitungsanlage Karlsruhe
WGU	Weapon-grade uranium
ZPPR	Zero Power Plutonium Reactor (Idaho, USA)

Conventions

g	gram
GW	gigawatt
GWd/t	gigawatt-days per tonne of fuel
GWe	gigawatt-electric
GWe (net)	gigawatt-electric (not including power consumed by the power station itself)
GWth	gigawatt-thermal
GWth-d	gigawatt-days of thermal energy
kg	kilogram
kW	kilowatt
kWh	kilowatt-hour
km	kilometre
kt	kiloton
m	metre
mg	milligram
MW	megawatt
MWd/t	megawatt-days per tonne of fuel
MWe	megawatt-electric
MWe-d	megawatt-days of electrical energy
MWth	megawatt-thermal
MWth-d	megawatt-days of thermal energy
s	second
SWU	separative work unit
t	tonne
THM	tonnes of heavy metal
TWhe	terrawatt-hour of electric energy
TWth-d	terrawatt days of thermal energy
y	year
..	Data not available or not applicable
–	Nil or a negligible figure
( )	Estimates

Part I
Introduction

1. Reasons, aims and sources

I. Introduction

In *World Inventory of Plutonium and Highly Enriched Uranium, 1992* (henceforth referred to as *World Inventory 1992*), the authors attempted to establish the quantities of plutonium and highly enriched uranium (HEU) which existed in the early 1990s and to identify where and in which forms they were held.[1] Capabilities for producing these materials were described, scenarios of possible future arisings were presented and policy issues were discussed, particularly in relation to the need for greater transparency and to the problems caused by growing material surpluses.

This revised and extended edition of the book has been prepared for three main reasons. The first is that much new information has become available, allowing more detailed and accurate assessments to be made of plutonium and HEU inventories, including a more precise definition of the uncertainties associated with them.

The second reason is that the world has moved on. To give just a few examples, in the four years since the first volume was prepared the North Korean plutonium production programme and the smuggling of nuclear weapon material from the former Soviet Union have become international issues; the dismantling of redundant nuclear warheads in the nuclear weapon states, and of Iraq's and South Africa's nuclear weapon programmes, has proceeded apace; the THORP (Thermal Oxide Reprocessing Plant) and UP2-800 reprocessing plants in Britain and France have begun operating; and programmes to utilize plutonium have both developed and been under reassessment, not least in Germany and Japan. Furthermore, significant policy initiatives have been taken or are being considered. In particular, international safeguards are being strengthened, commitments have been made to negotiate a treaty which would end the production of fissile materials for nuclear weapons (the Fissile Materials Cut-Off Treaty), and substantial attention is being given to the dismantling of nuclear warheads and the disposition of the materials and components extracted from them. This last issue has been addressed in two major studies published by the US National Academy of Sciences.[2]

---

[1] Albright, D., Berkhout, F. and Walker, W., SIPRI, *World Inventory of Plutonium and Highly Enriched Uranium, 1992* (Oxford University Press: Oxford, 1993).

[2] National Academy of Sciences, Committee on International Security and Arms Control, *Management and Disposition of Excess Weapons Plutonium* (National Academy Press: Washington, DC, 1994); and National Academy of Sciences, Committee on International Security and Arms Control, *Management and Disposition of Excess Weapons Plutonium: Reactor-Related Options* (National Academy Press: Washington, DC, 1995).

The third reason for revising the book is that many initiatives required in the fields of nuclear non-proliferation and disarmament policy now depend on the establishment of accurate and transparent material inventories. This is, for instance, a precondition for increasing confidence that the large quantities of excess weapon-grade plutonium and uranium in nuclear weapon states will not fall into the wrong hands and will eventually be placed under international safeguards. It is central to the efforts being made to strengthen and extend the safeguards system.

This volume takes account of these and other developments. The book follows the same broad structure as its predecessor. It should be stressed, however, that it is not simply an update. Most chapters are completely rewritten, two new chapters have been added (including an extended analysis of the Iraqi nuclear weapon programme) and the final chapter contains an extensive discussion of the policy agenda. Furthermore, the methods and definitions adopted are not everywhere the same. This book therefore *replaces* the *World Inventory 1992*— and it is accordingly retitled.

II. Four security contexts

Plutonium and highly enriched uranium are the essential materials in nuclear weapons. Their fissioning produces the enormous amounts of energy released in atomic bombs or used to ignite thermonuclear weapons. HEU is derived by 'enriching' uranium so that it contains a much higher proportion of the fissile isotope uranium-235 (^{235}U) than is found in natural uranium. Plutonium is derived by irradiating the abundant isotope uranium-238 (^{238}U) with neutrons in a nuclear reactor and then extracting the plutonium from the 'spent fuel' through a chemical technique known in the commercial nuclear industry as 'reprocessing'.

Plutonium and HEU were first produced in the 1940s by the USA and the USSR when they embarked on their nuclear weapon programmes. Since then, large quantities have been produced in many countries. Those with nuclear weapon programmes have acquired extensive stocks of HEU in order to manufacture warheads and to fuel naval reactors. In addition, HEU has been widely used, albeit in much smaller quantities, in civil research reactors. Many tonnes of plutonium have also been produced for nuclear weapons, but the largest amounts have arisen from the irradiation of uranium fuel in civil power reactors. There have long been plans to generate electricity by fuelling reactors with the plutonium extracted from the 'spent fuels' discharged from reactors after irradiation.

As nuclear production in both civil and military domains expanded, and as technologies diffused, knowledge of the scale and whereabouts of plutonium and HEU inventories became increasingly important to international security in the decades following World War II. Today there are four contexts in which this knowledge, or the lack of it, has assumed great significance.

The first is that of regional nuclear proliferation—the attempts by some countries to acquire materials for their fledgling nuclear weapon programmes. The monitoring and control of fissile materials and of the technologies for producing them have long been central to the workings of the nuclear non-proliferation regime. If anything, their importance has increased as knowledge of weapon design has diffused and as many countries have gained access to delivery vehicles. The clandestine activities of Iraq and North Korea have recently demonstrated how important it is to keep track of material and technology flows and to monitor developments in the countries involved. In both cases the international community has taken unprecedented measures to establish the extent of production capabilities and the amounts of material that have been acquired, in an effort to prevent them from being used to make nuclear weapons and to provide assurances that their nuclear weapon programmes have been halted.

The second context is that of nuclear arms reductions by the nuclear weapon states, and by the USA and former Soviet Union in particular. These developments have already involved the dismantlement of thousands of nuclear warheads and the extraction from them of large amounts of plutonium and HEU. Precise material inventories are required for managing the storage, disposal or recycling of the materials, for managing their entry into civil fuel cycles where that is envisaged, for providing confidence that they are well protected and will not again become available for making nuclear weapons, and for applying safeguards to them. Equally important, the quantities and condition of fissile material held outside weapons need to be precisely identified. The scale of this undertaking has only recently been appreciated as information about nuclear weapon production programmes, about the ways in which they were managed and about the condition of the materials, components and facilities linked to them, has become available. If complete nuclear disarmament is accepted to be the main objective, full disclosure of national inventories and their associated uncertainties will become even more important.

The third context is that of civil spent-fuel management, which is gaining in significance as the quantities of plutonium contained in fuels discharged from nuclear power reactors increase. Most of this plutonium is likely to remain in spent-fuel assemblies, which will be stored or eventually buried. However, the extraction of plutonium has been increasing significantly as a result of the expansion of commercial reprocessing in France and the United Kingdom, and may increase further early in the next century if the Rokkasho-mura reprocessing plant in Japan is constructed and operated according to plan. (Civil reprocessing also continues in India and Russia.) As the UK and France are reprocessing fuels from other European countries and from Japan, and as the plutonium is due to be returned to its owners for recycling in nuclear reactors, these activities will, if continued, lead to substantial growth in the international circulation of plutonium. This circulation would become even more pronounced if plutonium from dismantled warheads was burned in nuclear reactors outside its country of origin.

The fourth security concern is the theft of fissile materials and the development of black markets in them. Unauthorized trade in plutonium or highly enriched uranium could exacerbate nuclear weapon proliferation and increase risks of nuclear terrorism. If direct access were gained to fissile materials, one of the main barriers to proliferation—the construction of facilities for producing them—would have been circumvented. Although this problem may be most acute in the territories of the former USSR, it is not unique to that part of the world. It reinforces the need to establish effective national systems of accounts, to strengthen physical protection and to widen international safeguarding in the nuclear weapon states and in threshold countries where there are unsafeguarded materials and facilities.

To the above concerns should be added the environmental and safety risks attached to radioactive nuclear materials. The accumulation of spent fuels in both civil and military sectors, the extraction of plutonium and high-level wastes from them, the transport and recycling of plutonium, the dismantling of nuclear weapons, the storage and disposition of weapon components and materials—all these activities are hazardous and require strict monitoring and control. If such monitoring and control are not properly exercised, the confidence of publics and states in these activities will not be easily gained. Nuclear power's availability as one of the main options for meeting the expected growth of demand for electricity without adding to atmospheric pollution would be jeopardized.

III. The need for greater transparency

In each of these contexts, there is need for greater transparency with regard to inventories of nuclear materials. Knowledge of plutonium and HEU inventories is still very incomplete. Until recently, it was assumed that such knowledge should be kept under wraps by governments, industrial companies and international organizations. In countries possessing nuclear weapons, or trying to acquire them, information about HEU and plutonium produced and used for military purposes is still usually classified. The position is not much better in the civilian context. Information gathered by international agencies for safeguards purposes is held on a confidential basis and is not open to detailed public scrutiny, or even to the scrutiny of national authorities. The International Atomic Energy Agency (IAEA) and the European Atomic Energy Community (Euratom) only publish broad aggregates so as to protect the identity of the countries and industrial operators providing the information. In all areas, military and civil, the information that exists in the public domain is often inconsistent, scattered and incomprehensible to the layman.

Over the past four years, the situation has improved significantly. There is broad agreement that greater transparency is a desirable goal. More information has been released, not least by the United States about its military inventories and Japan about its civil inventories, and nine governments recently committed

REASONS, AIMS AND SOURCES 7

themselves to publishing civil plutonium inventories in a standard format.[3] Nevertheless, the situation remains far from satisfactory. In some countries and institutions, secrecy remains the norm. It is still the case that no official published collections of international statistics exist for these crucial materials.

Drawing on past work and recent publications, and on further research that has been carried out by the authors and others since *World Inventory 1992* was published, this book tries to bring together in one volume what is known and not known about the world's HEU and plutonium inventories. It seeks answers to five main sets of questions:

1. How much plutonium and HEU has been produced for military purposes in the nuclear weapon states (NWS), and at which locations? How much is held inside and outside nuclear weapons? How much may be released as weapons are dismantled? What is the scale of the excess stock?

2. Country by country, how much plutonium is contained in spent nuclear fuels arising from nuclear power programmes? How much more may be produced over the next two decades, and how much may be surplus to requirements?

3. How much plutonium has been separated from civil spent fuels, and where? How much may be separated and recycled in the next two decades, and how much may be surplus to requirements? How much HEU exists in the civil domain?

4. How much plutonium and HEU has been produced outside international safeguards in the non-nuclear weapon states (NNWS)? How much may have been assigned to nuclear weapon programmes? What is the status of key threshold states, such as Iraq and Iran?

5. What are the main gaps in information that need to be filled in order to gain a full picture of plutonium and HEU inventories?

In *World Inventory 1992*, the importance of establishing an international register of plutonium and HEU stocks was emphasized. A register of civil stocks in non-nuclear weapon states that are party to the 1968 Treaty on the Non-Proliferation of Nuclear Weapons (NPT) could easily be assembled (the possibility has been discussed in the IAEA and elsewhere) and readily extended to incorporate civil inventories in France, the UK and the USA. However, it is now recognized that an accurate and comprehensive register of plutonium and HEU stocks, embracing civil and military inventories in all countries, could not be established at this time even if the political will existed to do so. The simple and unpalatable truth is that the precise quantities of material that exist in a number of countries, in military and even in civil sectors, are probably not known to their governments. Although accurate figures on some stocks may be available to them, for others they may have only rough approximations, and for still others they may have little or no information. The necessary precursors to

---

[3] These countries are Belgium, France, Germany, Japan, the Netherlands, Russia, Switzerland, the UK and the USA.

the publication of inventories, and to their being brought under international safeguards, are the assembly of full histories of the production and disposition of fissile materials, and the establishment of comprehensive and effective national systems of accounts where they are still lacking. This is not a simple matter given changes in state boundaries, the lack of finance and authority required to put such systems in place, and the materials' differing forms and degrees of accessibility. It is, however, a vitally important one.

This issue is discussed further in chapter 15 and *passim*. It is evident that a two-pronged approach is required. Countries that have a firm grasp of the quantities of plutonium and HEU in their possession could submit the information to the IAEA or another accredited organization so that an annual register could be assembled. Those that do not have such information need to take immediate steps to establish, with international assistance where necessary, comprehensive national systems of accounts where they are lacking and to gain the fullest possible understanding of the sizes, locations and forms of material inventories within their countries. While progress is being made in this regard, particularly as a result of US–Russian bilateral initiatives, it is not yet sufficient. When precise data become available, they could gradually be incorporated in the international register.

IV. The limits to accuracy

Many hundreds of tonnes (t) of plutonium and HEU have accumulated since the 1940s. Ideally, the amounts held in any context should be known to the nearest few kilograms (kg), or even grams (g). Sophisticated modern nuclear weapons are typically estimated to contain on average 3–4 kg of plutonium and 15–30 kg of highly enriched uranium. The IAEA's Safeguards Division tries to ensure that inventories in specific material balance areas in non-nuclear weapon states are accounted for within annual error margins of 8 kg of plutonium and 25 kg of HEU. These are the 'significant quantities' recommended by the IAEA's Standing Advisory Group on Safeguards Implementation (SAGSI).

These levels of accuracy are not and could not be attempted in this book. The purpose of IAEA safeguards is to detect possible diversions of what may be very small quantities of material from civil to military use, or back into military use when its origin was in weapon programmes. The purpose of this study is instead to provide an empirical framework for policy analysis that is as nearly comprehensive as possible given the current state of knowledge, and to bring together and present information in ways that make it accessible to a public audience. This is something that the safeguards agencies would themselves have difficulty doing because many of the important stocks are outside their purview, and because they are constrained in what they can disclose.

The estimates presented here and the methods used to establish them have been cross-checked where possible with people and institutions with special expertise, and much effort has gone into making the figures as accurate as possible. Despite these best efforts, significant error margins are inescapable. Many of the figures presented here are derived from knowledge of reactor histories and fuelling arrangements and of enrichment, reprocessing and plutonium recycling programmes. Some of the statistical uncertainty is intrinsic to the subject and is shared by industries, governments and safeguards authorities. For instance, precise figures cannot be attached to the plutonium content in spent fuel unless and until it is separated during reprocessing, and there are losses of material in production processes which can usually only be estimated. However, much of the uncertainty stems from the public unavailability of information and could be eliminated by lifting the veil of secrecy.

Error margins vary from a few percentage points in the context of most civil programmes to the much larger margins encountered in particular in relation to the British, Chinese, French and former Soviet nuclear weapon programmes and to programmes in the threshold countries. The error margins presented in the following chapters are mainly indicative. They have not resulted from the application of formal statistical techniques, but instead suggest the degree of confidence that can be attached to the figures. Even greater uncertainty surrounds the scale, form and distribution of future plutonium and HEU inventories. Where appropriate, simple scenarios are presented, with little attempt to attach probabilities to them.

V. The scope of the book

The purpose of this book is to present information on plutonium and HEU, and on the capabilities associated with them. Broad policy implications are drawn in the final chapter, but analyses of consequences have elsewhere been kept to a minimum. It needs to be stressed at the outset that there is no necessary correlation between quantities of material and their political and strategic impact. In one location 50 kg or even 50 grams of weapon material may give rise to greater concern than 50 tonnes in another. Stocks of material in a country that appear 'safe' today may also appear 'unsafe' tomorrow, and vice versa. This is one reason why transparency and the universal and permanent application of safeguards, to the same high standards, are such important goals. The political significance of specific material inventories depends on the intentions and capabilities of the countries possessing them, the constraints and pressures upon them, their internal political conditions and their positions in the international order, all of which are prone to change. Weighing these factors is beyond the scope of the book.

Following this introduction, chapter 2 presents a brief explanation of the main technical features of plutonium and HEU and of their production processes. Part II is concerned with military inventories in the acknowledged nuclear weapon states. It opens in chapter 3 with an assessment of plutonium inventories, the estimates being derived mainly from information about military production reactors and from information divulged by governments (notably in the case of the United States). Chapter 4 considers their stocks of weapon-grade uranium, the estimates being derived from information on enrichment plants and programmes and on non-weapon uses of HEU, and again from information, albeit rather scanty information in this case, divulged by the US Government.

Part III is concerned with the principal civil inventories. Chapter 5 assesses, by country and region, the quantities of plutonium contained in spent fuels discharged from power reactors. Chapter 6 considers the amounts of plutonium separated from these fuels, mainly in France, Japan, Russia and the UK, where civil reprocessing continues on a large scale, and offers scenarios of future plutonium arisings. It also assesses the scale and location of current stocks of separated civil plutonium. The uses to which these inventories have been and may be put are examined in chapter 7. Chapter 8 covers the more limited civil inventories of HEU, particularly regarding the material's use in research reactors and efforts to reduce HEU inventories.

Part IV is concerned with inventories in 'threshold countries' which have attempted to gain access to the materials and technologies necessary for weapon production. Chapter 9 considers those among them with the most long-standing and developed nuclear weapon programmes (Israel, India and Pakistan). Chapter 10 covers North Korea, focusing both on the recent controversy and its plutonium inventory. Chapter 11 takes up the special case of Iraq, the only country that is forbidden to make or possess separated plutonium or HEU. Other countries which have received much international attention, Iran, Algeria, South Korea and Taiwan, are grouped together in chapter 12. Chapter 13 considers countries that have in recent years abandoned their nuclear weapon programme, with an assessment being provided of the situation in South Africa.

Part V concludes the book. Chapter 14 offers an overview of plutonium and HEU inventories, drawing on the statistics presented in the preceding chapters. Chapter 15 discusses the policy agenda with particular reference to the extension and deepening of international controls over fissile materials, and the disposition of excess stocks.

Inevitably, constraints on time, resources and access to information have resulted in less attention being given to some topics than to others. In particular it is regrettable that more could not have been said about HEU used in naval fuel cycles or the amounts, forms and whereabouts of material extracted from dismantled warheads in the nuclear weapon states.

VI. Sources

The findings in this book rest on information gathered from a variety of published and unpublished sources, from discussions with experts in government, industry, safeguards agencies, the nuclear community and academia, and from retired officials. A questionnaire was also sent to electricity utilities with operating nuclear power stations requesting up-to-date information about spent fuel and plutonium arisings, with excellent results. The authors are grateful for their cooperation.

In the process of writing this book, many nuclear facilities and experts in numerous countries were visited. These on-site visits were invaluable to gaining deeper insight into fissile material inventories and production capabilities. While some public information is very reliable, much is patchy and inconsistent. One of the main purposes of this book is to organize, weed and make sense of that which is available. The reader should nevertheless keep in mind that little of the information from the above sources can be taken for granted, particularly where secret programmes are involved. An effort is made throughout to explain the sources, check their reliability and indicate where doubts remain.

Attributions to individuals can be made only in a few cases. Usually information is imparted by persons employed by government, industry and international organizations, or by those formerly involved in nuclear weapon programmes, on the agreement that they remain anonymous. Ingrained bureaucratic caution often prevents people from talking on the record, even when the information sought has no special political or commercial value. This has not prevented us from seeking reliable, up-to-date information. Constraints on the disclosure of sources apply especially to the chapters on the nuclear weapon states and the threshold countries. Where information has been provided in confidence, we usually indicate broadly that it has come from 'an industry official', 'an intelligence source' or 'a government official'. There have been several occasions where this has not been possible because to indicate the source even in these vague terms might allow the person giving the information to be identified.

It should be stressed that the authors of this book have no affiliation to any government, commercial enterprise, security service or international organization. They are independent academic researchers, and the work has been funded by independent charitable foundations or donors.

2. Characteristics of highly enriched uranium and plutonium and their production processes

I. Introduction

Plutonium and highly enriched uranium have two features in common. The first is that they contain large proportions of fissile materials whose nuclei can break apart, or fission, when bombarded with neutrons, emitting more neutrons than they absorb. This gives rise to the possibility of sustained chain reactions, and thus to explosive or controlled releases of energy. The second common feature is that they are difficult and expensive to produce. Their acquisition requires heavy capital investment and mastery of a wide range of technologies. Were this not the case, many more countries would probably have nuclear weapons today—hence the great importance that nation states have attached to controlling the production and distribution of these materials and the technologies associated with them.

In other respects, the properties of plutonium and HEU, and the nature of their production processes, are very different. While the materials are used together in nuclear weapons, they present different kinds of problem to their producers and users and to those trying to exercise control over them. This chapter provides a brief introduction to their main characteristics.[1]

II. Highly enriched uranium

Uranium isotopes

Isotopes are forms of an element which have nearly identical chemical and physical properties but different nuclear properties. The chemical properties of elements are fixed by the number of positively charged protons in their nuclei and by the corresponding number of negatively charged electrons that they carry. The isotopes of an element have nuclei containing the same number of protons but different numbers of neutrons. Neutrons carry no electrical charge and can thus move with considerable freedom through atomic structures. They can penetrate an atomic nucleus and can cause the nucleus to fission, releasing a relatively large amount of energy.

Whereas most naturally occurring isotopes are stable, man-made isotopes such as those produced through the irradiation of uranium and its daughter products are unusually unstable and radioactive. The stability of an isotope is indicated by its half-life, which is the time taken for a quantity of an isotope to

---

[1] For a more detailed explanation see Kokoski, R., SIPRI, *Technology and the Proliferation of Nuclear Weapons* (Oxford University Press: Oxford, 1995).

halve through radioactive decay. Half-lives can vary from fractions of seconds to hundreds of millions of years. Radioactive isotopes emit three main kinds of radiation when they decay: alpha particles, which carry positive charges and consist of two protons and two neutrons (the helium-4 nucleus); beta particles, which are energetic electrons (negatively charged) or positrons (positively charged); and gamma rays, which have no charge and are the most penetrating. Neutrons and various sub-atomic particles may also be released.

Uranium (U) has 92 electrons and 92 protons (the atomic number). Of the 14 isotopes in the sequence ^{227}U to ^{240}U (the mass numbers), ^{235}U and ^{238}U are the most important.[2] With half-lives of 700 million and 4500 million years respectively, ^{235}U and ^{238}U are relatively stable isotopes. They are not strongly radioactive and can be handled by industrial workers without the need for substantial protection. In these respects, uranium contrasts with plutonium (Pu). Its principal isotopes have much shorter half-lives and workers handling this material require extensive protection.

Naturally occurring uranium consists of 99.28 per cent of ^{238}U and of 0.71 per cent of ^{235}U.[3] Moreover ^{235}U, like ^{239}Pu and ^{241}Pu, fissions when irradiated with relatively low energy ('thermal') neutrons, allowing heat to be released under controlled conditions in a class of reactor called 'thermal'.[4] In thermal reactors, neutrons are slowed down or 'moderated' by materials such as graphite and water.

For nuclear weapons, and for fuel burned in many types of nuclear reactor, it is necessary to increase concentrations of ^{235}U. This is the process known as 'enrichment'. The low-enriched uranium (LEU) used to fuel commercial power reactors generally contains 2–6 per cent ^{235}U. HEU is defined as uranium containing over 20 per cent ^{235}U. For fission-type nuclear weapons, ^{235}U concentrations of 90 per cent and over are usually desired. HEU at this level of enrichment is often referred to as 'weapon-grade uranium'. HEU with lower enrichments is also used in thermonuclear weapons.

The following five grades of uranium are commonly recognized:

1. Depleted uranium, containing less than 0.71 per cent ^{235}U.
2. Natural uranium, containing 0.71 per cent ^{235}U.
3. Low-enriched uranium, containing more than 0.71 per cent and less than 20 per cent ^{235}U.
4. Highly enriched uranium, containing more than 20 per cent ^{235}U.
5. Weapon-grade uranium, HEU containing more than 90 per cent ^{235}U.

---

[2] The isotope ^{233}U, which is derived from neutron capture in thorium-232, has chemical properties similar to ^{235}U, but a critical mass similar to that of ^{239}Pu. It has been evaluated as a nuclear weapon material in the USA and possibly elsewhere. However, the thorium fuel cycle has not progressed beyond the R&D stage (it received most attention in the 1950s and 1960s), and the quantities of ^{233}U that have been produced are very small. The thorium fuel cycle has been most actively researched in India, which has large deposits of ores containing thorium.

[3] It also contains 0.006% of ^{234}U.

[4] 'Thermal' implies neutron velocities akin to the velocities of molecules in gases at room temperature (i.e., 2200 m/s or 0.025 electron volts, eV). 'Fast' neutrons are usually defined as having energy within the range 20 keV to 10 MeV.

It should be stressed that a self-sustaining chain reaction in a nuclear weapon cannot occur in depleted or natural or low-enriched uranium and is only theoretically possible in LEU of roughly 10 per cent or greater. The critical mass of uranium which can give rise to explosive releases of energy can only be constructed in practice from materials containing high proportions of the fissile isotope ^{235}U. Thus the enriched uranium burned in conventional nuclear power reactors has no direct military value. For countries with enrichment plants, the possession of LEU can, however, reduce the time and cost involved in producing HEU since fewer separative work units are required to produce weapon-grade material if LEU is used as feed. Nevertheless, LEU's worthlessness as a weapon material is recognized in the two tiers of safeguards and physical protection measures applied to enriched uranium. Less stringent standards are applied when enrichment levels fall below 20 per cent. In contrast, a single set of regulations is applied to plutonium since it can be used in nuclear weapons in most available isotopic mixes.[5]

The corollary is that none of today's thermal power reactor designs require HEU fuels. The HEU-fuelled high-temperature reactor (HTR), prototypes of which were built in the Federal Republic of Germany (FRG) and the USA, is in abeyance. HEU is only used in submarine reactors, in a small number of breeder reactors and in a few large research reactors. Since the mid-1970s, many research reactors around the world have been converted to operate with uranium enriched to levels below 20 per cent (see chapter 8). In France and the former Soviet Union, some submarine reactors have also been designed so that they can be fuelled with LEU.

Enrichment techniques

The techniques of gaseous diffusion and centrifuge enrichment dominate today's enrichment industry.[6] However, they are not the only techniques available. The aerodynamic techniques applied in South Africa, and the electromagnetic separation technique applied in Iraq, raised considerable concern as the plants constructed there were not subject to international safeguards. Laser and chemical techniques have also been developed in recent years. The multiplicity of technical approaches to uranium enrichment has become one of the main problems in detecting clandestine enrichment programmes in countries such as Iraq, and in tracking the technology and equipment flows connected with them.

The separative work in an enrichment plant indicates the energy expended in separating the uranium feed into enriched product and depleted uranium waste, commonly called the tails. The tails assay is the concentration of ^{235}U left in this

---

[5] An exception is made for the isotope ^{238}Pu, which is a strong alpha-emitter and is used as a heat source in medical and space applications. Safeguards are not applied when concentrations of this isotope exceed 80%.

[6] For discussions of uranium enrichment techniques, see Krass, A. S., *et al.*, SIPRI, *Uranium Enrichment and Nuclear Weapon Proliferation* (Taylor & Francis: London, 1983); Tait, J. H., 'Uranium enrichment', ed. W. Marshall, *Nuclear Power Technology, Vol. 2: Fuel Cycle* (Clarendon Press: Oxford, 1983); and Kokoski (note 1).

CHARACTERISTICS OF HEU AND PLUTONIUM 15

waste. The unit of measurement is the kilogram separative work unit (kg SWU, usually abbreviated to SWU). The capacities of enrichment plants are expressed in SWU per year. It takes approximately 200 SWU to make 1 kg of weapon-grade uranium (uranium enriched to 90 per cent) using natural uranium feed and a tails assay of 0.3 per cent.

In brief, the following are the main enrichment processes. Further details of enrichment technologies and programmes appear in chapter 4 and appendix A.

Gaseous diffusion

This technique exploits the property of gases whereby heavy molecules travel more slowly than light molecules. If parts of the vessel containing the gas are made permeable, in the form of a barrier, the lighter molecules will pass through the diffusion barrier more rapidly, causing the escaping gas to be enriched in the lighter components. Thus the uranium hexafluoride (UF_6) gas emanating at the end of a diffusion stage will be slightly enriched in the isotope ^{235}U. The final degree of enrichment attained depends on the number of stages in the cascade and on the enrichment of the initial feed.

Gaseous diffusion accounts for approximately one-half of world enrichment capacity. Essentially all past HEU production in the USA, the UK, France and China has been based on this technique. As it is an electricity-intensive process, plants are usually sited by large hydroelectric power stations (as at Oak Ridge in Tennessee, USA) or by dedicated nuclear power stations (as at Tricastin in France).

Centrifuge enrichment

In this process the heavier molecules in a rotating gaseous mass move towards the outside of the fluid mass and are subjected to a counter-current flow. To first order, it is the same technique as that used in the separation of cream in the dairy industry. In the context of uranium enrichment, it requires high-precision engineering and sophisticated metallurgy because of the high speed of the centrifuges. Power consumption in a centrifuge plant ranges from 50 to 400 kilowatt-hours (kWh) per SWU, substantially less than the 2500 kWh per SWU that are typical of a gaseous diffusion plant.

The huge enrichment capacity that Russia has inherited from the Soviet Union is based on centrifuge techniques. They also form the basis of the Uranium Enrichment Company (Urenco) facilities in Germany, the Netherlands and the UK, and of the plant that recently began operating in Japan. There have long been plans to build a commercial-scale facility to this design in the USA, but overcapacity in the US enrichment industry has acted as a disincentive. As is discussed in chapters 9, 11 and 13, Pakistan, Iraq and Brazil have also followed this route when seeking to attain nuclear weapon capabilities.

Aerodynamic enrichment

This technique has two variants. The Becker jet nozzle, developed in the Federal Republic of Germany and only attempted on an industrial scale there and in Brazil (and since abandoned in both countries), exploits the mass dependence of the centrifugal force in a fast, curved flow of UF_6. The gas expands into a curved duct and the flow is split into heavier and lighter fractions by means of a skimmer. In the South African process, a mixture of hydrogen and UF_6 is allowed to swirl in a separating element which acts as a stationary-walled centrifuge. Neither aerodynamic technique has been shown to be commercially viable.

Electromagnetic separation

This technique, as recently applied in Iraq, is discussed in detail in chapter 11. In a device originally called a calutron, heavy and light uranium ions (atoms carrying electrical charges) follow trajectories with different curvatures in a strong magnetic field. This technique was used to produce HEU for the first US atomic weapons, but then rejected because of its very high capital and energy intensity.

Laser enrichment

There has been speculation since the early 1970s that laser enrichment will provide the basis of the next generation of enrichment plants. If made to operate effectively, it would have the lowest power consumption and the greatest mechanical simplicity among the physical techniques, although it would still be a complex and difficult process. The isotopes of uranium can be selectively excited by high-energy lasers. There are two technical approaches. In the atomic route, ^{235}U is selectively excited using tuneable lasers, and the resulting ionized atoms are separated electromagnetically. In the molecular route, selective infrared absorption of $^{235}UF_6$ gas is followed by further irradiation at infrared or ultraviolet frequencies, allowing the dissociation of the excited molecules or their chemical separation.

The first of the above techniques was adopted in the USA's Atomic Vapor Laser Isotope Separation (AVLIS) project. In 1985, the US Department of Energy (DOE) chose the AVLIS process when planning new enrichment capacity for the 1990s and beyond. However, the construction of AVLIS plants has been delayed owing to development problems and the mounting oversupply in the world enrichment market. Research programmes on laser enrichment are being carried out in several other countries.

Plasma separation and chemical exchange

Two other techniques are still at the research and development (R&D) stage. The first is the plasma separation process. In one version, uranium atoms are exposed to low-energy radio frequency waves resonating with the 'cyclotron frequency' of ^{235}U ions. If rotated, these ions can be collected on electrically charged plates. The second and more important technique is the chemical exchange process, which depends on a slight tendency of ^{235}U and ^{238}U to concentrate in different molecules when uranium compounds are continuously brought into contact. Catalysts are used to speed up the chemical exchange. Pilot plants using this technique have been built in France and Japan.

HEU recycling

For many years, HEU from spent submarine and research reactor fuels has been recovered by chemical reprocessing.[7] It has either been used as a fuel for plutonium production reactors, as in the cases of domestic and foreign spent HEU fuels reprocessed at the Savannah River and Idaho National Laboratory facilities in the USA; or it has been blended with depleted uranium, to produce low-enriched uranium which can be used in power reactors. The latter option has been followed by the former Soviet Union (FSU) in providing uranium fuel for the high-power channel-type reactors (RBMK—Reaktor Bolshoy Moshchnosti Kanalniy/Kipyashchiy).

What is now in prospect is that very large amounts of HEU, mostly containing more than 90 per cent of ^{235}U, will be released from dismantled nuclear weapons taken from the arsenals of the USA and the FSU. If this were diluted with depleted uranium, it would give rise to substantial quantities of low-enriched uranium, which could be used to fuel power reactors. A deal has been struck between the Russian and US governments whereby a substantial proportion of the former Soviet Union's stockpile of HEU will be purchased by the USA (Kazakhstan sent a much smaller stock of less than 600 kg of HEU to the USA in November 1994). After dilution, it will be introduced into the civil fuel cycle.

The point to be stressed here is that recycling HEU is relatively straightforward, at least in technical terms, providing it is sufficiently 'pure'. Steps have to be taken to limit the proportion of the ^{234}U isotope in the diluted product (as ^{234}U is also concentrated by the enrichment process, HEU tends to be rich in this isotope). In the case of the Russian HEU, this is being achieved by using 1.5 per cent enriched uranium rather than natural or depleted uranium as the diluting material. It is also necessary to ensure that the HEU has not been badly contaminated with other metals used in nuclear weapons. Otherwise, there are no technical obstacles to reducing stocks of this weapon material. Moreover, there is a ready commercial market for the resulting uranium fuel. If there is a problem, it is that introducing HEU to the world market when there is already a surfeit of enrichment supply will tend to depress the prices of natural and enriched uranium, or at least keep prices at their present low levels (see

---

[7] HEU extracted from spent fuels can only be re-enriched with difficulty, particularly because of the presence of the isotopes ^{232}U and ^{236}U. The ^{232}U increases radiation risks, while ^{236}U can form bubbles in the middle of a gaseous-diffusion cascade, impeding flows of the other isotopes.

chapter 15). While the uranium industry might lose, the electricity consumer would benefit from the low prices, although utilities would have to guard against the risk that under-investment in uranium mining would lead to a reduction in sources of supply and to future scarcity and price inflation.

As seen below, this is not the case for plutonium. Plutonium stocks cannot be extinguished in this way, and there will be economic disincentives to using them to fuel power reactors. Unless uranium prices rise very substantially their usage would have to be subsidized.

III. Plutonium

Unlike uranium, all but trace quantities of plutonium are manufactured material. Several of plutonium's isotopes are also highly radioactive, and its processing into weapon components or into fuel is far from straightforward. Plutonium is therefore a difficult and hazardous material to work with, but easier to make.

Plutonium isotopes and grades

Plutonium-239 is produced in a nuclear reactor when ^{238}U is irradiated with neutrons. Neutron capture turns ^{238}U into ^{239}U, which decays via neptunium-239 in a matter of days to ^{239}Pu. While ^{239}Pu has a half-life of 24 000 years, and is thus a relatively stable isotope, it is readily fissioned by both thermal and fast neutrons. It also absorbs neutrons, in addition to being fissioned by them, resulting in the formation of the isotope ^{240}Pu. Subsequent neutron captures lead to accumulations of the higher-numbered isotopes ^{241}Pu, ^{242}Pu and ^{243}Pu.

Plutonium-239 and ^{241}Pu are more susceptible to fissioning than the other plutonium isotopes, and are alone in being fissionable by thermal neutrons. They are therefore usually referred to as the 'fissile' isotopes of plutonium. In plutonium commerce, quantities are often expressed in terms of the amount of these fissile isotopes in a particular batch of material. The combined weight of ^{239}Pu and ^{241}Pu is then recorded as Pu_{fiss}, as distinct from Pu_{tot} which refers to the total weight of plutonium isotopes in the batch. For safeguards purposes, however, quantities are always measured in terms of total plutonium.

While the plutonium used in nuclear weapons usually contains very small quantities of ^{241}Pu, this is not the case with the plutonium derived from most power reactor fuels. The presence of ^{241}Pu can cause serious problems in plutonium handling since it decays to americium-241 (^{241}Am), which is an intense emitter of alpha particles and X- and gamma-rays. Plutonium-241 has a half-life of 13.2 years so that substantial quantities of ^{241}Am can quickly accumulate in plutonium separated from reactor fuels, leading to the need for heavier shielding to protect workers involved in the handling and safeguarding of plutonium. Limits have also been imposed on the acceptable amount of spontaneous neutron radiation from the even-numbered isotopes that are produced when fuels are irradiated for longer periods. It is usually best to recycle plutonium within three years of separation. If it is left longer, the costs of fuel fabrication can rise steeply because of the need to provide extra protection or to extract the americium by chemical means before fuel fabrication.

While ^{241}Pu is a problem in nuclear commerce, the even-numbered isotopes ^{240}Pu and ^{242}Pu are particular irritants for nuclear weapon designers.[8] Unlike ^{238}U, the isotopes ^{240}Pu and ^{242}Pu are readily fissioned by fast neutrons, even if less readily than the odd-numbered isotopes. Their presence does not prevent a chain reaction from occurring, although the critical mass required increases with their concentration. However, these isotopes fission spontaneously, producing energetic neutrons which can result in premature initiation of a chain reaction in the plutonium when the implosion of nuclear warhead material takes place. As a result, the weapon can 'fizzle', reducing its explosive yield. The heat emitted by the even-numbered isotopes is a further complication for the weapon designer. Reactor-grade material would also have to be used with more caution by a weapon manufacturer because of its higher radiotoxicity.

There are ways of avoiding these problems, even with plutonium containing substantial fractions of the even-numbered isotopes, without major compromises in yield, reliability, weight or efficiency. In the late 1950s and early 1960s, theoretical studies were carried out at the Los Alamos National Laboratory to assess the feasibility of using different grades of plutonium in nuclear weapons. The conclusion that operational weapons could be constructed from non-weapon-grade material was verified in a successful underground test carried out at the Nevada Test Site in the USA in 1962. Nevertheless, the weapon designer's preference is always for material with high concentrations of ^{239}Pu.

A distinction is therefore commonly made between three different grades of plutonium. The following definitions are widely used:

1. Weapon-grade plutonium, containing less than 7 per cent ^{240}Pu.
2. Fuel-grade plutonium, containing from 7 to 18 per cent ^{240}Pu.
3. Reactor-grade plutonium, containing over 18 per cent ^{240}Pu.

'Super-grade plutonium' is sometimes used to describe plutonium containing less than 3 per cent ^{240}Pu. The term 'weapon-usable plutonium' has no precise definition. It has been adopted on occasion to convey the message that most isotopic mixtures of plutonium can be used in nuclear weapons, or to imply that a given quantity of plutonium is in separated form and can thus be quickly introduced into weapon manufacture.

---

[8] The most authoritative discussion of weapon design and grades of plutonium is Mark, J. C., *Reactor-Grade Plutonium's Explosive Properties* (Nuclear Control Institute: Washington, DC, Aug. 1990). See also Sutcliffe, W. G. and Trapp, T. J., *Extraction and Utility of Reactor-Grade Plutonium for Weapons*, UCRL-LR-115542, Lawrence Livermore National Laboratory, Livermore, Calif., 1994; and National Academy of Sciences Committee on International Security and Arms Control, *Management and Disposition of Excess Weapons Plutonium* (National Academy Press: Washington, DC, 1994), pp. 32–33.

Table 2.1. Plutonium half-lives, and weapon-grade and reactor-grade isotopic concentrations, at given fuel discharges

Half-lives are given in years; concentrations are percentages.

Isotope	Half-life	Weapon-grade isotopic concentrations (typical)	Reactor-grade isotopic concentrations (typical)		
			PWR[a] (33 000 MWd/t[d])	Magnox reactor[b] (5000 MWd/t)	CANDU[c] (7500 MWd/t)
^{238}Pu	86.4	..	1.3	..	..
^{239}Pu	2.4×10^4	93.0	56.6	68.5	66.6
^{240}Pu	6.6×10^3	6.5	23.2	25.0	26.6
^{241}Pu	13.2	0.5	13.9	5.3	5.3
^{242}Pu	3.8×10^5	..	4.7	1.2	1.5

[a] Pressurized water reactor.
[b] Gas-cooled reactor with metallic fuel.
[c] Canadian deuterium–uranium reactor.
[d] Megawatt-days per tonne of uranium fuel.

Sources: Organisation for Economic Co-operation and Development, Nuclear Energy Agency, *Plutonium Fuel: An Assessment* (OECD: Paris, 1989), tables 2 and 4; and authors' data.

Plutonium production: irradiation of reactor fuels

The production of plutonium is carried out in two main industrial stages. The first involves the irradiation of uranium fuels by neutrons in nuclear reactors. The second involves the chemical separation of plutonium from the uranium, transuranic elements and fission products contained in discharges of irradiated fuel. The second technique is usually referred to as 'reprocessing' when applied commercially and 'plutonium separation' when carried out for military purposes. The need to construct both reactors and reprocessing facilities outside safeguards to acquire plutonium is one reason why some countries intent on acquiring weapon capabilities have favoured the use of HEU in recent years.

Although they can overlap, plutonium is produced in two different contexts. In the military context, the reason for irradiating nuclear fuel is to acquire stocks of weapon-grade material for use in nuclear warheads: plutonium supply is the *raison d'être*. In the civilian context, the purpose is to generate electricity, plutonium being a by-product which may or may not have further uses. The days have gone when it was considered imperative to accumulate plutonium as the starting fuel for a coming generation of advanced reactors. The isotopic content of discharged plutonium is a serious concern of nuclear weapon designers, but is less important to electricity producers.

As distinct from the production of uranium with high concentrations of ^{235}U, the technique of isotopic enrichment has not been used to produce weapon-grade plutonium from lower-grade material.[9] Research was carried out in the 1960s in the United States and former Soviet Union using calutrons (and centrifuges) to separate the plutonium isotopes. More substantial R&D programmes were launched in the 1980s to develop laser techniques for enriching plutonium, and plans were prepared, especially in the USA, to extract weapon-grade plutonium by these means. These plans came to nought because of the reduced demand for weapon-grade plutonium as weapon programmes were curtailed. However, it is recognized that the successful development of plutonium enrichment techniques would ease access to weapon-grade material and thus simplify the acquisition of high-performance weapons. Countries with nuclear power programmes could, in principle, obtain weapon-grade plutonium from their highly irradiated power reactor spent fuel if plutonium enrichment became a viable option.

Instead, the nuclear weapon producers have achieved the desired isotopic content of plutonium mainly by controlling the extent to which uranium fuel elements are irradiated with neutrons in nuclear reactors. This is known as the fuel burnup, whose unit of measurement is megawatt-days per tonne (MWd/t) of uranium fuel. Weapon-grade plutonium is produced by operating reactors at low burnups—400 MWd/t is typical—so that insufficient time elapses for a substantial buildup of ^{240}Pu and other plutonium isotopes.

The military fuel cycle is discussed in chapter 3. Civil power reactors are operated at higher burnups in order to optimize the energy output from a given amount of fissile material. Power reactors fuelled with natural uranium, such as the gas-cooled Magnox reactor developed in the UK and France, and the Canadian deuterium–uranium (CANDU) reactor, have burnups in the range 3000–8000 MWd/t. The most common type of thermal power reactor, the pressurized water reactor (PWR) which is fuelled with enriched uranium, is typically operated at 30 000–40 000 MWd/t. As table 2.1 shows, the concentrations of the even-numbered isotopes become substantial at these burnups. It should also be noted that the concentrations of total fissile plutonium (^{239}Pu plus ^{241}Pu) are not dissimilar for these reactor types, but that PWR fuel contains relatively low concentrations of ^{239}Pu and high concentrations of the troublesome ^{241}Pu.

The trend is towards still higher burnups, with many utilities aiming for 50 000–60 000 MWd/t for light water reactor (LWR) fuel in the coming decade. As the irradiation period is extended, the energy extracted from the fissioning of ^{239}Pu and ^{241}Pu increases as that from ^{235}U decreases. In effect, a high burnup strategy is a cheap and energy-efficient substitute for recycling plutonium from lower burnup fuels. These high burnup spent fuels will also have increasing concentrations of the isotopes ^{238}Pu, ^{240}Pu, ^{241}Pu and ^{242}Pu, which have detrimental consequences for the economics of plutonium recycling. In

---

[9] An indirect way of achieving the same end is to bombard a blanket of ^{238}U in a fast reactor. The reactor can be fuelled with low-grade plutonium, while weapon-grade material can be extracted from the blanket. Among the nuclear weapon states, this has only been practised by France (see chapter 3). This is an expensive route to follow unless fast reactors are already being operated efficiently to produce electricity, in which case it may be the least costly method.

general, the commercial attractions of reprocessing and plutonium recycling diminish with increasing burnups.

Plutonium production: spent fuel reprocessing

In the military fuel cycle, all plutonium is routinely separated from the irradiated fuels discharged from production reactors. In contrast, most of the spent fuel emanating from civil power reactors is today held in store. By the end of the century, one-fifth or less of world spent fuel arisings will have been reprocessed. Nevertheless, as shown in chapter 6, the reprocessing of spent fuels particularly at facilities in France, Russia and the UK is giving rise to large amounts of separated plutonium.

In contrast to enrichment, only one process is currently used to extract plutonium from spent reactor fuels.[10] This is the Purex (plutonium–uranium extraction) process developed in the USA in the late 1940s and early 1950s, the details of which were published at the first Geneva Conference in 1955. Plutonium separation occurs in three main stages. In the first, the spent fuel assemblies are dismantled and the fuel rods are chopped into short segments (after the cladding has been removed mechanically in the case of Magnox fuel). In the second stage, the extracted fuel is dissolved in hot nitric acid. In the third and most complex stage, the plutonium and uranium are separated from other actinides and fission products, and then from each other, by a technique known as 'solvent extraction'.[11] Tributyl phosphate is commonly used as the organic solvent in a kerosene-type diluent in the Purex process. The plutonium and uranium are usually taken through several solvent-extraction cycles to reach the required levels of purity.

In modern reprocessing plants, less than 1 per cent of the plutonium contained in spent fuel may end up in wastes. This high extraction efficiency means that there is a close relationship between capacity utilization and plutonium output. In assessing the quantities of plutonium separated, the most important variables are the characteristics of the spent fuel inputs—their types and irradiation histories—and the strategies adopted for reprocessing different batches of fuel. The separation process itself is not a major source of uncertainty, at least at the levels of accuracy attempted in this book.[12] Reprocessing also has no effect on isotopic concentrations of plutonium; in this respect, outputs are identical to inputs.

---

[10] The chemical properties of plutonium are discussed in Cleveland, J. M., *The Chemistry of Plutonium* (American Physical Society: La Grange Park, Ill., 1979). A general discussion of reprocessing technology can be found in Allardice, R. H., Harris, D. W. and Miils, A. L., 'Nuclear fuel reprocessing in the UK', ed. Marshall (note 6).

[11] Actinides are elements with atomic numbers 89–104, which display similar properties to actinium.

[12] Measuring material balances in bulk reprocessing plants at the kilogram accuracies required by safeguards authorities is, however, a serious problem. A good general discussion of safeguarding reprocessing plants is provided by Lovett, J. E., IAEA Department of Safeguards, 'Nuclear material safeguards for reprocessing', document no. STR-151/152 (International Atomic Energy Agency: Vienna, 1987).

Table 2.2. Neutron cross-sections
Cross-sections are expressed in barns.[a]

	Thermal neutrons			Fast neutrons		
Isotope	Fission cross-section	Capture cross-section	Fission/total cross-section (%)	Fission cross-section	Capture cross-section	Fission/total cross-section (%)
^{235}U	579	100	85	2.0	0.5	80
^{238}U	..	3	–	0.05	0.3	17
^{239}Pu	741	267	74	1.9	0.6	76
^{240}Pu	..	290	–	0.4	0.6	40
^{241}Pu	1 009	368	73	2.6	0.6	81
^{242}Pu	..	19	–	0.3	0.4	43
^{241}Am	3	832	0.4	0.4	1.9	17

[a] A barn is the unit of effective cross-sectional area of the nucleus equal to 10^{-28} m^2.

Sources: Farmer, A. A., 'Recycling of fuel', ed. W. Marshall, *Nuclear Power Technology, Vol. 2: Fuel Cycle* (Clarendon Press: Oxford, 1983), tables 9.1 and 9.2; and authors' calculations.

Plutonium recycling

Fissile plutonium (notably ^{239}Pu and ^{241}Pu) can be used as an alternative to fissile uranium (^{235}U) in reactor fuel. In the 1970s, the future expansion of nuclear power seemed predicated upon the recycling of plutonium, since it was believed that reserves of low-cost uranium would soon be depleted at the growth rates of nuclear electricity production then envisaged. Today, however, uranium is cheap and abundant, and the stock of nuclear power stations is not increasing rapidly. The real price of uranium is not expected to rise substantially for many years or even decades, while the economics of plutonium recycling are generally unfavourable (see chapter 7).

In terms of nuclear physics, plutonium is more suitable for recycling in fast reactors than in thermal reactors. The fission and capture cross-sections of an isotope are the technical terms used to indicate the probability of neutron absorption by an atomic nucleus. As they imply, the former indicates the probability that absorbed neutrons will fission nuclei, while the latter indicates the probability that neutrons will be captured without fissions occurring. Whereas the fission cross-sections of ^{239}Pu and ^{241}Pu irradiated with thermal neutrons are slightly higher than those of ^{235}U, this uranium isotope is the better fuel in thermal reactors because the isotopes of plutonium have considerably higher capture cross-sections (see table 2.2). In a thermal reactor, around 85 per cent of the neutrons absorbed by ^{235}U cause fissions (the remainder are captured to produce ^{236}U), while the proportion for ^{239}Pu is 74 per cent.[13] Moreover, ^{240}Pu and

---

[13] There are, however, compensating qualities of plutonium which increase its energy worth. Most ^{240}Pu becomes ^{241}Pu after neutron capture, whereas the isotope ^{236}U produced through neutron capture in ^{235}U requires 3 additional neutrons to reach the fissionable isotope ^{239}Pu.

^{241}Pu also have high capture cross-sections, as does ^{241}Am. As a result, a larger amount of fissile plutonium than fissile uranium is required for a given energy output in thermal reactors.

Plutonium is not used on its own to fuel reactors. It is typically blended with natural or depleted uranium in so-called mixed-oxide (MOX) fuels. The costs of fabricating MOX fuels for thermal reactors are today several times higher than the costs of fabricating ordinary uranium fuels (see chapter 7), so that the prices of MOX fuel are higher even when the plutonium used in them is regarded as a free good. The isotope ^{239}Pu also emits fewer delayed neutrons than ^{235}U, limiting the quantity of plutonium (typically to one-third core) that can be recycled in a reactor core designed for uranium fuelling without major modifications to the reactor control system. Utilities wishing to raise the burn-ups of new fuels such as MOX may also face licensing delays, causing them to lag behind those allowed with uranium fuels. In some contexts, spent MOX fuels may also have to be kept in cooling ponds at reactor sites for longer periods than spent uranium fuels because of their higher heat output, increasing the amount of storage capacity that is required there. For all these reasons, MOX fuels will not generally be competitive with uranium fuels when used in thermal reactors, unless uranium prices rise steeply and offset the higher costs associated with their manufacture and usage.

In fast reactors, plutonium is in physical terms a slightly better fuel than enriched uranium, although the fabrication of fast reactor fuels from plutonium is again more costly because of its radioactivity. Table 2.2 shows that, unlike in thermal reactors, all isotopes of uranium and plutonium are fissioned by fast neutrons, so that all contribute to the reactor's energy output. However, this is largely offset by the much lower fission and capture cross-sections exhibited by the fissile isotopes ^{235}U, ^{239}Pu and ^{241}Pu when they are bombarded by fast neutrons rather than by thermal neutrons. Table 2.2 also explains why higher enrichment levels are required in fast reactors. The total cross-section of ^{239}Pu is about 300 times that of ^{238}U in thermal reactors, but only seven times that of ^{238}U in fast reactors. Although the spread of neutron energies and resonances tends to reduce this differential, thermal neutrons are still 10 times more likely than fast neutrons to be absorbed by ^{239}Pu rather than by ^{238}U (the same applies to ^{235}U). Concentrations of ^{235}U or ^{239}Pu in the range of 15–25 per cent, compared to 3–6 per cent with LWR fuel, are therefore required in fast reactors. This is one reason why the start-up costs of fast reactors are so high.

Although fast reactor cores therefore require large initial inventories of plutonium or HEU, fast reactors also seemed attractive because the high neutron flux could be exploited to 'breed' plutonium. Once running, it was expected that they would produce more fissile fuel than they consumed, mainly through neutron capture in ^{238}U 'blankets' placed adjacent to the reactor core.

Despite these advantages, the prospects for fast reactors have diminished greatly in recent years. High capital costs, operational difficulties and doubts over safety have led all countries with fast reactor programmes to revise their plans. No government or utility now claims that it will construct fast reactors in significant numbers before the middle decades of the next century.

The heavy demand for plutonium to fuel fast reactors that was forecast in the 1970s has therefore largely evaporated. As shown in chapter 7, utilities in Europe and Japan have turned to MOX recycling in thermal reactors as a means of consuming the large quantities of plutonium that they will soon acquire as their spent fuels are reprocessed in France and the UK. The main reason for using MOX fuels today is to reduce stocks of plutonium arising from reprocessing contracts. Especially if they were located in or transferred to non-nuclear weapon states, the stocks would become political liabilities if left standing.

Plutonium from dismantled US and Soviet weapons might also become available for recycling early in the next century. As it is weapon-grade material, it is easier to handle than the plutonium from reactor spent fuels because of the relatively low concentrations of ^{241}Pu and of the other higher-numbered isotopes of plutonium. However, the USA has reservations about plutonium recycling, and it is questionable whether Russia, despite its desire to utilize its redundant weapon material, will have the technical or financial resources to use its large plutonium stock in this way except with foreign assistance.

The amounts of plutonium emerging from weapon dismantlement and civil reprocessing over the next two decades, and that will end up being recycled, are very uncertain. Possible approaches to the disposition of surplus materials are discussed in chapter 15. With strong economic disincentives to plutonium recycling, the majority of plutonium arisings from these sources might have to be stored and eventually treated as wastes. It should be noted, however, that plutonium storage also has its problems.

Both HEU and separated plutonium require extensive physical protection and safeguarding, the risks of criticality accidents have to be guarded against. In addition, special precautions are required so that plutonium does not endanger health and the environment.

In technical and environmental terms, dealing with surplus plutonium may thus be a more difficult problem than reducing stocks of HEU. The latter will have the greatest commercial impact. The quantities of HEU are much greater than those of surplus plutonium, and the HEU will be entering a market for enriched uranium that already suffers from over-supply. These policy dilemmas are discussed further in chapter 15.

Part II
Military inventories in the nuclear weapon states

3. Inventories of military plutonium in the nuclear weapon states

I. Introduction

This chapter assesses the amounts of plutonium produced for military purposes by the USA, the former Soviet Union, the UK, France and China.[1] Today, the stocks have reached a plateau as the production of weapon-grade plutonium has largely ceased. However, significant changes in the forms in which they are held are occurring as large numbers of nuclear weapons are decommissioned. There is also the prospect that inventories of weapon-grade plutonium will begin to be reduced as material is used as a nuclear fuel or is otherwise disposed of. As this is not expected to happen in the near term, inventories are likely to equate to the historical production and separation of weapon-grade material for several years to come.

Since *World Inventory 1992* was published, the US Government has declassified a considerable amount of information about the US weapon production system, and has released data on the plutonium produced and separated for warheads. These steps have yet to be reciprocated by the other nuclear weapon states (NWS). Knowledge of the British, French and former Soviet plutonium production programmes has increased, but not as a result of new information issued by the respective governments. Regrettably, there is little more to say at present about the inventories of military plutonium in China.

The more the authors have learned about military programmes in the NWS, the more they have come to appreciate the practical difficulties faced by their governments in establishing the precise scale and whereabouts of material

---

[1] Important texts on this subject are: von Hippel, F., Albright, D. H. and Levi, B. G., *Quantities of Fissile Materials in US and Soviet Weapons Arsenals*, PU/CEES Report no. 168 (Center for Energy and Environmental Studies, Princeton University: Princeton, N. J., July 1986); Cochran, T. B., *et al.*, NRDC, *Nuclear Weapons Databook, Vol. II: US Nuclear Warhead Production* (Ballinger: Cambridge, Mass., 1987); Cochran, T. B. and Norris, R. S., *Russian/Soviet Nuclear Warhead Production*, Nuclear Weapons Databook Working Paper NWD 93-1 (Natural Resources Defense Council: Washington, DC, 8 Sep. 1993); US Department of Energy, *Plutonium: The First 50 Years: United States Plutonium Production, Acquisition and Utilization from 1944 to 1994* (Department of Energy: Washington, DC, Feb. 1996); Gowing, M., *Independence and Deterrence: Britain and Atomic Energy, 1945–1952* (Macmillan: London, 1974); Simpson, J., *The Independent Nuclear State: The United States, Britain and the Military Atom* (Macmillan: London, 1986); and Lewis, J. W. and Xue, L., *China Builds the Bomb* (Stanford University Press: Stanford, Calif., 1988). A good history of the French weapon production programme has yet to be written. The nuclear statuses of Belarus, Kazakhstan and Ukraine (their stances on international agreements and treaties, the weapons held on their territories and the plans for transferring them to Russia, their nuclear production capabilities, and their nuclear export controls) are described in *Nuclear Successor States of the Soviet Union: Nuclear Weapon and Sensitive Export Status Report*, no. 2 (Carnegie Endowment for International Peace and Monterey Institute of International Studies: Washington, DC and Monterey, Calif., July 1995).

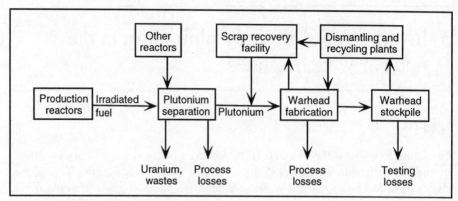

Figure 3.1. Plutonium in warhead production

inventories. Even in the USA, with its wealth of administrative resources, and where accurate material accounting has been practised since the 1960s, it has proved difficult to assemble unambiguous information about inventories and production histories at the various sites. Unlike the non-nuclear weapon states (NNWS), the NWS have not been required by international undertakings to keep track of every gram or kilogram of plutonium that has been produced and used. To get close to the truth, they are having to bring together whichever records were kept, reconstruct production histories and embark on the difficult task of assessing the quantities and conditions of materials contained in wastes, scraps and other forms.

While it is important that every effort should be made to pin down these inventories, no one should expect immediate results, especially in the former Soviet Union. Progress is nevertheless being made in delineating these inventories, through both the gradual declassification of information and the research being carried out inside and outside government. As shown in this chapter, this has resulted in significant changes to the estimates of the military plutonium inventories arrived at in *World Inventory 1992*.

II. The production process

This section gives a brief description of how the plutonium for nuclear warheads was acquired in the five declared nuclear weapon states and of the main differences in the approaches that they have adopted.

In each NWS, the production of plutonium for warheads was conducted under the wing of the state and amid strict secrecy. It involved in every case the construction of special 'production reactors' (see figure 3.1). The reactors were usually designed and operated to produce plutonium with a high concentration of the fissile isotope ^{239}Pu and a low concentration of other isotopes, notably ^{240}Pu, a strong neutron-emitter whose presence complicates the design and manufacture of nuclear weapons (see chapter 2). Weapon-grade plutonium is usually defined as plutonium containing less than 7 per cent of ^{240}Pu.

To avoid the build-up of the undesirable ^{240}Pu and of higher isotopes of plutonium, uranium fuel is discharged from production reactors after a shorter period than from commercial reactors. Typically, fuel elements would be submitted to burnups of hundreds of megawatt-days per tonne, compared to the thousands or tens of thousands of MWd/t in commercial operation. A consequence of these low burnups is that the discharged fuel was less radioactive, and did not have to be left for so long to cool before processing. The time which elapsed between the insertion of fuel in a reactor and the extraction of plutonium could thus be considerably shorter in the military than in the civil fuel cycle.

As figure 3.1 shows, the plutonium was extracted from irradiated fuel in plutonium separation plants, using chemical processes that either pre-dated or were identical to those applied commercially.[2] In the NWS, plutonium separation often occurred on the sites of production reactors (see table 3.1). After conversion into suitable metallic form at reprocessing or weapon fabrication sites, the plutonium was cast, machined and assembled into the 'pits' which form the fissile devices used either alone or in the first stages, or primaries, of thermonuclear weapons.

It is important to note that a small proportion of the plutonium discharged from production reactors is 'lost' during plutonium separation and warhead fabrication, as well as when plutonium is recovered from retired warheads. This occurs because small amounts of plutonium are left behind in chemical residues, or in scraps, some of which may be difficult or uneconomic to recover. The quantities of plutonium lost in this way depend on the efficiency of the production process and on the trouble weapon manufacturers take to recover materials from scrap. Plutonium stocks were also diminished by weapon testing, to an extent depending on the frequency of tests and the sizes and designs of devices being exploded.

As shown in this chapter, the plutonium production programmes of the NWS differed in a number of important respects:

1. *Reactor types.* The design most commonly used in the USA, the FSU and China was the light-water-cooled, graphite-moderated reactor (LWGR) (see table 3.1). The gas-cooled, graphite-moderated reactor (the GCR, or Magnox reactor) became the staple of the British and French weapon programmes, and subsequently of the North Korean weapon programme. Units of the heavy-water-cooled and -moderated reactor (HWR) were constructed by France, the FSU and the USA, partly to serve as tritium producers. This design has also been used by Israel and India to provide weapon-grade plutonium and possibly tritium.[3]

---

[2] The terms 'plutonium separation' and 'reprocessing' are used in the military and civil contexts, respectively, to indicate the same process.

[3] Tritium, the isotope hydrogen-3, is used to 'boost' the yield of the primary stages of nuclear warheads through a fusion reaction which produces copious amounts of high-energy neutrons that fission the plutonium or HEU more efficiently. It is produced mainly by irradiating lithium-6 in production reactors.

Table 3.1. Historical sources of weapon-grade plutonium

Country	Main production site/location	Reactor/plant number, type[a] (designation)	Other sources of weapon-grade plutonium
USA	Hanford Reservation, Washington	9 LWGR (B, D, F, DR, H, C, KE, KW and N); Purex reprocessing plant	Original Chicago and Clinton (Oak Ridge, Tenn.) piles; material imported from the UK; foreign research reactor fuel; blended super- and reactor-grade plutonium
	Savannah River, North Carolina	5 HWR (R, P, L, K and C); F and H reprocessing plants	
FSU	Kyshtym complex, Chelyabinsk, Urals	5 LWGR (A, IR, AV-1, AV-2 and AV-3); 2 HWR/LWR; RT-1 reprocessing plant	Original piles at Kurchatov Institute, Moscow
	Siberian Chemical Complex, Tomsk, Siberia	5 LWGR (I-1, I-2, ADE-3, ADE-4, ADE-5); unnamed reprocessing plant	
	Dodonovo, Krasnoyarsk, Siberia	3 LWGR (AD, ADE-1, ADE-2); unnamed reprocessing plant	
UK	Sellafield (Windscale), Cumbria	8 GCR (Calder Hall and Chapelcross (off-site)); B204 and B205 reprocessing plants	Original Windscale Piles; initial discharges from civil Magnox reactors
France	Marcoule, Côtes du Rhône	3 GCR (G1, G2 and G3); 2 HWR (Célestin 1 and 2); UP1 reprocessing plant	Civil Magnox reactors (esp. Chinon); uranium blankets in fast reactors
	La Hague, Normandy	UP2 reprocessing plant	
China	Jiuquan complex, Subei County	1 LWGR; unnamed reprocessing plant	
	Guangyuan, Sichuan	1 LWGR; unnamed reprocessing plant	

[a] GCR = gas-cooled, graphite-moderated reactor (Magnox reactor); LWGR = light-water-cooled, graphite-moderated reactor. The HWR (heavy-water-cooled and -moderated reactor) at Chelyabinsk was re-engineered in the 1980s to operate as an LWR (light-water-cooled and -moderated reactor).

Sources: Cochran, T. B., *et al.*, NRDC, *Nuclear Weapons Databook, Vol. II: US Nuclear Warhead Production* (Ballinger: Cambridge, Mass., 1987); Cochran, T. B. and Norris, R. S., *Russian/Soviet Nuclear Warhead Production*, Nuclear Weapons Databook Working Paper NWD 93-1 (Natural Resources Defense Council: Washington, DC, 8 Sep. 1993); and authors' data.

2. *Scale of production*. The US and Soviet programmes were many times the size of the British, Chinese and French programmes. The former entailed mass production, while the latter often involved batch production for limited weapon production runs. Whereas the US and Soviet reactors were mostly dedicated to serving military needs, the British and French reactors were operated as dual-purpose electricity and plutonium production facilities in order to spread costs and maximize revenues.

3. *Extent of recycling*. Particularly in the USA, where production of weapon-grade plutonium was cut back in the mid- to late 1960s, there has been heavy reliance on recycled material when constructing new warheads, allowing over 60 000 warheads to be manufactured over 45 years from a material stock sufficient for less than 30 000 warheads. Materials have been recovered by dissolving or oxidizing the plutonium pits and extracting the material by standard chemical methods.[4] The FSU relied less on recycled weapon material.[5] Instead, it was customary for warheads from old weapons either to be reallocated to new delivery vehicles, or to be placed in store. France and the UK both engaged in recycling, but less intensively than the USA. It is not known whether warhead materials have been recycled in China.

4. *Other sources of weapon materials*. The NWS did not only derive weapon-grade plutonium by producing it directly in production reactors or by recycling material from retired warheads. The USA, for instance, exploited stocks of lower-quality, fuel-grade plutonium by blending it with 'super-grade' plutonium containing 3 per cent or less of ^{240}Pu; France extracted plutonium from uranium blankets inserted in fast reactors; and Britain and France derived weapon-grade plutonium from 'civil' power reactors. The trade between the USA and the UK, whereby HEU and tritium have been exchanged for reactor- and weapon-grade plutonium, has been unique. No other NWS have exchanged weapon materials.

III. Methods of estimating military plutonium inventories

For domestic and foreign observers of weapon programmes, there have been three standard approaches to estimating the inventories of military plutonium possessed by an NWS: counting warheads; measuring krypton-85 emissions; and studying the design, fuelling and operating histories of production reactors. In principle, this last source of information has been available to the managers of weapon production programmes from material accounts and from the production records of reactors and separation plants. However, comprehensive

---

[4] This is carried out either in specialized facilities at the sites where warhead plutonium is processed, or by introducing batches of nitrate solution containing warhead material into the appropriate parts of plutonium separation plants.

[5] This appears to have reflected a policy towards innovation in nuclear weaponry different from that followed in the USA. In the FSU, warheads were standardized to a greater extent, with the delivery system being adapted to take the warhead. In the USA, many different designs of warheads and associated delivery vehicles were developed.

inventories were not usually compiled, especially in the early days of weapon programmes. Only now, as disarmament gathers pace, are governments in the NWS attempting to establish full inventories of their weapon materials.

Counting warheads

This approach involves taking the maximum number of warheads deployed by a given country, and making assumptions about the average quantity of plutonium per warhead. There are three difficulties here. The first is that only the number of strategic warheads is known with any precision, and only for France, the FSU and the USA. As a result of the Strategic Arms Reduction Treaties (START), the sizes of the former Soviet and US strategic arsenals became known,[6] and France announced the composition of its strategic arsenal in early 1994. For the UK, estimates of strategic warhead numbers are still subject to considerable uncertainty, while in the case of China estimating warhead numbers is still largely a matter of guesswork. Information on stocks of tactical nuclear weapons remains largely inaccessible. No NWS has published the numbers of tactical weapons in its arsenal, nor the numbers of redundant warheads that it holds in store.

The second problem is that the amounts of plutonium in nuclear weapons vary widely, depending on design specifications and the quantities of HEU that are used in the fission stages. Even the averages may differ across countries and historical periods. In the following pages the average range of 3–4 kg per warhead is used (except for older warheads, in which the plutonium content tends to be lower and the HEU content higher), with the caveat that the actual averages, or 'typical' amounts, for given batches could be above or below this range.

The third drawback with this approach is that it only allows estimates to be made of the inventory of plutonium in weapon arsenals. It tells nothing about quantities that are held in store or are locked up in the production process. These quantities can be substantial.

Measuring krypton-85 emissions

During the cold war, when little was known about Soviet production facilities, the US intelligence services tried to estimate plutonium output, and thus the rate of warhead production, by measuring the build-up of krypton-85 (^{85}Kr) in the atmosphere. By deducting the amounts of ^{85}Kr believed to have come from its own and from other countries' nuclear activities, the Soviet contribution could be estimated.

---

[6] The precise numbers allocated to bombers were not, however, declared. Instead, the numbers of these warheads controlled under the START I Treaty were estimated according to agreed formulae, depending on the carrying capacities of specific aircraft among other factors.

When the Soviet political system was opening up in the late 1980s and early 1990s, it was expected that more precise estimates would soon be provided by Russian officials, or could be derived as production histories were revealed. This has not proved to be the case. The Russian Government has either not known the extent of production or has been unwilling to release information about it. As a result, the US Government has recently returned to the krypton method as the most reliable means of estimating the amounts of plutonium that have been produced for the FSU's military programmes.

Krypton-85 is a product of the fissioning of ^{235}U or ^{239}Pu in nuclear reactors. It is an inert gas which is retained in reactor fuel until its release during reprocessing. Hitherto, it has been mostly vented into the atmosphere by the reprocessors. The concentration of ^{85}Kr in the atmosphere provides an indirect measure of the extent of fissioning in reactor fuel which has been reprocessed, from which estimates of plutonium arisings can be made.

There are three difficulties associated with this method of assessing plutonium inventories. The first concerns the accuracy of measures of ^{85}Kr concentrations in the atmosphere. As ^{85}Kr has a relatively long half-life (10.8 years), global releases can in principle be assessed from local sampling since the gas has time to diffuse widely through the atmosphere. However, measurements have to be taken far away from reprocessing sites, and account has to be taken of varying concentrations and rates of diffusion across the different layers in the atmosphere.[7]

The second difficulty is that the interpretation of ^{85}Kr releases still rests on assumptions about the type and operation of production reactors. The rates of fissioning of ^{235}U and ^{239}Pu, and thus of ^{85}Kr production, are functions of reactor fuelling and burnup. The ^{85}Kr approach does not obviate the need for some understanding of production reactors and their operating histories. Traditionally, calculations of plutonium arisings by this method relied on the very general assumption that US and Soviet production reactors had similar fuelling arrangements. They were thus first approximations. Moreover, account had to be taken of the significant quantities of ^{85}Kr released by other activities, notably the reprocessing of civil reactor fuels and nuclear weapon testing, and through leakage.

The third drawback is that the ^{85}Kr method only measures plutonium *separation*. It does not indicate the amounts of plutonium held in reactors and spent fuel stores. Although most of the spent fuel from military production reactors in each NWS was reprocessed soon after its discharge, some residual stocks may have remained (notably in the FSU).

Studying production reactors and operating histories

Where information is available, the best estimates of plutonium inventories come from studying the designs, fuelling arrangements and operating histories

---

[7] A full account of the measurement of ^{85}Kr concentrations is contained in von Hippel *et al.* (note 1).

of production reactors. While this knowledge is still incomplete, a fuller picture has emerged over the past decade.

In the USA details of the thermal power output (i.e., heat output) of production reactors were declassified by the Department of Energy in the early 1980s. Given an understanding of the relationships between thermal output and the fissioning of uranium and plutonium, it became possible to estimate past arisings of plutonium with greater accuracy. There was one important caveat: in some US reactors, uranium fuel was replaced on occasion by lithium targets in order to produce tritium for nuclear weapons. Wherever this occurred, plutonium output was reduced accordingly. Although much less precise, some information about tritium production also became available, so that adjustments to calculations of plutonium arisings could be made. These calculations, which were relied upon in *World Inventory 1992*, have now been superseded by the actual information on plutonium production at the main sites published by the DOE in 1993 and 1994 (see below).

In the cases of the UK and France, the main production reactors were dual-purpose electricity and plutonium producers. As their electrical output is on the public record, their thermal output can be calculated given knowledge of the efficiency with which heat is converted to electrical energy. As in the case of the USA, plutonium arisings can then be estimated, although, for reasons explained below, not with the same degree of accuracy. New information has also become available to the authors on the fuelling arrangements in British production reactors, and on the periods in which they were used to produce weapon-grade plutonium.

By comparison, little is yet known of the operating histories of Soviet production reactors, and still less about the Chinese reactors. With the exception of two water-moderated reactors, all Russian reactors are LWGRs, of the type used by the USA at Hanford (see table 3.1). For the FSU, estimates have largely been derived by Cochran and others by assuming that its LWGRs exhibited power ratings, burnups and load factors similar to their counterparts at Hanford.[8] (The extent of Soviet tritium production has also been estimated on the basis of the US experience.) The same approach can be adopted for China, but with still less confidence in its reliability. In both cases there is some justification in basing estimates on the US experience, since both the USSR and China learned from, and even copied, US reactor designs. However, for reasons that will become apparent, such estimates still cannot be regarded as reliable.

The estimates of military plutonium inventories for the NWS presented in sections IV–VIII below are thus derived mainly from the study of production reactors and their operating histories and, where available, from published information on the materials themselves. The one exception is the FSU, for which a combination of approaches is used. Each section also summarizes what is known about stockpiles of nuclear warheads in the country concerned. This provides a rough cross-check and allows gross estimates of the amounts of

[8] Cochran and Norris (note 1).

weapon-grade plutonium that are held outside nuclear weapons, and that will become available when arsenals of nuclear weapons are diminished through arms reductions. Estimates of plutonium arisings based on the measurement of ^{85}Kr are alluded to only in the context of estimating the FSU's total production of plutonium.

IV. The United States

Plutonium production history

US production of weapon-grade plutonium occurred in four stages.[9] During the first expansionist phase, lasting from the mid-1940s to the mid-1960s, 14 production reactors were constructed and operated: 9 at the Hanford Reservation in Washington State and 5 at Savannah River in North Carolina (see tables 3.1 and 3.2). Weapon-grade plutonium production reached its peak in the early 1960s, at a rate of around 6 tonnes per annum (figure 3.2 and table 3.3). By 1965 nearly 60 tonnes of weapon-grade plutonium, or approximately three-quarters of the final US inventory, had already been accumulated.

By and large, the rate of plutonium production was matched to the anticipated build-up of nuclear warheads. When their numbers peaked in the mid-1960s, and when plutonium began to be recycled from retired warheads, the demand for additional weapon-grade plutonium declined. Between 1965 and 1971 the original eight Hanford reactors were closed down, and the newly commissioned N reactor at Hanford was re-engineered and operated to produce electricity at minimum cost. Operating at higher burnups, it became a source of fuel-grade plutonium which was expected to be used by an expanding civil fast reactor programme. The Purex reprocessing plant at Hanford also ceased operating in 1972. In this second phase, from 1964 to 1972, the production of weapon-grade plutonium at Hanford was therefore gradually reduced to zero.

In the third phase, from 1973 to 1981, weapon-grade plutonium was produced, along with tritium, at Savannah River at an average rate of around 1 tonne per year. New warheads were manufactured mainly from existing stocks or from recycled materials.

The fourth and final phase lasted from 1982 to 1988. It began with ambitious plans to expand plutonium production, as part of the arms build-up initiated by the Soviet invasion of Afghanistan. In reality, the increases in plutonium output were comparatively modest. The N reactor at Hanford returned to the production of weapon-grade plutonium in 1982, with the Purex reprocessing plant reopening in 1983 to handle its fuel. The L reactor at Savannah River, which had ceased operating in 1968, was also restarted in 1985. In addition, the P, K and C reactors at Savannah River were adapted to produce super-grade plutonium which was then blended with fuel-grade material of US and British origin. Stocks of fuel-grade plutonium were thereby brought into play, in effect allow-

---

[9] See Cochran *et al.* 1987 (note 1) for details on US production reactors and separation facilities.

Table 3.2. US production reactors

Reactor site/ designation	Type and power rating (MWth)^a	Period of operation
Hanford Reservation		
B, D, F, H, DR, C	LWGR, up to 2 500	Start-up 1944–52, shut-down 1964–69
KW, KE	LWGR, up to 4 400	Start-up 1955, shut-down 1970–71
N	LWGR, 4 000	Produced weapon-grade Pu 1964–65, 1982–87; fuel-grade Pu 1965–82
Savannah River		
R	HWR, up to 2 260	Start-up 1953; shut-down 1964
L	HWR, up to 2 700	Start-up 1954; shut-down 1968; restarted 1985; shut-down 1988
P, K, C	HWR, up to 2 680 (P), 2 710 (K) and 2 915 (C)^b	Start-up 1954–55; C shut-down 1985, P shut-down 1988 and K maintained in cold standby as contingency for tritium production

^{a} MWth = Megawatt-thermal; LWGR = light-water-cooled, graphite-moderated reactor; HWR = heavy water reactor.

^{b} The P and K reactors were operated as producers of super-grade plutonium from 1981 (together with the restarted L reactor after 1985) until their closure in 1988. The C reactor was mainly operated as a dedicated tritium producer in the 1980s. Tritium was also produced in varying amounts by the Savannah River reactors during the 1970s.

Sources: Cochran, T. B., *et al.*, NRDC, *Nuclear Weapons Databook, Vol. II: US Nuclear Warhead Production* (Ballinger: Cambridge, Mass., 1987); and US Department of Energy, *Plutonium: The First 50 Years: United States Plutonium Production, Acquisition and Utilization from 1944 to 1994* (Department of Energy: Washington, DC, Feb. 1996).

ing the acquisition of three quantities of weapon-grade plutonium for every two quantities of super-grade plutonium extracted from the Savannah River reactors. Taken together, these steps led to an approximate doubling of the US output of weapon-grade material, the annual production rate approaching 2 tonnes during the mid-1980s.

By the end of 1988, however, all US plutonium production, as distinct from separation which continued albeit on a much reduced scale, had ceased. Concerns over reactor safety following the Chernobyl accident in 1986, and the *rapprochement* between East and West, led to the shut-down of the remaining reactors at Hanford and Savannah River. The US Department of Energy announced in early 1991 that it had no plans to produce plutonium 'in the foreseeable future'. The US determination to bring an end to the production of weapon-grade plutonium worldwide has since been evident in its commitment to achieving an international agreement on a Fissile Materials Cut-Off Treaty.

US plutonium production at Hanford Reservation and Savannah River

In 1993, the DOE embarked on its unprecedented Openness Initiative. The initiative has resulted in the declassification of much information on plutonium and HEU production, on their health effects and on the US nuclear weapon testing programme. Information was released in December 1993, June 1994 and February 1996. The aim has been to allow greater public debate on nuclear issues in the USA, to open up aspects of US nuclear history that have hitherto been kept secret and to encourage other nations to follow the US example in the belief that the arms control process would be strengthened as a result.

Details released by the DOE of annual plutonium production at Hanford Reservation and Savannah River, used to construct figure 3.2, are set out in table 3.3. The total quantities of weapon- and fuel-grade plutonium that had been produced before the closure of the last production reactor in 1988 are shown in table 3.3.

The data in table 3.3 have been assembled by the DOE from information contained in its Nuclear Materials Management and Safeguards System. The 'production numbers reflect production as measured during recovery of plutonium from reactor targets rather than theoretical production predicted by reactor physics models'.[10] They are *book inventories*, as distinct from *physical inventories*, based largely on measurements made in the accountancy tanks of plutonium separation plants.[11] These measurements cannot be regarded as totally accurate (see the discussion of inventory differences below): they always carry error margins, and accounting practices in the 1940s and 1950s were less precise than those developed in subsequent decades. When issuing its Fact Sheet on 27 June 1994 the DOE stated that the 'declassified quantities listed . . . are based on the evaluation of available records, some of which are very old. The quantities may be updated in the future after re-evaluation of the methodology originally used.'[12] Some of the records on production reactor operation were in fact destroyed.

Besides the plutonium production at Hanford and Savannah River, a small amount of plutonium was also produced as a by-product in 'non-production' reactors, including government research reactors and naval propulsion reactors. The total quantity was 0.6 tonne of plutonium, comprising 0.1 and 0.5 tonne of weapon- and fuel-grade plutonium respectively.[13]

---

[10] US Department of Energy, DOE Fact Sheet, Washington, DC, 27 June 1994, p. 42. Some of the numbers issued in Dec. 1993 were based on reactor physics models. This accounted for the discrepancy in the production of weapon-grade plutonium at Hanford reported in Dec. 1993 and June 1994.
[11] As defined by the DOE, in DOE Order 5633.3A, 'the "book inventory" is the quantity of material present at a given time as reflected by accounting records. The "physical inventory" is the quantity determined to be on hand by first physically ascertaining its presence and then using techniques that include measuring, sampling, weighing, and analysis'. DOE Fact Sheet (note 10), p. 107.
[12] DOE Fact Sheet (note 10), p. 2.
[13] US Department of Energy (note 1), p. 34.

Table 3.3. US Department of Energy total production of weapon- and fuel-grade plutonium (book inventory), 1947–89

Figures are in kilograms.

Year	Weapon-grade plutonium			Fuel-grade plutonium	DOE total
	Hanford Reservation	Savannah River	Total	Hanford Reservation	
1947	493	..	493	..	493
1948	183	..	183	..	183
1949	270	..	270	..	270
1950	392	..	392	..	392
1951	288	..	288	..	288
1952	662	..	662	..	662
1953	838	..	838	..	838
1954	1 113	..	1 113	..	1 113
1955	1 413	553	1 966	..	1 966
1956	2 074	1 151	3 225	..	3 225
1957	2 662	1 245	3 907	..	3 907
1958	3 303	672	3 975	..	3 975
1959	3 581	1 459	5 040	..	5 040
1960	4 266	1 734	6 000	..	6 000
1961	4 449	1 552	6 001	..	6 001
1962	4 169	1 578	5 747	..	5 747
1963	4 187	2 042	6 229	..	6 229
1964	4 247	2 123	6 370	256	6 626
1965	4 208	909	5 117	562	5 679
1966	3 130	1 302	4 432	800	5 232
1967	2 586	1 107	3 693	1 069	4 762
1968	1 494	1 253	2 747	1 530	4 277
1969	430	1 382	1 812	2 109	3 921
1970	977	872	1 849	707	2 556
1971	270	836	1 106	467	1 573
1972	..	1 028	1 028	414	1 442
1973	..	1 128	1 128	673	1 801
1974	..	1 226	1 226	607	1 833
1975	..	753	753	557	1 310
1976	..	1 400	1 400	429	1 829
1977	..	844	844	560	1 404
1978	..	835	835	559	1 394
1979	..	829	829	544	1 373
1980	..	1 010	1 010	413	1 423
1981	..	748	748	196	944
1982	..	793	793	449	1 242
1983	624	464	1 088	..	1 088
1984	294	809	1 103	..	1 103
1985	633	875	1 508	..	1 508
1986	934	935	1 869	..	1 869
1987	312	472	784	..	784
1988	−21	152	131	..	131
1989[a]	2	8	10	..	10
Total	54 463	36 079	90 542	12 901	103 433

[a] Adjustment to previous year's production to reflect the difference between theoretical calculations of plutonium produced in the reactors and the measured value of plutonium recovered in the chemical separation plants.

Source: US Department of Energy, *Plutonium: The First 50 Years: United States Plutonium Production, Acquisition and Utilization from 1944 to 1994* (Department of Energy: Washington, DC, Feb. 1996), tables 2 and 3.

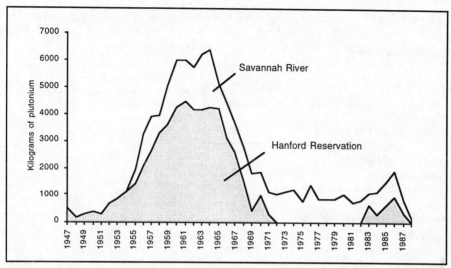

Figure 3.2. US Department of Energy weapon-grade plutonium production, 1947–88
Source: Table 3.3.

US military plutonium inventories

Table 3.3 displays the DOE's historical *production* of plutonium. In February 1996, the DOE announced that its current *inventory* consisted of 99.5 tonnes of plutonium, comprising 85.0, 13.2 and 1.3 tonnes of weapon-, fuel- and reactor-grade material respectively. There was no explanation of how these three totals were arrived at. Table 3.4 attempts to correct this omission, showing how various acquisitions and 'removals' gave rise to the amounts declared by the DOE. Figures recorded in brackets carry a margin of uncertainty, either because the DOE is uncertain over final quantities (regarding wastes and inventory differences), or because the proportions of weapon- and fuel- or reactor-grade plutonium in a given quantity have had to be estimated by the authors.

Plutonium acquisitions

Table 3.4 shows that the DOE acquired plutonium from four sources in addition to the production and non-production reactors. First, plutonium was transferred from the US civilian industry, involving 0.9 tonne from the Plutonium Credit Activity (whereby the Atomic Energy Commission provided 'credit' for plutonium produced in commercial reactors using uranium fuels purchased or leased from it) and 0.8 tonne from other sources, mainly in the form of spent fuel. Second, the DOE acquired 2.8 tonnes of weapon-grade plutonium (containing 6–7 per cent ^{240}Pu) by blending fuel-grade plutonium with super-grade plutonium (containing 3 per cent ^{240}Pu) produced in the Savannah River reactors after 1981. The increased inventory of weapon-grade material was balanced by a decrease in the fuel-grade inventory. Irradiated fuels from foreign research

Table 3.4. US Department of Energy inventory of plutonium, by grade, February 1996[a]

Figures are in tonnes.

Source	Weapon-grade	Fuel- and reactor-grade	Total
Acquisitions			
From US production reactors	90.5	12.9	103.4
From US non-production reactors	0.1	0.5	0.6
From US civil industry	0	1.7	1.7
Through blending	+ 2.8	– 2.8	0
Plutonium separated from foreign fuels	(0.3)	(0)	0.3
Imports from the UK	(0.1)	(5.3)	5.4
Sub-total: plutonium acquired	**93.8**	**17.6**	**111.4**
Removals			
Inventory differences[b]	(– 2.3)	(– 0.5)	(– 2.8)
Waste disposal (normal operating losses)[b]	(– 2.8)	(– 0.6)	(– 3.4)
Fission and transmutation	0	(– 1.2)	– 1.2
Transfer to civil industry	0	– 0.1	– 0.1
Transfer to foreign countries	0	– 0.7	– 0.7
Losses through weapon testing	– 3.4	0	– 3.4
Radioactive decay	(– 0.3)	(– 0.1)	– 0.4
Sub-total: plutonium removed from inventory	**– 8.8**	**– 3.2**	**– 12.0**
Classified transactions and rounding	0.1	0	0.1
Total inventory of plutonium	**85.0**	**14.4**	**99.5**

[a] Figures in brackets carry a margin of error, either because the DOE itself has difficulty providing precise measures, or because the authors have estimated the breakdown between weapon-grade and reactor- or fuel-grade plutonium.

[b] On the DOE's advice to the authors, the amounts of weapon-grade and fuel- and reactor-grade plutonium removed from the inventory have been broken down in proportion to the total production of these grades (81% and 19% respectively). These figures are still bracketed as the DOE seems uncertain about the actual amounts in these columns.

Sources: US Department of Energy, *Plutonium: The First 50 Years: United States Plutonium Production, Acquisition and Utilization from 1944 to 1994* (Department of Energy: Washington, DC, Feb. 1996); and authors' estimates.

reactors provided the third source of plutonium in the US military inventory. This quantity included 254 kg from Canada, 79 kg from Taiwan and 50 kg from 13 other countries. It is assumed here that this foreign plutonium was all weapon-grade.

Last but not least, the US Government announced in February 1996 that it acquired 5.4 tonnes of plutonium from the UK between 1962 and 1980 under the US–UK Mutual Defence Agreement. Of this amount, around 100 kg were weapon-grade, 4 tonnes were fuel- and reactor-grade plutonium produced in civil reactors, and the remaining 1.3 tonnes were fuel- and reactor-grade plutonium from the military reactors at Calder Hall and Chapelcross. Some or all of this last quantity was used in the 1980s to produce weapon-grade material for the US weapon programme in the blending operations discussed above.

The 5.4 tonnes of British plutonium were exchanged for US supplies of HEU and tritium. In February 1996, the DOE reported that it had transferred 6.7 kg of tritium and 7.5 tonnes of HEU to the UK in the period up to 1980. There were no subsequent transfers of plutonium from the UK to the USA, and the UK produced its own tritium in the Chapelcross reactor. Instead, British requirements for HEU for the weapon and naval submarine reactor programmes were met by 'swapping' British-produced low-enriched uranium for US-produced HEU or by purchasing HEU directly from the USA (see chapter 4).

Plutonium removals

When the DOE released information in February 1996, it reported on the 'removals' from its inventories that had occurred, again without distinguishing between grades of plutonium. These removals involved 'inventory differences' (see below), waste disposal (also see below), weapon-testing, the decay of ^{241}Pu to ^{241}Am, the fissioning and transmutation of fuel- and reactor-grade plutonium, and the transfer of US plutonium to civil industries at home and abroad. The latter involved transfers of 0.7 tonne to 30 countries. The countries receiving more than 1 kg were Australia (6.4 kg), Belgium (11.8 kg), Canada (3.5 kg), France (41.5 kg), Germany (518.1 kg), Italy (2.3 kg) and Japan (113.5 kg).[14]

'Excess' plutonium

Tables 3.5 and 3.6 show that, when not being used in operational weapons, the US plutonium inventory is held at eight sites. In regard to the 'excess plutonium', 'the Department of Energy and Defense performed an in-depth review of the fissile material required to support the nuclear weapons program and other national security needs. This was compared to available materials and as a result, 38 tonnes of weapon grade plutonium were declared excess to national defense needs'.[15] The DOE statement continues: 'In keeping with the President's policy, the Secretary of Energy announced on December 20, 1994, that plutonium and weapons usable highly enriched uranium that was separated and/or stabilized during the phase out, shutdown, and cleanout of weapons complex facilities would be set-aside as restricted use material and not used for nuclear explosive purposes.' The issue of 'excess' plutonium is discussed further in chapter 15, section VII.

---

[14] In 1982, it was reported that 1.2 t of plutonium had been exported from the USA, including 754 kg to Germany (87% ^{239}Pu), 159 kg to Japan (88%), 129 kg to Italy (79%), 58 kg to Belgium (81%) and 54 kg to the UK (84%). The balance between this 1.2 t, and the 0.7 t reported by the DOE in Feb. 1996, is believed to have come from private sources such as utility stocks.

[15] US Department of Energy (note 1), p. 75.

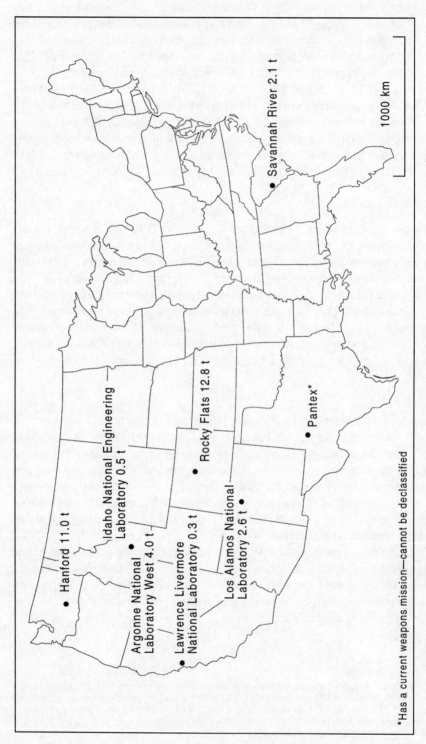

Figure 3.3. Plutonium inventories of the US Department of Energy, 31 December 1993.
Source: See table 3.5.

Table 3.5. US Department of Defense (DOD) and Department of Energy inventories of plutonium, by location

Figures are in kilograms unless otherwise indicated.

Site	Plutonium inventory (Dec. 1993)				Excess stock[a] (Dec. 1994)
	Weapon-grade	Fuel-grade	Reactor-grade	Total	Weapon-grade (t)
Pantex and DOD[b]	66 100			**66 100**	21.3
Rocky Flats	12 816[c]	..	..	**12 816**	11.9
Hanford	1 726	8 662[d]	640	**11 028**	1.7
Argonne National Laboratory-West	247	3 489[e]	212	**3 948**	< 0.1
Los Alamos National Laboratory	2 232[f]	349	12	**2 593**	1.5
Savannah River	1 512[f]	528	12	**2 052**	1.3
Idaho National Engineering Laboratory	162	1	341	**504**	0.4
Lawrence Livermore National Laboratory	236[f]	62	9	**307**	< 0.1
Totals	**85 031**	**13 091**	**1 226**	**99 348**	**38.2**

[a] Plutonium declared as excess by the US Secretary of Energy.
[b] Plutonium held at the Pantex site in Texas and in operational nuclear weapons.
[c] Includes plutonium metal (6.6 t), compounds (3.2 t), and mixtures and other forms (3.1 t).
[d] This mainly comprises plutonium in N reactor fuel elements and in the Fast Flux Test Facility.
[e] Most plutonium at Argonne is in Zero Power Physics Reactor plates (3.8 t).
[f] Mainly in the form of plutonium metals, compounds and scrap.

Sources: Communication by the DOE to the authors; and US Department of Energy, *Plutonium: The First 50 Years: United States Plutonium Production, Acquisition and Utilization from 1944 to 1994* (Department of Energy: Washington, DC, Feb. 1996).

US plutonium 'inventory differences' at seven nuclear sites

In addition to declassifying information about the plutonium produced or held at the seven nuclear sites discussed above, the DOE has also published the 'inventory differences' recorded at those sites. They deserve some attention here, not least because they have relevance to debates about the safeguardability of nuclear activities in the nuclear weapon states.

These 'inventory differences', defined as the book inventory less the physical inventory, were recorded at the sites over the periods in which their plants were in operation. Presumably annual physical inventories were taken at the sites and compared with the book inventories.[16] 'Inventory differences' are usually referred to as 'materials unaccounted for', or MUF, in the standard terminology

---

[16] The procedures followed when physical inventories have fallen short of book inventories are not described by the DOE.

Table 3.6. Total plutonium inventory differences at US production sites, cumulative to February 1996

Figures are in kilograms.

Site	Quantity[a]
Hanford Reservation	+ 1 266.0
Rocky Flats	+ 1 191.8
Savannah River	+ 232.0
Los Alamos National Laboratory	+ 47.5
Lawrence Livermore National Laboratory	+ 5.5
Argonne National Engineering Laboratory	– 3.4
Idaho National Engineering Laboratory	– 5.6
Other sites	+ 16.7
Total	**+ 2 750.5**

[a] A '+' or '–' denotes that the book inventory respectively exceeds or falls short of the physical inventory (a decrease from or increase from the book inventory in DOE terminology).

Source: US Department of Energy, *Plutonium: The First 50 Years: United States Plutonium Production, Acquisition and Utilization from 1944 to 1994* (Department of Energy: Washington, DC, Feb. 1996), figure 11.

of the IAEA. Those recorded at the seven DOE sites are shown in table 3.6. Detailed information about the movements in inventory differences over time has only been released for the Rocky Flats Plant. Between 1953 and 1993, the book inventory exceeded the physical inventory at this plant by 1191 kg, with the maximum recorded decreases and increases from the book inventory being 192 kg and 198 kg respectively (in 1971 and 1973).

According to the DOE, the factors which have contributed to these inventory differences include the following:

(1) high measurement uncertainty of plant 'holdup'; (2) measurement uncertainties because of the wide variations of the matrix containing the materials; (3) measurement uncertainties because of the statistical variations in the measurement itself; (4) in the early years, technology had not been developed to measure the material adequately; (5) measurements for waste are still very uncertain because often small quantities of plutonium or uranium are mixed in with a variety of other materials so variable that it is not possible to make accurate calibration sources for measurements; (6) losses from operations, such as accidental spills where accurate measurements were not made before the spill; (7) corrections of human errors in the input of data into accounting systems; and (8) rounding errors.[17]

Overall, 2750 kg of weapon-grade plutonium are thus formally unaccounted for at these seven DOE sites. This equates to 344 significant quantities (SQ) of plutonium, an SQ being defined by the IAEA as 8 kg for plutonium. Knowledge of the amounts of material unaccounted for at specific material balance

[17] DOE Fact Sheet (note 10), p. 109.

Table 3.7. US inventory of weapon-grade plutonium declared to be excess to weapon requirements
Figures are in tonnes.

Site	Metal	Oxides	Reactor fuel	Irradiated fuel	Other forms	Total
Pantex (future dismantlement)	21.3	..	..	..	..	21.3
Rocky Flats	5.7	1.6	..	..	4.6	11.9
Hanford	< 0.1	1.0	..	0.2	0.5	1.7
Los Alamos	0.5	< 0.1	< 0.1	..	1.0	1.5
Savannah River	0.4	0.5	..	0.2	0.2	1.3
Idaho National Engineering Laboratory	< 0.1	..	0.2	0.2	< 0.1	0.4
Other sites	< 0.1	..	..	< 0.1	< 0.1	0.1
Total	**27.9**	**3.1**	**0.2**	**0.6**	**6.4**	**38.2**

Source: US Department of Energy, *Plutonium: The First 50 Years: United States Plutonium Production, Acquisition and Utilization from 1944 to 1994* (Department of Energy: Washington, DC, Feb. 1996), table 15.

areas, facility by facility, would be required before judgements could be made on whether the DOE was meeting the accounting standards required by the IAEA where its safeguards are applied. This information is not publicly available. Most of the inventory differences arose from activities in the 1950s, 1960s and early 1970s, since when tighter measurement and accounting procedures have been applied. The DOE states that 'the total inventory differences for weapon-grade plutonium . . . are expected to be reduced by materials that will be recovered during decontamination of buildings and equipment'.[18] Submitting the facilities to IAEA safeguards might be problematic if the sources of the inventory differences were not more precisely identified.

Plutonium wastes

The Department of Energy reported in February 1996 that 3.4 tonnes of plutonium had been removed from the inventory as waste from 'normal operating losses'. These normal operating losses (NOL) are defined by the DOE as occurring 'when quantities of plutonium, determined by measurement or estimated on the basis of measurement, are intentionally removed from inventory as waste because they are technically or economically unrecoverable.' The reported quantities are drawn from the Nuclear Materials Management and Safeguards System (NMMSS).

---

[18] DOE Fact Sheet (note 10), p. 107.

Table 3.8. Plutonium in the DOE waste inventory
Figures are in kilograms of plutonium.

Location	Quantity	Description
Hanford site	455	High-level waste in the tank farms
	875	Solid waste in burial grounds
	192	Waste in cribs, trenches and ponds
Idaho, waste management	1 026	Solid waste in drums and boxes
Los Alamos National Laboratory burial ground	610	Solid waste in various forms
Savannah River site burial ground	382	Liquid waste in tanks
	193	Solid waste stored in containers
Idaho Chemical Processing Plant waste farm	72	Calcined waste stored in bins
	8	Solutions stored in tank farms
Rocky Flats	47	Solid waste in drums and crates
Oak Ridge National Laboratory	41	Particulate waste, dry solids and oxides, and solution and sludge in storage tanks
Nevada Test Site burial ground	16	Solid waste stored above and below ground
Argonne National Laboratory-West	2	In reactor assemblies stored underground
Total	**3 919**	

Source: US Department of Energy, *Plutonium: The First 50 Years: United States Plutonium Production, Acquisition and Utilization from 1944 to 1994* (Department of Energy: Washington, DC, Feb. 1996), table 16.

However, the DOE acknowledges that these quantities do not match precisely the quantities of plutonium wastes measured in burial sites and tanks, which amount to 3.9 tonnes (see table 3.8). The DOE seems unable to provide a persuasive explanation for this discrepancy. It states that:

In the early 1970s, sites began reporting details of plutonium in waste for the first time to NMMSS. At most sites the estimates of the amount of plutonium in waste were based on direct measurements of waste and provided confirmation of the NOL estimates of waste. In the case of Hanford, however, the 1974 estimate indicated 0.4 tonnes more plutonium in waste than in normal operating losses. This difference could be either: an accounting error at the site, such as reporting plutonium already included in the normal operating losses; or additional plutonium not captured by the normal operating losses tracking system, and therefore likely reported as 'inventory differences' ... Since 1974, the annual normal operating losses and waste inventories have tracked very closely ... the remaining 0.1 tonne inconsistency tracks closely to wastes received from sources outside the Department.[19]

---

[19] US Department of Energy (note 1), pp. 78–79.

MILITARY PLUTONIUM IN THE NUCLEAR WEAPON STATES

Table 3.9. The US plutonium inventory held by the Departments of Energy and Defense, February 1996

Figures are in tonnes.

Category	Quantity
Weapon-grade plutonium	85.0 ± 2%
Fuel-grade plutonium	13.2 ± 2%
Reactor-grade plutonium	1.3 ± 2%
Total	**99.5 ± 2%**
Inventory declared to be 'excess to national defense needs'	38.2

The figure of 3.4 tonnes of plutonium waste recorded in table 3.4, therefore, carries a substantial error margin. It also follows from the DOE's explanation that there may be double counting between this quantity and the 'inventory differences' recorded in the same table.

US plutonium inventories: the totals and their error margins

Table 3.9 presents a summary of the US plutonium inventories, as reported by the DOE. An error margin of ± 2 per cent is attached, reflecting continued uncertainties over plutonium wastes and inventory differences in particular. It should be noted that the 38.2 tonnes of plutonium that have been declared 'excess to national defense needs' are part of the reported total inventory of 99.5 tonnes.

Plutonium in US nuclear warheads

It has been estimated that the USA had 9250 nuclear warheads deployed in July 1994, 8300 in its strategic and 950 in its tactical arsenals.[20] A stockpile of 14 900 warheads in 1994 was later estimated, implying that about 5600 warheads are in reserve or await disassembly.[21] Numbers have fallen steadily from the peak of around 33 000 in the mid-1960s, the rate of decline accelerating in the early 1990s following the arms reductions agreements with the FSU. The average plutonium content per warhead rose from 2–3 kg in the 1960s to 3–4 kg in the early 1990s.

Because of the great variety in nuclear warhead size and design, it is not possible to make an accurate assessment of the plutonium inventory contained in operational US nuclear weapons. The 3–4 kg average leads, however, to an inventory of plutonium warheads in the range of 28–37 tonnes, with a midpoint of around 32 tonnes (assuming an inventory of 9250 operational war-

---

[20] Norris, R. S. and Arkin, W. M., 'Nuclear notebook: U.S. strategic nuclear forces, end of 1994', *Bulletin of the Atomic Scientists*, vol. 50, no. 4 (July/Aug. 1994), pp. 61–63.

[21] Norris, R. S. and Arkin, W. M., 'Nuclear notebook: U.S. strategic nuclear forces, end of 1994', *Bulletin of the Atomic Scientists*, vol. 50, no. 6 (Nov./Dec. 1994), pp. 58–59.

heads). Given that the DOE has declared that some 66 tonnes of weapon-grade plutonium are still assigned to the US nuclear weapon programme (see table 3.5), this would imply that 34 tonnes were held outside weapons in 1994.

As the numbers of warheads in the US arsenal are reduced to the few thousand currently envisaged, the quantities of excess plutonium will grow still larger. How much plutonium will be counted as excess to military requirements is currently the subject of discussion within the US Government, which has declared its intention of submitting excess stocks to IAEA safeguarding. Assuming an active arsenal of 3500 warheads, and a similar number held in reserve, and assuming that the 18.9 tonnes held at the seven nuclear weapon sites are also counted as excess, 45 or more tonnes of US weapon-grade plutonium might eventually be placed under IAEA safeguards.

V. The former Soviet Union

Early in 1992 Russia was accepted as the USSR's successor state, and thus as a nuclear weapon state, under the 1968 Treaty on the Non-Proliferation of Nuclear Weapons. Belarus, Kazakhstan and the Ukraine, the other former Soviet republics where strategic nuclear weapons were deployed, have acceded to the NPT as non-nuclear weapon states.[22] The former Soviet Union's inventory of weapon plutonium has become Russia's inventory. All nuclear weapons still located on the former Soviet territories are now Russia's property, as they are scheduled to be returned to Russia where the materials contained in them will remain, and as all significant nuclear production facilities are located in Russia. This inventory is regarded as a single entity in this chapter.

Besides the fissile material contained in warheads that have yet to be returned to Russia, there is today no weapon-grade plutonium assigned to military programmes in the FSU outside the frontiers of the Russian Federation. Quantities of HEU and plutonium are nevertheless held on the sites of a few research facilities in Belarus, Kazakhstan and Ukraine, although plutonium has only been recorded at the Aktau site in Kazakhstan. All of these stocks are due to be safeguarded by the IAEA.

It is evident from section IV that a much more accurate picture of military plutonium inventories in the USA can be assembled today than was possible when *World Inventory 1992* was prepared. This is not yet the case for the FSU. Information about weapon materials remains classified in Russia, and little more has been revealed about the operation of the former Soviet production facilities. No announcements have been made about the total historical production of weapon-grade plutonium, nor about the inventories held at different sites.

One reason may be that the Russian Government is itself struggling to establish the scale and locations of those inventories. It is believed that records were kept of materials entering and leaving sites. However, the records are probably

[22] See *Nuclear Successor States of the Soviet Union* (note 1).

incomplete, and few efforts were made to account for what happened to materials while they were being processed on site. In particular, it is unlikely that data exist on production losses and waste disposal. Some historical records may also be untrustworthy since, as was common practice in the Soviet economy, managers may have misrepresented annual production in order to justify their enterprises' incomes from the state.

Experience suggests that any reports on or estimates of the former Soviet inventories should be approached with great caution. Whereas the error margins attached to estimates of US inventories can be substantially reduced, large error margins are still unavoidable where the former Soviet inventories are concerned.

Since 1991, the possibility that nuclear materials are being smuggled out of the former Soviet Union, or are otherwise being diverted to unlawful purposes, has caused much international concern. The quantities of material that are known to have been transferred out of the FSU are mainly measured in grams or fractions of grams, and are thus extremely small parts of the inventories discussed here.[23] Significant quantities of material might be removed from the FSU without their making a dent on recorded inventories, let alone on the kinds of estimate presented here. The sources of the materials discovered in Germany and other European countries are still the subject of conjecture. It is possible that other batches of material have left the FSU by other routes.

Plutonium in the nuclear warheads of the FSU

The difficulties of assessing the scale of the former Soviet inventories of weapon-grade plutonium begin with the uncertainties surrounding the size of its arsenal of nuclear weapons. Estimates of the maximum number of warheads that were deployed at any one time vary widely from a 'low' of around 30 000 to a 'high' of around 50 000. Cochran and Norris report a comment made by Minister Viktor Mikhailov, of the Russian Ministry of Atomic Energy, to the effect that the Soviet nuclear weapon stockpile peaked at 45 000 warheads in 1986 and had declined to 32 000 warheads by May 1993.[24] This is consistent with a retirement rate of some 2000 warheads per year.

If each warhead contained 3–4 kg of plutonium, 45 000 warheads would have contained between 135 and 180 tonnes of plutonium. Taking production losses into account, the plutonium requirements would need to have been even larger (at least 150–200 tonnes). For reasons made clear below, it is unlikely that the output of weapon-grade plutonium approached these levels. This does not rule

[23] In Germany 40 cases of nuclear materials being smuggled were investigated by the police in 1991, 160 in 1992 and 240 in 1993. In 1994, the principal incidents that gave rise to concern were the discovery of 5.6 g of weapon-grade plutonium (99.7% ^{239}Pu) in Tengen-Weichs, Germany; 0.8 g of HEU (87.5% ^{235}U) in Landshut, Germany; 580 g of mixed-oxide fuel containing 300 g of plutonium (87% ^{239}Pu) at Munich Airport; 0.05 g of plutonium at Bremen railway station (Germany); 19 containers of nuclear material, one involving plutonium, in Sofia, Bulgaria; 750 g of HEU seized by Hungarian authorities on the border with Slovakia; and 2.27 kg of HEU seized in Prague.

[24] Cochran and Norris (note 1), p. 22.

out the possibility that 45 000 warheads were deployed in the mid-1980s, as substantial numbers of warheads might have contained little or no plutonium. They could have been constructed using HEU as the principal fissile material. Some low-yield tactical weapons may also have used relatively small quantities of HEU and plutonium.

No one claims that 45 000 warheads are still deployed in the former Soviet Union. The numbers of operational weapons are steadily being reduced as the provisions of the Strategic Arms Reduction Treaties, and the pledges given by Presidents Mikhail Gorbachev and Boris Yeltsin, are honoured. In June 1992, Presidents George Bush and Boris Yeltsin announced their ambition to reduce the US and former Soviet strategic arsenals to some 3000–3500 warheads each by January 2003. Norris suggests that Russian strategic forces after 2003 might comprise some 800 intercontinental ballistic missiles (ICBMs), 1700 submarine-launched missiles, and 1000 weapons assigned to bombers. Together with the deployment of perhaps 2750 tactical warheads, this would give a total arsenal of 6250 warheads with an additional 5000 warheads possibly being held in reserve.[25] Norris also estimates that the arsenal of operational weapons comprised around 11 000 warheads (7700 strategic and 3300 non-strategic) in mid-1994.

Until basic information is released about the size and composition of the former Soviet Union's nuclear arsenal, past and present, the warhead count cannot provide a reliable guide to the FSU's inventories of weapon-grade plutonium.

Production reactors, separation plants and the plutonium inventory

Although knowledge of the former Soviet plutonium production system has grown in recent years, much less is known than in the US case. The sites, numbers and types of production reactor have been established, but details of their power ratings and production histories remain, for the most part, undisclosed.

There are significant differences between the historical profiles of plutonium production for the US and Soviet weapon programmes. Production in the USSR began at around the same time as in the USA, but it developed more slowly. Significant difficulties were encountered in the 1950s, one symptom of which was the serious accident in 1957 at a nuclear waste storage site at Kyshtym in Siberia. Output peaked towards the end of the 1960s, a decade later than in the USA, and efforts were made to keep it at a high level until 1987 when the closure of production reactors began.

This is consistent with the picture of the lag in Soviet warhead production and deployment in the 1950s and 1960s, and the dogged efforts to match and even surpass the US weapon arsenal in the Brezhnev era. It also suggests a rather inflexible production system which, once in its groove, strove to maintain output at constant levels. Furthermore, it lends support to the view that warheads were not routinely constructed from recycled material.

[25] See Norris, R. S., 'Nuclear notebook', *Bulletin of the Atomic Scientists*, vol. 50, no. 5 (Sep./Oct. 1994), pp. 61–63.

MILITARY PLUTONIUM IN THE NUCLEAR WEAPON STATES 53

As table 3.1 shows, there were three production sites in the former Soviet territories (all in Russia): at Chelyabinsk on the eastern side of the Urals, at Tomsk in southern Siberia, and at Dodonovo (near Krasnoyarsk) in eastern Siberia.[26] The five LWGRs at Chelyabinsk were constructed in the late 1940s and early 1950s and were shut down between 1987 and 1990. Two water-cooled and -moderated reactors at Chelyabinsk are still operating, mainly to provide tritium for nuclear weapons and to maintain supplies of electricity to the local region. In the late 1970s the RT-1 reprocessing plant at Chelyabinsk was adapted to process spent fuel from the 440-MW water-water power reactors (VVER—*Vodo-Vodyanoy Energeticheskiy Reaktor*) in the Soviet Union, Eastern Europe and Finland, together with submarine and research reactor fuel. The intention was to use the plutonium arisings in fast reactors also planned for the Chelyabinsk site and to recycle recovered uranium in thermal reactors. By 1991, when the fast reactor programme had already come to a standstill, 25 tonnes of plutonium had been separated from VVER fuel and were held in store on the site.[27] At the end of 1994, the amount had increased to around 30 tonnes.

After 1978, the reprocessing of fuel discharged from the Chelyabinsk production reactors was switched to Tomsk, which also has five LWGRs, three of which have been shut down. Spent fuel from 1000-MW power reactors in the FSU (and one in Bulgaria) has been shipped to Krasnoyarsk in south-eastern Siberia where construction of a large new reprocessing plant began in the 1980s. Amid doubts whether its construction would be finished, President Yeltsin issued a decree in autumn 1994 committing the Russian Government to its completion., which may still be prevented by lack of finance (see chapter 6).

Three LWGRs are located at Krasnoyarsk, each constructed underground. Two were shut down in 1992 and one is still operating, ostensibly to maintain electricity supplies to the region. Along with the two operating reactors at Tomsk, this last reactor is due to re-optimized for electricity production so that it will no longer produce weapon-grade plutonium.[28]

Of the 13 plutonium production reactors at Chelyabinsk, Tomsk and Krasnoyarsk, the performance characteristics of only three reactors at Chelyabinsk are known in any detail. The main published assessments of plutonium output from the 13 reactors have been carried out by Cochran and Norris. The estimates contained in reports prepared by them in 1991 and 1993 increased by close to

---

[26] A detailed description of the Russian weapon production system, including assembly and dismantlement sites, can be found in *Nuclear Successor States of the Soviet Union* (note 1).

[27] Bukharin, O., 'Soviet reprocessing and waste management strategies', Paper presented at Princeton University's Workshop on Nuclear Weapons Materials Disposal, Berlin, 28–30 Nov. 1991.

[28] Under an intergovernmental agreement announced by Vice-President Al Gore and Prime Minister Viktor Chernomyrdin on 23 June 1994, the Russian Federation agreed to shut down its 3 remaining production reactors by the year 2000. Furthermore, the agreement stipulates that plutonium produced by these reactors would not be used in weapons while they still operated. However, their replacement became contentious and the current proposal is to convert their cores for electricity production.

Table 3.10. Estimated plutonium production by military reactors in the former Soviet Union, 31 December 1993

Figures are in tonnes.

Location	Plutonium
Ozersk (Chelyabinsk)	
A Reactor	8.4
IR-AI Reactor	4.2
AV-1 Reactor	11.6
AV-2 Reactor	11.5
AV-3 Reactor	11.1
Sub-total	**46.8**
Seversk (Tomsk)	
I-1 Reactor	10.7
I-2 Reactor	10.2
ADE-3 Reactor	13.6
ADE-4 Reactor[a]	12.8
ADE-5 Reactor[a]	12.4
Sub-total	**59.7**
Zheleznogorsk (Krasnoyarsk)	
AD Reactor	16.3
ADE-1 Reactor	13.5
ADE-2 Reactor[a]	13.4
Sub-total	**43.2**
Total	**149.7**

[a] Reactor still in operation.

Source: Anatoliy Diakov, Moscow, private communication.

50 per cent, from 116 tonnes to 171 tonnes of plutonium equivalent up to the end of 1990.[29]

In both reports, similar assumptions were made about the sizes of reactors and their up-rated power outputs, based upon experience at the Hanford reactors in the USA. However, different assumptions were made about the reactors' capacity factors (the ratio between average and full power levels, an indicator of operating efficiency). In the 1991 report, a capacity factor of 0.5 (0.4 in the first year of operation) was chosen 'so that the cumulative plutonium and tritium production was consistent with what we know from other sources'.[30] The reactors' operating performance was thereby made to tally with the 100–120 tonnes of plutonium reported to them by a Russian source. In the 1993 report, in contrast, a capacity factor of 0.8 (0.4 and 0.6 in the first and second years of operation respectively) was chosen 'to be reasonably consistent with the capacity factors achieved by the Hanford production reactors in the United

[29] See Cochran, T. B. and Norris, R. S., *Soviet Nuclear Warhead Production*, Nuclear Weapons Databook Working Paper NWD 90-3 (3rd rev.), (Natural Resources Defense Council: Washington, DC, Feb. 1991), tables 10, 12 and 13; and Cochran and Norris (note 1), tables 13, 15 and 16.

[30] Cochran and Norris (note 29), p. 60.

MILITARY PLUTONIUM IN THE NUCLEAR WEAPON STATES 55

States'.[31] Neither capacity factor was therefore based upon an understanding of how the reactors were actually operated. The decision to increase the capacity factor, and thus the estimate of plutonium output, may have been influenced by the report that the FSU had deployed as many as 45 000 nuclear warheads.

In spring 1995, a Russian expert, Anatoliy Diakov, provided the authors with what appeared to be more reliable estimates of the amounts of plutonium produced by the 13 production reactors up to the end of 1994. The estimates, which are shown in table 3.10, are based on more detailed knowledge of reactor design characteristics and operating practices than had previously been available. His estimated total production was 149.7 tonnes. Later in 1995, Diakov substantially reduced this amount to around 125 tonnes, in order to incorporate the following information that had become available to him:[32]

1. Reactor operations were on average halted six to seven times each year, for 6–10 days, to replace aluminium flow tubes which were prone to degradation. Around one-half of the reactor core was discharged in each shut-down.
2. Reactors were shut down for complete overhauls, lasting six months, every five years on average. Most reactors also experienced significant teething problems in their early years of operation, resulting in reduced plutonium output.
3. One-tenth of power output came from HEU fuel rods which were used to flatten the neutron flux, especially on the periphery of reactor cores.
4. The AV-3 reactor at Ozersk was used in the last 10 years of operation mainly for tritium production.

Two-thirds of the reduction in Diakov's estimate followed from his assumption that the presence of HEU would reduce plutonium production by an amount commensurate with the power output caused by its fissioning. This is surely incorrect. HEU is added to production reactors to maintain an even neutron flux, by raising the rate of neutron capture in ^{238}U atoms, and thus plutonium production, around the periphery. As the quantity of ^{238}U in the core will be little affected by the presence of HEU, the net effect may even be a modest increase in plutonium output. However, the reductions to Diakov's initial estimates because of the other factors mentioned above also need to be taken into account. A downward estimate of 5 tonnes is therefore being made here to the estimate in table 3.10, resulting in a rounded figure of 145 tonnes. It is apparent that only when full information is released about the design and operating histories of these reactors will it be possible to have confidence in estimates derived by these means.

Even if the figures in table 3.10 are accurate, three problems remain in assessing the amounts of plutonium that have been made available to the weapon programme: uncertainties over the substitution of tritium for plutonium production, over the rate and extent of plutonium separation, and over material losses.

[31] Cochran and Norris (note 1), p. 108.
[32] Diakov, A. S., 'Utilization of already separated plutonium in Russia and international security problems: consideration of short- and long-term options', Paper presented to the Fifth International Conference on Radioactive Waste Management and Environmental Remediation, 3–7 Sep. 1995, Berlin, p. 6.

Tritium is produced mainly by irradiating lithium targets. It can also be extracted from heavy water (deuterium oxide) which has been irradiated during its usage as a neutron moderator. If lithium targets are irradiated in plutonium production reactors, they substitute for uranium fuel elements and thereby reduce the amounts of plutonium that are produced.[33] The amounts of tritium that were acquired by this means in the LWGRs and in the two water-moderated reactors are not known. While an attempt has been made to take account of tritium production when compiling table 3.10, it is therefore difficult to estimate the proportion of plutonium-equivalent production (the total product of neutron irradiation) accounted for by weapon-grade plutonium production.

Uncertainties about the former Soviet reprocessing programmes also have to be taken into account. Most analysts have assumed that the separation of plutonium from spent fuels proceeded smoothly. There are grounds for believing that this was not the case. Many questions remain unanswered. How was the transfer from Chelyabinsk to Tomsk of production reactor spent fuels managed? What effects did accidents and technical mishaps at the reprocessing sites have on production schedules and on spent fuel storage? Why were East European governments instructed to expand storage capacity at the sites of their Soviet-supplied power reactors in the mid-1980s? Why did the rate of reprocessing at Krasnoyarsk increase in the early to mid-1990s? Is it possible that backlogs of spent fuel have built up from time to time as a result of difficulties in matching the rate of reprocessing to the rate at which spent fuels have been discharged from reactors?

Losses during plutonium separation and in the weapon production process also have to be subtracted. Those losses may have been high. In the USA, as much as 5 per cent of plutonium production was disposed of in wastes, is scattered in contaminated buildings and equipment, or is otherwise difficult or impossible to retrieve. The proportion lost in this way is unlikely to have been any lower in the FSU. Losses resulting from weapon testing also need to be subtracted (it is assumed below that 4 tonnes of plutonium were consumed in the test programme).

For all these reasons, knowledge of the former Soviet production system is not yet sufficiently advanced to estimate plutonium inventories reliably.

Estimating plutonium arisings using the krypton-85 method

As mentioned in section III, the US Government derives its principal estimates of the former Soviet plutonium inventories by applying the krypton-85 method. These estimates are updated annually using measures of ^{85}Kr concentrations in the earth's atmosphere, and by subtracting known and estimated ^{85}Kr releases from reprocessing occurring outside the FSU. The technique's great advantage is that it is based on actual measurements rather than on suppositions about the

---

[33] For this reason, the measure of 'plutonium-equivalent' which encompasses the total production of plutonium and other isotopes, including tritium, is often used. Approximately 1 kg of tritium can be produced with the same neutron irradiation as 72 kg of weapon-grade plutonium.

Table 3.11. Estimated Soviet military plutonium output, calculated from estimated krypton releases and plutonium arisings, 31 December 1983

	Million Curie (mCi)	Plutonium (t)
Krypton (^{85}Kr) releases		
Total measured krypton-85 release to atmosphere	144.8	
^{85}Kr releases from nuclear testing and from reprocessing in non-communist countries		
Reprocessing in the USA	47.4	
Reprocessing in the UK	18.5	
Reprocessing in France	12.3[a]	
Reprocessing in other West European countries and Japan	1.7	
Leakage from fuel outside communist countries	0.7[a]	
Nuclear weapon tests (communist and non-communist)	5.0	
Sub-total:	85.6	
Krypton releases from reprocessing in communist countries	59.2 ± 10	
Krypton releases from reprocessing carried out in China and for non-weapon purposes in the USSR	12.0[b]	
Krypton releases from reprocessing for weapon purposes in the USSR	47.2 ± 10	
Estimated Soviet production of weapon-grade plutonium		110 ± 25

[a] Note that the wrong figures were inserted here in the *World Inventory 1992*, table 3.5.

[b] This figure is based upon the assumption that 9 mCi had been released by Soviet reprocessing of power-, research- and submarine-reactor fuels, and 3 mCi by Chinese reprocessing of production reactor fuels, by the end of 1983.

Sources: von Hippel, F., Albright, D. H. and Levi, B. G., *Quantities of Fissile Materials in US and Soviet Weapon Arsenals*, PU/CEES Report no. 168 (Center for Energy and Environmental Studies, Princeton University: Princeton, N.J., July 1986), table 5.17; and authors' estimates.

performance of production facilities (although assumptions about fuel burnups are still required when calculating plutonium arisings from ^{85}Kr releases). The ^{85}Kr method has the added attraction that it is not affected by levels of tritium production, since the krypton is only produced by the fissioning of uranium and plutonium. Those carrying out these studies in government believe, perhaps erroneously, that weapon-grade plutonium arisings can be estimated by this method with an accuracy of plus or minus 5 per cent.

This said, the ^{85}Kr method still has its drawbacks. It does not indicate the amounts of plutonium that have been produced in reactors but are held in store prior to separation. Nor does it provide a measure of plutonium losses within production processes. It is also prone to inaccuracies in measuring krypton concentrations in the atmosphere and in matching them to their sources.

The US Government's analyses are classified. However, an authoritative independent study relying on this technique was carried out at Princeton University in the mid-1980s. It can provide a base-line. The figures derived for the end of 1983 are summarized in table 3.11.

Table 3.12. Total inventory of military plutonium of the FSU, 31 December 1993

Figures are in tonnes.

Source	Total inventory
Military plutonium production	145 ± 25
Losses in production processes and through waste disposal	10 ± 5
Losses through weapon testing	4 ± 1
Total available inventory of separated plutonium	**131 ± 25**

To the total of 110 tonnes of plutonium separated by the end of 1983 should be added a quantity of material that had been produced but was held in store, prior to reprocessing (and krypton release). In normal circumstances this would have amounted to between 2 and 3 tonnes, representing the six-months' worth of spent fuel that is placed in storage ponds for initial cooling. However, if a backlog of spent fuel did build up and reactor operation was not curtailed in response to it, several more tonnes may have been held in storage ponds prior to reprocessing. There are some grounds for believing that a backlog did build up in the early 1980s, possibly linked to the transfer of reprocessing from Chelyabinsk to Tomsk. A reference figure of 115 tonnes of plutonium arisings in production reactors is nevertheless being taken for the end of 1983.

This output implies an annual production rate of around 5 tonnes of weapon-grade plutonium, or an average of 385 kg per reactor, when all 13 reactors were operating at full power. Taking the closure of 10 reactors between 1987 and 1992 into account, this is consistent with an additional production of 35 tonnes of weapon-grade material up to the end of 1994. This implies a total plutonium-equivalent production of around 148 tonnes to the end of 1994, which is close to the figure derived from Diakov's original estimates.

There are still major uncertainties attached to the estimates arrived at through the krypton and reactor modelling methods discussed above. It has been decided to adopt 145 tonnes here as the central estimate for weapon-grade plutonium produced in the former Soviet reactors up to the end of 1994. In view of the great uncertainties that still surround the history of plutonium production in the FSU, an error margin of ± 25 tonnes is attached to this estimate.

This quantity does not, however, represent the amount that has been available to the weapon programme. To arrive at that inventory, two quantities need to be subtracted. The first involves losses incurred during plutonium separation, in the weapon production process. A central estimate of 10 per cent and an upper limit of 15 per cent for such plutonium losses are assumed here. Weapon testing is the second source of losses, which are assumed here to have absorbed 4 tonnes of plutonium.

It is also possible that several tonnes of spent fuel still await reprocessing. As there is no hard evidence that this is the case, they are assumed to be covered by the error margin.

MILITARY PLUTONIUM IN THE NUCLEAR WEAPON STATES   59

The results are shown in table 3.12. A central estimate of 131 ± 25 tonnes for the size of the FSU's inventory of potentially available weapon-grade plutonium at the end of 1993 seems the best that is possible today. If Norris's estimate of 11 000 operational weapons in 1994 is accurate, and each weapon contained 3.5 kg of plutonium on average, only 30 per cent of the FSU's military inventory of separated plutonium is now located in operational warheads. The remainder is contained in retired warheads, in stored plutonium pits taken from dismantled warheads and in stocks of separated material.

The final uncertainty concerns the grade of plutonium discharged from the former Soviet production reactors. While the great majority is likely to have been weapon-grade, it is possible that a proportion has been fuel- or reactor-grade.[34] No information is available on the burnups to which uranium fuels have been subjected nor, therefore, on the grades of plutonium produced.

VI. The United Kingdom

While the British, French and Chinese military plutonium inventories are much smaller than the US and Soviet inventories, reflecting the relatively modest scale of their nuclear weapon programmes, they are no easier to assess.

The British Government has made no announcements about the quantities of plutonium produced for nuclear weapons. The information remains classified. All estimates of Britain's military inventories rest upon calculations of plutonium arisings from production reactors, upon knowledge of their production histories, and upon the limited information available about Britain's exchange of nuclear materials with the USA.

Weapon-grade plutonium from Calder Hall and Chapelcross

The main plutonium production site in the UK is at Sellafield, formerly known as Windscale, in north-west England. Weapon-grade plutonium was derived from the two Windscale Piles between 1951 and 1957, and after 1956 from the Calder Hall and Chapelcross production reactors. The four Calder Hall reactors and the plutonium separation plants are located at Sellafield. The four Chapelcross reactors are situated at Annan near Dumfries in south-west Scotland.

Simpson has correctly estimated that the Windscale Piles produced 400 kg of plutonium in total.[35] Each pile produced 35 kg of plutonium per year on average. Between them, the two piles operated for about 11 years, giving a total output of 385 kg.

The output of weapon-grade plutonium from Calder Hall and Chapelcross is harder to assess. After further research, it is now apparent that the evidence presented in the *World Inventory 1992* was incorrect in a number of respects, although the totals were not far off the mark.

---

[34] The 3 operating reactors at Tomsk and Krasnoyarsk are evidently still operating at low burnups as their fuel elements have not been designed to withstand longer periods of irradiation.
[35] Simpson (note 1), p. 252, table A4a.

The Magnox reactors at Calder Hall were operated as suppliers of weapon-grade plutonium during three periods: between 1956 and 1964; in the late 1970s, when they provided material for the Chevaline warhead programme; and in the mid- to late 1980s, when they supplied plutonium for the planned modernization and expansion of the British nuclear arsenal, the most important component of which was the Trident submarine programme. In contrast, the Magnox reactors at Chapelcross are believed to have only produced weapon-grade plutonium in substantial quantities between 1959, when the first reactor was commissioned, and 1964. Chapelcross's main military function thereafter was to produce tritium. When not producing plutonium and tritium for military purposes, the Calder Hall and Chapelcross reactors were optimized to produce electricity: their fuel was submitted to much higher burnups in order to maximize power output at the lowest achievable cost.

Calculating the amounts of plutonium produced at the two sites between 1956 and 1964 is not straightforward. The reactors' main technical characteristics, and the dates on which they were commissioned and reached full power, were published long ago.[36] There is, however, limited detailed information about the reactors' operating histories on the public record. Matters are complicated by the raising of thermal and electrical capacities in stages in the 1960s, which had consequences for the rate of plutonium production and for the quality of plutonium that was separated from discharged fuel.

The calculations that follow are based on information gathered from two main sources. The first is an IAEA publication dating from 1971, which contains details on the characteristics and operating performance of the Calder Hall and Chapelcross reactors that are not available elsewhere.[37] In particular, this publication allows a reasonably coherent story of how and when the reactors were up-rated, and how they were operated thereafter, to be pieced together. Second, new information has been gained from private discussions with retired decision makers.

What the authors have learned can be summarized as follows:

1. The eight reactors were built to the same design. Their initial design ratings were 180 megawatt-thermal (MWth) and 42 megawatt-electric (MWe). These capacities were subsequently up-rated in the 1960s to 240 MWth and 54 MWe, and down-rated in the early 1970s to 48 MWe (Chapelcross) and 50 MWe (Calder Hall).

2. When producing weapon-grade plutonium, whole cores were discharged after the desired quantities and qualities of plutonium had been produced, and replaced by fresh ones. Typically a reactor was shut down for 70 days while this was carried out.

---

[36] For Calder Hall, the dates when the reactors reached full power were Oct. 1956, Feb. 1957, May 1958 and Feb. 1959. For Chapelcross, they were Feb. 1959, Aug. 1959, Nov. 1959 and Mar. 1960. A description of the engineering and operational improvements carried out at Calder Hall between 1956 and 1986 is contained in Ayres, G. P., 'An overview of thirty years of Calder Hall', *The Nuclear Engineer*, vol. 27, no. 6 (Dec. 1986), pp. 172–74.

[37] *Reactor Operating Experience up to 1970* (IAEA: Vienna, 1971), microfiche.

MILITARY PLUTONIUM IN THE NUCLEAR WEAPON STATES 61

3. British weapon designers demanded a very high quality of plutonium in the 1950s and 1960s. By submitting uranium fuels to burnups of 400 MWd/t, plutonium containing 97 per cent ^{239}Pu can be produced in Magnox reactors of these designs. The available evidence suggests that the Calder Hall reactors were operated at this burnup throughout the period 1956–64, before and after their up-rating in 1961–62. This implies that when producing weapon-grade material they operated on approximately yearly cycles before up-rating (300 days' operation plus 70 days' shut-down), and on approximately 300-day cycles (230 days' operation plus 70 days' shut-down) thereafter.

4. Calder Hall reactors provided lower electrical outputs than their counterparts at Chapelcross because steam from the reactors was also used in reprocessing and other industrial operations at the Sellafield site. In other respects, the Calder Hall and Chapelcross reactors were operated along similar lines, with fuel being submitted to burnups of 400 MWd/t when they were supplying plutonium for the weapon programme.

On the basis of these assumptions, the authors estimate that 1.33 tonnes and 0.87 tonnes of weapon-grade plutonium were produced at Calder Hall and Chapelcross respectively between 1956 and their re-optimization for electricity production in 1964.[38] To this amount should be added around 80 kg of weapon-grade plutonium which were separated from initial discharges after this re-optimization took place.[39] This gives a total of 2.28 tonnes. After rounding, our central estimate is now that 2.3 tonnes of plutonium were derived from Calder Hall and Chapelcross between 1956 and 1964 (compared with the 3.2 tonnes estimated in *World Inventory 1992*).

Another point about this early period deserves comment. As part of its declassification of information on the US production of weapon material, the US Department of Energy announced in June 1994 that the reactor-grade plutonium used in a nuclear weapon test conducted in 1962 was supplied by the UK.[40] A lively private debate on this issue ensued between the British and US governments. There are no records of reactor-grade plutonium being produced at Calder Hall and Chapelcross before 1962 (the fuel burnups were too low). For reasons that are hard to explain, a British Defence Minister nevertheless informed the House of Commons that the material did indeed come from those reactors.[41] The US Government has so far not provided documentary evidence to support its assertion. At the time of writing, the source of the material used in

---

[38] The calculations are as follows (see appendix A). In these Magnox reactors, a fuel burnup of 400 MWd/t yields 0.36 kg plutonium per tonne of fuel. At Calder Hall, 29 discharged reactor cores, each containing 127 t of heavy metal, therefore produced 1326 kg of plutonium. At Chapelcross, 19 cores irradiated at 400 MWd/t yielded 869 kg plutonium, according to the authors' calculations.

[39] Assuming that each initial discharge involved one-sixth of a reactor core, the initial discharges from 8 reactors provided the equivalent of 1.33 cores, resulting in 76 kg of plutonium at 0.45 kg Pu/t of fuel.

[40] In reponse to the question 'What was the source of reactor-grade plutonium?', it was stated that 'The plutonium was provided by the United Kingdom under the 1958 United States/United Kingdom Mutual Defense Agreement'. See DOE Fact Sheet (note 10), p. 190.

[41] Mr Aitken's responses to questions by Mr Llew Smith, *Parliamentary Debates, House of Commons [Hansard]*, parliamentary answers, 4 July and 15 July 1994.

this test had not been identified. It is unclear whether the material was produced in the USA or the UK, but it was definitely fuel- rather than reactor-grade.

As mentioned above, weapon-grade plutonium was also produced at Calder Hall on two later occasions. It is believed that close to 100 Chevaline warheads were produced to replace the original Polaris warheads.[42] The plutonium used in these warheads came partly from retired weapons and from existing stocks of separated material, and partly from Calder Hall. Although there is no hard information on this production run at Calder Hall, the evidence available to us is consistent with two of the four Calder Hall reactors being operated to produce weapon-grade plutonium in 1978 and 1979, their fuel being subjected to burnups of 400 MWd/t. Around 160 kg of weapon-grade plutonium would have been produced in these two years. Another 30 kg of weapon-grade plutonium would have been extracted from the parts of the previous two reactor cores that had experienced low burnups before the reactors were turned over to military production, and from the initial discharges after the reactors had again been optimized to produce electricity.[43] We therefore estimate that 190 kg of weapon-grade plutonium were produced in the late 1970s.

There is reason to believe that over twice this amount, or 500 kg, was produced for the Trident programme between 1986 and 1989. One source informed the authors that 'a few hundred kilograms', but 'less than 600 kilograms', of weapon-grade material were produced at this time. This would be consistent with the four Calder Hall reactors operating on military cycles for just over two years. The maximum number of warheads that could have been deployed on the Trident missiles was 512 (eight warheads on 16 missiles in four boats). The planned number may not have gone much above 250, in which case around 1 tonne of plutonium might have been sufficient for the whole programme. As in the case of Chevaline, the plutonium would have been drawn from stocks of separated plutonium and from recycled material as well as from Calder Hall.

Weapon-grade plutonium from 'civil' reactors

The last British sources of weapon-grade plutonium were the first discharges of fuel from the 'civil' Magnox power reactors constructed in the 1960s. A little history is required to explain the origin of this plutonium. Until 1969, reprocessing in the UK was conducted by the United Kingdom Atomic Energy Author-

---

[42] The British Government has never announced the numbers of warheads that it has deployed. After being re-equipped with Chevaline, Britain's Polaris missiles are believed to have been armed with 2 warheads each. If 3 operational submarines were each armed with 16 missiles, this gives rise to 96 warheads. For a discussion of the British nuclear weapon programme, see Norris, R. S., Burrows, A. S. and Fieldhouse, R. W., NRDC, *Nuclear Weapons Databook, Vol. V: British, French and Chinese Nuclear Weapons* (Westview Press: Boulder, Colo., 1994).

[43] 0.36 kg plutonium per tonne of heavy metal is produced at a burnup of 400 MWd/t. If two reactors were operated for two years on an annual refuelling cycle, four reactor cores would have been discharged. To this can be added 4 one-sixth cores, or two-thirds of a core, representing the low burnup fuel (500 MWd/t) extracted before and after the reactors were optimized to produce plutonium for military purposes. 4.66 cores, each containing 110 t of heavy metal, produce 190 kg of plutonium.

ity (UKAEA), which retained ownership of all British plutonium, whether derived from civil or military reactor fuels. In 1969 and in 1971 the two utilities operating nuclear power stations—the Central Electricity Generating Board, covering England and Wales, and the South of Scotland Electricity Board—gained title to the plutonium separated thereafter from the spent fuels discharged from their reactors.

Before 1969 no clear distinction was made between civil and military plutonium in the UK. The main distinction was drawn between weapon- and reactor-grade plutonium (the latter including fuel-grade in the British terminology).[44] Weapon-grade material was made available by the UKAEA to the Ministry of Defence, including plutonium derived from the initial low burnup discharges from the seven civil Magnox reactors commissioned before 1969. However, it appears that some of this material was retained in the civil sector. We estimate that 230 kg of weapon-grade plutonium, or the initial discharges from five of the reactors, came from this source.[45]

In total, it is therefore estimated that around 3.6 tonnes of weapon-grade plutonium have been produced in the UK and assigned to its nuclear weapon programmes since the 1940s. This consists of 230 kg from civil Magnox reactors, 400 kg from the Windscale Piles, 2.3 tonnes from Calder Hall and Chapelcross (1956–64), 190 kg from Calder Hall in 1978–79 and 500 kg from Calder Hall in 1986–89.

This quantity is not, however, the amount currently held in the British inventory. Losses in separation, in weapon production programmes and in weapon testing have to be taken into account. It is assumed here that the British (and French) programmes were slightly 'cleaner' than the US programme and that around 3 per cent of plutonium arisings were lost in separation and in weapon production. 300 kg are assumed to have been consumed in the 45 known British weapon explosions.[46]

Transfers of plutonium to the United States

In addition, substantial quantities of British weapon-, fuel- and reactor-grade plutonium were bartered for US highly enriched uranium and tritium under the US–UK Mutual Defence Agreement of 1958, as amended in 1959.[47] The precise terms of the exchange remain classified in both the UK and the USA. However, the US Government has revealed in 1996 that it acquired 5.4 tonnes of British plutonium in return for 7.5 tonnes of HEU and 6.7 kg of tritium.

[44] Most plutonium extracted from Magnox reactor fuels belongs in the category 'fuel-grade'.
[45] This assumes that the first one-sixth of a core, or 475 t of fuel in 5 reactors, is submitted to a burnup of 500 MWd/t, which gives rise to 0.45 kg Pu/t.
[46] This can only be a very rough estimate. A relatively large amount is assigned to testing because of the evidence that early British warhead tests involved tens to hundreds of kilograms of HEU and plutonium.
[47] A detailed examination of this agreement has been provided by Simpson (note 1). During the 1960s and 1970s the availability of sufficient and reliable reprocessing capacity to provide plutonium for the British weapon programme and for export to the USA, and to reprocess fuels from the growing stock of civil reactors, was the main problem faced by British authorities.

The export of plutonium to the USA has been controversial in the UK, not least because some critics have claimed that material from British civil reactors has been used in the US weapon programme.[48] Whereas the quantities exported by the UK were a drop in the ocean for the USA, they nevertheless formed a significant proportion of total British output. Which and how much material was dispatched when and from which source remains unclear. What is known may be summarized as follows:

1. Between 1959 and 1964, small quantities of weapon-grade plutonium from the UK production reactors were exported to the USA.

2. Between 1964 and 1971, approximately 4 tonnes of fuel- or reactor-grade plutonium produced in Britain's civil Magnox reactors were exported to the USA.[49] The British Government requested that this plutonium should not be used by the USA in its nuclear weapon programme.

3. Between 1962 and 1971, close to 1.3 tonnes of fuel- and reactor-grade plutonium produced in the Calder Hall and Chapelcross reactors were transferred to the USA. Some or all of this plutonium was blended with US supergrade plutonium in the 1980s to increase the stock of weapon-grade plutonium available to the US weapon programme.

4. Between 1971 and 1984 only 50 kg of fuel- or reactor-grade plutonium were transferred to the USA, of which 40 kg came from Chapelcross and Calder Hall.[50] During the Carter Administration's period in office, the agreements covering material exchanges remained in place, but the bartering of reactor-grade plutonium for HEU ceased owing to the Administration's concern that they might be perceived as breaching its non-proliferation policy. One consequence was that the British Government took steps to secure its supplies of weapon-grade uranium by constructing a new military enrichment plant at Capenhurst (see chapter 4), and its supplies of tritium by investing in new facilities at Chapelcross. The main transfers, particularly between 1976 and 1979, seem to have involved weapon-grade plutonium drawn from pre-1971 UKAEA stocks.

5. Besides the use of British fuel-grade plutonium in blending, the plutonium transactions between the USA and the UK ceased after 1980 (apart from warhead components being transferred to the USA as part of the British testing programme in Nevada). The main transactions in the 1980s involved purchases or 'swaps' of enriched uranium (see section IV and chapter 4).

---

[48] See the evidence submitted on 7, 8, 9 and 28 Nov. 1984 to the Sizewell B Public Inquiry. The debate was partly initiated by an analysis of plutonium arisings which suggested that some 2 t plutonium were 'unaccounted for'. See Barnham, K. W. J., et al., 'Production and destination of British civil plutonium', Nature, vol. 317, no. 6034 (19 Sep. 1985), pp. 213–17.

[49] In evidence given by Dr R. Hesketh to the Sizewell Inquiry, it is suggested that a small portion of one of the four 550 kg cores in the Fast Flux Test Facility (FFTF) at Hanford was manufactured from British material, and that the majority of reactor-grade material transferred to the USA between 1964 and 1971 ended up in the Zero Power Physics Reactor at Argonne National Laboratory-West. 200 kg of British reactor-grade plutonium may also have been used in a facility producing californium. Sizewell B Public Inquiry, 7 Nov. 1984.

[50] Evidence submitted by Mr J. W. Baker to the Sizewell B Public Inquiry, 9 Nov. 1984.

Table 3.13. Estimated British inventory of military plutonium, 31 December 1995
Figures are in tonnes.

Source	Inventory
Windscale Piles	0.4
Calder Hall and Chapelcross	3.0
Initial discharges from civil reactors	0.2
Sub-total	**3.6**
Losses in reprocessing and weapon production (c. 3%)	– 0.1
Losses through weapon testing	– 0.3
Transfers to the USA	– 0.1
Total, weapon-grade (central estimate)	**3.1 ± 0.5 (16%)**
Fuel- or reactor-grade plutonium in military inventory	8.7

The estimated British inventory of weapon-grade plutonium is summarized in table 3.13. According to the central estimate, 3.1 ± 0.5 tonnes are available to the British weapon programme, the main uncertainties being the level of production from the Calder Hall and Chapelcross reactors, and the scale of transfers to the USA. Until the British Government follows the example of the US Government in publishing information on its military inventories, a relatively high error margin must be attached to the estimate.

Fuel and reactor-grade plutonium from Calder Hall and Chapelcross

When optimized to produce electricity, the fuels discharged from the Calder Hall and Chapelcross reactors mainly gave rise to fuel- or reactor-grade plutonium.[51] As the Calder Hall reactors were completely refuelled on more than one occasion to produce fresh stocks of weapon-grade plutonium, the fuelling arrangements were complicated. The proportions of ^{240}Pu in discharged fuel were highly variable, reflecting the different burnups to which fuels were submitted. Taking this into account, our estimate is that 10.7 tonnes of fuel- or reactor-grade plutonium were produced at Calder Hall and Chapelcross from 1964 to the end of 1995. Of this amount, 0.7 tonne is known to have been exported to countries other than the USA.[52] In addition, 1.3 tonnes of the 5.4 tonnes of plutonium transferred to the USA under the Mutual Defence Agreement probably came from these reactors. The current inventory held in the UK is therefore estimated to be around 8.7 tonnes at the end of 1995. The fuel- or reactor-grade plutonium separated from Calder Hall and Chapelcross fuels that is held in stock in Britain is categorized as military material, and as such remains outside Euratom and IAEA safeguards.

---

[51] The cross-over between fuel-grade and reactor-grade plutonium (that is, between plutonium containing more or less than 18% ^{240}Pu) occurs at a burnup of around 3500 MWd/t.

[52] Evidence given by Mr J. Baker on 9 Nov. 1984 to the Sizewell B Public Inquiry. The precise amount is 663 kg.

Britain's future requirements for weapon-grade plutonium

A total of 3.1 tonnes of weapon-grade plutonium leaves the UK with a substantial margin above its weapon needs, given the reductions in present and planned warhead numbers that have been set in motion in the 1990s (around 100 warheads may have been withdrawn from service in the past three years). The size of the British nuclear arsenal is still a closely guarded secret, but is believed to lie in the range 200–300 warheads. 300 warheads may remain the upper bound despite the Trident submarine fleet being brought into service. At 3–4 kg per warhead, this arsenal could be serviced with less than one-half of the current stock of weapon-grade plutonium. As much as 1.5 tonnes of this stock could therefore be regarded as excess to requirements.

Nuclear material accounting in Britain

No information has been published on the accounting for military nuclear materials in the UK. It is not known how precisely the British Government can identify the quantities and locations of the plutonium in its military inventories. Nor has any announcement been made about the presence or absence of unresolved 'inventory differences' (see the discussion on the United States in section IV). As far as is known the British Government has yet to embark on the detailed checking of historical records that is required to give confidence that current figures are accurate, and that all acquisitions and removals are known about. The precise amounts of fissile material that have been consigned to waste (e.g., at Dounreay) may also not be known.

As in the other nuclear weapon states, accounting practices in the early period fell short of modern international standards. Unlike the civil nuclear materials held by the USA, the FSU and China, however, all those in the UK have been held under international (i.e., Euratom) safeguards since the early 1980s. The authors have been informed that identical accounting standards are now applied to civil and military nuclear materials in the UK. It is understood that preparations are currently being made to bring the Calder Hall and Chapelcross reactors under Euratom safeguards.

VII. France

France produced weapon-grade plutonium for nuclear weapons in a multiplicity of military and civilian reactors. This plutonium was separated in reprocessing facilities at Marcoule and La Hague, which also separated civilian plutonium.

The French Government has refused to release information about the size of its stock of military plutonium. However, President François Mitterrand in May 1994 declassified information about the size of the French nuclear arsenal (see below), which is generally considered a more sensitive subject.

Because France has made little distinction between military and civilian nuclear facilities, its theoretical plutonium production capacities have far exceeded its actual requirements. As a result, estimates of its total weapon-grade plutonium production (<7 per cent ^{240}Pu) are highly uncertain. Nevertheless, the estimated amounts of weapon-grade plutonium produced in France are set out in table 3.14. The figures are derived below.

Overview of plutonium production

The production of nuclear warheads is the responsibility of the Commissariat à l'Énergie Atomique (CEA). In 1976, the Compagnie Générale des Matières Nucléaires (Cogema) was formed as a subsidiary of CEA to carry out all fuel-cycle activities.

France first developed a source of plutonium for weapons in the late 1950s with the construction of the G1, G2, and G3 production reactors at Marcoule. Plutonium from these reactors was separated in the nearby UP1 reprocessing plant, which started operating in 1958 and was subsequently upgraded. The last of the G reactors was shut down in the mid-1980s.

From the mid-1960s until the early 1970s, France built six civil gas-graphite power reactors: Chinon-1, -2, -3; St Laurent-1, -2; and Bugey-1. Although they were principally for the generation of electricity, they also made weapon-grade plutonium. Because UP1 was unable to handle the fuel from all these reactors, the CEA constructed a second reprocessing plant, the UP2, at La Hague. The UP2 plant began operating in 1966 with a nominal capacity of 800 tonnes of gas-graphite fuel per year. This facility stopped processing gas-graphite fuel in the late 1980s, and now processes only light water reactor fuel. The last of the gas-graphite reactors was shut down in early 1994, and the last of their fuel will be reprocessed in 1997.[53]

By the time G2 was closed in early 1980, France had converted the twin Célestin heavy water reactors to plutonium production. Although these reactors started operation in the 1960s, they were originally dedicated to producing tritium for nuclear weapons. Tritium allowed the French to build smaller, more efficient nuclear weapons with more predictable yields. After conversion to plutonium production, these reactors made only small quantities of tritium.

France produced weapon-grade plutonium in the Phénix prototype breeder reactor. Earlier experimental breeder reactors might also have produced weapon-grade plutonium.

The CEA may have investigated the use of the laser isotope separation process to obtain weapon-grade plutonium from the spent fuel from its power reactors, including its light water reactors.[54] These reactors discharged reactor-grade plutonium containing about 55–65 per cent ^{239}Pu, and the laser process could purify this material to about 93 per cent ^{239}Pu. Although reactor-grade pluto-

---

[53] 'Marcoule: Bilan et perspectives', *Enerpress Document*, no. 5949 (15 Nov. 1993), p. 1.
[54] Office of Science and Weapons Daily Review, CIA, 'France: weapon-grade plutonium from spent-power reactor fuel,' *Science and Weapons Daily Review*, 20 Oct. 1981 (released under the Freedom of Information Act). An earlier declassification conducted under a request from the Natural Resources Defense Council in the early 1990s contains considerably fewer 'cut-outs' or deletions than a recent one received by one of the authors.

Table 3.14. Estimated French inventory of military plutonium, 31 December 1995[a]
Figures are in tonnes.

Source	Inventory
G1, G2, and G3 reactors	2.7–3.1
Célestin-1 and -2	0.5–1.5
Phénix	0.6–1.2
Chinon-1, -2, and -3; St Laurent-1 and -2; and Bugey-1	0.5–2.0
Sub-total	**4.3–7.8**
Losses through processing (c. 3–5%)	0.1–0.4
Losses through weapon testing	0.6–1.1
Total inventory of weapon-grade plutonium	**3.6–6.3**
Average total	**5.0 ± 1.4 (28%)**

[a] This table does not account for any plutonium from dismantled weapons transferred to civilian programmes.

nium can be used to make nuclear explosives, it is not desirable for use in modern nuclear warheads.

Production of weapon-grade plutonium for weapons has stopped. According to *Enerpress*, the Ministry of Defence informed Cogema at the end of 1991 that military plutonium separation would stop in 1994.[55] This source added that the military plutonium separation programme at the UP1 facility operated 16 weeks in 1992, only seven weeks in 1993 and was expected hardly to operate at all in 1994.

The G reactors

The G1 reactor

France's first military reactor went critical in early 1956 and soon afterward reached full power. Initially, the reactor's power was 38 MWth, but it was gradually increased to 42 MWth by 1962.[56] This reactor was moderated by graphite, cooled by air and fuelled by natural uranium fuel. It initially produced about 2 MWe of electricity, but needed about four times this amount of electricity to operate. The reactor was shut down in October 1968.

The core contained about 100 tonnes of fuel, which was loaded while the reactor was shut down. The average burnup of the fuel was low, approximately 100–200 MWd/t.[57] At an availability factor of 80 per cent, this reactor would produce about 12 260 megawatt-days of thermal energy (MWth-d) a year. At

---

[55] *Enerpress Document* (note 53), p. 1.
[56] IAEA, *Directory of Nuclear Reactors, Vol. 1: Power Reactors* (IAEA: Vienna, 1959), pp. 145–50; and the annual reports of the Commissariat à l'Énergie Atomique.
[57] IAEA (note 56). This reference states that no fuel rods had a burnup over 300 MWd/t.

the above burnups, this corresponds to an average of about 12 kg of weapon-grade plutonium (about 99 per cent ^{239}Pu) per year.[58]

Although the amount of plutonium would vary little over this range of average burnups, the amount of discharged fuel, and the concentration of plutonium in this fuel, would vary greatly. At 100 MWd/t, the reactor would need to discharge 120 tonnes of spent fuel per year. At 200 MWd/t, the reactor would need to discharge about 60 tonnes per year.

According to the CEA's *1962 Annual Report*, by 31 December 1962, G1 had produced in total about 59 630 MWth-d. At a burnup of 100–200 MWth-d, or a production rate of 0.98–1.0 grams of weapon-grade plutonium per megawatt-thermal-day, this reactor would have produced a total of about 58–60 kg of weapon-grade plutonium. During 1962, the reactor produced about 12 392 MWth-d of energy, or about 12 kg of weapon-grade plutonium. Assuming that it continued to produce at this rate from the beginning of 1963 through mid-1968, the reactor produced in total about 125 kg of plutonium. We assign an uncertainty of 10 per cent to this estimate. At an average burnup of 100–200 MWd/t, the plutonium would have been contained in a total of about 625–1250 tonnes of irradiated fuel.

The G2 and G3 reactors

These identical reactors reached full-power operation in 1959 with a power of 200 MWth and 40 MWe, and were refuelled while in operation. In the late 1950s, the reactors had a fuel loading of 120 tonnes of natural uranium, and the expected average burnup was 400 MWd/t.[59]

During the early 1960s, the power of each reactor was gradually increased to 260 MWth. The fuel loading is believed to have risen to about 150 tonnes of natural uranium fuel as the power was increased.

According to the CEA *1963 Annual Report*, the operators did their best to increase the reactors' thermal power as much as possible in order to increase plutonium production. The CEA *1969 Annual Report* stated that plutonium production had also been increased by using a fuelling technique using a certain portion of depleted uranium.

The G2 reactor was closed permanently in 1980, and the G3 reactor was shut down in 1984. Using gross electricity data from several issues of *Nucleonics Week*, the total gross electricity production of these reactors is seen to be 482 350 megawatt-days of electrical energy (MWe-d).[60] At an estimated average gross thermal efficiency of between 15.3 and 17 per cent, this corresponds to 2 994 980 MWth-d ± 157 630 MWth-d (or 5.3 per cent uncertainty

---

[58] Unless otherwise noted, plutonium data are from Turner, S. E., *et al.*, *Criticality Studies of Graphite-Moderated Production Reactors, Report prepared for the US Arms Control and Disarmament Agency*, SSA-125 (Southern Sciences Applications: Washington, DC, Jan. 1980).
[59] IAEA (note 56), p. 151.
[60] *Nucleonics Week*, 24 Feb. 1983, 24 Mar. 1983 and 31 Jan. 1985.

range).[61] At an average burnup of 300–400 MWd/t, total plutonium production was about 2780 kg ± 160 kg (5.7 per cent error). This plutonium would have been contained in an estimated 7100–10 500 tonnes of spent fuel, where the lower bound corresponds to the higher burnup.

This estimate is similar to one in a 1984 US Central Intelligence Agency (CIA) report.[62] As of 1984, this report says that UP1 had recovered over 2500 kg of plutonium for weapon use from these production reactors, and UP1 had reprocessed over 10 000 tonnes of spent fuel from these reactors.

The CFDT estimate

A publication of the Confédération Française Démocratique du Travail (CFDT), a major trade union in France, contains an estimate of the annual discharge of spent fuel from the G reactors and its average annual burnup until the end of 1977. According to this source, the three G reactors produced a total of 9910 tonnes of spent fuel during 1959–77.[63] From information on the associated annual burnup, an average burnup for all this fuel of about 400 MWd/t is derived. At this burnup, the spent fuel produced a total of 3 964 000 MWth-d of heat and contained a total of 3670 kg of plutonium.

This estimate, however, is inconsistent with the available information about the maximum power achieved by the G reactors. Assuming that G1, G2 and G3 operated at full power 100 per cent of the time, G1 is estimated to have produced at most about 161 000 MWth-d during its lifetime, and G2 and G3 would have produced at most 3 500 000 MWth-d by the end of 1977. The estimated upper bound for all three G reactors is thus 3 660 000 MWth-d. This value is less than the CFDT total energy estimate above, implying that this source's estimated burnup figures or spent fuel discharges are too high. At an average burnup of 300 MWd/t, however, the total energy output would be 2 973 000 MWth-d, close to our estimate.

Gas-graphite power reactors

France has acknowledged that it used its civilian gas-graphite reactors to make plutonium for its military programmes. For example, the CEA *1971 Annual Report* states that: 'The national civilian needs and military needs continued to be covered in 1971 by the production of the reactors of Marcoule and of Chinon.' In the 1973 report on its scientific and technical activities, the CEA added St Laurent-1 to this list.

The CIA has also provided some, possibly conflicting, information about the use of these reactors. In 1981, the CIA stated that France 'will continue to use'

---

[61] Both the thermal and the electrical power of the G2 and G3 reactors varied. This range of thermal efficiencies is estimated from various CEA annual reports.

[62] CIA, Directorate of Intelligence, *French Nuclear Reactor Fuel Reprocessing Program, An Intelligence Assessment*, Sep. 1984 (Released under the Freedom of Information Act).

[63] Syndicat CFDT de l'Énergie Atomique, *Le Dossier Électronucléaire* (Éditions du Seuil: Paris, 1980), pp. 186–87.

Bugey-1, Chinon-2, and Chinon-3 'for weapon-grade plutonium production'.[64] In 1984, a CIA report stated that France retains the option to use gas-graphite reactors 'to meet additional requirements for weapons, if they decide to produce more nuclear weapons and/or to modernize existing nuclear weapon systems'.[65] The report added that France 'may shift the power reactors Chinon-2 and/or Chinon-3 to plutonium production and could also adapt the St Laurent-1 and St Laurent-2 reactors to produce weapon-grade plutonium'.

There is little official information on the amount of weapon-grade plutonium produced in these reactors. However, the available information suggests that these reactors might have produced less weapon-grade plutonium than is commonly believed.

The Chinon-1 reactor

This 300 MWth or 68 MWe gas-graphite reactor was operated by Électricité de France (EdF), although it was also used to make military plutonium. The reactor went critical in 1962, but encountered problems, and did not reach full power until several years later. The reactor was closed down in 1973 because it was uneconomical.

According to the CEA annual reports, the core contained about 150 tonnes of fuel, which was reloaded while the reactor was shut down. Some of these shut-downs are recorded in CEA documents. According to the annual reports, major refuellings took place in 1967 (a five-month shut-down), 1968 (a four-month shut-down), 1969 (a two-month shut-down) and 1970 (a one and one-half month shut-down).

In addition, the annual reports contain information that implies earlier fuel discharges. The 1965 annual report says that the UP1 reprocessing facility at Marcoule processed Chinon fuel in December 1965. The 1966 annual report stated that substantial quantities of Chinon-1 fuel were reprocessed in that year for the first time.

By the time the Chinon-1 reactor was closed in 1973, it had produced in total 129 809 MWe-d (gross) of electricity.[66] Its gross thermal efficiency is estimated at about 27.6 per cent, and thus the reactor produced a total of about 470 000 MWth-d of energy.[67] Since the electrical power of this reactor probably varied, this estimate is uncertain.

By mid-1974, the reactor had discharged in total 53 625 fuel elements at an average burnup of 1215 MWd/t.[68] Each element weighed about 8 kg, for a total

---

[64] *Science and Weapons Daily Review* (note 54).
[65] CIA (note 62).
[66] *Nucleonics Week*, 24 May 1973, p. 8.
[67] *Nuclear Engineering International*, Aug. Supplement, 1982. This source lists an electrical output of 83 MWe-gross, 70 MWe-net, and 300 MWth.
[68] Arnaud, G., *et al.*, 'Résultats d'irradiation sur les éléments combustibles de la filière uranium naturalgraphite-gaz', CEA, *Bulletin d'Informations Scientifiques et Techniques*, no. 196 (Oct. 1974).

of about 430 tonnes of spent fuel.[69] The total energy extracted from this fuel is about 520 000 MWth-d, about 10 per cent higher than the above total energy estimate. Given the uncertainties in these estimates, this result implies that all or virtually all of the fuel had been discharged from Chinon-1 by 1974.

Gas-graphite fuel at an average burnup of about 1200 MWd/t contains about 1.0 kg of plutonium per tonne of spent fuel, where the plutonium contains about 92 per cent ^{239}Pu. The total plutonium output from Chinon-1 was therefore 430 kg. About 200 kg of this amount is estimated to be weapon-grade plutonium.

The Chinon-2, Chinon-3, St Laurent-1, St Laurent-2, and Bugey-1 reactors

These gas-graphite power reactors were larger than the Chinon-1 reactor and unloaded on-line. As a result, estimating any contribution from these reactors to the military stock is highly uncertain.

For example, if both the Chinon-2 and Chinon-3 reactors were dedicated to weapon-grade plutonium production, they could produce up to 600 kg per year.[70] However, indications are that these reactors were optimized to produce electricity and thus produced primarily fuel- or reactor-grade plutonium.

One indication of this conclusion is that the average burnup of the spent fuel discharged from these reactors by mid-1974 was higher than that for weapon-grade plutonium.[71] By mid-1974, Chinon-2 had discharged 89 170 rods, or about 936 tonnes of fuel with an average burnup of 2022 MWd/t. Chinon-3 had discharged 93 750 rods, or about 1070 tonnes of fuel, with an average burnup of 1425 MWd/t. By mid-1974, the St Laurent-1 reactor had discharged 71 093 fuel rods, or about 775 tonnes of spent fuel, with an average burnup of 1973 MWd/t. St Laurent-2 had discharged 63 006 rods or about 690 tonnes of fuel with an average burnup of 1990 MWd/t. The Bugey-1 reactor had discharged 20 370 rods with an average burnup of 1740 MWd/t.

How much weapon-grade plutonium was produced by the civil gas-graphite reactors?

The above information suggests that these reactors were not usually dedicated to making weapon-grade plutonium, but it is not sufficient to estimate the amount produced. Data about the reprocessing plants provide a method to make a crude estimate.

In the early years La Hague processed lower burnup fuel, much of which could have gone for weapons (see appendix C). During 1966–70, for example, La Hague processed a total of 740 tonnes of fuel from all the gas-graphite

---

[69] Simnad, M. T., *Fuel Element Experience in Nuclear Power Reactors* (Gordon and Breach: New York, 1971). This source says the Chinon-1 had 140 t of natural uranium fuel in 1148 fuel channels, each containing 15 rods.

[70] This estimate assumes that the reactors had a capacity factor of 80% and a conversion factor of about 0.9 g of weapon-grade plutonium per MWth-d.

[71] Arnaud *et al.* (note 68) and Simnad (note 69).

MILITARY PLUTONIUM IN THE NUCLEAR WEAPON STATES   73

reactors, including Chinon-1, with average burnups of roughly 500 to 1170 MWd/t, resulting in about 620 kg of plutonium (including losses). Roughly 350 kg of this plutonium could have been weapon-grade; the rest would contain more than 7 per cent ^{240}Pu.

According to the CFDT, Marcoule processed about 335 tonnes of civil reactor fuel by the end of 1977.[72] If the average burnup of this fuel was low enough, the fuel could have contained as much as 350 kg of weapon-grade plutonium. The result of this crude estimate is that up to 700 kg of weapon-grade plutonium could have originated from the civil gas-graphite reactors by 1977.

Available evidence suggests that civil gas-graphite reactors continued to produce weapon-grade plutonium at least into the 1980s, whenever some material was needed for military requirements. We therefore assign a total value of 500–2000 kg from these reactors to the military stockpile, although the basis of this estimate remains uncertain.

The Phénix breeder reactor

The 250-MWe, or 563-MWth, prototype Phénix breeder reactor started operation in 1973. According to a 1984 CIA assessment, the French had used Phénix to produce 'some weapons-grade plutonium'.[73] Phénix produces weapon-grade plutonium in a 'blanket' of depleted uranium that surrounds the reactor core.

France is believed to have separated the weapon-grade plutonium at UP1 at Marcoule. The first reprocessing campaign for material from the Phénix blanket was in 1978.

In theory, Phénix would discharge about 115 kg of weapon-grade plutonium (with about 97 per cent ^{239}Pu) a year in blanket material, when the reactor operates at full power 100 per cent of the time.[74] At this rate, this reactor would produce about 0.55 grams of weapon-grade plutonium in the blanket per megawatt-thermal-day of reactor output.

Nucleonics Week reports that Phénix had produced about 917 000 MWe-d by the end of 1995.[75] At a thermal efficiency of 44 per cent, the total energy output was 2 084 000 MWth-d. At the above weapon-grade plutonium production rate, Phénix would have produced about 1150 kg of weapon-grade plutonium by the end of 1995.

This estimate is probably an upper bound, since France might have dedicated some of this plutonium to civilian uses. We therefore assign a range of 570–1150 kg of weapon-grade plutonium to the military stockpile from Phénix.

The 40-MWth Rapsodie reactor at Cardarache might have also produced weapon-grade plutonium. Since its possible contribution is small relative to the uncertainty in the above estimate, it is ignored here.

[72] CFDT (note 63).
[73] CIA (note 62).
[74] Bussac, J. and Reuss, P., *Traite de Neutronique* (Hermann: Paris, 1978).
[75] *Nucleonics Week*, 8 Feb. 1996.

The Célestin reactors

The twin Célestin heavy water reactors at Marcoule were each initially rated at 190 MWth. The first started operating in 1967 and the second in 1968. Their initial purpose was to produce tritium for nuclear weapons, but they later produced military plutonium and civil isotopes such as plutonium-238, helium-3, cobalt-60 and carbon-14.

Initially, these reactors were fuelled with plutonium, but by the 1970s they were fuelled with highly enriched uranium. At start-up, the Pierrelatte enrichment plant was not finished and France therefore did not have a source of military highly enriched uranium (see chapter 4).

Starting in the mid-1970s, these reactors were converted to plutonium production for weapons. Evidence for this statement is a Cogema brochure, entitled 'Marcoule', from the late 1970s or early 1980s that says that in 1976 the Célestin reactors received a plutonium producing role, 'the ends of which are identical to the G reactors'.

By 1980, the reactors were reportedly producing primarily plutonium. Mary Davis has reported that, according to CFDT, the Célestin reactors were no longer operating in the tritium mode by 1980, although some tritium continued to be produced.[76] The 1984 CIA assessment mentioned above says that with the decommissioning of G1 and G2 the Célestin reactors were converted to plutonium production, although they probably also continued to produce small quantities of tritium.

At their rated power and an 80 per cent capacity factor, these reactors can produce about 100 kg of weapon-grade plutonium a year. This is equivalent to the production of about 1.5 kg of tritium per year, which appears far in excess of French military needs.

Little information is available about the amount of plutonium or tritium produced in the Célestin reactors. A rare reference is in the 1988 Cogema annual report, which says that in 1988 the production of plutonium and tritium in these reactors for the military programme was slightly higher than expected.

Since 1991, the reactors have been alternating in operation, and present plans call for their operation until the end of the century.[77] According to Davis, plutonium production in these reactors ceased in 1991, although they still make tritium.

Assuming that these reactors were dedicated to plutonium production from 1980 until 1990, or the equivalent of 10 full years, they could have produced about 1000 kg of weapon-grade plutonium. This estimate could have been as low as 500 kg or as high as 1500 kg.

---

[76] Barrillot, B. and Davis, M., *Les Déchets Nucléaires Militaires Français* [French military nuclear wastes],(Centre de Documentation et de Recherche sur la Paix et les Conflits: Lyon, 1994).

[77] *Enerpress Document* (note 53).

Losses through weapon tests

France conducted a total of 210 nuclear explosions. At an estimated average of 3–5 kg of plutonium per test, these tests could have consumed 630–1050 kg of weapon-grade plutonium.

Current nuclear arsenal

As of May 1994, France had a total of 492 nuclear weapons, of which 384 were on M4 submarine-launched ballistic missiles.[78] The rest were on land-based ballistic missiles, air-to-surface missiles and shorter-range tactical missiles. The number of warheads is now decreasing.

Little information exists about the amount of plutonium in French nuclear weapons. However, if each warhead contains on average 3–4 kg of plutonium, then the 1994 arsenal contained roughly 1500–2000 kg of plutonium.

With a total inventory of about 5.0 ± 1.4 tonnes, France appears to have a military plutonium surplus of more than 1.5 tonnes, even assuming that some plutonium is in the manufacturing pipeline. Whether France maintains a strategic plutonium reserve is unknown. It may have transferred some undisclosed amount of plutonium from dismantled weapons to its civilian programmes.[79]

Nuclear material accounting in France

As in the FSU and the UK, no information has been made public on the accounting for military materials in France. It is not known how precisely the French Government can identify the quantities and locations of the plutonium in the military inventories. Nor has any announcement been made about the presence or absence of unresolved 'inventory differences' (see the discussion on the United States in section IV).

As in the other nuclear weapon states, accounting practices in the military sector fell short of modern international standards, particularly in the early period. It is believed that France, like the FSU, restricted its accounting at military facilities to records of inputs and outputs. Unlike in the UK, the approaches followed in regard to civil and military activities were therefore not identical. Steps are apparently being taken in France to increase the accuracy of military plutonium inventories. However, as in the case of the UK, there are no firm indications that the French Government has yet embarked on the detailed checking of historical records that is required to give confidence that current figures are accurate, and that all acquisitions and removals are included.

---

[78] 'Intervention de M. François Mitterrand, President de la Republique sur le Thème de la Dissuasion', Palais de l'Élysée, Paris, 5 May 1994.
[79] Ministère de l'Industrie, L'Énergie nucléaire en 113 questions, Mar. 1996, pp. 35–36.

VIII. China

Considerable amounts of information became available about the Chinese nuclear weapon programme in the early 1990s, as a result of Chinese publications and the work of Lewis and Xue and, most recently, of Norris, Burrows and Fieldhouse.[80] Knowledge of plutonium production nevertheless remains hazy, and little information has become available since *World Inventory 1992* was published. It is believed that Chinese production of weapon-grade plutonium ceased in 1991.[81]

Before Soviet technical assistance was withdrawn in 1960, some design drawings and a few items of equipment for a production reactor were transferred to China. Priority was given to the enrichment programme, and the first production reactor began operating in 1966, two years after the first Chinese weapon test. This reactor, at the Jiuquan Atomic Energy Complex in Subei county (initially code-named Plant 404), experienced frequent technical difficulties, and interruptions because of the political turmoil in China during the Cultural Revolution, and was shut down for repair between 1973 and 1975. In the mid-1980s it was re-engineered so that it could be operated as a dual-purpose electricity and plutonium producer.

All fuels from this reactor were reprocessed on site from 1968 onwards. The main reprocessing plant began operating in April 1970.[82]

The other production site (Plant 821) is located at Guangyuan, in Sichuan Province. Its development began in the mid-1960s as a 'third-line' weapon manufacturing facility, further away than Jiuquan from China's frontiers. It is known that China's largest production reactor was constructed on this site. Although there are reports of other reactors in China, there is no evidence that there are plutonium production reactors other than the single units at Jiuquan and Guangyuan.[83]

In *World Inventory 1992*, the authors estimated that total weapon-grade plutonium production was between 1 and 4 tonnes, although this estimate was viewed as having a high degree of uncertainty and was referred to as a 'guesstimate'. Rather than derive a new estimate, the authors have decided to depend on a re-evaluation of the existing, sparse information about Chinese plutonium production conducted recently by a group of scientists at the Union of Concerned Scientists (UCS).[84]

The UCS group estimates that the Jiuquan reactor initially had a power of 250 MWth that was doubled by the early 1980s, giving a weapon-grade plutonium production by the end of 1991 of 1–2 tonnes. The Guangyuan reactor is

---

[80] The main sources on China are Lewis and Xue (note 1); Lewis, J. W. and Xue, L., 'Chinese strategic weapons and the plutonium option', *Critical Technologies Newsletter* (US Department of Energy: Washington, DC, Apr.–May 1988); and Norris, Burrows and Fieldhouse (note 42).
[81] Norris, Burrows and Fieldhouse (note 42), p. 350.
[82] Norris, Burrows and Fieldhouse (note 42), p. 348.
[83] Lewis and Xue (note 1), p. 113.
[84] Wright, D., Gronlund, L. and Liu, Y., 'Estimating China's stockpile of fissile material for weapons', draft, Union of Concerned Scientists Technical Working Paper, Washington, DC, Apr. 1996.

MILITARY PLUTONIUM IN THE NUCLEAR WEAPON STATES 77

estimated to have an initial power of 500 MWth that was later increased to 1000 MWth, producing about 1–4 tonnes of weapon-grade plutonium by the end of 1991.

Total production is therefore estimated as 2–6 tonnes of weapon-grade plutonium. Because of the lack of hard evidence on the production facilities, in particular on the power and operating histories of the reactors, the UCS group claims its estimates to be suggestive at best.

It is believed that China has deployed around 450 nuclear warheads.[85] Many of these warheads may have been constructed largely or entirely using HEU as the fissile material. A substantial modernization programme is under way in China which will probably entail the substitution of smaller warheads with plutonium being used as the main fissile material in weapon primaries. If the Chinese Government intends to maintain an arsenal of 400–500 warheads, their plutonium content could range between 1 and 2 tonnes if the warheads contained 3–4 kg on average. The plutonium requirement would be somewhat higher given the amounts needed in the production system and the desire to hold some material in reserve. The requirement could probably be met from a stock of 2–6 tonnes, although warhead considerations suggest 2 tonnes to be too small an amount.

As China seems prepared to accede to an indefinite cut-off of fissile material production, it is possible that its inventory of plutonium has been underestimated. Norris *et al.* suggest that the two production reactors could each have produced 300–400 kg of plutonium per year, giving an upper limit for the total inventory of 15 tonnes, but they provide no supporting evidence.[86] Besides being at variance with the information presented above, this high level of production seems very questionable in view of the reports from Chinese sources that the operating performance of the Chinese production reactors was often unsatisfactory.

The extent of plutonium losses is also unknown. How efficient were the reprocessing and weapon production facilities? Did accidents occur releasing plutonium into the environment? How much plutonium was consumed in the 45 Chinese weapon explosions that are known to have been carried out?

Until harder information about the production programmes becomes available, or the Chinese Government releases its own figures, any estimates of Chinese plutonium inventories must therefore be regarded as very tentative. A lower estimate of 2 tonnes and an upper limit of 6 tonnes are being attached here, giving a central 'guesstimate' of 4 tonnes (with an error margin of ± 50 per cent). One reason for keeping to a relatively low estimate is that it seems unlikely that Chinese production has far exceeded the British and French production of weapon-grade plutonium.

---

[85] Norris, Burrows and Fieldhouse (note 42), table 7.1, p. 359.
[86] Norris, Burrows and Fieldhouse (note 42), p. 350. These production rates were typical for the larger, up-rated US and Soviet production reactors.

Nuclear material accountancy in China

The authors are aware of no information on how weapon materials are accounted for in China. Nuclear materials at a number of civil facilities are eligible for safeguarding under China's voluntary offer agreement with the IAEA. In late 1994, one power reactor and one research reactor had been designated by the IAEA for inspection.[87]

---

[87] Civil materials and plants placed by China on a 'facilities list' can in principle be designated for safeguarding by the IAEA under the voluntary offer agreement concluded with the Chinese Government (INFCIRC/369).

4. Inventories of highly enriched uranium in the nuclear weapon states

I. Introduction

Highly enriched uranium in nuclear weapons usually has ^{235}U concentrations in excess of 90 per cent. This material is often referred to as 'weapon-grade uranium' (WGU).

HEU with enrichments between 20 per cent and 90 per cent has also been used in nuclear weapons. Although the nuclear weapon states used relatively small quantities of this material in their early fission weapons, they are believed to have used significantly larger quantities in thermonuclear weapons.

While more information exists now about military nuclear programmes than just a few years ago, the military inventories of HEU in the five declared NWS remain largely unknown. The one exception is the United States, where Secretary of Energy Hazel O'Leary has started an unprecedented effort to declassify information about military HEU stocks as a result of the end of the cold war and citizens' requests. This action is intended to encourage other nations to declassify similar information. Such declassified information is important to the current debate over the proper management and disposition of nuclear materials. As of early 1996 the DOE had released details of its total historical production of HEU for weapon, naval and civil purposes. The DOE anticipates declassifying additional information about its HEU stocks during 1996 and 1997.

This chapter provides an overview of the enrichment programmes in the NWS and presents estimates of the HEU inventories that are dedicated to nuclear weapons or associated stocks. The largest inventory is found in the former Soviet Union, and significant portions of this inventory continue to lack adequate physical protection against theft or diversion.

This chapter does not focus on HEU inventories in naval and civil programmes, although some preliminary naval inventory estimates are provided when applicable. Chapter 8 addresses civil stocks. However, HEU in the naval and civil fuel cycles also poses a risk of diversion.

II. Overview of enrichment programmes

The birth of the uranium enrichment industry occurred in the US Manhattan Project during World War II, when the USA developed several approaches to produce the HEU that destroyed Hiroshima in 1945. The only techniques that proved capable of producing HEU for nuclear weapons at the time were the

Table 4.1. HEU stocks dedicated to nuclear weapons, reserves and associated categories, including excess, as of 31 December 1995

Country	Main production period[a]	Halt to HEU production for weapons	Weapon inventory, reserves and excess weapon-grade uranium equivalent (t)[b]
USA	1950s–60s	1964	645 ± 10%
USSR	1960s–80s	1987–88	1 050 ± 30%
UK	1950s–60s	1963[c]	8 ± 25%[d]
France	1970s–80s	by early 1996	25 ± 30%
China	1970s–80s	1987–89	20 ± 25%
Total			1 750 ± 23% (± 400)[e]

[a] Period in which most weapon-grade uranium (WGU) is estimated to have been produced.

[b] Tabulated in terms of WGU equivalent, which is defined as the amount of 93% enriched weapon-grade uranium that could have been produced. In the US case, the stock is estimated at about 750 t of HEU with an average enrichment of 80%, and their number is converted to WGU equivalent. In all other cases, WGU-equivalent refers to the amount of WGU that could have been produced from the total estimated separative work output, assuming a typical tails assay of 0.3%. In these cases, the average enrichment is unknown although, as in the US case, the vast majority of the stock should be above 90%. See text for details of each estimate.

[c] Refers to domestic production only, does not reflect imports of WGU from the USA.

[d] Includes WGU imported from the USA (see also chapter 3).

[e] The total is rounded. If we assume the average enrichment of the HEU is 80%, 1750 t of WGU equivalent would correspond to about 2000 t of HEU (average enrichment 80% ^{235}U).

electromagnetic isotope separation (EMIS) and gaseous-diffusion processes. Because large-scale production using EMIS required enormous amounts of electricity and labour, the USA concentrated on building gaseous-diffusion plants after the war.

The Soviet Union launched its own enrichment programme right after World War II. It first deployed EMIS and gaseous diffusion at the end of the 1940s, and in the 1960s it started large-scale deployment of gas centrifuges. China received assistance from the USSR in building its first gaseous-diffusion plant, which began operation in the 1960s. When the UK and France started their nuclear weapon programmes, they also built gaseous-diffusion plants. The British plant started up in the mid-1950s and the French plant began operating in the mid-1960s. The UK has also received a significant amount of HEU from the USA.

Nuclear weapons provided the primary motivation for these enrichment programmes, summarized in table 4.1. In all the NWS these programmes were complemented by plutonium production programmes. The NWS have used both fissile materials, either alone or in combination, in their nuclear weapons; but weapon designers clearly desired HEU for boosted fission weapons with yields of hundreds of kilotons and for the second stages of thermonuclear weapons, where it is used in conjunction with the thermonuclear materials.

III. The United States

The USA curtailed production of HEU for nuclear weapons in 1964 after 20 years of production. Although the US Government planned to resume production in the early 1990s, claiming that its stockpile was running low, it cancelled these plans in early 1991. With the end of the cold war the US nuclear weapon arsenal is undergoing deep reductions, creating large excess stocks of HEU from dismantled weapons.

After 1964, the United States continued making large amounts of HEU for naval reactors, and much smaller quantities for space and research reactors. In November 1991 the USA announced that it would suspend production of HEU for any purpose.[1] Future requirements for naval and research reactors are being met from existing stockpiles.

The US enrichment complex

In the course of the Manhattan Project, the USA developed several enrichment processes, principally electromagnetic separation with calutrons and gaseous diffusion. After the war only gaseous-diffusion plants were built, because of their lower cost and greater energy efficiency in comparison to the other methods available at that time.

The first gaseous-diffusion plant was at Oak Ridge, Tennessee. By the mid-1950s, other plants were also located in Paducah, Kentucky and Portsmouth, Ohio. US production rose sharply in the 1950s, and reached a peak of 16.5 million SWU in 1961, out of a total US capacity of 17.2 million SWU per year. At peak production, the USA could produce over 80 000 kg of weapon-grade uranium per year.

From World War II until 1964 almost the entire US enrichment capacity was dedicated to the production of HEU for weapons, and most of this was weapon-grade uranium. Civil nuclear power was still in its infancy, and thus civil demand was small. After production for weapons was terminated in 1964, production dropped to a low of about 6 million SWU in 1970. The enrichment plants continued to produce HEU for naval reactors, but they produced mostly low-enriched uranium for civil purposes.

After 1970, production increased again in anticipation of large needs for civilian nuclear power reactors. The enrichment plants were expanded to a total capacity of 27 million SWU per year, but new reactor construction and enrichment sales were significantly lower than expected. In 1985, the Oak Ridge plant was shut down. Construction of the Gas Centrifuge Enrichment Plant at Portsmouth was also cancelled in 1985, after almost $3 billion had been spent.

---

[1] Dizard, W., 'Suspension of HEU production viewed favourably by friends, foes of UEE Bill', *Nuclear Fuel*, 25 Nov. 1991.

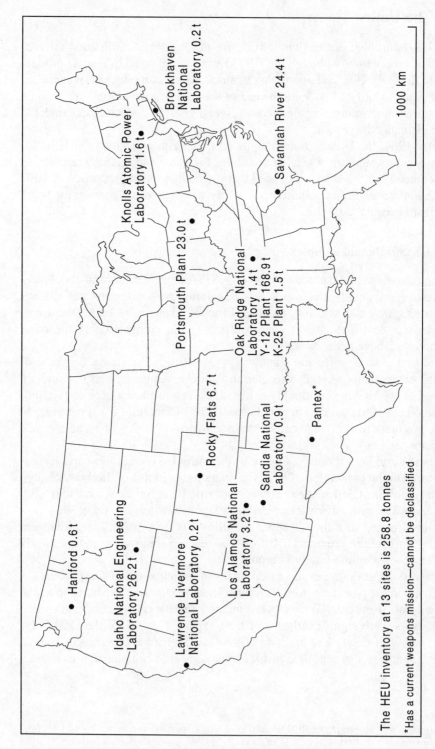

Figure 4.1. Highly enriched uranium inventories of the US Department of Energy, 31 December 1993

Declassification of HEU production information by the DOE

In June 1994, Secretary of Energy Hazel O'Leary declassified an estimate of the total amount of highly enriched uranium produced in the United States. The K-25 site in Oak Ridge made about 483 tonnes between 1945 and 1964. Portsmouth made another 511 tonnes from 1956 until 1992, for a total of about 994 tonnes of HEU. DOE officials say that this value remains uncertain for several reasons. There are poor records for the first two decades of the programme. In addition, an unexpected uncertainty is the difficulty in properly accounting for both the further enrichment of HEU to WGU at a later time and the blending down of HEU to LEU, causing the total HEU inventory to decrease as a result.

For example, from August 1945 until 1947, about 1 tonne of weapon-grade uranium was made in beta calutrons in the Y-12 Plant by further enriching over 4 tonnes of mostly 30 per cent HEU from the nearby K-25 gaseous-diffusion plant. (Before August 1945, alpha and beta calutrons produced just tens of kilograms of HEU.) This example shows that total production should be reduced by 3 tonnes. The DOE has said that the further enrichment of HEU to weapon-grade occurred often enough to severely complicate any cursory attempt to reduce the uncertainty in the estimate of total historical HEU production.

As of mid-1996, the DOE had not released information about the enrichment levels of the HEU it had produced or the amount that was dedicated to specific purposes. Most of it is expected to be weapon-grade, although DOE officials have said that a significant amount of this HEU was not weapon-grade and was produced before the 1964 cut-off in HEU production for weapons.

In February 1996, the Secretary of Energy announced that the DOE would produce a report detailing the production, use, disposition and inventories of HEU covering the past 50 years. She said the report would be completed in about a year. Because this report will be considerably more difficult to accomplish than the DOE's similar plutonium report (see chapter 3), the release date is expected to be delayed to at least late 1997.

Until production for weapons ceased, the enrichment complex produced about 150 million SWU, at an average tails assay of about 0.3 per cent.[2] About 4.6 million SWU of this total is estimated to have gone towards producing low-enriched uranium for plutonium production reactors, experimental reactors and early power reactors. The rest went to the production of HEU, resulting in the production of an equivalent of roughly 725 tonnes of weapon-grade uranium. As shown below, naval reactors have required in total about 100 tonnes of 97 per cent enriched uranium. Assuming that HEU production after 1964 was principally for naval reactors, then the equivalent of roughly 825 tonnes of weapon-grade uranium has been produced in total. This amount is equivalent to

---

[2] von Hippel, F., Albright, D. and Levi, B., *Quantities of Fissile Materials in US and Soviet Nuclear Weapons Arsenals*, PU/CEES Report no. 168 (Center for Energy and Environmental Studies, Princeton University: Princeton, N. J., July 1986); and Cochran, T. B., *et al., NRDC, Nuclear Weapons Databook, Vol. II: US Nuclear Warhead Production* (Ballinger: Cambridge, Mass., 1987).

170 million SWU, assuming that the tails assay of the naval enriched uranium was between about 0.2 and 0.25 per cent. This amount of SWU implies that the average enrichment of the 994 tonnes of HEU is about 80 per cent, assuming a weighted tails assay of 0.0029.

For the purposes of illustration, it can be assumed that the HEU is divided into two categories—weapon-grade uranium and non-weapon-grade HEU of a fixed average enrichment. Although the enrichment level of the non-weapon-grade material varied, the average is a convenient measure and we choose an enrichment level near the middle of the possibilities, or about 45 per cent. At an average enrichment of 45 per cent, the breakdown is about 280 tonnes of 45 per cent enriched material and 715 tonnes of weapon-grade uranium. If instead, 30 per cent is picked as the average enrichment level, the breakdown is 210 tonnes non-weapon-grade and 785 tonnes weapon-grade uranium.

In addition to revealing the total HEU production, the DOE has declassified the information that about 259 tonnes of HEU were stored at 13 sites as of 31 December 1993 (located as shown in figure 4.1). Although the inventories at many sites are relatively fixed, the inventory at the Y-12 Plant is growing as HEU is shipped there after weapons are dismantled at the Pantex weapon assembly and disassembly facility in Texas. DOE officials have said that a new inventory estimate is unlikely to be released, because of concerns that it could provide a method to derive classified nuclear weapon information.

The rest of the 994 tonnes produced, about 736 tonnes, is contained in two broad groupings. The first and most important category includes HEU dedicated to the nuclear arsenal or stored at the Pantex site. The other is HEU allocated to or consumed in the naval, civil and production reactor fuel cycle, nuclear tests or exported to other countries. This latter category is in essence a draw-down in the total HEU inventory. After subtracting the value from 736 tonnes, the resulting quantity is the amount dedicated to nuclear weapons.

Consumption and draw-downs of HEU for non-weapon purposes

HEU has been used for many purposes other than nuclear weapons. According to a US government official, determining all these draw-downs is a complicated and imprecise exercise. Nevertheless, our rough estimates of the consumption and major draw-downs of HEU are summarized below (see table 4.2 for the main results).

Savannah River reactors

The Savannah River reactors used HEU fuel. Before 1968, the reactors are believed to have used HEU fuel when making tritium but natural uranium fuel for plutonium production. Afterwards, the reactors used HEU fuel for both plutonium and tritium production in depleted uranium and lithium targets, respectively, and this HEU was about 60 per cent enriched.

Table 4.2. Estimates of US consumption of HEU, to 31 December 1994
Figures are in tonnes.

Category	HEU
Allocated to the Savannah River reactors	50–80
Sent to foreign and domestic civil reactors (minus amount currently stored at DOE sites)	35
Exported to UK and France for military purposes	10
Consumption in US nuclear weapon tests	8–17
Dedicated to naval reactors (excluding *c.* 20 t stored at DOE sites)	76–106
Processing losses (3% of total production)	30
Total	**209–278**
Mid-point	244

As part of early efforts to estimate the amount of weapon-grade uranium produced for weapons, two research groups, one at Princeton University and one at the Natural Resources Defense Council (NRDC) in Washington, DC, derived estimates of the amount of HEU consumed in these reactors.[3] Before 1968, both groups estimated that about 10 tonnes of ^{235}U in weapon-grade uranium were consumed during the production of tritium.

After 1968, the amount of ^{235}U consumed can be estimated from the total thermal output of the reactors. From the end of Fiscal Year 1968 to the end of 1988, when the reactors were shut down, the reactors produced in total about 31 terrawatt (thermal)-days (TWth-d) of heat.[4] About 90 per cent of this heat was produced by the fissioning of ^{235}U, or about 28 TWth-d. At 0.96 MWth-d per gram of ^{235}U fission, about 29 tonnes of ^{235}U fissioned. Another 7 tonnes were converted into ^{236}U, for a total consumption of 36 tonnes of ^{235}U. This estimate is similar to those given by the Princeton and NRDC groups.

The irradiated HEU fuel in these reactors was reprocessed at the Savannah River site, and the HEU was recovered and recycled back into the reactors along with that recovered from research reactor spent fuel at Savannah River and the Idaho National Engineering Laboratory (INEL) and from naval reactor fuel at INEL. To achieve the desired enrichment level for the fresh production reactor fuel, fresh HEU metal was also blended with recovered HEU. The inventory of HEU at the Savannah River site contains the remains of the fuel-cycle inventory when the reactors shut down in the late 1980s.

Because the fuel was continuously recycled, determining the actual draw-down of the HEU inventory would require detailed information which we do not have. However, a rough approximation of the amount of the draw-down in the HEU inventory is to assume that fresh weapon-grade uranium was provided

[3] von Hippel *et al.* and Cochran *et al.* (note 2).
[4] DOE, 'Production Reactor Information, Savannah River Plant MWD History', undated.

to make up for the ^{235}U consumed by the reactors. In this case, the amount of HEU consumed in these reactors is the equivalent of about 50 tonnes of weapon-grade uranium.

Because the feedstock for the production reactor fuel came not just from fresh HEU metal, but also from recovered HEU from naval and civil reactor fuel and recovered Savannah River reactor fuel, an estimate of 50 tonnes is a minimum estimate. If the enrichment level of the feedstock was less than weapon-grade, a greater quantity of HEU would be required to replace the consumed ^{235}U. We assign an upper bound of 80 tonnes to account for this possibility.

A draw-down greater than 50 tonnes of HEU implies that recovered HEU from naval and civil reactor fuel is included here. To avoid double-counting, later sections take account of this assumption.

Civil research reactors

The United States exported about 25 tonnes of HEU, at an average enrichment of 70 per cent, for use in overseas civil programmes (see chapter 8). The government also supplied about 30–35 tonnes of weapon-grade uranium to its own civil programmes and other domestic thermal, research and power reactors (see chapter 8).

Much of the spent fuel from these reactors was sent to DOE sites for reprocessing or storage. This HEU would be included in either the site inventories listed in figure 4.1 or consumed in the Savannah River production reactors. Inventories of HEU at DOE-owned reactors at the end of 1993 are also included in figure 4.1.

Based on estimates given in chapter 8, about 21–24 tonnes of HEU are already included in inventories at the DOE sites listed in figure 4.1 or consumed in the Savannah River reactors. The rest, about 34–36 tonnes, has been consumed in civil reactors, or is located at domestic and foreign civil fuel fabrication facilities, non-DOE civil reactors or overseas civil reactors. A midpoint of 35 tonnes is used in table 4.2.

Naval reactor programme

Until 1992, the USA produced large quantities of weapon-grade uranium (in this case 97 per cent enriched) for naval reactors. After the cut-off in HEU production for weapons in 1964, most of the HEU produced has been for naval reactors.

Since the *Nautilus* first began operation in 1954, US naval nuclear-powered ships have steamed over 100 million miles and have operated over 4300 reactor-years.[5] In March 1994, the US fleet had about 150 reactors in about 120

---

[5] Testimony of Admiral Bruce DeMars, Director, Naval Nuclear Propulsion, before the Subcommittee on Energy and Water Development, in *Energy and Water Development Appropriations for 1995*, Hearings before the Subcommittee on Energy and Water Development, Committee on Appropriations, US House of Representatives, 103rd Congress (US Government Printing Office: Washington, DC, 1994), Part 6, pp. 761–809; and Testimony of Admiral Bruce DeMars, Director, Naval Nuclear Propulsion, before the

submarines and surface ships, and in land-based prototypes.[6] In line with the end of the cold war, the current number of ships represents a decrease of over 26 reactors and 24 ships since 1991. In addition, the Defense Department has cancelled many orders for new ships; for example, it cancelled plans for 40 new nuclear-powered submarines in the past few years. It will shut down six of eight land-based prototype and training reactors.

The cores for new ships are expected to last sufficiently long that refuelling is either unnecessary or required only once during the life of the ship. This is a major advance since the initial Nautilus core, which lasted only about two years. The reactors to be placed in new attack submarines will last about 30 years—the expected life of this type of ship. The reactors for Trident submarines are expected to last 20 years, and those put in aircraft carriers will last 25 years—both types will need one refuelling.

Since the portion of the Portsmouth enrichment plant that can make weapon-grade uranium was closed in 1992, naval reactors now depend on the existing inventory of weapon-grade uranium. So far, the US Navy is demanding that virtually all of the DOE's current and projected excess WGU be reserved for possible use in naval reactors, despite the fact that this amount would meet naval requirements for hundreds of years.

The amount of HEU dedicated through 1994 to naval reactors can be derived from estimates of the number of reactor cores procured historically and the average amount of HEU per core.

Cores procured. In total, an estimated 600–750 cores had been procured by the Navy by 1995. This estimate is based on the following considerations.

According to congressional testimony by Admiral Rickover in the 1960s and 1970s: (*a*) by 1969, 297 cores had been procured and 66 refuellings had occurred; (*b*) by 1974, 409 cores had been procured and 124 refuellings had occurred; and (*c*) by 1979, 508 cores had been procured and 166 refuellings had occurred.[7] Extrapolating these numbers linearly implies that about 200 cores were procured between 1979 and 1989, when drastic down-sizing in the Navy began. Another 50 cores are estimated to have been procured since then, bringing the total to about 750 cores.

This estimate is probably an upper bound for the number of cores procured up to and including 1995. During this period, core lifetimes had been increasing, and therefore fewer refuellings had been needed. Because US Navy policy is to fabricate cores several years ahead of deployment, a core would be made five to seven years before it was needed in a ship and, thus, a core procured in the late 1980s would have an expected deployment date in the mid-1990s.

Subcommittee on Energy and Water Development, in *Energy and Water Development Appropriations for 1992*, Hearings before the Subcommittee on Energy and Water Development, Committee on Appropriations, US House of Representatives, 102nd Congress (US Government Printing Office: Washington, DC, 1991), Part 6, p. 871.

[6] Testimony of Admiral DeMars, 1994 (note 5), p. 762.

[7] Cochran *et al.* (note 2), p. 185, footnote 12.

Since this period has been one of drastic cut-backs in naval ship orders, it is assumed that fewer orders have occurred since the mid-1980s.

Core procurement can be better understood by considering information about the refuelling and construction of reactors. Until 1994, the Navy has completed over 300 reactor-core refuelling or defuellings, with about 35 cores defuelled from early 1991 until early 1994.[8] In 1994, about 28 reactors were being defuelled or refuelled. (The reactors currently being defuelled have operated between 15 and 20 years.[9]) At the end of 1993, the Navy had 156 operable reactors aboard ships, six prototype and training reactors, 16 reactors for ships being built and 60 deactivated reactors, for a total of 238 reactors built to the end of 1993.[10] In total, therefore, over 500 cores are known to have been procured. This estimate does not include cores procured for future refuellings, perhaps enough to last many years. This requirement is assumed to represent another 100 cores. A minimum estimate is therefore about 600 cores procured until 1994.

HEU per core. The NRDC has estimated that the first 200 cores each contained an average of about 80 kg of weapon-grade uranium, and subsequent ones had an average of 200 kg of weapon-grade uranium.[11] The latter cores would be expected to last an average of 20 years. Applying NRDC's estimates of core size, 600–750 cores would have required more than 96 tonnes and less than 126 tonnes of weapon-grade uranium. We take the mid-point of this range—111 tonnes—as the total amount of weapon-grade uranium dedicated to cores for naval reactors up until the end of 1994.

Spent fuel. Until the early 1990s, naval spent fuel was sent to the Idaho National Engineering Laboratory for chemical reprocessing. About 9 tonnes of ^{235}U were recovered from this fuel until 1984, and a few more tonnes were recovered afterwards.[12] Recovered HEU from the Idaho facility was sent to the Y-12 Plant for further processing and then used to fuel the Savannah River production reactors, but this practice has been suspended. The reprocessing plant is now shut down, and naval fuel will be stored, pending eventual geological disposal.

The 300 cores unloaded by the end of 1994 would have initially contained an estimated 36 tonnes of 97 per cent enriched uranium. If about 50 per cent of the ^{235}U was consumed, the spent fuel would contain about 17.5 tonnes of ^{235}U, in about 22 tonnes of HEU. Some of this was recovered and part of this quantity was consumed in the Savannah River reactors, but this contribution is accounted for above.

---

[8] Testimony of Admiral DeMars, 1994 (note 5), p. 863.
[9] Testimony of Admiral DeMars, 1994 (note 5), p. 765.
[10] 'US defense systems: nuclear reactors', *Defense and Economy World Report*, no. 1290/1051, Mar. 1994.
[11] Cochran *et al.* (note 2), pp. 71 and 185.
[12] Cochran *et al.* (note 2).

To avoid double-counting, we assume that about 20 tonnes of HEU from naval reactors are stored at INEL, Y-12, Knolls Atomic Power Laboratory and the Savannah River Site. Subtracting this quantity leaves a draw-down of 76–106 tonnes of HEU. Most of this HEU has been consumed, is stored in the form of fresh cores or at fuel fabrication facilities, located in naval reactors or stored as spent fuel at naval shipyards awaiting shipment to INEL.

Exports to the UK and France for military use

In 1959, the United States entered into a barter agreement with the UK to exchange HEU for British plutonium. Later, Britain entered into an enrichment service contract for weapon-grade uranium. The total estimated amount of HEU exported to Britain for weapons and naval reactor fuel by the end of 1995 is about 10 tonnes (see section V on the UK).

Also in 1959, the United States committed itself to supply about 0.5 tonnes of weapon-grade uranium to France for use in the development and operation of a land-based prototype submarine reactor.

Nuclear weapon tests

In total, the United States conducted 1054 atmospheric and underground nuclear explosions.[13] During the period 1963–92, the DOE stated that 63 of its tests involved more than one device, for a total of 158 separate explosions. Certain tests had as many as six nuclear explosive devices in a single vertical shaft. This practice was primarily a cost-saving measure and was allowed under the 1974 Threshold Test Ban Treaty. A test under this Treaty can involve two or more underground explosions as long as they are detonated within 2 km and 0.1 seconds of each other.

As of mid-1996, the DOE had not released information about the amount of HEU consumed in these test explosions. Because later tests have had smaller thermonuclear yields than earlier tests, and primaries have tended to shift to plutonium only, these tests probably consumed less HEU than would be expected by assuming a standard 15–30 kg of HEU per device, or about 17–34 tonnes of HEU in 1149 detonations. Based on preliminary information supplied by the DOE, which had determined the amount of HEU consumed in its tests, we lower our estimate by half to 8–17 kg per explosion. This lower result could possibly be explained by assuming that the roughly 500 tests officially listed as having yields less than 20 kt that occurred since the early 1960s contained little or no HEU and were essentially tests of plutonium primaries.

---

[13] US Department of Energy, 'Openness press conference: fact sheets', Revision 1, Washington, DC, 27 June 1994, pp. 140–61.

Processing losses

Relatively large quantities of HEU were lost during the production of HEU and subsequent fabrication into weapon components and reactor fuel. The losses are assumed to be 3 per cent of total HEU production or about 30 tonnes of HEU.

Other draw-downs

The USA has probably used HEU in more applications than discussed above. However, the additional amount cannot be estimated from the available information. In addition, re-enriching HEU to weapon-grade would in effect be another difficult-to-estimate draw-down.

Inventory of HEU in weapons or at the Pantex Plant

Subtracting the above draw-downs and the amounts at the sites shown in figure 4.1 leaves about 490 tonnes of HEU at the DOE's Pantex site and in nuclear weapons under the authority of the Defense Department. The uncertainty in this estimate is about 10 per cent.

Amount in active weapons

It has been estimated by Stan Norris and William Arkin that in mid-1994 the USA had approximately 9250 operational nuclear warheads, 8300 in its strategic and 950 in its tactical arsenals.[14] They later estimated a stockpile of 14 900 warheads in 1994, implying that about 5600 weapons are in a reserve or awaiting disassembly.[15]

In early 1995 they estimated a post-START II arsenal of about 4000 warheads for strategic forces and 450 warheads for non-strategic forces. In addition, they posit a reserve stockpile of about 3500 warheads that would allow a reconstitution of additional strategic forces, for a total of about 8000 warheads.[16]

Because warheads vary in their requirements for HEU, accurate assessment of the HEU inventory in weapons is not possible without more information. Nevertheless, assuming 15–30 kg of HEU per warhead, 9250 operational weapons would contain about 140–280 tonnes of HEU, with a mid-point of 210 tonnes. Subtracting this mid-point from the estimated 490 tonnes of HEU derived above and adding uncertainties, about 280 ± 120 tonnes are either in various reserves, in storage at the Pantex Plant or in weapons awaiting disassembly but not shipped to Pantex.

---

[14] Norris, R. S. and Arkin, W. M., 'Nuclear notebook: U.S. strategic nuclear forces, end of 1994', *Bulletin of the Atomic Scientists*, vol. 50, no. 4 (July/Aug. 1994), pp. 61–63.
[15] Norris, R. S. and Arkin, W. M., 'Nuclear notebook: U.S. strategic nuclear forces, end of 1994', *Bulletin of the Atomic Scientists*, vol. 50, no. 6 (Nov./Dec. 1994), pp. 58–59.
[16] Norris, R. S. and Arkin, W. M., 'Nuclear notebook: U.S. strategic nuclear forces, end of 1994', *Bulletin of the Atomic Scientists*, vol. 51, no. 1 (Jan./Feb. 1995), pp. 69–71.

Table 4.3. Estimated allocation of US highly enriched uranium, 31 December 1993
Figures are in tonnes.

Category	HEU
Total production (DOE estimate)	994[a]
Consumption and draw-downs	-245 ± 35
HEU inventory in weapons, reserve and associated categories, including excess	749 ± 50[b]
DOE sites (not including Pantex)	259
Active nuclear arsenal	210 ± 70
At Pantex, in reserve, awaiting disassembly	280 ± 120

[a] Average 80% ^{235}U.
[b] The range is increased to reflect additional uncertainties.

Amount at the Pantex Plant, in reserve or not yet disassembled

The DOE is dismantling between 1000 and 2000 warheads a year at the Pantex Plant. From the end of Fiscal Year 1989 until April 1994, when Pantex was closed temporarily for safety reasons, the DOE dismantled about 7100 warheads.[17] Since Rocky Flats produced its last pits in November 1989, which were assembled into weapons by the summer of 1990, few weapons have been built. This number of 7100 warheads should therefore represent a draw-down in the total number of weapons.

The DOE stated in late 1992 that the Pantex site stored about 3400–3800 pits from dismantled weapons at the beginning of the fourth quarter of 1992 and about 3900–4300 pits by the first quarter of 1993.[18] From the beginning of 1993 to April 1994, another 2482 weapons have been disassembled, bringing the total stored at that time to about 6400–6800 pits. Most of the pits from weapons dismantled since 1989 therefore appear to remain at Pantex.

HEU is used in both the fission primary and the thermonuclear secondary. Although HEU from dismantled secondaries is sent for storage at the Y-12 Plant, HEU in pits would remain at the Pantex Plant. Using the above estimate of average HEU per warhead, the 7100 warheads dismantled from the end of Fiscal Year 1989 to April 1994 would contain 105–210 tonnes of HEU, with a mid-point of about 160 tonnes. The Y-12 Plant had an inventory of 148 tonnes of HEU metal as of spring 1994, which should include a major portion of the HEU from weapons disassembled since 1989. This assessment therefore suggests that most of the 280 tonnes estimated as neither in operational weapons nor at Y-12 is at Pantex either in composite-core pits or awaiting disassembly, in weapons awaiting disassembly but not shipped to Pantex, or in the various reserves. Table 4.3 summarizes this discussion.

---

[17] DOE, 'Openness press conference, fact sheets', 27 June 1994, pp. 169 and 173.
[18] DOE, 'Environmental assessment for interim storage of plutonium components at Pantex', DOE/EA-0812, Dec. 1992, pp. 4–1 and 4–7.

Table 4.4. US highly enriched uranium inventories, declared excess, 6 February 1996

Location	Form of HEU	Excess (t)
Oak Ridge site	Metal	63.1
(Y-12, K-25. Oak Ridge National	Oxides	2.7
Laboratory)	Unirradiated fuel	10.6
	Irradiated fuel	0.6
	Other forms	7.9
	Sub-total	**84.9**
Idaho National Engineering Laboratory	Metal	1.6
	Oxides	1.7
	Unirradiated fuel	2.8
	Irradiated fuel	16.6
	Other forms	0.6
	Sub-total	**23.4**
Portsmouth Plant (GDP)	Oxides	7.3
	Other forms	15.2
	Sub-total	**22.5**
Savannah River Site	**Sub-total**	**22.0**
Pantex plus planned dismantlement	Metal	16.7
	Sub-total	**16.7**
Rocky Flats site	Metal	1.9
	Oxides	< 0.1
	Unirradiated fuel	< 0.6
	Other forms	< 0.4
	Sub-total	**2.8**
Hanford site (Richland)	Metal	< 0.1
	Oxides	0.1
	Unirradiated fuel	0.1
	Irradiated fuel	0.3
	Other forms	0.1
	Sub-total	**0.5**
Los Alamos National Laboratory	Metal	< 0.1
	Oxides	0.3
	Unirradiated fuel	0.1
	Irradiated fuel	< 0.1
	Other forms	< 0.1
	Sub-total	**0.5**
Brookhaven National Laboratory	Irradiated fuel	0.2
	Other forms	< 0.1[a]
	Sub-total	**< 0.3**
Sandia National Laboratories	Metal	< 0.1
	Oxides	0.1
	Unirradiated fuel	< 0.1
	Irradiated fuel	0.1
	Other forms	< 0.1[a]
	Sub-total	**0.2**

Location	Form of HEU	Excess (t)
Other sites	Metal	< 0.1
	Oxides	0.2
	Unirradiated fuel	0.2
	Irradiated fuel	< 0.1
	Other forms	< 0.1
	Sub-total	**0.5**
Total		**174.3**

^a Reactor fuel, which is characterized as unirradiated or fresh fuel at other sites.

Source: US Department of Energy, 'Openness Press Conference, Press Conference Fact Sheets' (Department of Energy: Washington, DC, 6 Feb. 1996).

Surplus HEU

The United States has a large excess of HEU. If all forms of HEU are considered, then the total estimated excess is about 540 tonnes, assuming an active arsenal of over 9000 operational nuclear weapons. Most of this HEU is in metal form, but a significant quantity is in forms that make recovery difficult. Ignoring this latter category and counting only the metal at the Y-12 Plant and Rocky Flats, the uranium hexafluoride at Portsmouth, and the 280 ± 120 tonnes estimated above, the USA has up to about 450 ± 120 tonnes of HEU that is potentially surplus and relatively easy to recover. How much of this material is weapon-grade is unknown as of mid-1996.

As part of its fissile material cut-off initiative launched in the autumn of 1993, the Clinton Administration offered in September 1994 to let the IAEA inspect about 10 tonnes of HEU at the Y-12 Plant. This offer is part of a US determination to begin demonstrating its commitment to the irreversibility of the nuclear disarmament process. The USA has said that this material will not be withdrawn from safeguards for any nuclear weapon or other nuclear explosive purpose. This pledge, however, does not forbid withdrawing the HEU from safeguards for use as naval reactor fuel, although such a withdrawal is currently seen as unlikely.

On 1 March 1995, President Clinton announced that another 200 tonnes of fissile material would be permanently withdrawn from the US nuclear weapon stockpile. In February 1996, the DOE provided detailed information about this excess stock. About 174 tonnes of this material is HEU, which is in many chemical forms, including metal, reactor fuel elements, and chemical compounds and solutions (see table 4.4). About 33 tonnes are enriched over 92 per cent, and 142 tonnes are enriched to between 20 and 92 per cent.[19] According to a DOE official, the reason for the relatively low percentage of WGU is that the

[19] US Department of Energy, 'Openness Press Conference: Press Conference Fact Sheets' (Department of Energy: Washington, DC, 6 Feb. 1996).

US Navy insisted that WGU be reserved for its potential needs. Table 4.4 provides a list of the location and amounts of this excess HEU.

The vast majority of the HEU comes from the stock of 259 tonnes declared at DOE sites (see figure 4.1). About 50 tonnes of the excess HEU are scheduled to be given to the US Enrichment Corporation where it is expected it will be blended down to LEU. These 50 tonnes will come from inventories at the Portsmouth and Y-12 plants.

The DOE expects to increase the amount of fissile material declared excess to weapon requirements. As weapons are dismantled, more excess HEU could be removed from the weapon inventory. Under current US plans, however, this quantity is not expected to increase significantly, at least this century.

Clinton's 1995 announcement means that the United States is turning over a large proportion of its non-weapon-grade HEU stock to civil uses and preserving most of its WGU for naval reactors. However, any naval requirement, which amounts to at most a few tonnes of WGU a year, is far less than the rate at which WGU will become surplus.

During 1995, the DOE was considering three main options for its excess HEU: (a) continued storage as HEU for up to several decades and then blending it down to LEU; (b) blending down the HEU into LEU for commercial reactors; and (c) blending down to LEU and disposing of the LEU as low-level nuclear waste.

The DOE has not decided the quantity of HEU that will be blended down into LEU and used commercially within the next decade. Of the 174 tonnes declared excess already, a total of 103 tonnes could be blended down for commercial use.[20] The exact amount will be determined by many factors, including available industrial infrastructure to blend down HEU, legislative and policy decisions, and future market conditions.

IV. The former Soviet Union

Despite more openness in other nuclear areas, the Russian Government remains unwilling to release information on its historical enrichment production. Although a considerable amount of information about its enrichment facilities has been released in the past few years, most of it concerns the current gas centrifuge programme. Unlike the US case, there is little information about historical Russian enrichment capacity and output. The focus of this section is to gain a better understanding of the history of the Soviet enrichment complex in order to derive estimates of both the historical capacity and total HEU production.

Energy Minister Viktor N. Mikhailov said in the autumn of 1993 that Russia had an inventory of 1250 tonnes of HEU. This statement has been challenged by many US and Russian officials, and Mikhailov has been unwilling to clarify or add support to his statement. As shown below, Russian enrichment capacity

---

[20] DOE (note 19).

was large enough to have produced this quantity of HEU, particularly if some of the HEU was not weapon-grade.

The HEU is in many chemical forms, although the vast majority of it is probably in metallic form. Other forms are expected to include uranium oxide solids, uranium fluoride gas, and irradiated and fresh fuel. The HEU is located at scores of locations within the former Soviet Union. US and Western officials regularly question whether Russia has an accurate knowledge of the quantity of HEU at many of these sites.

As part of an offer to the USA to stop fissile material production for weapons, the Soviet Union announced in October 1989 that 'this year it is ceasing the production of HEU'.[21] The Soviet Union actually stopped making HEU for defence purposes in 1988.[22] It is unclear how much more HEU was produced for non-weapon purposes after this date, although the total amount of HEU is probably small. Production ceased by the early 1990s.

Besides freeing a considerable amount of enrichment capacity for civil uses, including exports, this step represented the end of a massive effort to make HEU that began immediately following World War II. By the early 1990s the Russian enrichment complex had a total capacity of about 20 million SWU per year, all derived from gas centrifuges.[23]

The development of the Soviet enrichment complex

In 1945, following the US detonation of atomic bombs over Japan, the Soviet Union launched a coordinated effort to research and develop several uranium enrichment methods, principally gaseous-diffusion and electromagnetic methods. In 1946, the Soviet Union also started developing the gas centrifuge.

In its early enrichment efforts, the Soviet Union received significant assistance from German and Austrian scientists who had either been captured at the end of World War II or who had volunteered to work in the Soviet Union. Many of them worked at institutes near Sukhumi, on the Black Sea and in the former Soviet republic of Georgia.

Gernot Zippe, one of the captured scientists, said that the initial leader of the German scientists, Manfred von Ardenne, was summoned to a meeting with high officials in Moscow soon after the USA dropped atomic bombs on Japan.[24] At this meeting, Ardenne was told that the USSR wanted the captured scientists to help them make nuclear weapons. Ardenne reasoned that if they worked on

[21] Petrovsky, V. F., Deputy Head of the USSR Delegation to the 44th UN General Assembly, 'Statement on the item entitled Report of the IAEA', 25 Oct. 1989.

[22] Mikerin, E., Bazhenov, V. and Solovjov, G., 'Directions in the development of uranium enrichment technology', presented at the 1993 International Enrichment Conference, 13–15 June 1993, Washington, DC, Sponsored by the US Council for Energy Awareness (an earlier, more detailed version of this paper was also used in this section); and NUEXCO, 'Conversion and enrichment in the Soviet Union', *NUEXCO Monthly Report*, no. 272 (Apr. 1991). This latter source says that HEU production was halted in 1987.

[23] Hart, K., 'USEC claims AVLIS ready by 2000, MINATOM pushes centrifuge technologies', *Nuclear Fuel*, 11 Sep. 1995, pp. 3–4.

[24] Zippe, G., Private communication to one of the authors, July 1993.

the weapon itself, they would never be allowed to go home. He then proposed to the Soviet officials that the German teams design and develop uranium enrichment technologies, arguing that isotope separation would be harder to accomplish than the construction of atomic weapons. The officials apparently agreed and several German groups were created to pursue specific enrichment technologies.

Although the German teams initially dominated the Soviet enrichment effort, parallel Soviet R&D teams were set up that soon developed sophisticated enrichment expertise. The German and Soviet teams competed, but the competition was one-sided. Although the German teams regularly transferred their knowledge and experience to these parallel Soviet teams and the Soviet members of their own teams, the Germans received no reports from their Soviet counterparts and little or no information about their progress or accomplishments.[25] The Scientific–Technical Council of the USSR Council of Ministers reviewed the work of both sets of teams, decided which was the best method and issued new assignments.

The Scientific–Technical Council also decided when to withdraw Germans from a particular project. For security reasons, the initial idea was to withdraw German groups from a nuclear project after the laboratory stage. But this policy was not always followed, particularly early in the enrichment programme. For example, after Peter Adolf Thiessen developed a gaseous-diffusion barrier in his laboratory at Sukhumi he was transferred to Moscow to work on the production of barriers for the gaseous-diffusion plants.

Gaseous diffusion

The Soviet Union first successfully developed the gaseous-diffusion process, and the first industrial-scale facility started full operation in 1949 near Verkh Neyvinsk outside the city of Ekaterinburg (formerly Sverdlovsk).[26] The site is called the Ural Electrochemistry Kombinat (UEK), or Sverdlovsk-44.

The first plant, called D-1, was relatively inefficient, and required additional development work during 1949 and part of 1950.[27] Major problems were encountered in stopping corrosion by uranium hexafluoride gas, with resulting unacceptable losses, and in developing the compressors which raise the pressure of the gas to the required level. The inefficiency of these compressors was a principal motivation for accelerating the development of gas centrifuges in the early 1950s.[28]

---

[25] US Central Intelligence Agency, *The Problem of Uranium Isotope Separation by Means of Ultracentrifuge in the USSR*, Report no. DB-G-3,633,414, 8 Oct. 1957 (unclassified version). This report is essentially a summary of an extensive debriefing of one of the leading captured participants in the centrifuge programme.

[26] Sinev, N., 'Enriched uranium for nuclear weapons and power: History of developments of industrial technology and production of HEU (1945–52)', Central Institute for Atomic Information, Moscow, 1991 (in Russian); and Chernov, A., 'Uranium enrichment in the USSR', Paper presented at the Annual Symposium of the Uranium Institute, London, 6–8 Sep. 1989.

[27] NUEXCO (note 22).

[28] NUEXCO (note 22).

Initially, the plant produced only 75 per cent enriched material, and reaching this level required batch recycling of the product.[29] Further enrichment to weapon-grade was accomplished by EMIS. Following more improvements, the D-1 plant started producing tens of kilograms of weapon-grade uranium per year in 1950. Its capacity reached no more than about 7500 SWU per year, or less than about 50 kg of weapon-grade uranium per year, at a tails assay of 0.5 per cent.

The larger D-3 and D-4 plants were added during the next two years. (There is no mention in the sources of a D-2 plant.) The D-3 plant was several times larger than D-1. The D-4 plant, which began operation in late 1952, is estimated to have had a nominal capacity of about 100 000 SWU per year.[30] At this rate, it could produce about 800 kg of weapon-grade uranium a year at a tails assay of 0.5 per cent. The D-5 plant was subsequently added. Other plants were also built and stages improved. Details about these efforts, however, are sketchy.

The D-1 plant was closed in 1955 because it was too inefficient. Heinz Barwich, a captured German scientist who worked in the early Russian gaseous-diffusion programme, and then settled in East Germany before defecting to the West in 1964, told Zippe that the first facility was dismantled and sent to China in the late 1950s.

During this first decade, the main Soviet focus was on improving the performance of the diffusion stages. Sixteen different models were manufactured, with the last model having a separation capacity of 850 SWU per year, or 6500 times more separation capacity than the first-stage model.[31] The electricity consumption rate decreased from 35 000 to 3500 kilowatt-hours per SWU, and the flow rate of uranium hexafluoride increased considerably.

Three more gaseous-diffusion sites were established during the 1950s: the Electrochemistry Kombinat at Krasnoyarsk; the Electrolyzing Chemical Kombinat at Angarsk, about 50 km north-west of the Siberian city of Irkutsk; and the Siberian Chemical Kombinat, near Tomsk.[32] These three facilities are also known as Krasnoyarsk-45, Angarsk and Tomsk-7, respectively. These gaseous-diffusion plants in combination with the Urals Electrochemistry Kombinat worked as one technological unit to produce HEU.

Beginning in the 1960s, gaseous-diffusion equipment was replaced by gas centrifuges as it became obsolete. According to Evgeni Mikerin, Head of Fuel Cycle Directorate, Ministry of Atomic Energy of the Russian Federation, the change to the new gas-centrifuge technology that occurred from 1966 to 1992 allowed Russia, with a minimum of additional construction, to increase the

---

[29] Sinev (note 26). The description of the early gaseous diffusion experience is from this source.

[30] Sinev (note 26), pp. 96 and 101. This source says that D-4 contained 672 T-49 stages (75 SWU per stage per year), 1420 T-47 stages (45 SWU/y), 928 T-45 stages (9.1 SWU/y), 1098 T-44 stages (2.0 SWU/y), and 1368 OK-19 stages (0.6 SWU/y). This information implies that D-4 had a maximum capacity of less than 125 000 SWU per year, since not all the stages would work in tandem.

[31] Mikerin *et al.* (note 22); and Sinev (note 26), p. 96.

[32] 'Interview with Evgeni Mikerin', *Atom Press*, vol. 56, no. 4 (1993). Unofficial translation.

enrichment capacity by a factor of 2.4 and to decrease the electricity consumption by a factor of 8.2.[33]

Russia has not released information about the total capacity of these four gaseous-diffusion plants. Nevertheless, applying Mikerin's information to the total estimated capacity of 20 million SWU per year in the early 1990s implies that the gaseous-diffusion plants had a nominal capacity of 8.3 million SWU per year in 1966.

This estimate is consistent with anecdotal information about Russian enrichment capacity. Many references in the late 1970s and 1980s cite the Soviet enrichment capacity as being 7–10 million SWU per year. This value originated from the nuclear scientist Manson Benedict, who published this estimate in the mid-1970s. He heard it while he was Chairman of the US General Advisory Committee in the mid- to late 1960s.[34] At the time, according to Benedict, this value was considered in a number of government circles as a reliable estimate of total Soviet capacity.

Gas centrifuges

Parallel to the gaseous-diffusion programme, the USSR also developed gas centrifuges. The initial R&D effort was accomplished mainly by captured German and Austrian scientists.

Because so little is published about the early Russian gas-centrifuge enrichment effort, some historical information is provided here.[35] This information will later be the starting point for an estimate of historical Russian gas-centrifuge enrichment output.

The German team. The German team, comprised of about 60–65 German and Russian workers, was led by the theorist Max Steenbeck, who had been a director of the German company Siemens during World War II. At first, he led an effort to develop a 'droplet' enrichment method, but by late 1946 he was also working on the development of a gas centrifuge.

Steenbeck himself directed the theoretical work. In mid-1946, Zippe joined the team and led the experimental group for mechanical development and separation tests at Sukhumi. Rudolph Scheffel was in charge of the experimental electrical work for the team.

When problems were encountered in the gaseous-diffusion process in 1949, Steenbeck proposed directly to L. P. Beriya that 10-metre centrifuges be used to further enrich material from the gaseous-diffusion process up to weapon-grade.[36] His proposal to build a cascade composed of tall centrifuges was

---

[33] 'Interview with Evgeni Mikerin' (note 32).
[34] Interview with Manson Benedict, Dec. 1986.
[35] This information is largely based on a series of interviews by one of the authors with Gernot Zippe which occurred in 1993, 1994, 1995 and 1996; Sinev (note 26); and CIA (note 25).
[36] Zippe, G., Beams, J. W. and Kuhlthau, A. R., 'The development of short bowl ultracentrifuges', University of Virginia/Ordnance Research Laboratory, UVA/ORL-2400-58, Progress Report no. 1, 1 Dec. 1958.

accepted, and his group at Sukhumi received substantially greater resources than before. Later, when informed that such a topping cascade was no longer necessary, the team shifted its focus back to a machine suitable as the basis for a complete enrichment plant.

The German and Soviet teams had few prior results upon which to base their work. They were armed only with unclassified publications from the 1930s by the US researcher Jesse Beams, the unclassified Smyth report on the Manhattan Project and some early German centrifuge work. Nevertheless, effectively combining theory and experiment, the German experts, particularly Steenbeck and Zippe, and later Soviet centrifuge experts revolutionized the mechanical design of centrifuges, incorporating many of the most important features of the modern centrifuge. Based on this initial work, the Soviet Union was the first nation to deploy centrifuges industrially. Russia currently has the largest gas-centrifuge programme in the world.

Results of the German and Soviet centrifuge research and development. From 1946 until 1953, the German and Soviet research teams developed both a supercritical and a subcritical machine. Although Steenbeck had originally told the Russians he intended to build a 10-m high, thin-walled supercritical machine, his team realized after a few years of development that a 3-m long machine was the limit of their capabilities and sufficient for their purposes. The rotor tube had a diameter of 58 mm and a peripheral velocity up to 240 metres per second. The rotor was composed of about 10 short tubes connected by 9 metal bellows. A bellows acts like a spring and enables the entire tube to pass through critical bending frequencies that would otherwise tear apart a solid rotor.

Earlier, they had developed a subcritical machine with a diameter of 58 mm and 30 cm in length. A subcritical machine uses one short rigid tube without any bellows that spins more slowly than the first critical bending frequency. An essential feature of the design was a needle-point bottom bearing with a damper that was designed so that the rotor acted as a self-stabilizing spinning-top. Achieving this result and developing the top bearing, according to Zippe, were the hardest parts of the German scientists' research effort in the Soviet Union.

The rotors were made of aluminium. According to Zippe, Russian industry pressed the precise aluminium tubing for 58-mm experimental centrifuges on an old Siemens machine imported into Russia in 1913.

In late 1952 and early 1953, the Russians moved the German team to the Test Design Bureau, or OKB-133, located at the Kirov Plant in Leningrad and which had earlier developed the compressors for the gaseous-diffusion cascade. Compressor work continued there after the German team started centrifuge work at OKB-133. The Soviet Union decided to cancel any plans to build supercritical machines and to concentrate instead on a subcritical design. Looking back, Zippe believes that the Soviet leaders had already made the decision to deploy the subcritical centrifuge industrially.

This shift to subcritical machines was fought by Steenbeck, who still wanted to build a cascade of supercritical machines. Nevertheless, Zippe agreed with the Soviet decision. He said that he even encouraged it, since they could demonstrate a subcritical machine faster than a supercritical one, and the Soviet authorities had promised that they would release him after the successful demonstration of a machine.

Although Soviet experts had started to suggest their own ideas of ways to improve centrifuge designs, Zippe believes that the Russians could not at the time have deployed supercritical machines on their own. He said that the Soviet teams did not work on supercritical machines, and only one Russian on the German team had any knowledge of how to produce and operate supercritical machines. Zippe said that each supercritical rotor tube had to be balanced by hand, and required specialized knowledge to assemble.

The last major act of the German team was to demonstrate in the summer of 1953 a subcritical centrifuge incorporating many new features that are common in modern centrifuges. (For a diagram of a modern centrifuge see appendix A.) Zippe says that the demonstrated design had a diameter of 58 mm, a length of 25 cm, and spun at 340 metres per second (m/s).

The demonstrated design incorporated many innovative features that have become standard in Russian and Western centrifuges. The design used a non-contacting top bearing composed of ring magnets, allowing the pipes that carry the uranium hexafluoride gas to enter the rapidly spinning rotor tube without physically contacting the tube. At Sukhumi, the Germans had developed a magnetic bearing, but they developed ring magnets in Leningrad. At the suggestion of the Soviet experts, they used a molecular pump, which is a simple way both to create a good vacuum inside the centrifuge housing by the action of the spinning rotor and to prevent uranium hexafluoride from leaking out of the top bearing into the region surrounding the rotor. Following a Soviet suggestion, the German team applied stationary 'scoops' within the spinning gas that amplified the counter-current gas flow and, more importantly, created a pressure difference that extracted the depleted and enriched uranium from the rotor tube without the need for compressors. The German team continued to use a hysteresis-type motor that acted directly on the rotor tube.

In addition to contributing to a simple mechanical design, the molecular pump and scoops allowed the power consumption per SWU for the gas centrifuge to be drastically lower than it is for the gaseous-diffusion process. The molecular pump greatly reduced frictional losses. The scoops, by eliminating the need for compressors, drastically reduced the power requirement for centrifuges. Both the molecular pump and the scoops had been used in other applications, and the Soviet teams probably learned about these ideas from the literature. Nevertheless, they deserve credit for their suggestion to apply these ideas to a gas centrifuge. Zippe in particular deserves credit for realizing the value of these suggestions and demonstrating them in a centrifuge.

After demonstrating the 58-mm machine to the satisfaction of the Soviet team, the Germans were removed from the centrifuge project. Shortly before leaving OKB-133 in October 1953, however, Zippe worked on a subcritical machine with a longer rotor—about 45 cm long and 100 mm in diameter. He said that about 3–4 samples were made on a lathe before he left OKB-133. From theory, the Germans realized that the separative capacity could be increased by increasing the length and diameter of the rotor in a geometrical ratio that maintains a subcritical design. The final length of a production centrifuge would be determined by manufacturing considerations. Zippe said that OKB-133 officials planned to construct more 45-cm machines, although he was unaware of Soviet accomplishments with these larger machines.

The German team stopped at single-machine trials, and never deployed a centrifuge cascade. Steenbeck proposed doing so with a cascade of replaceable machines, but the Soviet Union was not interested in having the Germans work on the idea.

According to Zippe, the Germans were released after they had demonstrated how to apply the new subcritical centrifuge design industrially and their Soviet counterparts knew enough details about the theory and operation of centrifuges to continue on their own. After sitting in a hotel in Leningrad for a few months, Zippe and Steenbeck were transferred to Kiev in March 1954 where they worked only on unclassified projects. In July 1956, they were allowed to leave the country, Steenbeck settling in the GDR and Zippe returning to Austria.

On returning home, Zippe, Scheffel, and Steenbeck started filing for Western patents on their centrifuge design. In January 1958, Zippe received oral permission from Soviet Minister of Atomic Energy V. S. Emelyanov to work freely on centrifuges in the West. In the 1960s, Zippe recreated the 100-mm machine and spent several years optimizing it, something he had not had the opportunity to do in Leningrad. This new machine spun at about 360 m/s and had a separative length of 43 cm (about 48 cm in total). At its peripheral speed of 360 m/s, the machine had an effective separative capacity of about 0.57 SWU per year at an efficiency of 30 per cent. It could spin at 400 m/s, but the aluminium in the rotor 'creeps' at that speed, drastically shortening the rotor lifetime.

By this time, Steenbeck had stopped working on centrifuges. Zippe continued, contributing significantly to the development of advanced supercritical Western centrifuges and to the formation of the commercial European enrichment consortium Urenco.

Soviet objections. Following the decision to reveal details of the Soviet centrifuge programme in the early 1990s, N. M. Sinev, who was head of OKB-133 when the German team was there and one of the holders of the Certificate of Authorship of the initial Soviet centrifuge design, claimed that the Soviet scientists did not use the German team's design, except for one feature, namely the bottom-bearing support needle and oil damper.[37] He writes that the Russian team 'demonstrated the imperfection of the Sukhumi centrifuge', particularly its prospects for industrial production. Work on this centrifuge was stopped in 1953. Sinev quotes Steenbeck, who wrote in his memoirs published in East

---

[37] Sinev (note 26). Certificate of Authorship no. 23286, 20 Apr. 1953 was given to N. M. Sinev, Kh. A. Murinsom, S. A. Arkin, G. V. Kudryavtsev, I. K. Kikoin, P. F. Vasilevskii, I. B. Starobin and A. S. Voznyuk.

Germany that the Soviet scientists found an original solution to the problem that the German team had not solved—the optimal method of connecting centrifuges into a cascade.

Sinev charged that Steenbeck and Zippe had 'free access to all works' of the Leningrad Test Design Bureau on centrifuges, and patented the Soviet design in the West. However, this charge appears to be exaggerated. First, it appears to mix up the supercritical and subcritical designs, and the roles of Steenbeck and Zippe. Second, it contradicts the Soviet policy of denying information about Soviet advances to the captured scientists. Third, it ignores the fundamental importance of the bearings. Finally, it ignores the main point that in Leningrad the development of the centrifuge resulted from teamwork between the Soviet and German teams but depended on the German team's ability to implement these ideas experimentally.

Sinev says the Soviet leadership decided not to challenge Zippe's patents in order not to draw attention to their own centrifuge programme. The Soviet Union wanted the world to believe it possessed only gaseous-diffusion plants.

After 1953 the Soviet centrifuge teams worked alone to develop their own industrial designs and to build centrifuge plants. Unlike many other centrifuge programmes, the Soviet programme remained with subcritical machines, although it significantly improved their design and output.

Deployment of centrifuges. The first Soviet gas-centrifuge pilot plant went into operation at Verkh Neyvinsk in October 1957, and contained 2500–3500 centrifuges.[38] This plant tested both the industrial-design centrifuges and the operation of cascades. It was followed by an industrial-scale plant in 1959 that contained 'several tens of thousands of centrifuges.'[39] A full-scale plant with three modules was built there between 1962 and 1964.[40] It contained several hundred thousand subcritical centrifuges with aluminium alloy rotors.[41]

Subsequently, the capacity at this site was increased and centrifuges were installed at the other three enrichment facilities. In addition, centrifuges in all four plants have been updated with new models.

By 1992, centrifuges had completely replaced gaseous-diffusion equipment at the four sites. Russia has not published the total capacity of these sites at that time, or revealed whether the plants were still being expanded in capacity. In 1993, the US Department of Energy listed the capacity of the four sites as roughly 20 million SWU per year.[42] As mentioned above, the total capacity was 20 million SWU in September 1995. This information implies that the total capacity has remained constant for several years. However, the actual annual output during the past several years is believed to be significantly lower, about 10 million SWU per year.

[38] NUEXCO (note 22) and Sinev (note 26).
[39] NUEXCO (note 22).
[40] Chernov (note 26).
[41] NUEXCO (note 22).
[42] See, e.g., Starr, D., 'Enrichment', in US Department of Energy, *Nuclear Non-Proliferation Seminar*, Washington, DC, 7–9 Dec. 1993.

Table 4.5. Russian enrichment capacity in the early 1990s

Fraction of capacity is expressed as a percentage; estimated capacity is given in million SWU/y.

	Fraction of capacity[a]	Estimated capacity
Ural Electrochemistry Kombinat (UEK) Krasnoyarsk	49	9.8
Electrochemistry Kombinat (EK)	29	5.8
Siberian Chemical Kombinat (SCK) Tomsk	14	2.8
Electrolyzing Chemical Kombinat (ECK)	8	1.6
Total	*100*	*20*

[a] The fractions are from 'Interview with Evgeni Mikerin', *Atom Press*, vol. 56, no. 4 (1993). Unofficial translation.

Mikerin has provided a breakdown of the capacity at each site on a percentage basis. This information is listed in table 4.5, and used to derive an estimated capacity for each of the four sites. Little is publicly available about the historical growth in capacity. The sections below attempt to estimate this growth by evaluating Soviet centrifuges in more detail.

Centrifuge models. The USSR developed eight centrifuge models over the course of 35 years.[43] The separation capacity of the latest model is about 10 times higher than the first model, although the last two to three models have not been deployed as of 1995. The fifth-generation machine has a separative capacity 40 per cent higher than the fourth-generation model, and 4.5 times greater than first-generation machines.[44] In general, the separative capacity of the first several centrifuge models increased by about 50 per cent per generation.

The industrial plant built near Verkh Neyvinsk between 1962 and 1964 used machines that lasted 10–12 years, at which time they were replaced with newer models.[45] The bottom bearing initially lasted about three years, although current ones last as long as the machine. This increase in lifetime implies that the design of the bearing may have shifted from a hard metal pin rotating in a socket to a spiral-grooved design in which the pin pumps oil between itself and the socket, dramatically reducing friction and extending the bearing's lifetime.

As of late 1991, the Urals plant employed only fourth- and fifth-generation centrifuge machines. Over 98 per cent of the fourth-generation machines were over 5 years old with about 85 per cent of them between 5 and 15 years old. In contrast, about 60 per cent of the fifth-generation machines were less than 5 years old and over 99 per cent of them were less than 10 years old. A sixth-generation machine may have been installed either there or in other plants. The electricity requirement for the machine itself is 120 kWh per SWU for the

---

[43] Mikerin *et al.* (note 22).
[44] Mikerin *et al.* (note 22) and 'Interview with Evgeni Mikerin' (note 32).
[45] NUEXCO (note 22).

fourth-generation machine, 80 kWh per SWU for the fifth-generation and 50 kWh per SWU for the sixth-generation.[46]

The first industrial centrifuge models took about three to four years to develop and start producing serially. Development of later models took about seven years, and the launching of serial production required another three years.[47]

The first few centrifuge machines were independently fixed to platforms, but later models have been interchangeable, easing repair and replacement. The centrifuges are stacked three or four high in order to occupy the available volume of the old gaseous-diffusion buildings. For example, four buildings at UEK currently contain centrifuges. Of these, three originally had gaseous-diffusion equipment. (The first gaseous-diffusion building is no longer used for enrichment.) The fourth one was build specifically to house centrifuges, and is the one often visited by Western experts and officials. According to some of these officials, the newest building is 1 km long and contains about 660 000 centrifuges in about 500 rows, each row split in half by a long central corridor. Each row is composed of about 16 stacks of centrifuges, where each stack has four shelves of centrifuges. Each shelf has 20 machines, 10 on each side. According to one US official, each shelf is replaced as a unit and connected into the cascade piping as one unit.

Unlike Urenco, Russia did not deploy supercritical machines industrially. One reason given recently for initially choosing subcritical machines is that machine failure could be decreased more easily in subcritical than supercritical machines.[48] The bellows apparently caused special problems.

Russia has subsequently been unable to achieve a 20-year lifetime for supercritical machines—the minimum lifetime currently desired for industrial deployment—finding that after 10 years of operation the centrifuges were difficult to repair or restart after a halt in operation. Nevertheless, development work on supercritical machines continues. Russia has produced several hundred experimental supercritical models, but they were never utilized in industrial plants. As early as 1957, each subcritical generation was accompanied by a supercritical analog.[49] Russia currently uses a small number of supercritical machines to separate stable isotopes.

As expected, many of the features of the modern Russian centrifuges are similar to those used in Urenco machines, implying a common ancestry or 'genetic code'. Both types use a magnetic top bearing, a molecular pump, scoops, a needle bearing and an oil dampener to absorb the energy of rotor vibrations and a disk-type hysteresis motor.[50]

As elsewhere, advances in Russian centrifuges have been influenced by the development of rocket engines and missiles. This implies that Russian centri-

[46] Mikerin *et al.* (note 22).
[47] Mikerin *et al.* (note 22).
[48] Mikerin *et al.* (note 22).
[49] Mikerin *et al.* (note 22).
[50] NUEXCO (note 22).

fuges moved from aluminium alloys to materials with a greater ratio of tensile strength to density. Russia has so far not published details of the material used in its centrifuges, but it has stated that its centrifuges have used a thin-walled rotor reinforced with composite material.[51] It has also stated that it has improved the quality of its composite rotors.

Based on Western experience, the Russian rotor could be composed of an aluminium alloy tube with an overwrap of glass fibre or carbon fibre. Newer Russian machine designs indicate that speeds may approach twice that of machines already deployed. This increase implies that Russia is using advanced composite materials, perhaps high-strength carbon fibre.[52]

According to Mikerin, the higher separative capacity of the Russian machines following increased rotor speeds resulted from the 'application of materials with high specific strength, optimal use of mechanical property structure of the material, and improvement of the gas dynamics characteristics'.[53]

Estimated centrifuge capacity. Although the estimates of the separative capacity of Soviet centrifuges below are speculative, they conform to standard expectations of methods to improve output and are consistent with the above information.

The starting point is the 100-mm machine Zippe proved in Germany in the 1960s that had a separative capacity of about 0.57 SWU per year. This machine is assumed to be a fair representation of the first centrifuges industrially deployed in the Soviet Union.

The most straightforward methods by which to increase a machine's separative capacity are to spin the rotor faster and to increase the rotor's length. This simplistic point of view, however, has severe limits at higher speeds and lengths. In addition, improvements in the efficiency of isotope separation are harder to achieve at higher speeds.

Until missile developments led to new materials in the 1960s and 1970s, special aluminium alloys were the most common material available for centrifuges. Early Soviet rotors used duraluminum. We assume that the first several generations of Soviet centrifuges used aluminium alloys, and the next several used maraging steel or aluminium with an overwrap of composite material. The most recent ones, the seventh- and eighth-generation machines, might utilize a carbon-fibre rotor or a composite material over another material besides aluminium. The change from an aluminium rotor to an overwrapped aluminium one would allow increases in peripheral speed from roughly 350 m/s to over 550 m/s. The use of maraging steel or titanium as a liner would allow peripheral speeds of 600 m/s. Carbon-fibre rotors can have speeds of over 600 m/s.

At low speeds, an increase in the rotor's peripheral speed increases the separative capacity in proportion to the fourth power of the tangential velocity. At higher speeds, however, the separative capacity grows in proportion to the

---

[51] NUEXCO (note 22).
[52] Mikerin *et al.* (note 22).
[53] Mikerin *et al.* (note 22), and NUEXCO (note 22).

Table 4.6. Estimated increase in the separative capacity of Soviet centrifuges
Capacity is expressed in terms of SWU/y.

Generation	Period of deployment	Separative capacity	Comments
First	1950s	0.4	Diameter = 10 cm, speed = 340 m/s, efficiency = 0.3
Second	1950s–1960s	0.6	Machine optimized, perhaps diameter = 12 cm and speed = 360 m/s
Third	1960s–1970s	1.0	Same except speed = 425 m/s
Fourth	1970s–1980s	1.4	Same except speed = 475 m/s
Fifth	1980s–1990s	1.9	Same except speed = 530 m/s
Sixth	1990s	2.5	Same except speed = 580 m/s
Seventh	?	3.2	Same except speed = 630 m/s
Eighth	?	4.2	Same except speed = 690 m/s

cube of the velocity. This occurs in regimes where the speed is greater than about 300 m/s but below roughly 600 m/s. At higher speeds, separative capacity grows in proportion to the square of the speed, or less. If everything else is held constant, increasing the speed from 360 m/s to 450 m/s would increase the separative capacity by a factor of 1.96.

Since separative capacity varies linearly with length, it is standard to increase the rotor length. In this case, however, the increase must be done so that the machine remains subcritical. In practice, this condition is satisfied if the ratio of the length to diameter remains smaller than four.

However, there are economic and technical limits to increasing the diameter. With an increase in diameter, the peripheral speed of the rotor increases, possibly exceeding the specific strength of the rotor material. In addition, increasing rotor speed means that the outer casing must be thicker to contain the increased energy of a crashed rotor. Costs associated with thicker outer casings limit rotor diameters and thus the length of subcritical centrifuges.

Nevertheless, if the diameter of the above model is increased from 10 cm to 12 cm, the separative length would become about 52 cm, and the separative work for each machine would increase by a factor of 1.2 to about 0.68 SWU per year, assuming tangential speed remains at 360 m/s. The angular velocity would decrease from about 1150 revolutions per second (rps) at a diameter of 10 cm to 950 rps at a diameter of 12 cm.

Although this increase in separative output is not large, the larger diameter would permit faster rotor speeds, assuming the specific strength of rotor material is not exceeded. If the angular velocity in the example above remains fixed, a diameter of 12-cm would translate into a rotor speed of 430 m/s.

Western experts involved in the commercial uranium business with Russia have said that each centrifuge is about the size of a diver's compressed air tank, or about 1 m in total length and about 15 cm in diameter. They say that the rotors currently have an angular velocity of about 1500 rps.

Another Western expert has said that the centrifuges located in the newest of the UEK buildings have a total length, including the top and bottom bearing assemblies, of roughly 70 cm. The rotor itself is about 46 cm long.

Table 4.6 presents an estimate of the growth in the separative capacity of Soviet centrifuges. This approximation assumes that the increase in separative output resulted essentially from increases in peripheral velocity.

The estimates in table 4.6 have a high degree of uncertainty, and other choices of key dimensions, speed and efficiency would result in estimates that also fit the available data. Despite these ambiguities, this table nevertheless provides a reasonable estimation of the improvements in Soviet centrifuges that fit the available information. For the first five generations, each model has about 40–50 per cent more separative work than the previous one. Afterwards, the increase in capacity is limited by the difficulty of mastering higher speeds.

Historical plant capacity

Little has been published about the historical separative capacity of the enrichment plants. There is the estimate of 7–10 million SWU per year during the mid-1960s mentioned above. According to a former senior US official, the USA estimated the capacity of the Soviet enrichment plants at about 12 million SWU per year during the 1970s.

In the mid-1980s, a US official said that the Soviet enrichment capacity was a matter of intense controversy within the US Government. Some believed correctly that gaseous-diffusion equipment was being replaced by centrifuges in the same buildings. Others believed, apparently incorrectly, that the Soviet Union had built secret, undiscovered facilities.

As discussed above, from 1966 to 1992 gas centrifuges gradually replaced gaseous-diffusion equipment in the enrichment buildings as this equipment became obsolete. Little information, however, exists about the deployment or the decommissioning history of the diffusion plants. NUEXCO has reported that the plants operated reliably with repair intervals of seven to nine years.[54]

According to Russian nuclear officials, improvements in centrifuge lifetimes and manufacturing capabilities permitted the intensive replacement of diffusion equipment after 1976. How this was done in practice is shown below.

Mikerin *et al.* have stated that the centrifuge manufacturing plants produced only one model at a time.[55] In contrast, there were two gaseous-diffusion manufacturing plants working independently to produce gaseous-diffusion stages. Mikerin *et al.*, however, said that in the case of centrifuges, the emphasis was mass production with a high level of production technology standardization and quality control.

[54] NUEXCO (note 22).
[55] Mikerin *et al.* (note 22).

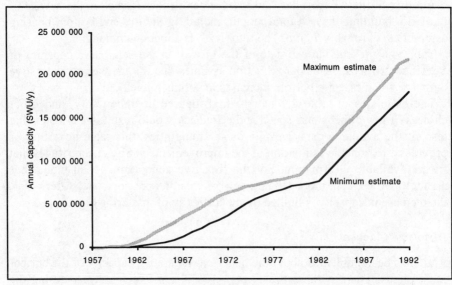

Figure 4.2. Maximum and minimum estimates of annual Soviet gas-centrifuge capacity, 1957–92

The first centrifuge designs ran for 10–12 years with a failure rate of less than 1.5 per cent per year.[56] Existing designs operate for over 15 years with a failure rate of a few tenths of 1 per cent per year.

Because of the current financial crisis gripping Russia, Mikerin is pessimistic regarding the likelihood of Russia manufacturing more powerful centrifuge designs on schedule.[57] He added that since the last centrifuge design 'utilized most engineering, design, and construction models, future models will require more time for design, more money for materials, and longer production times'.[58]

Historical enrichment capacity estimate. Estimates of the historical enrichment capacity are summarized below and shown in figures 4.2 and 4.3.[59]

The gaseous-diffusion capacity is assumed to have increased to about 7–8.5 million SWU per year by 1966 and then to have decreased to zero by 1992 under two different scenarios. The first scenario is a maximal estimate and assumes the capacity grows linearly to 8.5 million SWU per year in 1966, remains level until 1978 and then decreases linearly to zero by 1991. The second is a minimal estimate and assumes the capacity grows more slowly, increasing linearly to 1 million SWU per year in 1956, then increasing faster to

---

[56] NUEXCO (note 22).
[57] 'Interview with Evgeni Mikerin' (note 32).
[58] 'Interview with Evgeni Mikerin' (note 32).
[59] More complete information is available from the Institute for Science and International Security, Washington, DC..

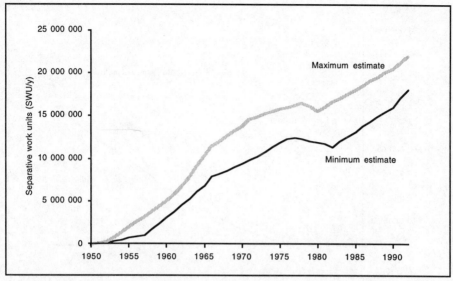

Figure 4.3. Maximum and minimum estimates of annual Soviet uranium enrichment capacity, 1950–92

a peak of 7 million SWU per year in 1966, decreasing slowly to 6 million SWU in 1977 before decreasing linearly to zero by 1990. The minimal estimate assumes that until 1956 most of the Soviet effort went into developing more efficient enrichment stages, and that afterwards it went into large-scale industrial deployment of improved stages.

Our model of the centrifuge capacity is more complex. It is based on an estimate of the lifetime of each centrifuge model, a fixed centrifuge manufacturing rate where only one model is manufactured at a time, and a constant replacement of centrifuge models as they reach the end of their design lifetime. The centrifuges are assumed to have the separative capacities shown in table 4.6.

Under these assumptions, Soviet centrifuge capacity increased in two main stages. The first two or three centrifuge models had lifetimes of about 10 years. In this case, the increase in total capacity resulted from a one-to-one replacement of an old machine with a new one with a higher capacity. After all older machines are replaced over a 10-year period, total capacity would increase by the same factor as the growth in separative capacity from one model to the next one. For example, the fourth-generation machine replaced the third-generation machine and had a separative capacity 40 per cent larger than the third. The total capacity will therefore increase by a factor of 1.4 after a 10-year replacement period.

If the centrifuge lifetime is 20 years, which we assume to be the case for the fourth- and succeeding generations, then the total capacity would increase significantly faster. For example, if all fourth-generation machines were

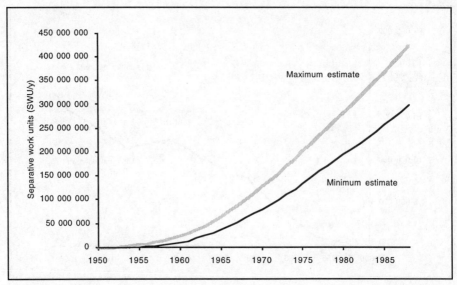

Figure 4.4. Maximum and minimum estimates of cumulative Soviet separative output, 1950–88

deployed over a 10-year period, none of them would need replacement for another 10 years; but if the fifth-generation machine goes into production when production of the fourth-generation halts, the capacity will grow by a factor of about 2.4 over the next 10 years. This rapid an increase would have enabled the USSR to phase out the gaseous-diffusion plants after 1976 while still increasing total capacity.

Two centrifuge projections are presented. The maximal case assumes that the successive generations of centrifuges were deployed faster and in larger numbers than the minimal estimate. For example, the maximal estimate assumes that the fourth and fifth generations of centrifuge were deployed two years earlier than the minimal estimate (1971 vs 1973 and 1981 vs 1983). In both cases, the centrifuge manufacturing rate is assumed to reach an equilibrium in the 1960s. For the maximal and minimal cases, the equilibrium rates are 600 000 and 550 000 centrifuges per year, respectively. These values are chosen so as to achieve a total enrichment capacity of about 18–22 million SWU per year in 1992, assuming the centrifuge capacities shown in table 4.6.

Separative output. Using the above estimates, the USSR produced about 284–404 million SWU by the end of 1987. If production continued throughout 1988, this range would increase to 299–423 million SWU. We average these estimates and arrive at 291–414 million SWU, as shown in figure 4.4. About half of this output came from gaseous-diffusion plants and the other half came from gas-centrifuge plants.

Non-weapon requirements

A large fraction of this total would have gone to produce enriched uranium for various non-weapon uses:

1. About 40 million SWU were exported as low-enriched uranium to Western power reactors. (This represents deliveries up to the end of 1988, which we assume were all enriched before the end of 1987.)
2. Domestic, East European and Finnish power reactors required roughly 35 million SWU to the end of 1987.
3. If Soviet plutonium production reactors used slightly enriched uranium and the same type of fuel cycle as the US Hanford production reactors, the enrichment of the fuel would have required about 5 million SWU. A US official has stated that the three remaining production reactors use natural uranium fuel in combination with WGU fuel spikes. This combination is used in these reactors to flatten the neutron flux, permitting the production of plutonium with a more exact fraction of ^{240}Pu. Each of these reactors contains almost 85 kg of weapon-grade uranium, for a total of about 250 kg of weapon-grade uranium in the three reactors. Using similar information, Oleg Bukharin has derived total historical estimates of WGU requirements for the production reactors, assuming that they all used natural uranium fuel with WGU fuel spikes. He estimates that the total requirement was about 23.5 tonnes of WGU through 1994, most of which was used by the late 1980s.[60] This amount corresponds to about 4.7 million SWU.
4. The Soviet Union was reported to have used LWRs and HWRs to produce tritium.[61] Without specific information, we assume that up to the end of 1987 the Soviet Union produced about 200 kg of tritium (not corrected for decay). This is similar to the estimated US production.[62] We roughly estimate SWU requirements in two different ways.

(a) We first consider a 1000 MWth LWR (water-cooled and graphite-moderated). Such a reactor could produce about 4 kg of tritium a year.[63] At this rate, 50 reactor-years of operation would be required to make 200 kg of tritium. Each year, such a reactor would require about 140 000 SWU.[64] Production of 200 grams of tritium would therefore require about 9 million SWU.

(b) Next, we consider the Savannah River reactors (heavy water-moderated and -cooled). The NRDC estimated that the production of 1 kg of tritium in a Savannah River reactor would require the consumption of about 96 kg of ^{235}U.[65] Assuming that the fuel is 60–90 per cent enriched and about 40 per cent of the

---

[60] Bukharin, O., 'Analysis of the size and quality of uranium inventories in Russia', presented at the Nuclear Energy Institute's International Uranium Fuel Seminar, Williamsburg, Va., 8–11 Oct. 1995.
[61] Cochran, T., Norris, R. and Bukharin, O., *Making the Russian Bomb: From Stalin to Yeltsin* (Westview Press: Boulder, Colo., 1995).
[62] Cochran *et al.* (note 2).
[63] Cochran, Norris and Bukharin (note 61)
[64] Assuming 80% capacity and a 50% fuel burnup, which implies an annual requirement of about 700 kg WGU per year.
[65] Cochran *et al.* (note 2), p. 185.

Table 4.7. Estimated Soviet consumption of separative work units, to 31 December 1987

Area of consumption	SWU (x 10^6)
LEU exports	40
LEU for domestic, East European and Finnish power reactors	35
Fuel for production reactors, plutonium production	5
Tritium production	7
Fuel for naval propulsion reactors	25
Fuel for domestic and foreign research	4
Consumption in nuclear weapons tests	4
Process losses (3% of original enrichment output)	10
Working inventory	5
Total	**135**

initial ^{235}U is consumed, then about 10 million SWU would have been needed. This estimate assumes no recycling of the spent fuel, which is unlikely. If up to half the separative work was saved through recycling into other reactors, then as little as 5 million SWU would have been required. As can be seen, this estimate is highly uncertain, but the above shows that the requirement for tritium production is unlikely to have exceeded 10 million SWU.

As can be seen, this estimate is not very sensitive to the two reactor types. We assign 5–10 million SWU, or a mid-point of 7 million SWU, to the production of tritium.

5. An estimated 470 reactors have been installed on about 260 Soviet naval vessels and icebreakers.[66] We assume crudely that enriched uranium enough for two to four times that number of reactor cores had been produced by the end of 1987. Soviet naval reactors are reported to use various enrichments from 20 per cent to over 90 per cent.[67] In any case, we will assume that an average reactor requires between 50 and 100 kg of ^{235}U, requiring between 47 and 190 tonnes of ^{235}U to produce all the cores. If the fuel was 10 per cent enriched, its production would have required about 9–36 million SWU. If the fuel was all fully enriched, it would have required about 11–42 million SWU. We pick the mid-point of these ranges, or roughly 25 million SWU.

6. A total of over 30 research reactors have generated an estimated 3 million MWth-d of energy up to the end of 1987.[68] Generation of this amount of heat would require roughly 10 tonnes of 90 per cent enriched fuel. Soviet space reactors are estimated to have required about 1 tonne of weapon-grade uranium. Critical facilities and other uses are likely to have raised this total to 20 tonnes.

[66] Cochran, Norris and Bukharin (note 61), pp. 286–87; and Bukharin (note 60).
[67] Handler, J., 'Preliminary report on Greenpeace visit to Vladivostok and areas around the Chazma Bay and Bolshoi Kamen submarine repair and refuelling facilities, 9–19 October 1991', Greenpeace, Washington, DC, 6 Nov. 1991; and Bukharin (note 60).
[68] IAEA, *Nuclear Research Reactors in the World* (IAEA: Vienna, 1991), table 14. We assume an average capacity factor of 70% for Soviet research reactors.

In total, this amount of fuel would require about 4 million SWU. Exports of enriched uranium fuel for foreign research reactors would not add significantly to this total (see appendix D).

7. About 10–30 tonnes of HEU are estimated to have been consumed in almost 1000 nuclear weapon explosions, or the equivalent of 2–6 million SWU.

8. Process losses might have consumed about 3 per cent of the amount of enriched uranium produced. This would entail the loss of an equivalent of about 10 million SWU.

9. A working inventory of several months' worth of production, roughly 5 million SWU, might have been needed in the enrichment complex in order to meet fluctuations in demand and provide a stock of various enrichment levels.

Table 4.7 summarizes the above draw-downs. In total, about 135 million SWU would have gone to non-weapon and testing uses. We assign an uncertainty of ± 15 per cent.

HEU inventory for weapons

Subtracting these requirements from the total enrichment effort leaves about 155–279 million SWU. We are unaware of the average tails assay achieved in Soviet enrichment plants, but an average tails assay much greater than 0.3 per cent or less than 0.25 per cent was unlikely to have been sustained during the period when HEU was made in large quantities. At a tails assay of 0.3 per cent, the weapon-grade uranium equivalent is about 775–1395 tonnes. If the average tails assay was 0.25 per cent, the weapon-grade uranium equivalent would be 720–1290, or about 7 per cent lower. We assign a range of 720–1395 tonnes of weapon-grade uranium equivalent, with an average of 1050 tonnes. The uncertainty is about 30 per cent. If, like the United States, the Soviet Union produced non-weapon-grade HEU for its weapons, this estimate could be greater by several hundred tonnes.

The total Soviet nuclear arsenal is claimed to have peaked in 1986 at 45 000 warheads, and then declined to about 32 000 warheads for the FSU by May 1993.[69] These numbers represent total warheads and include many removed from active service and stored or awaiting disassembly. According to Norris at the NRDC, Russia did not recycle fissile materials from retired warheads as efficiently as the United States. If a weapon was removed from the field, Russia did not always take it apart and reuse the fissile material.

Norris and Arkin estimated that Russia's total arsenal was about 29 000 warheads in mid-1994, and about 2000–3000 warheads were being dismantled each year.[70] This total includes many weapons that were not on active service and were either awaiting disassembly or held in reserve. The size of the active

---

[69] Cochran, T. B. and Norris, R. S., *Russian/Soviet Nuclear Warhead Production*, Nuclear Weapons Databook Working Paper NWD 93-1 (Natural Resources Defense Council: Washington, DC, 8 Sep. 1993), pp. 22–24.
[70] Norris and Arkin (note 15), pp. 58–59.

arsenal is largely unknown, although Russia has about 9000 strategic warheads, most of which are considered active. We arbitrarily assign 11 000 warheads to Russia's active arsenal.

We have little information about the amount of HEU in Soviet warheads. Assuming a simple model of about 15–30 kg of HEU per warhead, the estimated 45 000 warheads would have contained about 675–1350 tonnes of HEU, which approximates our estimated range of HEU produced for weapons, and implies little recycling of HEU from old weapons into new ones.

The estimated 11 000 warheads in Russia's active arsenal would contain about 165–330 tonnes of HEU. The mid-point of this estimate is 250 tonnes in active nuclear weapons, implying a mid-point of about 770 tonnes in the production pipeline, in storage, in weapons awaiting disassembly or in weapons held in reserve.

With such a large excess, Russia has agreed to sell 500 tonnes of weapon-grade uranium from dismantled weapons to the United States over 20 years. Russia will blend down this weapon material into low-enriched uranium before sending it to the USA. The USA will receive about 4.4 per cent enriched uranium and pay about $12 billion.

During 1995, Russia sold the USA 6 tonnes of HEU (in the form of LEU) and agreed to sell 12 tonnes in 1996.[71] These amounts are far below expectations and are the result of disagreements over price and other issues between the USA and Russia. These disagreements are gradually being resolved as of mid-1996.

The above HEU inventory estimate does not include any HEU dedicated to the naval fuel cycle. The amount could total up to hundreds of tonnes of HEU, depending on the enrichment level of the fuel. The amount of WGU in the naval fuel cycle is on the order of tens of tonnes. Much of this HEU is in the form of irradiated fuel at tens of facilities, including docked ships or decommissioned submarines.

Location and form of HEU

Little information is available about the location of Russia's major HEU stocks, and considerable uncertainty remains about the amounts of HEU at various sites. Moreover, few details exist about the form of the Russian HEU at specific sites. Although most of Russia's HEU is expected to be metal, significant quantities could be in the form of scrap, oxides, fresh and spent fuel, and other compounds.

Although Russia has sharply reduced the number of sites with nuclear weapons, it has not similarly consolidated the sites involved in processing or storing fissile materials. US officials have stated that scores of military and civil sites in Russia contain significant quantities of fissile materials. Two US

---

[71] 'Sale to US of Russian uranium and blended-down HEU from weapons', *Nuclear Fuel*, 25 Mar. 1996, p. 12.

research groups have attempted to catalogue many of these sites.[72] Because so little information is available about the inventory of material at these sites and much of the information is contradictory, we do not try to assign HEU inventories to specific sites, as the DOE has done for its major military inventories.

Public and international concern has focused most intensely on HEU (and plutonium) stocks held at research and fuel fabrication centres. These sites are believed to have weaker physical security and safeguards than the main nuclear weapon production sites. Nevertheless, the vast bulk of Russia's HEU stock is at military sites, which are not believed to have adequate protection or safeguards, even if they are marginally better than civil sites.

Some HEU remained in the nuclear weapons on the territories of the former Soviet republics after the collapse of the USSR. Kazakhstan had returned all the warheads to Russia by April 1995, however, and on 1 June 1996 the Ukrainian Government announced that all nuclear weapons on Ukrainian territory had been returned to Russia. The last of the nuclear warheads based in Belarus were scheduled to be returned to Russia by the end of 1996.

In addition, former Soviet republics possess relatively small but significant stocks of HEU at research or fuel fabrication sites. Currently, civil research reactors in Kazakhstan, Latvia, Ukraine and Uzbekistan have unknown quantities of fresh HEU fuel, and in addition they possess stocks of spent HEU fuel (see chapter 8 and appendix D). In total, these countries possess on the order of a few hundred kilograms of WGU, but significantly more HGU.

In early 1996, the US DOE and the Latvian Nuclear Research Centre (the site of a 5 MWth research reactor) finished the first materials protection control and accounting upgrades in the former Soviet Union consistent with international standards. Other civil research reactor sites in the former Soviet Union are also involved in similar bilateral security upgrades.

Belarus possesses a few tens of kilograms of WGU for one critical facility and a few hundred kilograms of non-weapon-grade HEU for another critical facility. It may also possess kilogram quantities of HEU in fresh fuel and significant quantities of HEU in irradiated fuel left over from a 4 MWth research reactor that closed in the late 1980s.

Georgia had a research reactor that used HEU, but it was shut down in 1990. Fresh and spent HEU fuel remains there, however.

At Semipalatinsk in Kazakhstan, Russia is reported to control about 200 kg of HEU in rocket reactor fuel.[73] Kazakhstan also has a large fast reactor that annually uses hundreds of kilograms of 20–25 per cent enriched uranium and stores tonnes of spent HEU fuel (see chapter 8). Kazakhstan sold to the USA an estimated 650 kg of HEU, containing about 580 kg of ^{235}U, left over at the Ulba Metallurgical Plant at Ust-Kamenogorsk. This site made naval reactor fuel, and

---

[72] Cochran and Norris (note 69) and Carnegie Endowment for International Peace and Monterey Institute of International Studies, *Nuclear Weapon and Sensitive Export Status Report: Nuclear Successor States of the Soviet Union*, no. 4, May 1996.

[73] 'Sale near of homestake material: Benton applauds customs action; more Kazakh HEU?', *Nuclear Fuel*, 4 Dec. 1995.

Table 4.8. Estimated uranium-235 content of the HEU at the Ulba Plant
Figures are total kilograms of ^{235}U (average enrichment $c.$ 90 per cent).

Form of material	Total ^{235}U
HEU metal	168.7
Uranium oxides	29.7
Uranium-beryllium, alloy rods	148.6
Uranium oxide-beryllium, alloy rods	1.6
Uranium-beryllium alloy, scrap	231.5
Other scrap	0.9
Total	**581**

Source: Riedy, A. W., 'Project Sapphire', Martin Marietta Energy System viewgraph presented at the workshop on 'Weapon dismantlement and material controls in the Former Soviet Union; sponsored by the Institute for Science and International Security (ISIS), Washington, DC, 1 Feb. 1995.

the HEU was left over from an earlier order. The material was in various forms, including metal, oxides, uranium-beryllium fuel rods, and scrap (see table 4.8). An exact accounting of the scrap was unfinished.

V. The United Kingdom

The UK has obtained weapon-grade uranium for its nuclear arsenal from a gaseous-diffusion enrichment facility at Capenhurst and from the USA. Little is known about the amount produced at Capenhurst and questions remain about the quantity acquired from the USA. It is assumed that the UK's stock of military HEU is almost all weapon-grade uranium. However, if this is not the case, the estimate should be treated as the quantity of weapon-grade uranium equivalent.

Most of the Capenhurst gaseous-diffusion plant was shut down in 1962, ending British production of HEU. The reason for closing the plant was related to the appearance of an alternative cheaper US supply of weapon-grade uranium for military purposes and low-enriched uranium for civil reactors.

The UK signed a Mutual Defence Agreement with the USA in the late 1950s (see chapter 3). Under this agreement, the UK traded a growing surplus of plutonium for weapon-grade uranium and tritium from the USA. Later, it entered into an enrichment services contract to acquire HEU from the USA. The current contract is in force until 2001.

Weapon-grade uranium supply

Capenhurst

Britain built the gaseous-diffusion plant at Capenhurst in the 1950s, at the time the largest industrial building under one roof in Europe. The first buildings were commissioned in 1952–53. The plant was constructed so that, while construction was proceeding, it could produce slightly enriched uranium for the Windscale plutonium production reactors and re-enrich depleted uranium left over after reprocessing production reactor fuel.

The plant was expanded in 1957 with the installation of more efficient enrichment stages in additional buildings. After the expansion was completed in 1959, one source put the plant's total capacity at about 325 000 SWU per year. Another source states that three years' output of Capenhurst in 1960 was equivalent to the production of 5 tonnes of 93 per cent enriched uranium.[74] At a tails assay of 0.3–0.4 per cent, 5 tonnes of production is equivalent to about 300 000–330 000 SWU per year. This range closely matches the above estimate and implies that the average tails assay for Capenhurst was in the above range.

Before the expansion, Capenhurst's nominal capacity is estimated at about 100 000–150 000 SWU per year, or slightly less than half its maximum capacity. When Capenhurst reached this capacity is unknown, but it appears to have sustained this level over a few years.

Our estimate of HEU production is in two parts. The first period extends to 1959, when the expansion was completed. The UK is reported to have started producing HEU after about 1954. Production appears to have been relatively small, and much of the weapon-grade uranium was used in nuclear explosive tests.

In 1957, Britain had available about 250 kg of weapon-grade uranium for use in high-yield test devices during Operation Grapple, conducted on Christmas Island in May and June 1957.[75] These tests involved designs for high-yield boosted fission weapons and two-stage thermonuclear weapons.

Nevertheless, Britain faced a shortage of weapon-grade uranium at the time of Operation Grapple, implying that production at Capenhurst was only of the order of a few hundred kilograms per year. The estimated 100 000–150 000 SWU per year output during this early period would imply a considerably higher level of production, about 500–750 kg of weapon-grade uranium a year. We believe that an average production of 150–200 kg of weapon-grade uranium per year from 1955 through 1958 is a better estimate. At this latter rate, Britain produced 600–800 kg of weapon-grade uranium during this first period. The higher rate would result in the production of 2000–3000 kg, which contradicts the available information.

---

[74] Memo from John Hill (later Sir John Hill, Chairman of the UK Atomic Energy Agency) in the file AB38.101, 'Purchase of U^{235} from USA' (from the papers of J. C. C. Stewart, 1960–61). These files are available at the Public Record Office, London.

[75] Brotherhood, W. R., 'Operation Grapple: Details of megaton warheads and order of firing', 16 Jan. 1957 (declassified version), unpublished document available at the Public Record Office, London.

The second period lasts two years, ending in 1961, when Capenhurst operated at a capacity of about 325 000 SWU per year. Assuming an average tails assay of 0.3–0.4 per cent, the plant would have produced about 3.2–4.1 tonnes of weapon-grade uranium during this second period.

In total, it is estimated that Capenhurst produced about 3.8–4.9 tonnes of weapon-grade uranium until 1961. The mid-point is 4.4 tonnes.

About 90 per cent of Capenhurst was closed in 1962. Some of the older stages were kept running to provide low-enriched uranium for non-military reactors, although the record is unclear about the plant's operation after 1962. In 1982, the British began decommissioning, dismantling and decontaminating the plant.

Britain decided in the late 1970s to build a military centrifuge plant at Capenhurst, called A3, to produce HEU because the Carter Administration decided to end the practice of bartering British plutonium for US HEU (see chapter 3). In 1982, however, Britain announced that the plant would produce intermediate-level enriched uranium and this material would be sent to the USA for final enrichment to the level required for defence purposes.[76] After starting in 1984 or 1985, according to one British official, the plant produced 4.5 per cent enriched uranium for shipment to the USA (see below). This plant is estimated to have a capacity of about 200 000 SWU/y.

In 1991 the British Ministry of Defence decided to terminate its enrichment contract with the Capenhurst A3 plant, as of 1993.[77] Following the end of the military's use of the A3 plant, Urenco, the European enrichment consortium, took over the operation of the plant to produce LEU for civil purposes. Since January 1993, Urenco has operated the A3 plant under IAEA and Euratom safeguards.

HEU from the USA

The USA has provided the UK with HEU for use in nuclear-powered submarines and nuclear weapons since the early 1960s. The nature of the supply, however, has changed over time. Currently the USA can supply HEU to the UK under a contract to provide uranium enrichment services that runs from 1991 to 2001. It is not known if the USA is actually supplying or planning to supply additional HEU to the UK under this arrangement.

The original basis for supplying HEU was a 'barter' provision in the 1958 US–UK Mutual Defence Agreement. According to recently declassified information from the US Department of Energy (with the agreement of the British Government), from 1960 to 1970 the USA supplied 6700 kg of HEU in exchange for 6.7 kg of tritium and 5400 kg of plutonium (see chapter 3).[78]

[76] 'In Parliament', *Atom*, 31 Aug. 1982.
[77] Marshall, P., 'BNFL will cut 400 jobs at Capenhurst as a result of UK's axing of contract', *Nuclear Fuel*, 1 Apr. 1991.
[78] US Department of Energy, *Plutonium: The First Fifty Years: United States Plutonium Production, Acquisition, and Utilization from 1944 to 1994* (Department of Energy: Washington, DC, Feb. 1996), p. 42; and DOE (note 19), pp. 81–86.

The DOE decided in 1995 not to declassify information about the amount of HEU the USA sent to the UK after this barter arrangement ended. In fact, there are few details about the more recent HEU transfers, although, as mentioned above, the arrangement centres on the purchase of enrichment services, or SWUs, from the USA. However, the DOE was reconsidering its decision in the spring of 1996, and was seeking British agreement for additional disclosures.

Britain contracted with the United States for up to 100 000 SWU per year from 1981 until 1986 for military purposes.[79] We have no information about the amount of separative work actually used, but the main reported objective was to obtain HEU for submarine reactors.[80]

It is known that once the Capenhurst A3 plant started the UK sent LEU (less than 5 per cent enriched) from this plant to the DOE under its enrichment services contract. According to a 1987 press report, the initial arrangement involved the USA sending natural uranium to Britain, where it was enriched at Capenhurst to an intermediate level and then sent back to the USA for enrichment to weapon-grade, before being sent to Britain for defence uses.[81] In 1987, this arrangement was replaced by one in which the UK provided LEU but the British material was 'swapped' for an equivalent amount of weapon-grade uranium. The HEU was reported to be in metallic form suitable for weapons, and was taken from the US weapon stockpile that had been produced before 1964.

We can only estimate very roughly the total amount of weapon-grade uranium imported from the USA under the more current arrangements. As shown below, we estimate that about 2–4 tonnes of weapon-grade uranium were imported for naval reactors during this period. Additional shipments of weapon-grade uranium for weapons may have occurred, and we estimate that 0–2 tonnes were acquired for this purpose.

Demand for weapon-grade uranium

Britain had several non-weapon requirements for weapon-grade uranium:

1. Weapon-grade uranium is used to fuel the UK's nuclear-powered submarines. Some of this material would have been taken from the British nuclear weapon inventory, but not all of it. The British Government announced in 1982 that it needed to obtain additional HEU for its submarines and that it planned to obtain the material from the USA.[82] We do not know when the UK first received this enriched uranium from the USA. In any case, we estimate that the UK has fabricated a total of about 60–80 cores.[83] We estimate that each core

---

[79] *Nuclear Fuel*, 13 Apr. 1981, p. 9.
[80] John Nott, Written Parliamentary Answer, *Parliamentary Debate, House of Commons [Hansard]*, 23 June 1982, cols. 128–29; and John Lee, Written Parliamentary Answer, *Hansard*, 17 July 1985, cols 181–82.
[81] Associated Press, 'Oak Ridge uranium headed for Britain', newspaper reports, 28 Nov. 1987.
[82] Nott (note 80).
[83] Fishlock, D., 'New design for submarine reactors', *Financial Times*, 12 June 1981. Fishlock wrote that by mid-1981, 37 cores had been built for British submarines.

Table 4.9. Estimated British inventory of HEU, 31 December 1995
Figures are central estimates in tonnes.

Inventory	Estimate
Supply	
Capenhurst WGU equivalent production	4.4
From the USA	
Barter provision	6.7
Under contract	4.0
Sub-total	**15.1**
Draw-downs	
Nuclear submarines	5.8
Nuclear tests	1.0
Processing losses (3% of supply)	0.5
Sub-total	**7.3**
Remainder for weapons, the pipeline and reserve	7.8 ± 25%

required about 70–100 kg of weapon-grade uranium, where the first 40 cores each contained 70 kg, or a total of 2.8 tonnes, and the next 20–40 cores each contained 100 kg, or a total of 2–4 tonnes. The 60–70 cores therefore required about 4.8–6.8 tonnes of weapon-grade uranium. We assume the first 40 cores came from pre-1980 stocks. The rest, an estimated 2–4 tonnes, would probably have been obtained from the USA under the enrichment services contract.

2. Civil research reactors might have required some HEU from the military inventory. We believe that the total quantity was small, and we ignore it here. The main reason is that in addition to receiving military shipments of weapon-grade uranium from the USA, the UK also received over 2 tonnes of HEU from the USA for civil purposes. Some of the imported material was used in British civil reactors, and some of it was re-exported after fabrication into fuel. The largest research reactor using HEU, the 65 MW Dounreay Fast Reactor (DFR) which operated from 1959 until 1977 did not use weapon-grade uranium fuel. The reactor started with about 300 kg of 46 per cent enriched uranium (about 25 000–30 000 SWU), and then used 75 per cent enriched uranium.[84] When the reactor started, HEU was in short supply. This shortage apparently led to an early decision to reprocess the spent fuel, lowering further the amount of enriched uranium dedicated to this reactor.[85] At any one time, no more than 1 tonne of HEU is estimated to be committed to the DFR fuel cycle (see chapter 7). Since the production of this amount of non-weapon grade HEU would have required a relatively small number of separative work units, its effect is ignored here.

---

[84] 'The Dounreay fuel cycle facilities of AEA Fuel Services', *Atom*, Jan. 1992. Originally published under 'Flexible reprocessing at Dounreay', *Nuclear Engineering International*, Nov. 1991.
[85] See note 84.

3. By the end of 1991, the UK's 45 nuclear explosions would have required about 650–1100 kg of HEU, where we assume 10–20 kg per weapon, except for two large-yield fission tests that each required over 100 kg of weapon-grade uranium. We use about 1.0 tonnes.

4. Processing losses of 3 per cent of the total amount of HEU produced or acquired would amount to about 400 to 500 kg.

These draw-downs of weapon-grade uranium total about 6.2–8.3 tonnes. Subtracting them from the above estimate of 13.5–17.6 tonnes leaves a stockpile of about 6.3–9.3 tonnes of weapon-grade uranium dedicated to weapons, the pipeline and strategic inventories.

The central estimates are summarized in table 4.9, giving a current weapon-grade uranium inventory of about 7.8 tonnes. We assign an uncertainty of ± 25 per cent—or 2 tonnes—to this estimate.

Britain is estimated to have between 200 and 300 nuclear weapons. If each weapon has about 15–20 kg of weapon-grade uranium, about 3–6 tonnes would be in those weapons. Britain is bringing into service its fleet of Trident submarines, which have new warheads. It is, however, also reducing the number of weapons, following the end of the cold war. Therefore, the total number of weapons is expected to remain constant or decline to about 200 warheads. Nevertheless, building significant numbers of new warheads would require a significant amount of weapon-grade uranium in the pipeline, perhaps as much as a few tonnes of material.

VI. France

France started producing HEU for nuclear weapons in 1967 at the Pierrelatte gaseous-diffusion plant. In the end, it constructed a far larger capacity to produce weapon-grade uranium than it needed for military purposes.

France has stopped making HEU for nuclear weapons, although exactly when this occurred is unclear. In February 1996, President Jacques Chirac said that France had sufficient stockpiles of fissile materials to meet its future defence needs and that France would no longer make HEU for weapons. He also announced the impending closure of the Pierrelatte plant.[86] It was closed by 30 June 1996, and the cascades will be emptied by about mid-1997.[87]

Soon after Chirac's announcement, the French Defence Minister Charles Millon implied that the French supply of fissile material for defence purposes was large when he said, 'France today holds stocks of fissile material [sufficient] for the next fifty years' and that 'beyond these fifty years, we know how to recycle the materials currently used in our weapons.'[88] Nevertheless, France has not disclosed the size of its HEU stock.

[86] 'France to close HEU production plant', *Nuclear Fuel*, 26 Feb. 1996, p. 17.
[87] MacLachlan, A., 'HEU production ends at Pierrelatte', *Nuclear Fuel*, 18 July 1996.
[88] 'Pierrelatte ferme mais les salaires seront réorientés' [Pierrelatte closes but the salaries will be reoriented], *Midi Libre*, 6 Mar. 1996.

Pierrelatte enrichment plant

The core of the original Pierrelatte plant comprises four main buildings called the low-, medium-, high-, and very-high-enrichment buildings. Combined, these four buildings house several thousand separation stages and occupy an area of 220 000 square metres. The low-enrichment building took natural uranium up to about 2 per cent enrichment, the medium went from there to about 6–8 per cent, the high took that material up to 25 per cent, and the very-high-enrichment building continued the enrichment to 90 per cent and above. The low-enrichment building was inaugurated in 1965, and the other three were commissioned over the next two years. The entire plant started operating as a unit in early 1967, and the first weapon-grade uranium was produced in April 1967.

The first HEU was used in nuclear tests at Muroroa and Fangataufa in the South Pacific during the summer of 1967.[89] These eight tests investigated the use of HEU in high-yield boosted fission explosions and thermonuclear devices. The number and type of tests, which included boosted explosions of about 500 kt, suggest that Pierrelatte produced several hundred kilograms of weapon-grade uranium within several months of the commencement of the very-high-enrichment building.

During the rest of the 1960s, the French concentrated on improving operations and throughputs at the Pierrelatte plant. According to the Commissariat à l'Energie Atomique annual reports, the plant operated satisfactorily since then.

The nominal capacity of the plant is secret, although numerous estimates exist. Most estimates of the plant's nominal capacity when fully operational are between 400 000 and 600 000 separative work units a year. At a capacity of 500 000 SWU per year, Pierrelatte could produce about 2.5 tonnes of weapon-grade uranium a year, assuming a tails assay of 0.3 per cent.

Pierrelatte was first dedicated exclusively to military needs. During the 1960s and early 1970s, French civilian needs for enriched uranium had to be covered by sending natural uranium to the United States for enrichment. According to the 1970 CEA annual report, the contract concluded in 1969 with the United States was extended through 1971–76 and involved 200 000–500 000 SWU. In the 1970s, France also contracted with Russia for low-enriched uranium for its power reactors.

In the early 1970s, the Pierrelatte plant began producing enriched uranium for civil purposes. According to the 1972 CEA annual report, although the Pierrelatte plant remains essentially a military plant, a 'not negligible part of its activity will be directed from now on to civilian ends'.[90] One reason for this excess capacity might have been France's development of two-stage thermonuclear weapons, which eliminated the need for high-yield boosted fission

---

[89] Norris, R. S., Burrows, A. S. and Fieldhouse, R. W., *Nuclear Weapons Databook, Vol. 5: British, French, and Chinese Nuclear Weapons* (Westview Press: Boulder, Colo., 1994).

[90] Commissariat à l'Énergie Atomique, *Rapport Annuel 1972, Tome 2* (CEA: Paris, 1972), p. 13.

weapons and their large HEU requirement. Norris *et al.* estimate that France deployed only one type of high-yield boosted fission weapon—about 35 MR-41 warheads, each with a yield of about 500 kt, for submarine-launched ballistic missiles.[91] Assuming 75–125 kg of weapon-grade uranium per weapon, the MR-41 warheads required about 2.6–4.3 tonnes of weapon-grade uranium, or about 525 000–860 000 SWU at a tails assay of 0.3 per cent. This weapon was retired in the late 1970s and the HEU was recycled.

Reduced need for Pierrelatte

France formed Eurodif in 1972 in partnership with Belgium, Iran, Italy and Spain. The Eurodif plant first started operating in 1979 and reached a nominal capacity of 10.8 million SWU per year. With the operation of this plant the need for Pierrelatte was sharply reduced, and its low-and medium-enrichment buildings were closed in 1982.

The closing of the low- and medium-enrichment buildings evidently did not interfere in meeting military needs. The CEA said that France had available large stocks of nuclear materials both for the needs of its missile programme and to meet the fuelling needs of its nuclear submarine programme.[92] In addition, the Eurodif plant was contracted to provide LEU to replace the output of these cascades.[93]

In 1984, Pierrelatte was switched to a seasonal schedule. To take advantage of cheaper summer electricity rates, Pierrelatte operated on a seven-month, April–October schedule. Because the plant needed to produce as much in seven months as it produced in 12 months, France restarted a portion of the medium-enrichment building.

The closing of Pierrelatte brings to an end 30 years of operation and ushers in many years of work to decommission and dismantle this facility.

Historical production

Based on the above information, Pierrelatte's enrichment output can be roughly estimated. Production is divided into several periods. The first is from 1967 to the end of 1970, when production is assumed to be about 200 000 SWU per year, for a total of 600 000 SWU. The second period lasted from 1970 to the end of 1982, when we estimate full production of 500 000 SWU per year, or a total of 6 million SWU. The third period stretches from 1982 until the end of 1995, when production averaged 200 000 SWU per year, or a total of 2.6 million SWU. (Production in the third period also includes the separative work units to produce the LEU feedstock for the Pierrelatte plant.)

[91] Norris *et al* (note 89), p. 187.
[92] Augereau, J. F., 'L'usine militaire d'enrichissement de l'uranium de Pierrelatte ferme deux de ses ateliers' ['The military uranium enrichment plant at Pierrelatte closes two of its plants'], *Le Monde*, 15 Dec. 1982.
[93] MacLachlan, A., 'French defense enrichment plant to hibernate five months a year', *Nuclear Fuel*, 3 Dec. 1984.

Total estimated production is therefore 9.2 million SWU. This estimate is assigned an uncertainty of 30 per cent.

Non-weapon uses

Pierrelatte supplied several non-weapon purposes, primarily fuel for civilian, production and naval reactors, and nuclear weapon tests.

Civilian reactors

By the early 1970s, Pierrelatte started supplying enriched uranium to civilian reactors. According to CEA annual reports, the first civilian reactors provided with enriched uranium were the Phénix prototype breeder and French light water reactors. In total, Phénix received highly enriched uranium worth a total of 43 000 SWU.[94]

The CEA *1973 Annual Report* states that excess capacity allowed the CEA to offer low-enriched uranium to foreign customers who would later be covered by production at Eurodif beginning in 1978. The first foreign utilities included Synatom in Belgium, which received 425 000 SWU, and K. K. Kaiser in Switzerland, which received 396 000 SWU.[95] The shipment to K. K. Kaiser was transferred to the South African utility ESCOM in 1981, and perhaps about one-quarter of the sale to Synatom was similarly sent to South Africa.[96]

In summary, we have established that Pierrelatte produced at least 864 000 SWU for civilian purposes during the 1970s. Undoubtedly, it produced additional amounts for the EdF light water reactors. We estimate that EdF reactors could have received up to 900 000 SWU. In total, about 1 760 000 SWU are estimated to have been used for civilian power reactors.

Production reactors

Some plant capacity might have also been dedicated to re-enriching depleted uranium up to natural uranium assay for use in French gas-graphite reactors.[97] We ignore this contribution here.

Célestin reactors

The 200 MWth Célestin reactors, which have produced plutonium and tritium for nuclear weapons, switched from plutonium to HEU fuel in the early 1970s (see chapter 3). As a result of fission and neutron capture, each reactor consumes roughly 73 kg of ^{235}U per year at a capacity factor of 80 per cent.[98] If

---

[94] Nuclear Assurance Corporation, *Enrichment Status Report*, Oct. 1982, Section E, p. 2.
[95] Nuclear Assurance Corporation (note 94), Section E, p. 2
[96] Nuclear Assurance Corporation (note 94), Section C, pp. 94, 148, and 160.
[97] 'The low- and medium-enrichment sections', *Nuclear News*, mid-Sep. 1982, pp. 29B and 116A.
[98] 1 g of fission of ^{235}U produces about 0.96 MWth-d of heat. In addition, neutrons are captured in ^{235}U, resulting in ^{236}U. The ratio of the capture to fission cross-sections is about 0.2, therefore 1.2 g of ^{235}U are consumed to produce 0.96 MWth-d of heat, or 1.25 g ^{235}U/MWth-d.

93 per cent enriched uranium is used as fuel, and the spent fuel has a burnup of 50 per cent, the two reactors would require about 290 kg of weapon-grade uranium per year. This corresponds to about 58 000 SWU per year at a tails assay of 0.3 per cent.

Although the Célestin reactors are believed to have achieved this high capacity factor, they are unlikely to have sustained such high values historically. We assume that the historical capacity factor of the Célestin reactors averaged 60 per cent from the early 1970s until the early 1990s. At this capacity factor, the reactors would have required an average of 220 kg of weapon-grade uranium per year. Assuming they both operated 20 years at this capacity factor, they would have required in total about 4400 kg of weapon-grade uranium fuel. The production of this amount of material required 44 000 SWU per year, or a total of 880 000 SWU.

Since 1991, only one Célestin reactor has operated at a time, and this pattern is expected to continue until the end of the century (see chapter 3). If each reactor attains a capacity factor of 60 per cent, the annual HEU requirement for one reactor is about 60 kg of weapon-grade uranium or about 12 000 SWU per year. For 10 years of operation at this level the reactors would need in total about 600 kg of HEU fuel, or about 120 000 SWU.

Naval reactors

France has built nuclear-powered ballistic missile submarines, attack submarines and prototype reactors, and is building a nuclear-powered aircraft carrier. We estimate that France has procured a total of 40–60 cores for its naval reactor programme.

France's first ballistic missile nuclear-powered submarines might have initially used HEU fuel. Its attack submarines have always used LEU fuel with an estimated average enrichment of 7 per cent.

For the sake of simplicity, we assume that each core required roughly 100 kg of ^{235}U in LEU. If all this material had an average enrichment of 7 per cent, the cores would have required about 57–86 tonnes of material. Production of this amount of material would have required about 640 000–960 000 SWU, at a tails assay of 0.3 per cent.

Nuclear weapon tests

France conducted 215 nuclear explosions. If each device required an average of 10–20 kg of WGU, they would have required 2150–4300 kg of highly enriched uranium, or about 430 000–860 000 SWU.

Processing losses

We assume that about 3 per cent of total HEU production was lost during the many processing steps. This rate corresponds to 270 000 SWU, or the equivalent of 1.4 tonnes of weapon-grade uranium at a tails assay of 0.3 per cent.

Inventory of HEU for weapons

Subtracting these requirements, which total 4.0–4.8 million SWU, we are left with about 4.4–5.2 million SWU dedicated to weapon stocks. This is enough for roughly 22–26 tonnes of weapon-grade uranium at a tails assay of 0.3 per cent. Some of the separative work might have produced highly enriched uranium that was not weapon-grade for thermonuclear weapons, but this effect is ignored here. We assign an uncertainty of 30 per cent to the estimate.

As of May 1994, France had a total of 492 nuclear weapons, of which 384 were on M-4 submarine-launched ballistic missiles. The rest were on land-based ballistic missiles, air-to-surface missiles and shorter-range tactical missiles. At 15–30 kg per weapon, a total of 492 missiles corresponds to a total of 7400–14 800 tonnes of HEU. France thus appears to have an HEU surplus, particularly in light of its recent decision to reduce its nuclear arsenal.

VII. China

China first produced highly enriched uranium in early 1964 at the Lanzhou uranium enrichment plant, located in Gansu province (central China). During the next several months, the plant produced enough material for China's first atomic test in October 1964.

US intelligence agencies thought as late as August 1964 that the enrichment plant was incomplete.[99] At that time the CIA believed that 'on balance' China would not have enough fissile material for a test until after the end of 1964.[100]

During the late 1970s and early 1980s, China redirected its nuclear infrastructure to serve civilian goals. China is reported to have stopped producing weapon-grade uranium for weapons in 1987.[101]

Gaseous-diffusion plants

As a result of Soviet–Chinese collaboration in the 1950s, the Soviet Union assisted China in the initial construction of a gaseous-diffusion enrichment plant at Lanzhou.[102] Construction started in 1957, but the rift between China and the Soviet Union led to the withdrawal of all Soviet aid by mid-1960. Nevertheless, China received a significant amount of the necessary technology

---

[99] Director of Special Intelligence, 'The chances of an imminent Communist Chinese nuclear explosion', Special National Intelligence Estimate, 13-4-64, 26 Aug. 1964, declassified version in Ruffner, K. C. (ed.), *CIA Cold War Records: America's First Satellite Program* (CIA Center for the Study of Intelligence: Washington, DC, 1995).

[100] Note 99.

[101] MacLachlan, A. and Hibbs, M., 'China stops production of military fuel: all SWU capacity now for civil use', *Nuclear Fuel*, 13 Nov. 1989. The 1987 data are from a personal communication from M. Hibbs, who was told this by the head of the China Nuclear Energy Industry Corporation.

[102] The description of the Chinese development of gaseous diffusion plants is based on *China Today: Nuclear Industry*, Apr. 1987. Selections translated by the Foreign Broadcast Information Service, JPRS-CST-88-002, 15 Jan. 1988; and JPRS-CST-88-008, Washington, DC, 26 Apr. 1988.

INVENTORIES OF HEU IN THE NUCLEAR WEAPON STATES 127

and components for a gaseous-diffusion plant.[103] This included some of the diffusers (with separation membranes) and much of the specialized equipment necessary to monitor and run the plant.

Much work remained, however. Only in mid-January 1964 did the plant start producing weapon-grade uranium. The capacity of the plant during the mid-1960s is unknown, although we estimate that its initial nominal capacity was between 10 000 and 50 000 SWU per year. At this capacity, the plant could produce about 60–300 kg of weapon-grade uranium each year at a tails assay of 0.5 per cent. The Chinese encountered problems in learning to use and duplicate the Soviet-supplied components and in manufacturing missing equipment. This delayed both the opening of the plant and its expansion.

In 1972, the US Defense Intelligence Agency estimated that the Lanzhou facility was producing weapon-grade uranium at a rate of 150–330 kg per year.[104] At a tails assay of 0.5 per cent, this rate is equivalent to 24 000–53 000 SWU per year. At a tails assay of 0.3 per cent, the plant would produce the equivalent of 30 000–66 000 SWU per year. The first major increase in capacity occurred in the mid-1970s, when the plant was renovated. It was reported in 1978 that the plant's capacity was 180 000 SWU per year.[105]

A second period of expansion of the plant occurred during the early 1980s, when China achieved an 'enormous breakthrough in separation membrane technology'.[106] This development led to a further increase in separation efficiency. These improvements during the 1970s and 1980s resulted in a 'manyfold' increase in production capacity over initial output.[107] Afterwards, the plant is believed to have reached a capacity of about 300 000 SWU per year.[108]

China has a second gaseous-diffusion plant at Heping in Sichuan, although details of it are sketchy, including its size and its operational status. Citing Western sources, *Nuclear Fuel* reported that the capacity of the second plant is a little larger than that of the first plant.[109] In 1972, several years before the plant started operating, the Defense Intelligence Agency estimated that this plant would be capable of producing 750–2950 kg of weapon-grade uranium per year.[110] This corresponds to 120 000–470 000 SWU per year at a tails assay of 0.5 per cent, or 150 000–590 000 SWU per year at a tails assay of 0.3 per cent.

The plant is believed to have started operating in the mid-1970s and to have been built as a 'third-line' weapon manufacturing site. Starting in 1964, the Chinese leadership ordered the construction of strategic facilities in the interior

---

[103] *China Today* (note 102).
[104] US Defense Intelligence Agency, *Soviet and Peoples' Republic of China Nuclear Weapons Employment Policy and Strategy*, TCS-654775-72, Mar. 1972.
[105] 'Mainland China talking to French, Germans, about nuclear power', *Nucleonics Week*, 12 Jan. 1978.
[106] *China Today* (note 102), p. 14, in translation.
[107] *China Today* (note 102), p. 15, in translation.
[108] MacLachlan and Hibbs (note 101).
[109] MacLachlan and Hibbs (note 101).
[110] Defense Intelligence Agency (note 104).

of the country, far from the more vulnerable coasts and borders. This pattern was also followed in the case of plutonium production (see chapter 3).

Gas-centrifuge plants

China has conducted R&D on gas centrifuges since 1958. For years, Chinese officials have stated that China might build a gas-centrifuge or laser enrichment plant to replace the gaseous-diffusion plants.[111]

Recently, Russia agreed to supply China with a gas-centrifuge plant to produce low-enriched uranium. The first part of the plant, currently being built in Shaanxi province, is scheduled to be operational in late 1996 or 1997, producing 200 000 SWU per year.[112] Eventually, the plant could reach 500 000 SWU per year. The centrifuges are reportedly not the most advanced Russian machines but are older, surplus models.

Inventory of weapon-grade uranium

Historical production

China's inventory of weapon-grade uranium can be roughly estimated from the above information. We assume that production in the first plant averaged about 50 000 SWU per year from 1964 until the end of 1975, at which time the capacity was increased to 180 000 SWU per year over a three-year period. During an expansion which we assume started in 1980 and lasted until 1985, the plant's capacity linearly increased further to 300 000 SWU per year. Under these assumptions the first gaseous-diffusion plant is estimated to have produced about 3.1 million SWU to the end of 1987.

The output of the second plant is highly uncertain. We assume that it started operating in 1975 at a low capacity and increased linearly to 200 000–400 000 SWU per year in 1987. Total output of this plant is estimated at 1.2–2.4 million SWU to the end of 1987.

Since uncertainties about output at both plants are high, we estimate total production as a range of 4–6 million SWU.

Non-weapon uses

The enrichment programme would have supplied about 1 million SWU of enriched uranium for several non-weapon purposes, primarily fuel for research, production and naval reactors, and for nuclear weapon tests:

---

[111] MacLachlan and Hibbs (note 101).
[112] 'China finds Russian nuclear services cheaper than western counterparts', *Post-Soviet Nuclear Complex Monitor*, 28 June 1994; and Hibbs, M., 'Russian centrifuge plant in China to be finished, operating next year', *Nuclear Fuel*, 25 Sep. 1995.

1. China's research reactors would have required 165 000–230 000 SWU by the end of 1987, when HEU production for weapons ceased.[113]

2. Production of LEU for five nuclear-powered submarines and one land-based prototype would have required roughly 150 000–200 000 SWU.[114]

3. About 900–1350 kg of weapon-grade uranium would have been consumed in 45 nuclear weapon tests, where we assume an average of 20–30 kg per test. We assume that China's nuclear weapons depended more on HEU than other weapon states. At a tails assay of 0.3 per cent, this amount of weapon-grade uranium would correspond to 180 000–270 000 SWU. We use the mid-point of 225 000 SWU.

4. China's graphite-moderated, water-cooled production reactors are believed to use a fuel cycle similar to US production reactors. If China produced 4 tonnes of weapon-grade plutonium, fuel for these reactors would have required about 160 000 SWU where slightly enriched fuel is used.[115] Alternatively, natural uranium could have been used as fresh fuel and the uranium recovered after reprocessing the irradiated fuel could have been enriched up to weapon-grade uranium. In this case, the enrichment level of the feed would have been less than that of natural uranium, about 0.65 per cent for this type of production reactor and a burnup of about 600 MWd/t. Production of weapon-grade uranium (93 per cent ^{235}U) would have therefore required about 5 per cent more SWU per kilogram of product. In total the production of weapon-grade uranium using all the recovered uranium as feed would have required roughly an additional 85 000 SWU assuming a tails assay of 0.30 per cent.

5. Process losses of 3 per cent would have consumed about 150 000 SWU. A working inventory, equivalent to five months' supply, would have required another 250 000 SWU.

Inventory for weapons

Subtracting these requirements, we are left with about 3–5 million SWU, or enough for between 15 and 25 tonnes of weapon-grade uranium equivalent at an average tails assay of 0.3 per cent. This number remains very uncertain.

China is estimated to have stockpiled roughly 300 deployed weapons on launchers and another 150 tactical weapons available but not deployed.[116] This estimate is uncertain and the actual number could be significantly larger. Its weapons are believed to depend heavily on weapon-grade uranium, so we esti-

---

[113] IAEA (note 68). Combined, the research reactors could have produced about 270 000–380 000 MWth-d of heat, assuming a 50–70% capacity figure. Two-thirds of this heat was produced by the 125 MWth reactor at the Reactor Operation Institute that uses weapon-grade uranium fuel. Assuming that about 40% of the ^{235}U is fissioned or converted into ^{236}U, fuel for these reactors would require about 165 000–230 000 SWU.

[114] According to *China Today*, Part II (note 102), Chinese submarine reactors use low-enriched uranium fuel. We assume that a typical naval core contains about 1400 kg of 5% enriched uranium fuel (70 kg of ^{235}U). Assuming that a total of 15–20 cores have been fabricated by the end of 1987, total requirements would be 21 000–28 000 kg of 5% enriched uranium, or 150 000–200 000 SWU.

[115] Cochran *et al.* (note 2).

[116] Norris *et al.* (note 89).

mate that each weapon contains on average 20–30 kg of weapon-grade uranium. China would therefore need about 9.0–13.5 tonnes of weapon-grade uranium to build about 450 nuclear weapons.

Part III
Principal civil inventories

5. Plutonium produced in power reactors

I. Introduction

Plutonium production is the inevitable consequence of irradiating uranium in a nuclear reactor. It arises primarily through neutron capture by ^{238}U to form ^{239}U and then, via decay, ^{239}Pu (see chapter 2). Heavier isotopes of plutonium, from ^{240}Pu to ^{243}Pu, are produced by further neutron captures. Only ^{239}Pu and ^{241}Pu are fissile when irradiated with thermal neutrons and have an energy value as fuel in today's conventional thermal power reactors.

The rate at which plutonium is produced in nuclear fuel depends on the reactor type and its characteristics (moderator, coolant and fuel type), the enrichment of the fuel, where the fuel was located in the reactor and how the reactor is operated. In general, the longer fuel is irradiated, the greater the amount of plutonium contained within it. Fuel in commercial power reactors is typically irradiated for longer periods than fuel in military plutonium production reactors. Increased irradiation times change the isotopic composition of the plutonium, with lower proportions of the fissile isotopes ^{239}Pu and ^{241}Pu in more highly irradiated fuel. Plutonium in fuel discharged from a commercial reactor will typically have a fissile content of about 70–74 per cent (see table 2.1). Weapon-grade material is defined as containing less than 7 per cent ^{240}Pu.

Detailed information on reactor fuelling is not routinely published in most countries. Moreover, even with such information, there will always be some uncertainty about the amount of plutonium produced in irradiated fuel. This is because of errors in measuring the heat output of a reactor and uncertainties in modelling plutonium production. For instance, operator estimates of plutonium production in British Magnox reactors have an accuracy in any one year of no better than plus or minus 5 per cent.[1] Only when a fuel element has been dissolved at a reprocessing plant and the plutonium within it assayed can an accurate measurement (that is, with an error margin of below plus or minus 0.5 per cent) be made of its plutonium content.

Estimates presented in this chapter closely follow the methods developed by the Systems Studies Section of the Department of Safeguards at the IAEA in a series of reports produced during the early and mid-1980s.[2] Fuelling, burnup and plutonium production histories have been modelled for each of the world's civil power reactors. In this chapter only summary data are presented, together with some impression of their reliability. Rather than producing annual

---

[1] Hinkley Point 'C' Inquiry, UK, Transcript of evidence day 147, UK, 1989, pp. 19–25.
[2] Bilyk, A., *Forecast of Amounts of Plutonium at Power Reactors Subject to Safeguards (1981–1990)*, STR-125 (IAEA Department of Safeguards: Vienna, June 1982); Mal'ko, M., *Estimation of Plutonium Production in Light Water Reactors*, STR-226 (IAEA Department of Safeguards: Vienna, Nov. 1986).

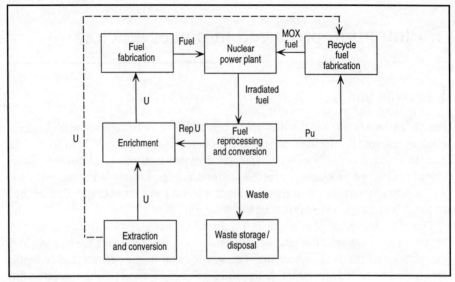

Figure 5.1. The nuclear fuel cycle including reprocessing

plutonium discharge figures, the emphasis is on decadal estimates and forecasts of discharges from 1970 to 2010.

For a number of countries, estimates of historical fuel and plutonium discharges have been verified in a new survey conducted by the authors in 1994. Utility operators were sent estimates of plutonium discharges prepared for *World Inventory 1992*, and asked to respond with actual data for the period up to the end of 1993.[3] Through this benchmarking exercise, and by using published data where they are available, it has been possible to verify about 55 per cent of historical world spent-fuel and plutonium discharges.[4] In addition the authors have good confidence in estimates of fuel and plutonium discharges from British reactors. These account for a further 18 per cent of world fuel discharges and about 7 per cent of plutonium discharges up to 1993. Since the survey demonstrated that at the national level the 1990 estimates were typically less than 10 per cent in error, the authors are now confident that the historical world estimates presented here are less than 5 per cent in error.

---

[3] Utilities and agencies from the following countries responded to the survey with data: Argentina, Brazil, Canada, the Czech Republic, Finland, France, Hungary, Italy, the Netherlands, Slovakia, Spain, Sweden, Switzerland and Taiwan.

[4] For US data see US Department of Energy/Energy Information Agency (EIA), *Spent Nuclear Fuel Discharges from U.S. Reactors 1992*, EIA Service Report, SR/CNEAF/94-01 (Department of Energy: Washington, DC, May 1994); and US DOE/EIA, *World Nuclear Capacity and Fuel Cycle Requirements 1993*, DOE/EIA-0436-93 (Department of Energy: Washington, DC, Nov. 1993). A variety of sources were used for British figures: Barnham, K. W. J. et al., 'Production and destination of British civil plutonium', *Nature*, vol. 317 (19 Sep. 1985), pp. 213–17; *Parliamentary Debates, House of Commons [Hansard]*, parliamentary answers; and the British Department of Energy and Department of Trade and Industry's 'Annual Plutonium Figures', published annually (1987–96).

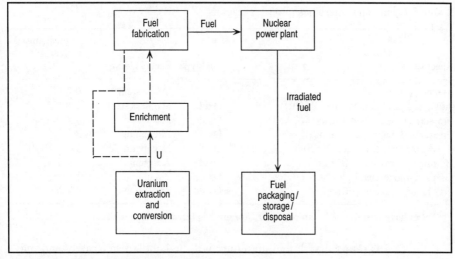

Figure 5.2. The once-through nuclear fuel cycle

With the passage of time, it also becomes clearer that the stagnation of nuclear capacity in many parts of the world is a deep-seated trend. Greater confidence can therefore also be attached to forecasts for future fuel and plutonium discharges provided in this chapter. The combined effects of increasing fuel burnups and reactor shut-downs mean that annual discharges of spent fuel and plutonium peaked in 1994–95. Thereafter a slow decline set in. The primary uncertainties associated with forecasts are over the rate of reactor decommissioning and the rate of reactor commissioning in Asia.

II. The fuel cycle in civil reactor systems

Nuclear fuel in reactors has to be replaced when its fissile content becomes depleted, or when the fuel matrix deteriorates as a result of the effects of heat or irradiation. In some gas-cooled, CANDU and RBMK reactors fuel may be replaced while the reactor is operating. This is known as 'on-load refuelling'. For most reactors, including the dominant LWR, each irradiation campaign is followed by a shut-down when a proportion of the fuel core is replaced.

On discharge, the fuel contains uranium, including a residual amount of fissile ^{235}U, plutonium, and waste fission and transuranic products. It is hot and highly radioactive. Fuel is normally stored under water at the reactor and allowed to cool, for at least one year at Magnox stations and at least three years at LWRs. There are then two options. The fuel may be sent for reprocessing, or it can continue to be stored pending final disposal as a waste or reprocessing in the longer term (see figures 5.1 and 5.2). For metallic fuels such as magnox fuels there is generally held to be no alternative to reprocessing. Ceramic oxide fuels with corrosion-resistant cladding can be safely stored under water or in air for periods of decades.

Table 5.1. Fuel characteristics of power reactors

Reactor	Fuel type	Fuel cladding	Typical initial enrichment (% ^{235}U)
Magnox (GCR)	Metallic	Magnesium	0.7
Advanced gas-cooled reactor (AGR)	Oxide	Stainless steel	2.1–2.5
Pressurized heavy water reactor (CANDU)	Oxide	Zirconium	0.7
Pressurized water reactor (PWR)	Oxide	Zirconium	3.2–3.8
Pressurized water reactor (VVER)	Oxide	Zirconium	3.6
Boiling water reactor (BWR)	Oxide	Zirconium	2.6–2.8
Graphite-moderated light water reactor (RBMK)	Oxide	Zirconium	2.4–3.0
Fast breeder reactor (FBR)	Oxide	Zirconium	15–20[a]

[a] FBRs have been fuelled with either HEU or mixed plutonium/uranium fuel.

If the fuel is reprocessed, the plutonium will be handled separately and may be recycled as fuel. If the fuel is not reprocessed, the plutonium will remain locked in the fuel matrix.

III. Fuelling strategy and fuel burnup

The fissile content of uranium is exploited differently across the range of conventional thermal reactor types. A common feature of these reactor designs is that neutron speeds are slowed to thermal velocities in the reactor core through their interaction with a moderator. This sets them apart from 'fast' reactors in which neutron velocities are not reduced. At slower velocities the likelihood of fission occurring in fissile atoms is enhanced (see chapter 2, section III). The main reactor systems treated here are listed in table 5.1 together with their typical fuel characteristics.

Commercialized nuclear reactor systems use either natural or enriched uranium. The chief difference between these reactor systems is their power density—the amount of heat produced per unit volume of the core. LWRs (fuelled with enriched uranium) typically have power densities five times greater than heavy water reactors (natural uranium) which, in turn, have higher power densities than gas-cooled Magnox reactors (natural uranium). LWRs are therefore more compact, with smaller fuel cores. A pressurized water reactor (PWR) with an electric power rating of 1000 MWe typically has a core of uranium fuel weighing about 90 tonnes, whereas an equivalent Magnox station would have a core of about 1000 tonnes. Per unit of electricity produced, about 10 times as much fuel is discharged from a Magnox reactor as from an LWR.

Differences in power density have an important effect on fuel design and on plutonium production. Lightly irradiated natural uranium fuel (discharged from gas-cooled reactors) contains less plutonium, per kilogram, than more highly irradiated enriched uranium fuel. It is also less contaminated with the higher-

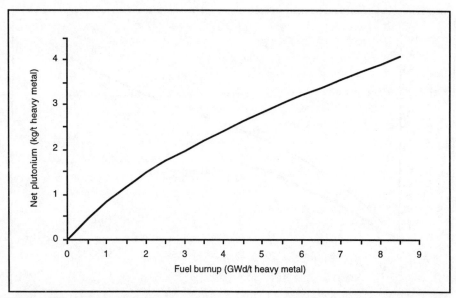

Figure 5.3. Specific plutonium production as a function of fuel burnup, natural uranium fuel

Sources: Bilyk, A., International Atomic Energy Agency, *Forecast of Amounts of Plutonium at Power Reactors Subject to Safeguards (1981–1990)*, STR-125 (IAEA Department of Safeguards: Vienna, June 1982); and Turner, S. E. et al., *Criticality Studies of Graphite-Moderated Production Reactors*, Report prepared for the US Arms Control and Disarmament Agency, SSA-125 (Southern Sciences Applications Inc.: Washington, DC, Jan. 1980).

numbered isotopes of plutonium. The standard measure of fuel irradiation is burnup. Fuel burnup is the cumulative thermal power generated per unit weight of fuel (usually measured in megawatt-days per tonne of fuel, MWd/t). A simple relationship exists between burnup and the fissile content of irradiated fuel. About 1 gram of fissile material is required to generate 1 megawatt-day of thermal power (see appendix B). This can be translated into the relationship between fuel burnup and the amount of plutonium remaining in irradiated fuel, as shown in figures 5.3 and 5.4.

Natural uranium-fuelled reactors generally operate at lower burnups than those fuelled with enriched uranium: magnox fuel burnups typically range from 3000 to 5000 MWd/t; CANDU burnups from 6000 to 8000 MWd/t; and AGRs, fuelled with slightly enriched fuel, from 18 000 to 21 000 MWd/t. Fuel in a typical PWR is enriched to 3.25 per cent ^{235}U and irradiated to 30 000 MWd/t during the course of three annual campaigns (one-third of the reactor core being replaced each year). Boiling water reactor (BWR) fuel typically has an initial enrichment of 2.6–2.8 per cent and achieves burnups of 28 000 MWd/t.

In recent years PWR and BWR operating experience and improved fuel technology have led utilities to increase fuel burnups. This may be achieved in two

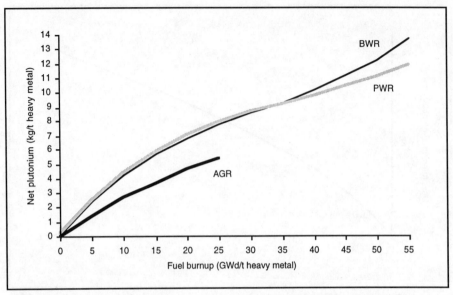

Figure 5.4. Specific plutonium production as a function of fuel burnup, enriched uranium fuel[a]

[a] BWR = boiling water reactor; PWR = pressurized water reactor; AGR = advanced gas-cooled reactor.

Source: Mal'ko, M., International Atomic Energy Agency, *Estimation of Plutonium Production in Light Water Reactors*, STR-226 (IAEA Department of Safeguards: Vienna, Nov. 1986).

ways: by increasing the length of time between refuelling or by reducing the number of assemblies replaced during each campaign (one-quarter, one-fifth, or one-sixth core, rather than one-third). Initial enrichments must be raised in each case, although by slightly less in the second option. Depending on the approach followed, average PWR burnups may reach 50 000 MWd/t by early next century, while for BWRs they could increase to 40 000 MWd/t. By 1995 mean PWR burnups in the United States and Europe were nearly 40 000 MWd/t, while BWR burnups averaged about 32 000 MWd/t.

IV. A sketch of methods

Information on reactor capacities, core weights, nominal thermal efficiencies and annual electrical outputs is widely available for most of the world's reactors.[5] Usually the only available reliable time-series evidence on reactor operation is the monthly, annual and cumulative electrical output. This can be used to calculate total heat output by the reactor, from which plutonium pro-

[5] Nuclear Engineering International, *World Nuclear Industry Handbook 1995* (Reed Business Publishing Group: London, 1994) publishes reliable data on reactor characteristics; *Nucleonics Week* (McGraw-Hill: New York) publishes tables of monthly power outputs for all operating commercial reactors, bar those in Lithuania.

duction can be estimated. In general, it is assumed here that reactor operators have sought to optimize their use of fuel by maximizing fuel irradiation over the long term. Heat output is calculated by dividing electrical output by the efficiency of the reactor (usually around 30 per cent). If a reactor is at equilibrium, fuel burnup can be inferred by dividing this heat production (measured in megawatt-days of thermal energy, MWth-d) by the amount of fuel discharged from the reactor (either by using published information or by making an assumption about the rate of fuel discharge—between one-sixth and one-half of the core per annum, depending on the reactor type and utility fuelling strategy). By using the conversion factors shown in figures 5.3 and 5.4, plutonium production can then be deduced.

A more detailed description of the range of methodologies used to derive plutonium discharge figures is given in appendix B. While plutonium discharges were modelled for all reactors, not all estimates could be corroborated against actual discharges.

The best situation is where modelled estimates of fuel and plutonium discharge could be corroborated and adjusted, as for countries participating in the 1994 Science Policy Research Unit (SPRU) Fuel and Plutonium Survey. Information from an earlier survey conducted by the authors of East European fuel-management policy was also used. In the absence of survey data, full use was made of published information. Nominal design equilibrium burnups for reactors in non-nuclear weapon states are published in IAEA publications and elsewhere, and these have been supplemented with country-specific knowledge of plans to increase fuel burnups.[6] Since *World Inventory 1992* appeared, information on reactor and fuel characteristics and on power outputs has been published regularly for most reactors in Central and Eastern Europe, as well as those in Russia and most other countries in the Commonwealth of Independent States (CIS). This has greatly improved confidence in fuel and plutonium discharge estimates for these countries.

For one reactor type (the RBMK) few reliable data exist on fuel burnup and the relationship between fuel burnup and plutonium concentration. For these reactors it has been necessary to fall back on a cruder method based on standardized conversion factors which match electrical power output to plutonium production (for example, RBMK reactors are assumed to generate 300 kg of plutonium per GWe(net)-year).

Clearly, the degree of confidence in estimates varies depending on the method used and whether estimates have been reviewed by operators. In most cases, estimates for countries not participating in the survey are believed to be correct to within plus or minus 10 per cent.

In making forecasts, the authors have assumed that classes of reactor (PWR, BWR, RBMK, and so on) operate at standard load factors. We have chosen to

---

[6] Nominal design fuel burnups are published in a number of places, e.g., IAEA, *Operating Experience with Nuclear Power Stations in Member States in 1992*, STI/PUB/951 (IAEA: Vienna, 1993).

Table 5.2. Past discharges of spent fuel from nuclear power reactors, to 31 December 1993

Figures are in tonnes of spent fuel. *Denotes benchmarking in the SPRU Fuel and Plutonium Survey 1994.

Country	1960–70	1971–80	1981–90	1991–93	Total
Argentina*	0	360	1000	450	1 810
Armenia	0	10	260	80	350
Belgium	0	200	1080	380	1 660
Brazil*	0	0	30	30	60
Bulgaria	0	110	580	170	860
Canada*	50	3 790	10 780	4 330	18 950
China	0	0	0	0	0
Czech Republic*	0	0	180	140	320
Finland*	0	50	770	210	1 030
France*	370	5 550	10 610	4 100	20 630
Germany	20	940	4 040	1 420	6 420
Hungary*	0	0	310	170	480
India	0	190	560	300	1 050
Italy*	460	640	600	80	1 780
Japan	200	1 260	6 230	2 890	10 580
Korea, South	0	20	1 350	930	2 300
Lithuania	0	0	440	430	870
Mexico*	0	0	0	60	60
Netherlands*	0	90	140	50	280
Pakistan	0	50	50	30	130
Russia	260	1 390	5 730	1 670	9 050
Slovakia*	0	0	420	150	570
Slovenia	0	0	120	50	170
South Africa	0	0	160	100	260
Spain	0	900	1 780	810	3 490
Sweden*	0	310	1 840	650	2 800
Switzerland	0	320	790	260	1 370
Taiwan*	0	70	940	340	1 350
Ukraine	0	50	1 510	840	2 400
United Kingdom	4 200	9 720	9 010	3 110	26 040
United States*	440	6 590	15 150	6 390	28 570
Total	**6 000**	**32 610**	**76 460**	**30 620**	**145 690**

Source: SPRU Spent Fuel and Plutonium Database 1994.

adopt the same factors used in the IAEA Actinide Database.[7] To calculate spent-fuel and plutonium discharges, assumptions have to be made about the evolution of fuel burnups in different countries. This affects both the amount of spent fuel and the amount of plutonium which is forecast to be produced.

[7] Chantoin, P. and Pecnik, M., 'Actinide database and fuel cycle balance worldwide', Paper presented at the American Nuclear Society Summer Meeting, San Diego, Calif., 20–24 June 1993.

Utilities in some countries, such as France and Germany, look set to pursue higher burnups aggressively (leading to lower spent-fuel and plutonium discharges), while those in others, such as Japan, will move more slowly. Where specific information about burnup planning is to hand (as for France) it has been used. For those countries with plans to increase fuel burnup but for which no specific data are to hand, a set of standard increases is assumed.[8] All data are presented in this chapter as single figures. No sensitivity analysis has been conducted on estimates, whether pertaining to historical or future discharges.

V. Discharges of spent fuel and plutonium from civil reactors

At the end of 1995 installed world nuclear capacity was 356 GWe, having tripled over the previous 15 years. By then, the world's commercial reactors had discharged a total of some 166 000 tonnes of spent fuel, containing nearly 990 tonnes of total plutonium. Spent fuel is currently arising at a rate of around 10 300 tonnes each year, containing about 70 tonnes of plutonium. It is estimated that at the end of 1995 an additional 120 tonnes of plutonium were held in partially irradiated fuel in reactor cores. In the first quarter of 1996 the discharge of plutonium from the world's power reactors therefore passed the 1000-tonne mark.

Past spent-fuel discharge estimates up to the end of 1993 are set out country-by-country in table 5.2. Past discharges of plutonium are set out in table 5.3. Plutonium had been generated in power reactors in 31 countries (excluding the former German Democratic Republic) by the end of 1993. Of these, 19 countries had discharged more than 5 tonnes of plutonium from power reactors. By far the largest amount had been discharged by power reactors in the USA, some 236 tonnes. Only minor changes have been made to the historical fuel discharge figures presented in *World Inventory 1992*. A more significant change has been made to bring down plutonium discharge estimates. This is almost entirely explained by a correction made to plutonium discharge rates for reactors in the former USSR.[9]

Plutonium is currently held in a variety of forms and at several sites in each country. Most plutonium remains embedded in spent fuel and is stored at reactor sites. Of the material removed from reactor sites, a large proportion has crossed international boundaries to be transported to reprocessing plants in France, the UK and Russia. Much of this plutonium is subsequently separated from the fuel, and may then have been recycled as fuel. Table 5.3 is

---

[8] Burnup increases in other countries are assumed to happen only in reactors which operate throughout a given decade (i.e., 2001–10). Belgium, Germany, Japan, Sweden and Switzerland are assumed to raise mean PWR fuel burnups to 40 GWd/t in 1991–2000 and to 45 GWd/t in 2001–10. Mean BWR fuel burnups are assumed to rise from a mean of 30 GWd/t in 1991–2000 to 35 GWd/t in 2001–10. Mean LWR burnups in South Korea, Spain and Taiwan are taken to come into line after 2000. Before that, design burnups are assumed.

[9] Reactors in Armenia, Lithuania, Russia and Ukraine are now estimated to have discharged about 63.6 t of plutonium to the end of 1990, rather than 78.1 t as estimated in *World Inventory 1992*.

Table 5.3. Past discharges of plutonium from nuclear power reactors, to 31 December 1993

Figures are in kilograms of total plutonium. *Denotes benchmarking in the SPRU Fuel and Plutonium Survey 1994.

Country	1961–70	1971–80	1981–90	1991–93	Total
Argentina*	0	1 100	3 350	1 520	**5 970**
Armenia	0	90	1 770	180	**2 040**
Belgium	0	1 770	9 730	3 750	**15 250**
Brazil*	0	0	250	270	**520**
Bulgaria	0	940	4 960	1 460	**7 360**
Canada*	170	12 630	39 220	15 210	**67 230**
China	0	0	0	0	**0**
Czech Republic*	0	0	1 530	1 170	**2 700**
Finland*	0	330	6 640	1 900	**8 870**
France*	430	11 900	67 820	30 830	**110 980**
Germany	140	8 190	37 280	13 690	**59 300**
Hungary*	0	0	2 260	1 500	**3 760**
India	0	1 050	2 390	1 060	**4 500**
Italy*	1 060	2 210	2 070	330	**5 670**
Japan	400	7 840	49 000	22 760	**80 000**
Korea, South	0	150	8 240	6 280	**14 670**
Lithuania	0	0	3 150	1 890	**5 040**
Mexico*	0	0	0	130	**130**
Netherlands*	30	720	1 220	390	**2 360**
Pakistan	0	150	170	90	**410**
Russia	1 660	8 930	36 800	11 390	**58 780**
Slovakia*	0	0	3 030	1 070	**4 100**
Slovenia	0	0	1 030	430	**1 460**
South Africa	0	0	1 450	890	**2 340**
Spain	0	3 280	10 570	5 520	**19 370**
Sweden*	0	2 540	15 370	5 390	**23 300**
Switzerland	0	2 540	6 820	2 440	**11 800**
Taiwan*	0	390	7 540	3 050	**10 980**
Ukraine	0	290	10 940	7 260	**18 490**
United Kingdom	6 650	20 480	26 100	9 120	**62 350**
United States*	2 210	45 660	131 310	57 290	**236 470**
Total	**12 750**	**133 180**	**492 010**	**208 260**	**846 200**

Source: SPRU Spent Fuel and Plutonium Database 1994.

therefore not a picture of the physical distribution of plutonium at the end of 1993, although it does represent a rather good picture of the distribution of plutonium.

Making forecasts of spent-fuel and plutonium production in the future is inevitably prone to some uncertainty, even with a relatively stable number of reactors operating. Table 5.4 shows that, by the end of the year 2000, nearly 1400 tonnes of plutonium are expected to have been discharged from power reactors (850 t plus 540 t). This will grow to a little over 2100 tonnes of plutonium by the end of 2010. These forecasts are similar to those presented in

Table 5.4. Estimated discharges of spent fuel and plutonium from nuclear power reactors, 1994–2000 and 2001–10

Figures are in tonnes of spent fuel and kilograms of total plutonium.

Country	Spent-fuel discharges 1994–2000	Plutonium discharges 1994–2000	Spent-fuel discharges 2001–10	Plutonium discharges 2001–10	Total spent-fuel discharges 1994–2010	Total plutonium discharges 1994–2010
Argentina	1 460	4 980	2 040	7 220	3 500	12 200
Armenia	0	0	0	0	0	0
Belgium	870	8 750	920	9 180	1 790	17 930
Brazil	170	1 070	320	3 330	490	4 400
Bulgaria	400	3 400	770	7 520	1 170	10 920
Canada	10 070	35 500	17 830	63 770	27 900	99 270
China	400	2 760	650	5 830	1 050	8 590
Czech Republic	320	2 730	350	3 490	670	6 220
Finland	520	4 680	510	5 170	1 030	9 850
France	9 250	76 830	11 620	120 390	20 870	197 220
Germany	3 550	59 300	3 810	32 810	7 360	92 110
Hungary	480	4 200	370	3 720	850	7 920
India	1 560	5 060	4 750	16 060	6 310	21 120
Italy	370	1 340	0	0	370	1 340
Japan	8 070	61 290	9 080	84 660	17 150	145 950
Korea, South	2 880	17 500	5 660	32 370	8 540	49 870
Lithuania	1 440	4 410	1 290	5 880	2 730	10 290
Mexico	300	2 120	250	2 580	550	4 700
Netherlands	100	880	80	580	180	1 460
Pakistan	60	220	120	560	180	780
Russia	4 110	28 830	6 820	34 780	10 930	63 610
Slovakia	450	3 110	750	7 430	1 200	10 540
Slovenia	110	1 000	110	1 180	220	2 180
South Africa	230	2 090	340	3 510	570	5 600
Spain	1 760	13 560	1 320	12 860	3 080	26 420
Sweden	1 510	12 590	2 150	19 340	3 660	31 930
Switzerland	690	5 900	560	4 960	1 250	10 860
Taiwan	990	8 380	980	10 010	1 970	18 390
Ukraine	1 860	16 450	2 670	22 320	4 530	38 770
United Kingdom	6 700	21 320	8 120	24 080	14 820	45 400
United States	13 400	127 190	19 000	195 730	32 400	322 920
Total	**74 080**	**537 440**	**103 240**	**741 320**	**177 320**	**1 278 760**

Source: SPRU Spent Fuel and Plutonium Database 1994.

World Inventory 1992, although changes in detail have been made. World plutonium discharges for the decade 2000–10 have risen by about 4 per cent owing to higher fuel burnups in LWRs.

In making these forecasts, all reactors are assumed to operate smoothly for their design lifetimes. This projection could be a slight underestimate. On the one hand, reactor lives may be extended. On the other, it is conceivable that currently unplanned reactors could be operating by 2010, especially in East and South Asia. Against this, it is possible that some reactors will be retired

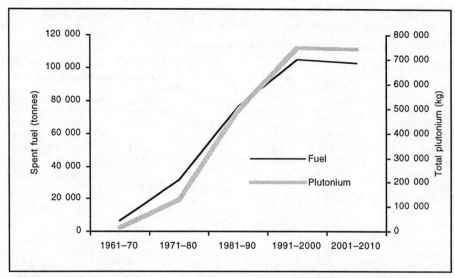

Figure 5.5. World spent-fuel and plutonium discharges from power reactors by decade, 1961–2010

early, especially in the United States on economic grounds and in the former Soviet Union and Eastern Europe on safety grounds. Overall, neither effect is likely to have a significant impact, and unless there is a dramatic change of nuclear policy the authors have good confidence in the forecasts presented here. The overall trend in spent-fuel and plutonium discharges from power reactors is shown in figure 5.5. World annual fuel discharges from power reactors reached a peak in 1995–96 and are set to decline slowly thereafter. Plutonium discharges show a similar but gentler decline from the late-1990s as plutonium concentrations in fuel rise with increasing burnups, so compensating for slightly lower fuel discharges. In the post-2000 period, the combined effect of reactor decommissioning and higher fuel burnups in LWRs will lead to a sharpening decline in spent-fuel and plutonium discharges. This overall pattern is shown in figure 5.5.

The geography of spent-fuel discharges

Spent-fuel discharges have historically been dominated by just six countries, and this looks set to continue in the next two decades. In two of these, nuclear power systems are based on natural uranium-fuelled reactors—Canada and the UK—while the rest are those with the largest LWR programmes—the USA, France, Japan and Russia. To the end of 1993, spent-fuel discharges in these countries accounted for about 78 per cent of the world total (they will still account for over 70 per cent of the world cumulative total in 2010). Spent-fuel discharge profiles by decade for these countries are given in figure 5.6.

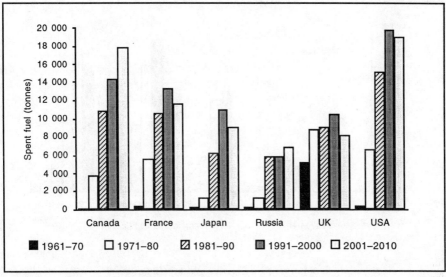

Figure 5.6. Spent-fuel discharge profiles by decade for Canada, France, Japan, Russia, the UK and the USA, 1961–2010

Figure 5.6 shows that, individually, countries show distinct fuel discharge patterns. LWR-based programmes show the greatest similarity with a sharp peak in the 1990s, followed by a decline. The scale of Canadian spent-fuel discharges is explained by the operation of heavy water, natural uranium-fuelled CANDU reactors which generate large quantities of spent fuel. In the decade 1981–90, over 40 per cent of North American spent fuel came from CANDU reactors (10 780 t out of 25 900 t), although they account for only about 10 per cent of the region's nuclear-generating capacity. The future Canadian profile is tied more rigidly to total capacity since there is limited scope for improvements in CANDU fuel burnups.

The significance of British fuel discharges is also the result of having a long-established nuclear programme based on reactors which use natural uranium. While gas-cooled reactors represented just 9 per cent of installed nuclear capacity in 1980 worldwide, they had generated over 55 per cent of the world's spent fuel (about 20 400 t).[10] Most of these reactors (nearly 80 per cent of gas-cooled capacity) were located in the UK. As other nuclear programmes based on water-cooled reactors have grown, so the overall significance of magnox fuel discharges has fallen.

The Russian fuel discharge profile is explained by the scale of its nuclear programme and the relatively large fuel requirements of its RBMK reactors. Future trends are highly sensitive to reactors which are currently under construction coming on-line. Without this new capacity, Russian fuel discharges will also begin to fall in 2000–10, not rise as shown in figure 5.6.

[10] AGRs had discharged just 250 t of spent fuel by 1980.

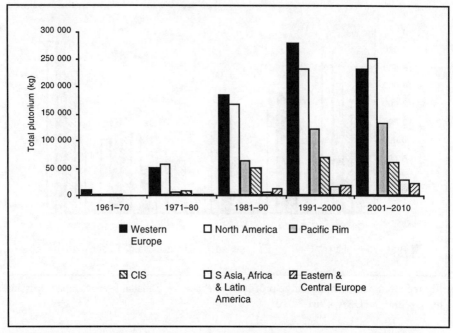

Figure 5.7. Plutonium discharges from power reactors in six regions, 1961–2010

The geography of plutonium arisings

Because plutonium production is more closely tied to the production of thermal power in nuclear reactors, world plutonium arisings better reflect the distribution of nuclear power capacity. This means that plutonium arisings are more evenly distributed than are spent-fuel arisings.[11] In figure 5.7 plutonium discharges from power reactors are shown aggregated into six regions: Western Europe, Central and Eastern Europe, North America, the Pacific Rim, the CIS, and South Asia, Africa and Latin America.

The figure shows that Western Europe and North America (including Mexico) have been and will remain the main producers of power-reactor plutonium. Up to the end of 1993 about 74 per cent of cumulative world production had been generated in these two regions. Asian countries (Japan, the Republic of Korea, Taiwan, China) have generated smaller amounts which will remain below 20 per cent of the world total in 2010. The other main future trend is a sharp decline in plutonium discharges in Western Europe—resulting primarily from the expected closures of British, German and Swiss power reactors in the decade 2001–10.

---

[11] The 6 countries with 78% of spent fuel discharges account for about 65% of plutonium discharges.

VI. Conclusions

Once discharged, spent fuel must be stored under safe and secure conditions. Typically fuel is stored in a water pool at the reactor site. There are then two options for the operator: to continue to store the fuel by extending storage capacity on-site or by gaining access to capacity 'away from the reactor' (AFR); or to send it for processing to a reprocessing plant. Historical and forecast rates of reprocessing are assessed in chapter 6. The AFR option is available in Sweden, and is being planned in several other countries. In many countries, the provision of extended spent-fuel storage capacity elicits sharp political controversy, usually because no consensus has been reached over nuclear or nuclear waste policy. In countries where fuel is to be stored the policy is eventually to dispose of spent fuel as a waste in geological repositories. The Netherlands is currently the only exception, having implemented an 'isolate, contain, monitor' (ICM) policy as an alternative to long-term disposal. In chapter 6 we estimate that over two-thirds of fuel discharged from power reactors up to 2010 will be left in storage and not reprocessed.

While in store, spent fuel continues to pose safety, environmental and proliferation risks. Accidents or diversions of material are possible up to the point at which the fuel is finally disposed of underground and could even occur after that. Institutional supervision of spent-fuel wastes above ground is likely to last several decades and could go on for much longer. The main justification for postponing disposal is that the decay in heat generated by the spent fuel will make geological disposal less hazardous and cheaper. Throughout the long periods in which spent fuels may be held in store, they should be subjected to strict safeguarding and physical protection. Indeed, as the fuel cools and radiation decays, the diversion risk could increase, as fissile materials become more accessible.

Spent fuel will continue to accumulate at a rate of 9000–10 000 tonnes per year for the next two decades. Each year a further 75 tonnes of plutonium will be added to the world total. The overwhelming majority will be produced in Europe, North America and in Asia, although the amounts now being generated in transitional and less developed countries are reaching the hundreds of tonnes of plutonium. Maintaining control over this material presents a major challenge for safeguards regimes.

6. Reprocessing programmes and plutonium arisings

By the end of 1995 about 990 tonnes of plutonium had been discharged from the world's power reactors. Most of this material remained fixed in spent fuel, but a substantial and growing amount had been separated. In this chapter estimates are presented of how much plutonium has been separated at reprocessing plants from power-reactor fuel, and how much is due to be separated over the period to 2010.

I. Reprocessing in the nuclear fuel cycle

Reprocessing is the chemical separation of plutonium (0.2 to 1 per cent by weight of the irradiated fuel) and uranium (over 95 per cent by weight of the fuel) from the fission products and transuranic wastes contained in spent nuclear fuel. Historically, three main justifications have been put forward for spent fuel reprocessing.

First, it makes available fissile materials which can be recycled as fuel in thermal and fast reactors. The recycling of plutonium has often been seen as desirable on grounds of energy security.

Second, reprocessing is claimed to be a safer way of dealing with spent fuel in the long term. This may be especially so with metallic fuels whose cladding corrodes rapidly when stored under water. Splitting the fuel from its cladding and dealing with the fission and transuranic waste products while separating plutonium and uranium from the waste stream may be safer than trying to deal with the untreated fuel rods.

Third, fuel is taken off the reactor site where it would otherwise accumulate. In practice this has become the major driving force behind most utilities' reprocessing policies. Reactor operators have been under legal and logistical pressure to solve the problem of managing spent fuel through resort to reprocessing, so long as the cost does not greatly affect the overall cost or public acceptability of nuclear electricity. Typically, a utility reactor operator will have a contract with a fuel services company (such as Cogema in France or British Nuclear Fuels (BNFL) in the UK) for a specified amount of spent fuel to be reprocessed. The components of the fuel (plutonium, uranium and waste products) normally remain the property and responsibility of the utility, and in due course they are returned.

If plutonium and uranium are to be reused, they must be in a chemically pure form, free of fission products and other materials. In reprocessing, fuel cladding is first removed from the fuel, either mechanically (as with metallic fuel) or chemically (as with oxide fuel) and the fuel is dissolved in nitric acid. It is at this stage that the first accurate measurement can be made of the amount of plutonium contained in the fuel. Radiations emitted by spent fuel can be measured by non-destructive assay methods, but these do not yield the same accuracy as weighing and analytical techniques applied to chemical solutions in the process areas of a reprocessing plant. The plutonium content in spent fuel cannot be measured directly, but is estimated by utilities using their knowledge of reactor operating histories and fuelling strategy.

In reprocessing, the dissolved fuel is put through a series of solvent extraction phases in which the fission products and transuranics are separated out first, followed by the uranium and plutonium (the reprocessing products). Uranium is usually made available to the customer as a nitrate, whereas plutonium is normally returned as an oxide.

There are inevitably losses of material during decladding and chemical processing. In the overall material balance of the Eurochemic plant in Belgium, which operated from 1966 until 1974, some 2.3 per cent of the uranium and 4.8 per cent of the plutonium were lost to waste streams.[1] At Sellafield it has been estimated that between 2 and 3.5 tonnes of plutonium are contained in a variety of plutonium-contaminated solid wastes which have accumulated at the site over the past 40 years.[2] A 1986 report by the British Government showed that about 1.8 tonnes of plutonium were held in stripped fuel cladding and other plutonium-contaminated materials (PCM) produced in magnox reprocessing.[3] These figures suggest losses of 3–6 per cent in magnox reprocessing.[4]

Losses have generally been reduced with improved technology, better housekeeping and tighter regulatory control on waste management. Today overall plutonium losses at oxide fuel reprocessing plants lie between 0.1 and 0.5 per cent of the input total. Modern oxide reprocessing plants are more efficient. In its first four years of operation the UP3 plant at La Hague achieved a separation efficiency of 99.88 per cent of uranium and plutonium.[5] In making estimates here losses of 0–2 per cent are assumed for oxide fuel reprocessing (depending on the plant) and of 2–5 per cent for magnox fuel reprocessing.[6]

---

[1] Detilleux, E., 'Operation of the plant and the period after shutdown', ed. W. Drent and E. Delande, *Proceedings of the Seminar on Eurochemic Experience, 9–11 June 1983*, ETR-318 (Eurochemic: Mol, Belgium, Apr. 1984), p. 59.
[2] Hinkley Point 'C' Inquiry, Transcript of evidence day 98, UK, 1989, p. 17.
[3] *Radioactive Waste Management Advisory Committee Seventh Annual Report* (HMSO: London, Sep. 1986), figure 3.
[4] These large losses are chiefly the result of the mechanical decladding of magnox fuel. Selective chemical decladding used for oxide fuels tends to reduce plutonium losses to <1%.
[5] Laurent, J. P. *et al.*, 'Reprocessing and plutonium recycling: The French view', Paper presented at the International Conference on the Nuclear Power Option, IAEA, Vienna, 5–8 Sep. 1994.
[6] For a discussion of magnox reprocessing losses see Barnham, K. W. J., 'Calculating the plutonium in spent fuel elements', ed. F. Barnaby, *Plutonium and Security* (Macmillan: London, 1991), pp. 110–32.

II. The evolution of fuel-cycle strategies

The 1960s

Irradiated fuel was first discharged from commercial power reactors in the early to mid-1960s in the USA (Dresden 1 discharged fuel beginning in 1961), the UK, the Soviet Union and France. In these early days, nuclear fuel remained the property of state authorities in charge of nuclear power development—the US Atomic Energy Commission (AEC), the United Kingdom Atomic Energy Authority (UKAEA), the Soviet Ministry of Atomic Power and Industry (MAPI)[7] and the Commissariat à l'Énergie Atomique (CEA). In all these cases power-reactor fuel was sent for reprocessing from the earliest date, with France and the UK leading the way.

In the UK and France reprocessing lines originally devoted to plutonium production for weapons were used to process fuel from commercial Magnox power reactors. In the early days, magnox fuel could not be stored for long periods, and rapid reprocessing was therefore an operational necessity. Some plutonium contained in this fuel was appropriated for weapon purposes (see chapter 3). By the early 1970s, as utilities began to assume greater control over their reactor programmes, ownership of spent fuel was passed to them in most Western countries. Since then a clearer distinction has gradually come to be drawn between civil and military inventories, although under existing safeguards arrangements nuclear weapon states are still permitted to move fissile materials into and out of safeguards.

To those few countries that purchased gas-graphite reactors from Britain and France, reprocessing services were offered as part of the reactor supply agreement. Under the British agreements plutonium and uranium were returned, if requested, to the customer (Italy and Japan) while, under a French 'take-back' agreement with Spain, fissile material and reprocessing wastes were retained by France. Up to the mid-1970s radioactive wastes remained the responsibility of the reprocessor. Plutonium arisings from commercial reactors in non-nuclear weapon states have generally been kept out of military programmes in nuclear weapon states.[8]

In parallel with the conversion of military reprocessing capacity to civil uses in the nuclear weapon states, a multinational venture to build a new oxide commercial reprocessing plant was being launched in Europe. Thirteen countries, all non-nuclear weapon states except for France, set up the Eurochemic company in the late 1950s, which rapidly established a reprocessing plant at Mol in Belgium. As a model for providing fuel management services, the appeal of multinational centres continues to be debated to the present day. The eventual failure of the Eurochemic venture, as France and Germany set out to establish

---

[7] Sometimes called the Ministry of Nuclear Power (MNP) and today called Minatom.
[8] The only exception may be Spain's Vandellos 1 Magnox reactor, which operated between 1972 and 1990. Plutonium separated at Marcoule from early fuel discharges from this reactor may have been used in the French nuclear weapon programme.

national reprocessing capacities in the early 1970s, demonstrates that maintaining the industrial and political coordination necessary for such centres is extremely difficult.

During the 1960s and early 1970s, with the anticipated rapid growth of nuclear power and a consequently growing need for reprocessing services, many countries other than the five nuclear weapon states developed plans for national civil reprocessing facilities. Such plans typically included a commitment to future large-scale plutonium recycling in fast reactors. In some cases they were bound up with a policy of establishing or keeping open a nuclear weapon option. Of these programmes, only four were successful in setting up reprocessing plants—those of Germany, India, Italy and Japan.

The 1970s

Between 1970 and 1980 world installed nuclear capacity rose from around 10 GWe to 120 GWe. During this period an assumption prevailed that fuel would be reprocessed, and most reactors were built with limited on-site fuel storage capacity. A typical Westinghouse pressurized water reactor, for instance, was supplied with storage capacity sufficient for about four years'-worth of fuel discharges, while the equivalent Soviet VVERs were built with five to six years' storage capacity. Provision therefore had to be made to remove fuel from the reactor site, usually to a reprocessing plant with large spent-fuel storage capacity. The exceptions were the Canadian CANDU pressurized heavy water reactors and the Soviet graphite-moderated light water (RBMK) reactors, for which long-term fuel storage was seen as the preferred option. India alone has reprocessed CANDU fuel.

Several important trends in the structure and politics of reprocessing services around the world were established during the 1970s. The first was that a European–Japanese fuel management regime was consolidated around plans to build large new oxide reprocessing plants at La Hague in France and Windscale (now Sellafield) in the UK. This became possible as the ambitions of smaller European countries to set up national or independent multinational reprocessing schemes foundered, and as the Japanese, German and US programmes were postponed or cancelled. Cogema in France and BNFL in Britain therefore became the sole providers of commercial reprocessing services, and utilities throughout Europe and Japan contracted to have their fuel reprocessed at La Hague and Windscale. Although both Britain and France sought to convert part of their magnox fuel reprocessing capacity to handle oxide fuel, only France succeeded, and thereby established a commercial and technological lead over the UK. The return of plutonium and wastes to the country of origin became a standard feature of reprocessing contracts during this period.

The second trend was the creation of a separate fuel management regime orchestrated by the Soviet Union. Nuclear fuel was supplied by MAPI to all reactors in the FSU and Eastern Europe, as well as two VVERs supplied to Finland. All VVER spent fuel was sent to Chelyabinsk (another converted military plant) for reprocessing following initial storage at the reactor. Under intergovernmental agreements this 'take-back' system for spent fuel was provided free of charge.[9] The plutonium separated was not returned, and was to have been used to fuel the Soviet fast-reactor programme. Substantial amounts of reprocessed uranium (RepU) have been recycled in Russia, as in the UK.[10] RepU was used to fabricate fuel for the Soviet RBMK reactor programme.

The third trend was a growing concern over the risks and rationale of reprocessing. As it became clear that civil reprocessing plants were emitting large amounts of radioactivity to the environment, concerns about local and regional health effects became the focus of a continuing controversy. Perhaps even more important, a growing and soon dominant body of opinion in the United States held that the separation of plutonium in civil reprocessing presented an unacceptable risk of nuclear proliferation. This led to the Carter Administration's policy calling for a world-wide ban on reprocessing.

At the same time, the model of a plutonium economy in which thermal reactors would make way for fast reactors fuelled with plutonium became increasingly discredited by industrial and economic reality. From the mid-1970s onwards, reduced expectations for the growth of nuclear electricity generation, the abundance of uranium and delays in commercializing fast reactors undermined the rationale for rapid and complete reprocessing. Doubts were also raised over the commercial viability of reprocessing by the failure of a privatized US reprocessing industry to get off the ground in the mid-1970s. Reprocessing of commercial reactor fuel was effectively abandoned in the United States after 1976.[11] A similar non-reprocessing policy was adopted by Sweden in 1980. Reactor operators in these countries have adapted to the change in policy by extending fuel storage capacities, either by stacking fuel more tightly into existing ponds, or by making available new capacity at and away from reactors.

The 1980s and 1990s

During the 1980s Japanese and European reactor operators persisted, in most cases, with the policies laid down in the 1970s. For a mixture of commercial, strategic, political and legal reasons, programmes which were increasingly regarded as uneconomic and burdensome were sustained. As a result, a majority of Japanese and West European spent fuel discharged by the year 2000 (c. 75 per cent of fuel containing about 40 per cent of discharged plutonium) is due to be reprocessed—most of it at La Hague and Sellafield, which have

[9] A 1966 fuel supply agreement with Hungary did not contain a 'take-back' provision. Ravasz, K., 'Russians agree to take back Paks spent fuel, but protests abound', *Nuclear Fuel*, 9 May 1994, pp. 9–10.

[10] Over 15 000 t of RepU have been fabricated into AGR fuel. Forsey, D. C. and Gresley, J. A. B., 'An outline of the requirements for successful uranium recycle', Paper presented at the *RECOD 87* conference, Paris, Aug. 1987.

[11] Three US reprocessing facilities, at West Valley, Barnwell and Morris, closed down or failed to come into operation during the mid-1970s. Rochlin, G., *Plutonium, Power and Politics* (University of California Press: Berkeley, Calif., 1979), pp. 104–105.

become major hubs in the global fuel service industry. The strong connection between European and Japanese fuel-cycle policies was reinforced. Almost two-fifths of LWR reprocessing capacity available to non-British and non-French utilities during the 1990s will be taken up with Japanese fuel.[12] At the same time, Japan is continuing to pursue a policy of independence in the nuclear fuel cycle, although construction of its first commercial reprocessing facility at Rokkasho-mura may not be completed until well into the next century.

Nevertheless, priorities within the European–Japanese fuel management regime have changed. Many utilities have partly or totally switched to a policy of fuel storage, rather than reprocessing. The legal change which enabled this in Germany was finally achieved in 1994, and this is only one of a succession of policy changes in Europe over the past five years. Storage and 'direct disposal' of spent fuel are now seen as safe and technically proven alternatives to reprocessing, while the costs of reprocessing and the value of plutonium and reprocessed uranium as a fuel have been widely questioned. A growing proportion of European and Japanese fuel will therefore be left in store at reactors or in central stores in future years.

The state of the Soviet–East European fuel management regime became far more precarious towards the end of the 1980s. Space constraints had already led to delays in the transfer of spent fuel from Eastern Europe to Chelyabinsk during the early part of the decade. In Czechoslovakia this led to the construction of a large centralized spent-fuel store during the mid-1980s. Nevertheless, the assumption held that fuel would be sent back to the Soviet Union. The major shock to the regime came with the unravelling of the Council for Mutual Economic Assistance (Comecon) trading system in 1988–90 and the breakup of the USSR. Nuclear trading relations were severely disrupted by these events. In 1988 the USSR began demanding dollar payments for fuel management services from its clients, while in 1991 the Russian Parliament passed an environmental protection law which effectively prohibited the import of foreign spent fuel into the country. Little fuel from non-Russian reactors has therefore been transferred to Chelyabinsk in recent years. As a result, East European and Ukrainian reactor operators have been forced to implement plans to rapidly increase spent-fuel storage capacities at reactors and at central stores.

The situation remains in flux. The Russian Ministry of Atomic Energy (Minatom) is keen to maintain and expand the reprocessing industry in Russia, especially by attracting business from West European and Asian reactor operators. Great political weight has been thrown behind the expansion of Russian reprocessing capacity. If political uncertainty in Russia subsides, and if attractive offers are made to potential customers (including, for instance, the retention of reprocessing wastes for disposal in Russia), Western utilities may

---

[12] Berkhout, F., Suzuki, T. and Walker, W., 'The approaching plutonium surplus: a Japanese–European predicament', *International Affairs,* vol. 66, no. 3 (July 1990), pp. 523–45.

Table 6.1. National spent-fuel management policies, 1960–2000 and beyond[a]

Country	1960s/1970s	1980s	1990s	Post-2000
Argentina	S + R(D)?	S	S	S
Armenia	R(D)	R(D)	TB	..
Belgium	R(D)	R(F+D)	R(F) + S	S + R(F)?
Brazil	S + R(D)?	S	S	S
Bulgaria	..	TB	TB + S	S + R(F)?
Canada	S	S	S	S
China	..	..	S	S + R(D)?
Czech Republic	..	TB + S	TB + S	S
Finland	..	TB + S	S + TB	S
France	R(D)	R(D)	R(D) + S	R(D) + S
Germany	R(F+D) + TB	R(F+D) + TB	R(F) + S	S + R(F)
Hungary	..	S + TB	S + TB	S
India	R(D)	R(D) + S	R(D) + S	R(D) + S
Italy	R(F)	R(F)	R(F)	..
Japan	R(F)	R(F+D)	R(F+D)	R(F+D) + S
Kazakhstan	S + R(D)	R(D)	R(F)	R(F) + S
Korea, South	..	S	S	S + R(F)?
Lithuania	..	S	S	S
Mexico	..	..	S	S
Netherlands	R(F)	R(F)	R(F) + S	S
Pakistan	S + R(D)?	S	S	S + R(D)?
Romania	..	..	S	S
Russia	R(D) + S	R(D) + S	R(D) + S	R(D) + S
Slovakia	..	S + TB	S + TB	S?
Slovenia	..	S	S	S
South Africa	..	S	S	S
Spain	TB	R(F) + TB	R(F) + TB + S	S
Sweden	R(F)	S	S	S
Switzerland	R(F)	R(F)	R(F) + S	S
Taiwan	..	S	S	S
Ukraine	..	S + R(D)	S + R(F)?	S + R(F)?
United Kingdom	R(D)	R(D)	R(D) + S	R(D) + S
United States	R(D) + S	S	S	S

[a] S = interim storage (either at or away from the reactor); R(D) = reprocessing (domestic); R(F) = reprocessing (foreign); TB = fuel returned to supplier under a 'take back' arrangement. For states that were Soviet republics before 1990, R(D) indicates reprocessing domestic to the Soviet Union.

be tempted to conclude reprocessing contracts with Minatom. A gradual merging of the two formerly separate fuel management regimes could then take place. Both Cogema and BNFL have been active in providing nuclear assistance to Russia since 1992. Chelyabinsk would then become the third node in a globalized fuel management regime potentially spanning the whole of Europe and Asia. In 1996 it seems unlikely that this will happen.

A summary of fuel management strategy in countries with major nuclear programmes is given in table 6.1. It shows the general shift towards storage policies, although in several countries there is uncertainty about the policies that will be followed in future.

III. A sketch of methods

Making estimates of the amount of plutonium separated at reprocessing plants is more problematic than calculating plutonium production in reactors. Complete historical information on fuel throughputs, plutonium concentrations and material losses at reprocessing facilities is usually not in the public domain. Instead, a number of assumptions have to be made, often based on partial or aggregated information. Three different approaches have been used.

1. In the best cases, annual throughputs of fuel and its mean burnups are known. For no plant is a full set of data available, but for some—UP1, UP2 and UP3 in France, Tokai-mura in Japan, and B205 at Sellafield in the UK (since 1981)—a good approximation of plutonium separation can be made. In some cases, reprocessors have published cumulative totals of plutonium separated which may be used as a benchmark.

2. Information on totals of spent fuel reprocessed at a particular plant up to a certain date, together with inferences about burnups, can be used to estimate plutonium separation. These estimates can be corroborated with estimates made for discharged fuel from individual reactors (see chapter 5) in some cases.

3. There may be information about the capacity of a given reprocessing plant and the reactors which it is servicing (as with B205 before 1981, and Russian and Indian reprocessing plants). In this case, rates of plutonium separation can be estimated using information about spent-fuel and plutonium discharges from reactors or about reprocessing contracts.

The basic data used in this section are set out in appendix C. As far as possible all final quantities are stated in terms of total plutonium. When converting from fissile to total plutonium quantities, the assumption is made that the material was 70 or 80 per cent fissile if produced in LWR or Magnox reactors respectively.[13] This does not take into account the subsequent decay of ^{241}Pu (half-life: 14.4 years).

IV. Overview of power-reactor fuel reprocessing

Power-reactor fuel has been reprocessed at numerous facilities around the world, several of them small plants and laboratories. In this chapter only the most important in terms of plutonium separation are considered, namely those which have separated at least 500 kg of total plutonium. These facilities are described in table 6.2, which presents design annual throughputs and the period of operation.

---

[13] The fissile isotopes ^{239}Pu and ^{241}Pu make up 70.42% of PWR uranium spent fuel with a burnup of 33 000 MWd/t; and 67.28% at a burnup of 43 000 MWd/t. OECD/Nuclear Energy Agency, *Plutonium Fuel: An Assessment* (OECD: Paris, 1989), table 9, p. 41.

Table 6.2. World industrial-scale reprocessing plants
Capacity is in tonnes of heavy metal per year.

Country	Location	Owner/ operator[a]	Facility[b]	Fuel[c]	Design capacity	Years of operation
Belgium	Mol	Eurochemic	Eurochemic	oxide+metal	30	1966–75
France	Marcoule	Cogema	UP1	metal	400	1958–97
		CEA	APM/TOR	oxide (FBR)	6	1988–
	La Hague	Cogema	UP2	metal	400	1966–87
				oxide	400	1976–93
			UP2/HAO	oxide (FBR)	..	1979–84
			UP2-800	oxide	800	1994–
			UP3	oxide	800	1990–
Germany	Karlsruhe	KfK/DWK	WAK	oxide	35	1971–90
India	Tarapur	DAE	PREFRE	oxide	100	1982–
	Kalpakkam	DAE	KARP	oxide	100–200	1996–
Japan	Tokai-mura	PNC	Tokai	oxide	100	1977–
	Rokkasho-mura	JNFS	Rokkasho	oxide	800	2003?
Russia	Chelyabinsk-65	Minatom	RT-1	oxide	600	1976–
	Krasnoyarsk-26	Minatom	RT-2	oxide	1000	2005?
UK	Windscale/	BNFL	B205	metal	1500	1964–2014?
	Sellafield		B204/B205	oxide	300	1969–73
			THORP	oxide	700	1994–
	Dounreay	UKAEA	D1206	oxide (FBR)	7	1958–97
			D1204	oxide (MTR)	< 1	1959–97?
USA	West Valley	NFS	West Valley	oxide+metal	300	1966–72

[a] Cogema = Compagnie Générale des Matières Nucléaires; CEA = Commissariat à l'Énergie Atomique; KfK = Kernforschungszentrum Karlsruhe; DWK = Deutsche Gesellschaft für Wiederaufarbeitung von Kernbrennstoffe; DAE = Department of Atomic Energy; PNC = Power Reactor and Nuclear Fuel Development Corporation; JFNS = Japan Nuclear Fuel Service Company; Minatom = Ministry of Atomic Power; BNFL = British Nuclear Fuels; UKAEA = UK Atomic Energy Authority; NFS = Nuclear Fuel Services Company.

[b] APM = Atelier Pilote Marcoule; WAK = Wiederaufarbeitungsanlage Karlsruhe; PREFRE = Power Reactor Fuel Reprocessing; THORP = Thermal Oxide Reprocessing Plant.

[c] LWR (oxide) or magnox (metal) fuels unless otherwise stated. FBR = fast breeder reactor; MTR = materials test reactor.

Within the industry, daily throughputs are more usually cited for these plants, together with an estimate of how many days the plant is expected to operate in a given year. Down-times between reprocessing campaigns for maintenance and retrofitting are relatively lengthy (a typical plant may be operable for 150–200 days each year) and are the main factor determining annual throughput. A reduction in down-times allows annual throughputs to be raised.

V. Commercial reprocessing programmes

In many cases, reprocessing plants have not achieved design fuel throughputs. They have frequently incurred technical problems which have either restricted production or forced lengthy closures while changes were made to the plant. The most consistent load factors have been achieved in plants handling low burnup fuel from gas-graphite reactors. Lower than expected fuel throughputs have meant lower than expected rates of plutonium separation. Modern commercial plants appear to have matched design capacity more closely. The French UP3 plant which began operating in 1990 reached full capacity in 1995. In this section a detailed analysis is given of the operating histories of commercial-scale reprocessing plants. Summary tables of the information presented here are given in section VI.

The United Kingdom

Magnox fuel

Separation of plutonium for the British nuclear weapon programme began at Windscale/Sellafield in 1951 at the Butex B204 plant. Fuel from two plutonium production reactors (the Windscale Piles) was reprocessed until they were shut down in 1957. From 1956 on, fuel from the eight dual-purpose plutonium- and electricity-producing Magnox reactors at Calder Hall and Chapelcross was also reprocessed at the B204 plant. Maximum throughput at the plant was about 750 tonnes of low-burnup fuel per year until it was shut down in 1964 (see chapter 3).

Magnox power-reactor fuel has been reprocessed at the follow-on plant (B205) since 1964—one year after fuel was first discharged from the Magnox station at Berkeley in the west of England. Processing of fuel from Calder Hall and Chapelcross was also transferred to B205, where it was 'co-processed' with power-reactor fuel (that is, in mixed reprocessing campaigns) until 1986. Since then civil and military spent fuel have been processed separately.

The B205 plant has always been a critical element of the Magnox reactor programme. Not only were fuel storage capacities at reactors limited, but the onset of cladding corrosion made the fuel hazardous to handle and transport if stored in ponds for over 12 months. The smooth transfer of fuel to Windscale/Sellafield and its rapid processing there was therefore vital to continued reactor operation. When production at B205 was interrupted by problems in fuel storage ponds in the early 1970s, Magnox reactor operation was put in jeopardy.

Two factors have significantly reduced this logistical pressure in the past decade. First, Magnox reactors have operated at higher burnups, and have thus discharged less fuel. Second, the control of pond water chemistry at reactor sites and at Sellafield has improved, thereby reducing rates of cladding corrosion. The typical lag between fuel discharge and reprocessing today is over two years.

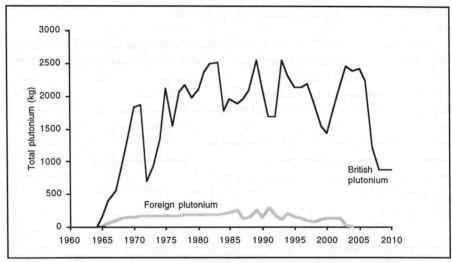

Figure 6.1. Past and projected plutonium separation from power-reactor magnox fuel at the British B205 reprocessing plant, 1960–2010

Sources: Parliamentary Debates, House of Commons [Hansard], Official Report, 1 Apr. 1982, cols 168–69; 27 July 1983, col. 440; 25 Jan. 1985, cols 545–46; 23 July 1985, col. 413; 21 July 1986, col. 10; British Department of Energy, Press Releases, 'Annual plutonium figures', 16 Dec. 1987; 13 Oct. 1988; 5 Dec. 1989; 18 Oct. 1990; 17 Oct. 1991; Department of Trade and Industry, Press Notice, 4 Feb. 1993, 1 Mar. 1994, 19 July 1994; Barnham, K. W. J. *et al.*, 'Production and destination of British civil plutonium', *Nature*, vol. 317 (19 Sep. 1985), pp. 213–17; and SPRU Spent Fuel and Plutonium Database 1995.

A small part of the capacity at B205 has been taken up by fuel discharged from two Magnox reactors sold to Italy and Japan in the mid-1960s. When operating normally, these reactors typically discharged about 90 tonnes of fuel annually. This fuel has been routinely shipped to Sellafield after about a year's cooling at reactor sites. The Italian Latina reactor was closed down in 1987, and the Tokai unit in Japan is due to follow suit in 1996. Reprocessing of foreign magnox fuel is therefore set to end in 2002/3.

Basic operating information on fuel throughput and plutonium separation for the B205 plant has been published since 1982. For the period 1964–81 it has been necessary to make certain assumptions about the rate of reprocessing. The work of Barnham and colleagues has provided good fuel discharge estimates for the Magnox programme.[14] These can be brought together with fuel discharge estimates derived for the Tokai and Latina reactors to produce a picture of reprocessing at B205. An axiom in deriving plutonium separation estimates is that all magnox fuel was reprocessed soon after discharge from the reactor. By assuming a universal one-year lag between fuel discharge and reprocessing before 1981, and by adopting the mean fuel burnups calculated for these reac-

[14] Barnham, K. W. J. *et al.*, 'Production and destination of British civil plutonium', *Nature*, vol. 317 (19 Sep. 1985), pp. 213–17.

tors (ranging from 3000 MWd/t before 1980 to 4900 MWd/t at present), reliable estimates can be derived for plutonium separation. Inevitably the accuracy of estimates for individual years may vary in the pre-1981 period (for instance, no documented account is available of the effect of the 6-month halt in production at B205 during 1973 on fuel throughputs at the plant), but we have good confidence in the cumulative figures.

For the period 1981–86 annual declarations were made by the British Government of the plutonium stocks held at Sellafield, and these can be approximately matched to fuel throughput estimates.[15] Since 1987, annual declarations have been made on the fuel throughput at B205 and the amount of plutonium separated there.[16] Figure 6.1 gives estimates of plutonium annually separated from magnox fuel at Windscale/Sellafield (B205).

In the years since 1970, plutonium separation has fluctuated between 0.5 and 3.0 tonnes per year, with a mean of 2 tonnes. For the period 1964 to 1993 British material made up just over 90 per cent of the cumulative total of power reactor plutonium separated at B205 (47.6 t Pu out of 52.2 t Pu). Plutonium production is expected to fall away (from a mean of 2.35 t/y in the 1980s to about 1.9 t/y in the 1990s), principally as a result of reactor closures. By the end of 1993 a little over 24 500 tonnes of magnox fuel had been reprocessed at B205. Some 2400 tonnes of this was foreign fuel. By the end of 1995 some 26 800 tonnes of fuel had been processed at B205 from which a total of nearly 59 tonnes of plutonium had been separated.

Magnox reprocessing is not expected to end before 2015 if current life extension plans for British Magnox reactors are realized. Following privatization of the AGR and PWR reactors in mid-1996, Magnox reactors will be operated by a new state-owned company, Magnox Electric. Since the decommissioning and waste management liabilities for these reactors now fall on government and not the consumer, there are strong financial incentives to operate these stations for as long as possible. The B205 reprocessing plant will need to operate so long as British Magnox reactors operate, and for between 5 and 10 years beyond in order to process final cores. Long-term interim storage of magnox fuel is not seen as a safe alternative. Nevertheless, B205 has now been operating for over 30 years. Whether the plant can operate smoothly for a further 20 years remains an open question. Assuming that it does operate steadily, by 2010 a total of about 86 tonnes of plutonium may have been separated from power-reactor magnox fuel at B205, 5.6 tonnes of which will have been discharged at the Japanese and Italian reactors.

Aggregate totals for plutonium separation based on these figures are set out in table 6.3. All figures are given for the end of the financial year (31 March), as is the British convention. Published figures for the British plutonium

---

[15] John Moore, Written Answer (WA), *Parliamentary Debates, House of Commons [Hansard]*, 1 Apr. 1982, col. 169; Giles Shaw, WA, *Hansard*, 27 July 1983, col. 438; Mr Goodlad, WA, *Hansard*, 25 Jan. 1985, cols 545–46; and Mr Goodlad, WA, *Hansard*, 23 July 1985, col. 475.

[16] Department of Energy, News Release, 16 Dec. 1987, 13 Oct. 1988, 5 Dec. 1989, 18 Oct. 1990 and 17 Oct. 1991; and Department of Trade and Industry, Press Notice, 4 Feb. 1993, 1 Mar. 1994, 19 July 1994 and 13 July 1995.

Table 6.3. Cumulative past plutonium separation at Sellafield and Dounreay, at the end of 1970, 1980, 1990 and 1993

Figures are in kilograms of total plutonium.

Site	31 Dec. 1970	31 Dec. 1980	31 Dec. 1990	31 Dec. 1993
Sellafield				
Magnox				
British	5 200	22 100	43 800	49 700
(Since 1971)		(16 800)	(38 600)	(44 500)[a]
Foreign	600	2 200	4 100	4 600
Total	**5 800**	**24 300**	**47 900**	**54 300**
Oxide[b]				
Foreign	200	400	400	400
Dounreay				
FBR			3 000	3 400
Total	**6 000**	**24 700**	**51 300**	**58 100**

[a] The British Department of Trade and Industry reported that on 31 Mar. 1994 some 41.5 t of plutonium separated from British power-reactor fuel were held in store at Sellafield. This is material separated since Mar. 1971 which has not been used for other purposes.

[b] Note that THORP began operating in early 1995.

Source: SPRU Spent Fuel and Plutonium Database, 1996.

inventory take March 1971 as the base point. In 1971 title to nuclear fuel, including plutonium, was moved from the UKAEA to the English and Scottish utilities. A little over 5 tonnes of plutonium had been separated from British power-reactor fuel before then and it is presumed that most of this material was bartered with the USA for highly enriched uranium (see chapter 3).

The figure derived here for plutonium separation between 1971 and 1993 (44.5 t) compares well with figures published by the British Government. The Department of Trade and Industry (DTI) has announced that as of 31 March 1994, 39.5 tonnes of separated plutonium belonging to the British nuclear-generating companies were in store at Sellafield and 5 tonnes had been sold or leased to AEA Technology for fast-reactor research (albeit since 1969, not 1971 which is used as the datum here).[17] This gives a total of 44.5 tonnes.

At current rates of production, and assuming that THORP is brought into operation as scheduled, there will be a stock of some 65 tonnes of British power-reactor plutonium at Sellafield by 2000 (assuming that material currently held by AEA Technology is returned to the British stockpile held by British Energy), and this will have risen to about 90 tonnes by 2010.

---

[17] In addition, 1.5 t of plutonium separated from Italian and Japanese fuel were in store at Sellafield.

Oxide fuel

Oxide-fuel reprocessing at Windscale/Sellafield began in 1969. A new head-end plant (HEP) at which oxide fuel was chopped up and dissolved was commissioned at the shut-down B204 plant. The HEP fuel solution was passed through one solvent extraction cycle at B204, before being fed on a campaign basis into the B205 plant.[18] In all, 78 tonnes of power-reactor oxide fuel and about 35 tonnes of research-reactor fuel were processed in this way before an accident caused the B204 plant to be shut down permanently in 1973. Some 360 kg of plutonium were extracted.

Large-scale thermal-reactor oxide fuel reprocessing was restarted at Sellafield in 1994 when commissioning of the new Thermal Oxide Reprocessing Plant (THORP) began. After construction delays and a lengthy review process, the go-ahead for THORP was given by the British Government on 15 December 1993. Commissioning was then further interrupted when this decision was challenged in the courts. This challenge was unsuccessful and full commissioning began in April 1994. The first commissioning phase, due to last up to 18 months, is planned to involve the reprocessing of some 200 tonnes of fuel, although there have been delays in the start-up phase.[19] Early production at THORP has been hampered by technical and licensing problems, so that by late 1996 commercial operation had not yet begun.[20]

About 70 per cent of THORP capacity in the first 10 years of operation will be dedicated to handling foreign spent fuel. The contractual position as it stood in August 1995 is set out in table 6.4. 'Baseload' and 'options' contracts for 6618 tonnes of fuel due to be processed in the first 10 years of THORP's operation are binding and, barring accident or other technical problems, most of this fuel will almost certainly be reprocessed (only Swedish fuel is unlikely to be reprocessed). Assuming timely commissioning and good operating performance, baseload contract work should be completed by 2005.

Contracts for reprocessing beyond 2005 are less secure. New contracts entered into with German utilities in 1990 for work beyond the baseload period have been the subject of debate since they were signed. These contracts allow utility customers to withdraw from their obligations if political or economic circumstances change. Before 1994 German utilities were not legally entitled to adopt storage in favour of reprocessing as a spent-fuel management strategy. In May 1994, an amendment of the 1960 Atomic Law left utilities free to choose their preferred strategy.[21] Many German utilities would now like to opt for extended fuel storage at reactor sites and at central fuel stores. Their main constraint is the uncertain availability of spent-fuel storage capacity as a result of continued political opposition to away-from-reactor stores at Gorleben and

---

[18] Hudson, P., 'Developing technology to reprocess oxide fuel', *British Reprocessing, Nuclear Engineering International,* Special Publication, Oct. 1990, pp. 17–20.
[19] Marshall, P., 'Accident at THORP delays shearing of spent fuel for several months', *Nuclear Fuel,* 23 May 1994, pp. 11–12.
[20] *Whitehaven News,* 'More delays ahead for THORP plant', 31 Oct. 1996.
[21] Roser, T., 'Changing Germany's nuclear law', *Nuclear Engineering International,* Aug. 1994, p. 9.

Table 6.4. Existing contracts for fuel reprocessing at THORP, 1995
Figures are in tonnes of heavy metal.

	Fuel contracted	
Country	1994–2004[a]	Post-2004
UK		
Nuclear Electric	1 540[b]	1 520
Scottish Nuclear	618[b]	1 080
Germany	884[c]	703[d]
Italy	143	–
Japan	2 673	–
Netherlands	53	–
Spain	145	–
Sweden	140[e]	–
Switzerland	422	–
Total	**6 618**	**3 303**

[a] Figures include both 'baseload' contracts and 'options' subsequently taken up.

[b] Contracts for a total of 4758 t of AGR fuel are reported to have been signed. Lifetime arisings for British AGRs are put at 7400 t, leaving a balance of 2642 t not covered by reprocessing contracts..

[c] BNFL states that German utilities have contracted for 969 t of capacity in the first 10 years, and 1600 t of 'new' contracts. The figures given here are those published by the German utilities themselves.

[d] BNFL claims to have secured 1344 t of post-baseload contracts. The figure used here represents 'fixed-commitments-type' and 'requirements-type' contracts.

[e] The Swedish contracts were signed in 1979. Swedish utilities subsequently abandoned their policy of reprocessing, and no fuel has been sent to Sellafield under this contract. Swedish spent fuel sent to Sellafield under earlier contracts is being returned to Sweden.

Sources: BNFL, *The Economic and Commercial Justification for THORP,* Risley, July 1993, p. 15; and Schmidt, U., 'Problems concerning accumulation of isolated plutonium', Paper presented at the IAEA Advisory Group Meeting on Problems Concerning the Accumulation of Separated Plutonium, Vienna, 26–29 Apr. 1993.

Ahaus. If this uncertainty were removed, some or all of the 'post-baseload' reprocessing contracts would be cancelled. At the end of 1994 two sets of new German post-baseload contracts were withdrawn, and other utilities had been expected to follow suit.[22] Given the prevailing impasse over AFR storage capacity, German utilities have begun to reconsider a pull-out from reprocessing. Since the Japanese reprocessing programme and the US spent fuel management programmes are also in trouble, future clients may yet be found for European commercial reprocessing.

[22] European Energy Report, 'German utilities cancel nuclear reprocessing contracts with THORP in favour of storage option', no. 426, 6 Jan. 1995, p. 2; and Hibbs, M., 'German utilities begin pull-out of post-2000 reprocessing deals', *Nucleonics Week,* 5 Jan. 1995, pp. 1, 10–11.

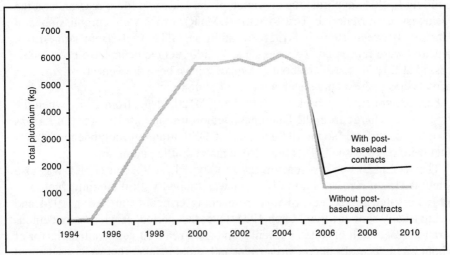

Figure 6.2. Projected rates of plutonium separation from oxide fuel at the British Sellafield THORP reprocessing plant, 1994–2010
Sources: Hudson, P., 'Developing technology to reprocess oxide fuel', *British Reprocessing, Nuclear Engineering International,* Special Publication, Oct. 1990, pp. 17–20; and table 6.4.

Further new THORP reprocessing contracts were agreed in 1991/92 between British Nuclear Fuels and the British nuclear generating companies, Nuclear Electric (NE) and Scottish Nuclear (SNL), as part of a large fuel supply agreement. Under this agreement the British Government was to have underwritten financial risks from past and future liabilities. In November 1992 the government refused to accept these liabilities, and so threw the contracts into doubt. However, in February and March 1995 both Scottish Nuclear and Nuclear Electric agreed new fuel supply and fuel management contracts with BNFL covering future requirements. In total, two-thirds of anticipated AGR fuel arisings are covered by reprocessing contracts.

Projected rates of plutonium separation from oxide fuel at Sellafield are shown in figure 6.2. Two scenarios are shown: one including German post-baseload contracts; the other omitting these. Assuming that THORP operates as planned, there will be a rapid increase in fuel throughputs; full capacity being reached in 1998–99. By early next century, THORP is expected to be separating nearly 6 tonnes of plutonium per year, about twice the rate ever achieved at B205. By 2006, when the baseload contracts have been serviced, THORP is expected to have separated about 46 tonnes of plutonium, of which about 39 tonnes will have come from foreign LWR fuels. All of this material is due to be returned to the countries of origin.

Fast-reactor and materials test-reactor fuel

Fast-reactor and materials test-reactor (MTR) fuel has been reprocessed at Dounreay in northern Scotland since July 1958. MTR fuel reprocessing at building D1204 supported the operation of the Dido and Pluto reactors and the Demonstration Materials Test Reactor (DMTR), which were commissioned at Harwell between 1956 and 1958. In addition, MTR fuels from non-British reactors were processed. To date, some 11 500 fuel elements have been reprocessed at D1204, of which about one-quarter have been of overseas origin. This activity has yielded only small amounts of plutonium.[23]

Fast-reactor fuel reprocessing at building D1206 dates from 1961, when the first fuel discharge from the Dounreay Demonstration Fast Reactor (DFR) was processed. In total, some 10.06 tonnes of DFR uranium-molybdenum driver fuel were reprocessed.[24] This fuel also contained little plutonium.

The second British fast reactor, the Prototype Fast Reactor (PFR), was also sited at Dounreay. Unlike the DFR, it was fuelled with plutonium fuel. The PFR operated for 20 years between reaching criticality in March 1974 and being finally shut down in March 1994. Over that period, as with many demonstration reactors, it operated unevenly, and achieved a cumulative load factor of about 22 per cent. Both core driver fuel and radial blanket fuel were discharged from the reactor, and a total of 93 000 fuel pins were irradiated.[25] This fuel was reprocessed at Dounreay. Building D1206 was shut down between 1974 and 1979 for refurbishment. Since 1979 it has processed fuel discharged by the PFR. By the end of 1993, some 19.3 tonnes of PFR fuel had been processed, containing about 3.68 tonnes of plutonium.[26] Estimating the total weight of driver and blanket fuel irradiated at PFR is complicated because a variety of different fuel pin designs were used to fuel the reactor. Assuming a mean cooling time for the driver fuel of three years, it is estimated that by the end of 1993 about 20.5 tonnes of driver fuel assemblies had been discharged from PFR. This fuel contained about 4.3 tonnes of plutonium. Added to this is the final fuel core of 4.1 tonnes of fuel—containing some 800 kg of plutonium[27]—which will be discharged by the end of 1996. In total, therefore, we estimate that some 5.1 tonnes of plutonium will be discharged from the PFR. Current plans are that all this fuel will be reprocessed by the end of 1997.

The future of MTR fuel reprocessing at Dounreay is uncertain at present, although AEA Technology has been successful in attracting some further work. Demand for this service developed during 1990–93, following the refusal of the US Department of Energy to accept back fuel from US-supplied and licensed research reactors. Several European research reactor operators considered sending fuel to Dounreay instead. However, the USA has subsequently reversed its policy on research reactor fuel, and has agreed to take the fuel back (see

[23] Barrett, T. R., 'Specialist reprocessing needs', Paper presented at the Management of Spent Nuclear Fuel Conference, IBC Technical Services, London, 29–30 Apr. 1991.
[24] Mégy, J. et al., 'The fast breeder reactor fuel cycle in Europe—present status and prospects', *Nuclear Technology*, vol. 88 (Dec. 1989), pp. 283–89.
[25] Broomfield, T., 'PFR builds firm foundations for fast reactors', *Atom*, no. 433 (Mar./Apr. 1994), pp. 22–26.
[26] Some 17.3 t of this fuel was driver fuel from which was separated a total of 3.66 t of total plutonium (an average of 21.2%). In 1991–93 a further 2.04 t of breeder fuel was reprocessed, yielding another 18 kg of total plutonium. Personal communication, Doug Gordon, UKAEA Dounreay, 31 Oct. 1994.
[27] Assuming a 10% burnup.

chapter 8). No future MTR fuel reprocessing at Dounreay is expected to yield significant quantities of plutonium.

France

The distinction between military and civil nuclear materials has been less clear in France than in the UK, even though, as in the UK, all civil material is subject to Euratom safeguards.[28] Indeed, France has not sought to hide its use in weapons of plutonium produced in power reactors operated by Électricité de France (EdF) (chapter 3, section VII). This section traces the reprocessing of fuel from all power reactors, whether or not it has ended up in military or civil use. An estimate is given in chapter 3 of the small amounts of power-reactor plutonium which were transferred into the French military stockpile.

Gas-graphite reactor fuel

Plutonium separation began in France in November 1949 when French chemists at the Le Bouchet centre succeeded in isolating 15 milligrams from a fuel rod irradiated in the Zoé heavy water reactor. This was followed in 1954 with the commissioning of the first pilot reprocessing plant at Fontenay-aux-Roses.

An industrial-scale reprocessing plant (Usine de Plutonium 1, UP1) began operating at Marcoule in July 1958 and handled fuel discharges from three plutonium production reactors on the same site until about 1985, soon after the last of three production reactors was shut down. The plant also reprocessed initial fuel discharges from the three Chinon power reactors in 1965 and 1966.

Reprocessing of power-reactor fuel was switched to a new site—La Hague in Normandy—in the mid-1960s. At La Hague the UP2 plant began operating in 1966, taking fuel from French and Spanish Uranium Naturel Graphite Gaz (UNGG) reactors. Processing of metallic fuel continued until 1987. In total about 4894 tonnes of UNGG fuel were processed at UP2. Meanwhile, reprocessing of UNGG power-reactor fuel restarted at UP1 in the mid-1970s. All UNGG reprocessing was gradually moved back to Marcoule, leaving facilities at La Hague dedicated to LWR fuel processing. Co-processing of production reactor and power-reactor fuel ended at UP1 in about 1986. It is estimated that

---

[28] The main references used in this section besides those noted later are: Commissariat à l'Énergie Atomique (CEA), 'Le retraitement des combustibles irradiés', *Industrie Nucléaire Française*, Paris, 1982, pp. 154–64; Couture, J., 'Status of the French reprocessing industry', Paper presented at the American Nuclear Society Conference, *Fuel Processing and Waste Management*, 26–29 Aug. 1984, Jackson, Wyo.; Delange, M., 'Operating experience with reprocessing plants', *Atomwirtschaft*, Jan. 1985, pp. 24–28; Delange, M., 'LWR spent fuel reprocessing at La Hague: ten years on', Paper presented at the *RECOD 87* conference, Paris, 1987, pp. 187–93; 'Reprocessing and waste management, country: France, pt 1', *NUKEM Market Report*, no. 3 (1988), pp. 15–18; Lewiner, C. and Gloaguen, A., 'The French reprocessing programme', *Atomwirtschaft*, May 1988, pp. 227–29; CEA, *Cycle du combustible nucléaire: retraitement* (CEA: Paris, Mar. 1989); 'Reprocessing and waste management: review 1989', *NUKEM Market Report*, no. 2 (1990), pp. 14–23; EdF, 'Retraitement recyclage', Paper by Service des Comustibles, Paris, 6 Mar. 1990; Ledermann, P., 'Operating UP3: Three years of experience', *Nuclear Engineering International*, Jan. 1994, pp. 46–49; and Laurent et al. (note 5).

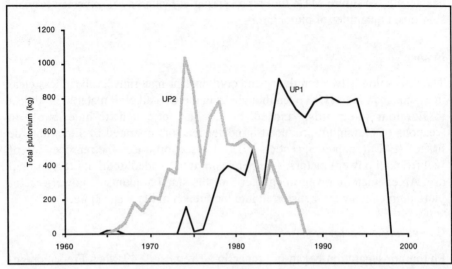

Figure 6.3. Past and projected plutonium separation from gas-graphite power-reactor fuel at the French Marcoule UP1 and La Hague UP2 reprocessing plants, 1960–2000

Sources: Syndicat CFDT de l'Énergie Atomique, *Le dossier électronucléaire* (Éditions du Seuil: Paris, 1981), pp. 186–91; Hirsch, H. and Schneider, M., 'Wiederaufarbeitung in Europa: Wackersdorf ist tot es—lebe La Hague?' [Reprocessing in Europe: Wackersdorf is dead—is the Hague alive?], *Rest-Risiko*, no. 6 (Greenpeace: Hamburg, Apr. 1990); and Ricaud, J. L., 'Main issues of the fuel cycle back end', Paper presented at the *RECOD 94* conference, London, 24–28 Apr. 1994.

by the end of 1993 almost 5300 tonnes of power-reactor UNGG fuel had been processed at UP1. The rate of plutonium separation from UNGG fuel at La Hague and Marcoule is shown in figure 6.3.

UNGG fuel reprocessing at UP2 increased gradually over the 10 years from 1966 onwards, reaching a peak of just over 1 tonne of plutonium separated in 1974. Thereafter, with the restart of power-reactor reprocessing at Marcoule's UP1, reprocessing of UNGG at UP2 fell. By 1987, UP1 was handling all French and Spanish UNGG fuel, separating about 750 kg of plutonium annually. French UNGG plutonium separation was at its peak between 1974 and 1987, when mean annual production was above 900 kg. This is shown in figure 6.4. According to a 1988 agreement between Cogema and EdF, UP1 will continue to process gas-graphite reactor fuel until 1997.[29] The last UNGG reactor, Bugey 1, was finally shut down in May 1994. When UNGG reprocessing ends in France, a total of about 11 200 tonnes of UNGG fuel will have been processed, leading to the separation of a little under 20 tonnes of plutonium.

---

[29] Barrillot, B. and Davis, M., *Les déchets nucléaires militaires français*, Études du CDRPC (CDRPC: Lyons, Feb. 1994), p. 129.

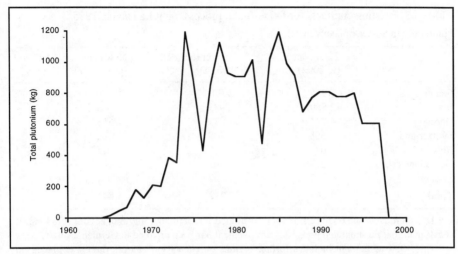

Figure 6.4. Total plutonium separation from gas-graphite power-reactor fuel in France, 1960–2000

Sources: Syndicat CFDT de l'Énergie Atomique, *Le dossier électronucléaire* (Éditions du Seuil: Paris, 1981), pp. 186–91; Hirsch, H. and Schneider, M., 'Wiederaufarbeitung in Europa: Wackersdorf ist tot es—lebe La Hague?' [Reprocessing in Europe: Wackersdorf is dead—is the Hague alive?], *Rest-Risiko*, no. 6 (Greenpeace: Hamburg, Apr. 1990); and Ricaud, J. L., 'Main issues of the fuel cycle back end', Paper presented at the *RECOD 94* conference, London, 24–28 Apr. 1994.

Oxide fuel

Oxide-fuel reprocessing began in France in 1976 at the UP2 plant after a modified head-end plant (Haute Activité Oxyde, HAO) was installed. Capacity was gradually raised as UNGG fuel processing was phased out at UP2, and reached 400 tonnes of oxide fuel per year by the late 1980s. With the commissioning in 1989–90 of the new UP3 oxide-fuel reprocessing plant (800 t fuel per year), foreign fuel reprocessing at UP2 was phased out and moved to UP3. By 1992 only French fuel was being handled at UP2. In 1994 construction was completed of two new workshops at the UP2 plant, so allowing the expansion of capacity at the plant (now known as UP2-800) up to 850 tonnes fuel throughput per year. The commissioning of UP2-800 brings to an end a major programme of plant extension at La Hague begun in 1979.

By the end of 1993 some 5702 tonnes of oxide fuel had been reprocessed at La Hague.[30] Of this, about 40 per cent (2370 t) was fuel discharged from French PWRs, the rest being primarily from foreign LWRs.[31] The contractual situation for reprocessing oxide fuel at La Hague is shown in table 6.5. This shows that at the end of 1993 Cogema had processed exactly one-third of the fuel for which it holds service contracts.

[30] Of this 4078 t had been processed at UP2 and 1624 t had been handled at UP3.
[31] Some 4.5 t of LWR–MOX fuel and 10 t of FBR–MOX fuel had been processed. A. Gloaguen, personal communication, 26 May 1994.

Table 6.5. Existing contracts for oxide-fuel reprocessing at La Hague, 1995
Figures are in tonnes of heavy metal.

	Total contracts (to 2002)	Fuel reprocessed (to 31 Dec. 1993)	New contracts (post-2002)
France	7 965	2 370[a]	..
Japan	2 869	933	..
Germany	4 731	1 672	1 616
Switzerland	558	208	..
Belgium	670	310	225[b]
Netherlands	219	162	..
Sweden	57	57	..
Total	**17 069**	**5 712[c]**	**1 841**

[a] This includes some 667 t of EdF fuel substituted for German utility fuel at UP2, 1984–88. It is assumed that Pu separated under this arrangement has been credited to German utilities.

[b] This 1991 agreement includes further options on capacity. A moratorium was placed on new reprocessing contracts by the Belgian Government in 1990, and no final decision has been taken over whether to pursue the 1991 agreement.

[c] A further 1276 t of fuel were processed in 1994, and 1650 t in 1995, giving a total cumulative oxide fuel throughput at La Hague of 6988 t.

Sources: Reprocessing News, no. 15 (United Reprocessors Group (URG): Hanover, Feb. 1990), p. 3; Ricaud, J. L., 'Main issues of the fuel cycle back end', Paper presented at the *RECOD 94* conference, London, 25 Apr. 1994; Schmidt, U., 'Problems concerning accumulation of isolated plutonium', and Bay, H., 'Problems concerning the accumulation of isolated plutonium—Swiss situation', papers presented at the IAEA Advisory Group Meeting on Problems Concerning the Accumulation of Separated Plutonium, Vienna, 26–29 Apr. 1993; Belgische Kamer van Volksvertegenwoordigers, *Resolutie: betreffende het gebruik van plutonium- en uraniumhoudende brandstoffen in Belgische kerncentrales en de opportuniteit van de opwerking van nucleaire brandstofstaven* [Resolution: about the use of plutonium and uranium fuel in Belgian nuclear reactors and the opportunity for reprocessing nuclear fuel rods], -541/6-91/92 (B.Z.), Brussels, 16 Dec. 1993, pp. 5–7; Gloaguen, A., EdF, personal communication, 26 May 1994; and Giraud, P., Cogema, 'Reprocessing–recycling: the industrial stakes', Konrad Adenauer Stiftung, Bonn, 9 May 1995.

Rates of plutonium separation from oxide fuel at UP2 and UP3 are shown in figure 6.5. In generating these figures, no distinction has been made between boiling water reactor (BWR) and PWR fuel. In addition we assume that post-baseload contracts will not be fulfilled at UP3 in the short term. The figure shows a very gradual buildup of LWR fuel reprocessing at UP2-400 starting in 1976. Full capacity was reached in 1987 when UNGG reprocessing ceased. With the commissioning of UP3 in 1989–90 and the shift of foreign fuel processing away from UP2, fuel throughputs at UP2 fell. EdF insisted that French plutonium separation be brought in line with demand. Following completion of the Melox mixed-oxide (MOX) fabrication plant at Marcoule in 1995, EdF has agreed to increase throughputs of its fuel at UP2 to coincide with the commissioning of the new expanded capacity. Existing EdF contracts are likely to be

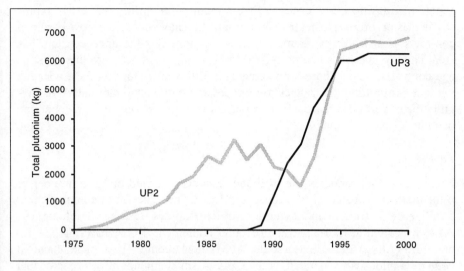

Figure 6.5. Plutonium separation from oxide-fuel reprocessing at the La Hague UP2 and UP3 reprocessing plants in France, 1975–2000

Sources: Hirsch, H. and Schneider, M., 'Wiederaufarbeitung in Europa: Wackersdorf ist tot es—lebe La Hague?' [Reprocessing in Europe: Wackersdorf is dead—is the Hague alive?], *Rest-Risiko*, no. 6 (Greenpeace: Hamburg, Apr. 1990); 'Active commissioning of the UP3 plant: the first days ...', *Reprocessing News*, no. 15 (United Reprocessors Group (URG): Hanover, Feb. 1990); 'UP3 in figures', *Nuclear Engineering International*, June 1992, p. 4; and Ricaud, J. L., 'Reprocessing–recycling: the industrial stakes', Paper to the Konrad Adenauer Stiftung, Bonn, 9 May 1995.

completed by 2003. No new contracts have yet been agreed, and their scale will depend on the performance of existing facilities and the general strategic position of plutonium in EdF fuel policy. If plants do function as planned, it seems safe to assume that reprocessing of French PWR fuel will continue for at least another 10 years or so. We assume in our projections that 850 tonnes of fuel will be reprocessed per year. EdF strategy assumes that some 350 tonnes of spent fuel will be placed into storage per year from around 2000 (105 t of UO_2 fuel, 110 t of RepU fuel and 135 t of MOX fuel). EdF plans to recycle plutonium and reprocessed uranium just once and to store all spent RepU and MOX fuel indefinitely.

After the first fuel was sheared at UP3 in 1989, commissioning of the plant proceeded rapidly. Full capacity was reached in 1995. By the end of 1989, about 7500 tonnes of foreign fuel covered by baseload and 1990s options contracts remained to be processed. At planned rates of production, UP3 is expected to complete this task by 2001. The future of existing post-2001 contracts is uncertain. As with post-2000 contracts at THORP, the main customers are likely to be German utilities who signed new contracts in 1989–90. In addition, Cogema has signed a 'new' contract with the Belgian fuel-cycle company Synatom. However, in both Germany and Belgium utility back-end policy is in a state of transition. In Germany there remains some utility interest in reprocessing as an option. Whereas German utilities have cancelled post-baseload contracts with BNFL, Cogema has taken a more flexible approach and has negotiated to convert some post-2000 reprocessing contracts into interim storage contracts. Under these arrangements, utilities may store fuel at La Hague without committing themselves to reprocessing until some later date.[32] In Belgium the government has decided to place a moratorium on the 'new' contract, pending a final decision in 1997.

Fast-reactor fuel

Fast-reactor fuel reprocessing in France has been carried out at a number of pilot facilities located at Marcoule and La Hague. The first dedicated plant (Atelier de Retraitement des combustibles Rapides—Atelier Traitement 1, AT1) was sited at La Hague, the second (Traitement d'Oxydes Pilote, TOP), and third (Atelier Pilot de Marcoule, APM) at Marcoule. These plants have all been on a pilot scale, operated by the CEA. A small amount of fast-reactor fuel was also processed at UP2 between 1979 and 1984.

Laboratory-scale reprocessing of fast-reactor fuel began in the Cyrano laboratory at Fontenay-aux-Roses in 1968. About 100 kg of fuel from the Rapsodie and Phénix reactors were processed there, producing a total of 15–20 kg of plutonium.

The decision to build the AT1 plant at La Hague to reprocess fuel from the Rapsodie fast reactor was made in 1964. AT1 was designed to have the capacity to reprocess one fuel core (134 kg total)[33] per annum and began active operation in 1969 when it processed 220 fuel rods weighing 25 kg.[34] Rapsodie was fuelled with both plutonium and HEU. In all, some 910 kg of Rapsodie plutonium fuel were processed at AT1 by the time it was shut down at the end of 1979.[35] Estimating the amount of plutonium contained in this fuel is difficult because initial fuel enrichments are likely to have varied and because of the wide range of burnups exhibited in the fuel (up to a maximum of 200 000 MWd/t). However, if it is assumed that all of the processed fuel initially had a plutonium enrichment of 30 per cent (see table 7.1), and was irradiated up to a mean burnup of 20 per cent, then the plutonium inventory in the reprocessed fuel would be about 220 kg Pu_{fiss} (about 290 kg Pu_{tot}) per tonne of heavy metal. A small amount of fuel from the Phénix fast reactor (about 0.18 t) was also reprocessed at AT1. Assuming a mean plutonium enrichment in fresh fuel of 27 per cent (see table 7.1), and a mean burnup of 20 per cent, this fuel would have contained about 40 kg of Pu_{fiss} (about 50 kg Pu_{tot}).

[32] Hibbs, M. and MacLachlan, A., 'Cogema and Preag float new deal for German spent fuel after 2000', *Nuclear Fuel*, 4 July 1994, pp. 1, 15–17.

[33] The initial core was 30% enriched with plutonium (about 40 kg). Groupement Centrale Nucléaire Européene à Neutrons Rapides (NERSA), *The Creys-Malville Power Plant*, brochure, EdF, Direction de l'Équipement, Alpe Lyons, 1987, p 8. For a listing of more technical information about French fast reactors, see table 7.1.

[34] CEA, 1989 (note 28), p. 164.

[35] Mégy *et al.* (note 24), p. 285.

Fast-reactor fuel reprocessing at Marcoule began in 1974 when the Service de l'Atelier Pilote (SAP), which had previously treated UNGG fuel, was converted to take fuel discharged by Rapsodie. The refurbished TOP plant had a design capacity of 10 kg of fuel per day, although this appears not to have been achieved. Between 1974 and 1976 fuel discharged from Rapsodie (0.05 t of MOX), the EL4 heavy water reactor and the German KNKI thermal reactor (an HEU-fuelled heavy water reactor, shut down for installation of a fast core in 1974) was reprocessed. In all, 1.65 tonnes of KNK HEU fuel were processed. From 1977 to 1983, when the TOP plant was shut down for refurbishment, only fuel from Phénix was processed there. A total of 9.04 tonnes of Phénix fuel were handled, of which 6.74 tonnes were FBR MOX fuel. The remainder (2.3 t) was HEU fuel. Some 1305 kg of fissile plutonium (about 1650 kg total plutonium) were separated from this fuel.[36]

In 1979 experiments began with fast-reactor fuel reprocessing at UP2. The fuel was declad and dissolved in the HAO facility, and then sent for chemical separation, diluted in UNGG fuel solution. A total of 10.07 tonnes of Phénix fuel were treated in this way at UP2 up to 1984.[37] Assuming a fissile plutonium inventory in spent Phénix fuel of 20 per cent, this would have yielded a total of some 2000 kg of fissile plutonium (about 2500 kg of total plutonium).

In 1978 the CEA decided to expand the capacity of the TOP facility. The new APM facility was brought into operation in 1988 with a design throughput of 6 tonnes of spent fuel per year. APM serves as the head-end for the Traitement d'Oxyde Rapides (TOR, formerly TOP) separation and materials-finishing workshops. Between January 1988 and January 1991 about 5 tonnes of fuel from the Phénix and the German KNKII fast reactors were processed at APM/TOR.[38] Some LWR MOX fuel has also been processed.[39] It is assumed here that about 4 tonnes of Phénix fuel were processed, together with the complete first plutonium core of KNKII. If the Phénix fuel had a plutonium enrichment of about 20 per cent and the KNK fuel an enrichment of 18 per cent, some 900 kg of fissile plutonium (about 1200 kg total plutonium) would have been separated at APM by the end of 1991. Since then, no further Phénix fuel has been reprocessed.

In total, therefore, the authors estimate that about 4.5 tonnes of fissile plutonium (about 5.7 t of total plutonium) had been separated from fast-reactor fuel in France by the end of 1993. About 90 kg of this material were owned by

---

[36] Barrillot and Davis (note 29), p. 137; Davis, M. D., *The Military–Civilian Nuclear Link: A Guide to the French Nuclear Industry* (Westview Press: Boulder, Colo., 1988), pp. 76–77 and 104; and Mégy et al. (note 24), p. 285.

[37] CEA, 1982 (note 28), p. 165, states that by 1982, 2.1 t had been processed at UP2. CEA, 1989 (note 28), p. 17, states that of the 20 t of fast-reactor fuel reprocessed in France by 1989, 19 t had been discharged by Phénix.

[38] Barrillot and Davis (note 29), p. 138. Two cores of MOX fuel weighing 689 kg each, the first enriched to 22% with plutonium and the second to 60%, were loaded at KNKII between 1977 and 1991 when the reactor was closed down. It is assumed that the first core was reprocessed by 1991 and that the second is currently in store. 'KNK II—an experimental power station equipped with a fast core', *Nuclear Engineering International*, Jan. 1979, pp. 41–44; and *Nucleonics Week*, 5 Sep. 1991, pp. 9–10.

[39] CEA, *Rapport Annuel 1989* (CEA: Paris, 1989), p. 39.

Table 6.6. Cumulative past plutonium separation at La Hague and Marcoule, at the end of 1970, 1980, 1990 and 1993

Figures are in kilograms of total plutonium.

	31 Dec. 1970	31 Dec. 1980	31 Dec. 1990	31 Dec. 1993
Magnox	650	7 900	16 700	18 900
LWR	..	1 900	25 000	40 700
FBR	..	1 800	4 600	4 700
Total	**650**	**11 600**	**46 300**	**64 300**

Source: SPRU Spent Fuel and Plutonium Database, 1994.

Germany. Given the effective halt in the French fast-reactor programme between 1992 and 1995, we assume that little more fast-reactor fuel has been reprocessed at APM since 1993.

Future separation of plutonium from fast-reactor fuel in France is dependent primarily on the operation of the Phénix and Superphénix reactors (see chapter 7, section IV). Phénix operated for 10 days at 65 per cent power between late 1990 and October 1994, having been ordered to shut down following generally critical safety reports by the French nuclear regulator. Superphénix was shut down in July 1990, and was not restarted until August 1995, following a French Government refusal to issue a restart licence in July 1992.

The character of the French fast-reactor programme has been completely altered since the Curien Report on the future of Superphénix and the public debate that followed it in 1993.[40] Phénix and Superphénix have been transformed into research reactors, primarily developing fuel technology for actinide transmutation in fast reactors. Under the Consommation Accrue de Plutonium dans les Rapides (CAPRA) and SPIN (separation–incineration) projects both reactors are to become consumers of plutonium and other actinides, rather than plutonium producers.[41] Conversion of the Superphénix reactor to a plutonium consumer is not expected to be completed until 2001. Radial fuel assemblies will be replaced with steel sub-assemblies while the initial MOX core and a second that has already been fabricated are both fully irradiated. With insertion of a newly fabricated third core, plutonium will be burnt at a rate of about 100 kg per year, assuming a 70 per cent capacity factor. No prospective fuelling strategy for Phénix has been published. It is assumed that all Superphénix and

[40] Ministère de la Recherche et de L'Espace, *Le traitement des produits de la fin du cycle electronucléaire et contribution possible Superphénix* [The treatment of nuclear spent-fuel products and the possible contribution of Superphénix], 17 Dec. 1992, Paris.

[41] 'Le projet CAPRA pour que les "rapides" consomment le plutonium' [The CAPRA project for the consumption of plutonium in fast reactors], *RGN-Actualités*, no. 5 (Sep./Oct. 1993), pp. 359–60; Lacroix, A. and Gloaguen, A., 'Use of the Creys-Malville Reactor as a plutonium and/or actinide burner', Paper presented at the The Uranium Institute Annual Symposium 1993, London, Sep. 1993; and Anzieu, P. and del Beccaro, R., 'Plutonium burning and actinide transmutation in Superphénix', Paper presented at the International Conference on Evaluation of Emerging Nuclear Fuel Cycle Systems, Versailles, 11–14 Sep. 1995.

Phénix fuel will be reprocessed, given the long-term objective of the transmutation programme.

Taking only current and known fabricated fuel into account, the two reactors would be expected to discharge some 8.3 tonnes of fissile plutonium over the next 10 years.[42]

Summary information for plutonium separation in France is presented in table 6.6.

Russia

As elsewhere, a closed fuel cycle stood at the heart of nuclear power policy in the former Soviet Union. The basic concept was that all fuel discharged from VVER reactors in the USSR and among its client states in Central and Eastern Europe would be reprocessed. Plutonium would be recycled in Soviet fast reactors, while recovered uranium would be recycled in light-water-cooled, graphite-moderated RBMK reactors. Fuel discharged from RBMKs would not be reprocessed on economic grounds, but stored at the reactor sites.[43]

Only the first phase of this strategy was implemented—the construction and operation of the RT-1 reprocessing plant at the Mayak Chemical Combine (MCC) at Chelyabinsk-65 which has handled primarily fuel from the smaller VVER-210 and VVER-440 power reactors.[44] Construction of the second reprocessing plant, RT-2, at the Mining and Chemical Combine (Sibkhimstroy) at Krasnoyarsk-26 to handle fuel from the larger VVER-1000 reactors was begun in 1978. While a spent-fuel storage facility was brought into operation at Krasnoyarsk in 1985, construction of the 1000 tonnes/year chemical separation facility has been much delayed.[45] Uranium recovered at RT-1 has been used to fabricate fuel for the BN-350 and BN-600 reactors, as well as for RBMK, and more recently VVER-1000 reactors. The recycling of plutonium in fast reactors is still in its infancy.

Oxide and fast-reactor fuel

The RT-1 reprocessing plant is based on a chemical separation plant which first started processing spent fuel from Soviet plutonium production reactors in 1956

---

[42] Assuming a 20% initial enrichment for Phénix fuel and 15% enrichment for Superphénix fuel, and a 20% burnup for all discharged fuel, we would expect a total of about 700 kg of fissile plutonium to be discharged in a single Phénix core, and 7.6 t of fissile plutonium in 2 Superphénix cores.

[43] At a burnup of about 18 GWd/t RBMK-1000 fuel contains about 5 kg/t of plutonium with a fissile content of 53%. Kudriavtsev, E. G. and Mikerin, E. I., 'Russian prospects for plutonium utilization', *Proceedings of the 1993 International Conference on Nuclear Waste Management and Environmental Remediation*, Prague, 5–11 Sep. 1993, pp. 639–43.

[44] For the sake of simplicity the older names of Russian facilities are used here. All fuel-cycle facility sites have now been renamed: Chelyabinsk-65 has become Ozersk; Chelyabinsk-70, Shezhinsk; Krasnoyarsk-26, Zheleznogorsk; Tomsk-7, Seversk. Carnegie Endowment Center for Russian and Eurasian Programs, *Nuclear Successor States of the Soviet Union*, Nuclear Weapon and Sensitive Export Status Report, no. 2, Moscow, Dec. 1994.

[45] Bukharin, O., 'Nuclear fuel cycle activities in Russia', ed. T. B. Cochran and R. S. Norris, *Russian/Soviet Nuclear Warhead Production*, Nuclear Weapons Databook Working Paper NWD 93-1 (Natural Resources Defense Council, Washington, DC, 8 Sep. 1993), p. 127.

Table 6.7. VVER-440 fuel dispatched to RT-1, 1976–93

Spent fuel is given in tonnes of heavy metal. Figures in brackets are estimates; others are validated data.

Country	Spent fuel	Percentage of total VVER-440 discharges
Armenia[a]	(350?)	(100?)
Bulgaria	(600)	(75)
Czech Republic	140	45
Finland	280	70
Germany (formerly GDR)	140	18
Hungary	84	18
Slovakia	84	15
Russia	(1 500)	(80)
Ukraine	200	70
Total	**(3 400)**	**(60[b])**

[a] The Armenia 1 and 2 VVER-440 reactors were shut down in early 1989 following an earthquake. It is estimated that by then some 270 t of fuel had been discharged by the two reactors. We assume that the 84 t of partially irradiated fuel held in the reactor cores were discharged following reactor closure and that all this fuel was subsequently transferred to Chelyabinsk.

[b] Total fuel discharges to the end of 1993 from VVER-440 reactors are estimated at 5900 t.

Sources: SPRU Survey of Reactor Operators, 1994; SPRU Spent Fuel and Plutonium Database, 1994; MacLachlan, A., Silver, R. and Hiruo, E., 'Ukraine on way to ending storage crisis; construction slated to begin this year', *Nuclear Fuel*, 17 Jan. 1994, pp. 16–17; Petchera, I., 'Independent spent-fuel storage facilities licensing process in Ukraine', IAEA-SM-335/38; and Velkov, L. M., 'Licensing of spent fuel storage in Bulgaria—regulations and problems', IAEA-SM-335/32, papers to the International Symposium on Spent Fuel Storage—Safety, Engineering and Environmental Aspects, Vienna, 10–14 Oct. 1994.

(see chapter 3). In 1971 a new head-end plant was commissioned to enable the reprocessing of stainless steel and zircalloy-clad fuel. By 1976 production at RT-1 was shifted from processing military reactor fuel to processing spent fuel from VVER-440 reactors, fast reactors (BN-600 at Sverdlovsk and BN-350 Mangyshiak Peninsula in Kazakhstan), and the propulsion reactors of icebreakers and submarines.[46]

RT-1 has a design capacity of 400 tonnes fuel per year, sufficient to service all the currently operating VVER-440s.[47] However, the plant has never operated at this capacity, for several reasons. Under Soviet fuel supply agreements, it was expected that VVER fuel would be stored at reactor sites for about three years (later extended to five years) before being shipped either to Chelyabinsk/Ozersk (VVER-440 fuel) or to Krasnoyarsk/Zheleznogorsk (VVER-1000 fuel). Since the height of Soviet reactor commissioning in the USSR and in other

[46] Bibilashvili, Yu. K. and Reshetnikov, F. G., 'Russia's nuclear fuel cycle: an industrial perspective', *IAEA Bulletin*, no. 3 (1993), pp. 28–33.

[47] In 1993, 26 VVER-440s were operating, discharging about 360 t of spent fuel each year.

countries was during the early 1980s, fuel from these reactors would not have been expected at Mayak until the mid- to late 1980s. By then, with the unravelling of the Comecon trading system, the intergovernmental agreements covering take-back of fuel of Soviet origin were beginning to come apart. Continued political instability since then, and the adoption by the Russian Parliament in 1992 of a law which effectively prohibited the import of spent fuel into Russia, have meant that shipments of fuel to Chelyabinsk and Krasnoyarsk from outside Russia have been far less regular than planned. This law was subsequently overridden by a decree from President Yeltsin in April 1993, and shipments of fuel restarted in 1995.[48]

One report suggests that up to 1989 the average annual throughput of fuel at RT-1 was about 200 t.[49] The authors estimate that up to the end of 1993 about 3200 tonnes of VVER-440 fuel had been reprocessed at RT-1.[50] In addition to this an amount of fast-reactor fuel was processed. It is estimated that some 200 tonnes of fast-reactor fuel would have been reprocessed.[51] Considering that the fuel throughputs at RT-1 in 1991, 1992 and 1993 were 160 tonnes, 120 tonnes and 100 tonnes respectively, this leaves an average annual throughput according to our estimates, in the period up to 1990, of 195 tonnes of fuel per year. In the absence of any other data, it is not possible to estimate how much naval propulsion and other fuel has been handled at the facility. It is assumed here that the amounts have been relatively small.

By using information collected in our survey of reactor operators and knowledge of the fuel cycle policies of Bulgaria and the former USSR, it is possible to reconstruct a picture of the VVER-440 fuel management policies across the former USSR and in Eastern Europe. Table 6.7 shows the estimated amount of fuel sent to Chelyabinsk by each of the countries which operate VVER-440s. Apart from Russia, VVER-1000s were built only in Ukraine and Bulgaria. We assume that some 200 tonnes of fuel remained in store at Mayak.

A variety of figures have been published for the amount of plutonium which has been separated at RT-1 and placed in storage at Chelyabinsk. Taken together these suggest that by the end of 1993 a total of about 26.5 tonnes of plutonium had been separated from all types of fuel since 1976.[52] The material was of variable quality, ranging from plutonium separated from VVER fuel

---

[48] Spent fuel from the Paks nuclear station in Hungary was shipped to Mayak on 18 Jan. 1995 after the Russian Government overcame parliamentary opposition to the shipment. Uranium Institute, *News Briefing*, no. 95/5, London, 31 Jan. 1995, p. 1.

[49] 'Soviet Union postpones completion of Siberian reprocessing plant', *Nuclear Fuel*, 16 Oct. 1989, pp. 1–2.

[50] It is assumed here that the storage time for fuel at Mayak is short, of the order of a year, since the rate of reprocessing appears to be constrained by shipments of fuel, rather than reprocessing capacity.

[51] Each year BN-600, which started operating in 1980, discharges about 6 t of fuel, and BN-350 (commissioned in 1973) discharges about 7.4 t. Kudriavtsev and Mikerin (note 43).

[52] Dzegun, E. G., 'Experience with the management of fissile materials at "Mayak"', Paper presented at the Workshop on the Future of the Chemical Separation of Plutonium and Arrangements for the Storage and Disposition of Already Separated Plutonium, Moscow, 14–16 Dec. 1992; Kagramanyan, V. S., 'Utilization in BN-800 fast reactors of isolated plutonium being accumulated in the Russian Republic', Paper to the IAEA Advisory Group Meeting on Problems Concerning the Accumulation of Isolated Plutonium, Vienna, 26–29 Apr. 1993; and Kudriavtsev and Mikerin (note 43).

with a range of burnups to material separated from the blanket fuel of fast reactors.[53] Most of it (estimated at about 25.5 t) had been separated from VVER-440 fuel, while the rest was extracted from fast reactor, naval propulsion and test reactor fuel.[54] At present fuel production rates of 100–200 tonnes per year, 600–1200 kg of total plutonium are being separated each year.

In 1975 it was decided to construct a second reprocessing facility (RT-2) at Krasnoyarsk to handle fuel from the larger 1000 MWe VVER and 'other' reactors.[55] Nominal design capacity is put at 1000 tonnes of fuel per year, yielding about 8 tonnes of plutonium. Construction of a large central fuel store was also authorized. Work began at Krasnoyarsk in 1978 and by December 1985 the 6000-tonne capacity fuel store was commissioned. At that stage, the reprocessing plant was just 30 per cent complete, and following a sharp decrease in funding for the project in 1985, construction was first interrupted and then halted in 1989 as a result of public opposition. The project looked in danger of perishing until in September 1994 President Yeltsin issued a decree ordering construction of the plant. Implementation of the decree appears to depend on foreign capital being secured through reprocessing contracts with non-Russian nuclear utilities. The prospects appear to be gloomy.

Japan

Magnox fuel

Spent fuel from the single Japanese Magnox reactor, Tokai 1, has been sent for reprocessing at Windscale/Sellafield since the reactor started discharging fuel in 1967. Under a commercial fuel supply arrangement agreed when the UK sold the reactor, provision was made for separated plutonium to be returned to Japan. By the end of 1993 a little over 1100 tonnes of Tokai 1 fuel had been reprocessed at B205. Some 2.1 tonnes of total plutonium had been separated from this fuel. The last shipment of plutonium to Japan from the UK took place in 1981, by which time some 800 kg of plutonium had been sent back by BNFL.[56] By the end of 1993, 1286 kg of Japanese plutonium were held in store at Sellafield.[57]

---

[53] Nikipelov, B. V. et al., 'Technological aspects of U and Pu recycle as most important element of closed nuclear fuel cycle', *Proceedings of the 1993 International Conference on Nuclear Waste Management and Environmental Remediation*, Prague, 5–11 Sep. 1993, p. 647.

[54] Solonin states that 'about 4 tonnes' of plutonium had been separated from BN-350 and BN-600 fuel by the beginning of 1992, but we have been unable to corroborate this estimate. Solonin, V. N., 'Utilization of nuclear materials released as the result of nuclear disarmament', Paper presented at the International Seminar on Conversion of Nuclear Warheads for Peaceful Purposes, Rome, 15–17 June 1992, table IV.

[55] Cochran and Norris (note 45), p. 101.

[56] Table 1 in 'Plutonium—do we really need it?', *Nuke-Info Tokyo*, no. 16 (Mar./Apr. 1990), shows that 660 kg of fissile plutonium were returned by the UK to Japan between 1970 and 1981 in a total of 13 shipments, 8 of them by air. To derive a total plutonium figure we assume that the material was 83% fissile.

[57] In Mar. 1994, 1.5 t of foreign plutonium were stored at Sellafield. This includes Italian material. See Department of Trade and Industry, *News Release*, DTI P/94/439, 19 July 1994; and 'Atomic energy White Paper unveils conditions of Japan's plutonium inventory', *Atoms in Japan* (Tokyo), Nov. 1994, pp. 4–7.

Oxide fuel

Reprocessing has long been regarded in Japan as a prerequisite for a strong and independent nuclear power programme. Political and technical obstacles have delayed the establishment of a large capability, but in the 1980s government and utilities embarked on an ambitious plan to turn Japan into a major producer and user of plutonium by the beginning of the next century.

Japan's first reprocessing plant for oxide fuel was constructed at Tokai-mura during the 1970s. Although completed in 1974, it did not begin full operation until 1981. The Tokai-mura reprocessing plant is unique in that plutonium and uranium are not separated from each other. A plutonium–uranium nitrate solution is used as a direct feed to mixed-oxide fuel fabrication plants on the same site. This was a concession to US concerns in the 1970s that Japan should not acquire a plutonium separation capability.

The plant's design capacity was set at 210 tonnes of fuel per year (0.7 t per day), but it has operated at well below this level, processing a total of 716.9 tonnes of Japanese PWR, BWR and advanced thermal reactor (ATR) fuel by the end of 1993.[58] By then some 5.2 tonnes of total plutonium had been separated (see table 6.8 and figure 6.6). The plant's comparatively poor performance has been the result of technical problems as well as the insistence of utility companies on running separate campaigns for their batches of fuel.

Japanese plutonium production has rather erratically increased to a level of 500–700 kg per year. This is comparable, for instance, with current French UNGG plutonium production. Japan has slowly established a domestic reprocessing capacity during the 1980s, and today it has separated more material than any other non-nuclear weapon state. It is also the only NNWS party to the NPT with plans to continue separating plutonium.

A large new reprocessing facility is planned to start operation in 2003 at Rokkasho-mura in northern Honshu.[59] The plant is being built by the Japan Nuclear Fuel Services Company (JNFS) and is financed by the electric utilities. Like the Tokai plant, it will be based on French technology, and is modelled on the UP3 plant at La Hague. Its design capacity is 800 tonnes of fuel (6–7 t of plutonium) per year (4 t per day).[60] The facility is being built to satisfy only domestic demand. There are no plans for JNFS to become a competitor to BNFL and Cogema in the international fuel services market. Projections of plutonium separation at Rokkasho-mura are given in table 6.12.

---

[58] 'Tokai marks reprocessing of 500 tonnes of fuel', *PNC Review*, no. 17 (spring 1991), p. 6; and Citizens' Nuclear Information Center, Tokyo, personal communication, 17 Oct. 1994.

[59] Official publications state that the Rokkasho plant will start operating 'sometime after 2000'. Japanese Ministry of International Trade and Industry, Nuclear Subcommittee, Advisory Committee for Energy, *Interim Report*, Tokyo, 6 June 1994. 'Construction costs for JNFL's reprocessing plant reach Y1.88 trillion ($18 billion)', *Atoms in Japan*, June 1996, p. 39.

[60] Atomic Energy Commission, *Nuclear Fuel Recycling in Japan*, Report by the Advisory Committee on Nuclear Fuel Recycling, Tokyo, 2 Aug. 1991, provisional translation, Aug. 1991.

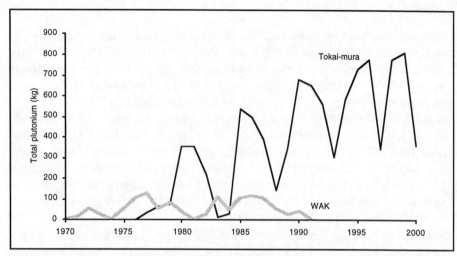

Figure 6.6. Past and projected quantities of plutonium separated at the German WAK and Japanese Tokai-mura reprocessing plants, 1970–2000

Sources: Dr P. Lausch, Deutsche Gesellschaft für Wiederaufarbeitung von Kernbrennstoffe (DWK), Karlsruhe, Private communication with the authors, 20 Jan. 1992; 'Tokai reprocessing plant completes 91-1 campaign', *PNC Review*, no. 19 (autumn 1991), p. 10; and Dr Jin Ta Kagi, personal communication, Nov. 1994.

Germany

Oxide fuel

Reprocessing of oxide fuel began in Germany in 1971. Apart from a number of laboratory rigs, reprocessing was centred at the WAK plant at the Karlsruhe Nuclear Research Centre (KfK). WAK was a pilot facility with a design throughput of 35 tonnes of fuel annually, although this was never achieved because of technical problems. About half of the fuel processed was from German light water reactors, while the rest was discharged by the materials testing heavy water reactor (MZFR) also located at Karlsruhe.[61]

Plans to extend German domestic reprocessing capacity were first unveiled in the early 1970s and went through a number of stages before foundering in 1989 when utility support was withdrawn from a facility being built at Wackersdorf in Bavaria. Soon after, a decision was taken jointly by federal and state governments to end funding for reprocessing research, and this led to the closure of WAK on 31 December 1990. A total of 208 tonnes of fuel (93 t from commercial LWRs) were reprocessed at WAK, 1164 kg of plutonium being separated.[62] The WAK plant is now being decommissioned.

---

[61] The MZFR operated between 1966 and 1984.
[62] Personal communications with Dr P. Lausch, Deutsche Gesellschaft für Wiederaufarbeitung von Kernbrennstoffe (DWK), Karlsruhe, Jan. 1992.

Figure 6.6 shows the rate of plutonium separation at the WAK and Tokai plants. It reveals that German plutonium separation for the 20 years 1971–91 was at a relatively low level, never going above 130 kg per year.

Belgium

The Eurochemic company was set up at Mol in Belgium in December 1957. It was an OECD Nuclear Energy Agency (NEA) venture through which European countries could collaborate in developing reprocessing technology. The first project was to construct and operate a demonstration reprocessing facility capable of handling fuels with a variety of claddings and burnups. Based on the technical and operational experience gained at this plant, the plan was to build an industrial-scale reprocessing plant to service fuel from OECD countries.

Construction of the demonstration plant with a design capacity of 300 kg uranium per day (50 t of fuel per year) was carried out between 1958 and 1966. The plant was then operated for almost nine years, processing fuel from gas-graphite reactors, heavy water reactors, PWRs and BWRs. A small amount of blanket fuel from Rapsodie (FBR) was also processed. In total 181.7 tonnes of fuel were reprocessed, yielding 677.7 kg of plutonium.[63] The plant was finally closed down in early 1975.

Although widely regarded as a technical success, the Eurochemic venture failed because of the growing commercialization and concentration of the European reprocessing industry during the late 1960s and early 1970s. While the Eurochemic plant was designed as a demonstration facility, it had to be operated commercially because of an emerging surplus of reprocessing capacity. The cheap reprocessing services being offered at Windscale by the UKAEA using capabilities and infrastructure developed for military reprocessing, and the lack of sustained government funding, meant that Eurochemic had to compete economically in order to continue operating. The plant's final demise was signalled in the late 1960s when two of the major shareholders, France and Germany, decided to develop national civil reprocessing capacities, thus undermining the multinational concept of Eurochemic.[64] Although there were later attempts to revive the project, these failed, and the plant was turned into a demonstration decommissioning facility.

The United States

Besides the military plutonium separation facilities at Hanford in Washington State and the Savannah River site in South Carolina (see chapter 3), three commercial reprocessing plants were built in the USA: West Valley, Morris and Barnwell.

---

[63] von Busekist, O., Detilleux, E. and Olivier, J. P., 'Fuel reprocessing and radioactive waste management: Twenty years of experience in the OECD Nuclear Energy Agency and the Eurochemic Company', IAEA-CN-42/327, in *Nuclear Fuel Cycle* (Vienna), vol. 3 (1983), pp. 83–100.

[64] Heinz, W. and Randl, R. P., 'Eurochemic: a challenge or a lost opportunity?', in *Proceedings of the Seminar on Eurochemic Experience*, 9–11 June 1983, Mol, Belgium, pp. 203–208.

Only one, the Nuclear Fuel Services (NFS) plant at West Valley, New York, ever operated. This was a Purex process plant with a design capacity of 300 tonnes of oxide fuel per year. West Valley operated between 1966 and 1972. During this time, 2058 kg of plutonium were separated from the 676 tonnes of fuel reprocessed. Over half of the fuel reprocessed (375 t) came from the N-reactor at Hanford, although this had a low burnup, yielding a total of only 553 kg of plutonium. A total of 1505 kg were separated from 301 tonnes of commercial power-reactor fuel.

The Midwest reprocessing plant at Morris, Illinois, built by General Electric (GE) with an annual capacity of 300 tonnes of fuel, ran into serious technical difficulties during cold testing and was declared inoperable in 1974. GE subsequently abandoned the project.[65] Beginning in the late 1960s, Allied-General Nuclear Services (AGNS) built a reprocessing facility at Barnwell, adjacent to the Department of Energy Savannah River site in South Carolina. It was due to begin operation in 1974 with a nominal capacity of 1500 tonnes per year. Following delays in construction and licensing, it had still not been finished in 1977 when President Carter decided to defer indefinitely all commercial reprocessing in the United States. Federal funding was cut, and licensing and construction suspended. The Barnwell facility was also eventually abandoned and the plant decommissioned.

No power-reactor fuel has been reprocessed in the USA since the West Valley plant closed in 1972, and there are no plans to revive domestic commercial reprocessing. Recent reports that nuclear utilities are considering reprocessing fuel in Europe also appear unlikely to come to anything.

India

India has long had a policy of developing a closed fuel cycle with plutonium recycling in fast reactors. Uniquely it has done this on the basis of a power-reactor programme based on natural uranium-fuelled CANDU reactors. India is the only country committed to reprocessing CANDU power-reactor fuel.

The first Indian reprocessing plant at the Bhabha Atomic Research Centre (BARC) at Trombay began operating in 1964, processing fuel from the Cirus and Dhruva research reactors primarily for its nuclear weapon programme (see chapter 9). A second reprocessing plant, the Power Reactor Fuel Reprocessing (PREFRE) facility, dedicated to reprocessing fuel from power reactors, was brought into operation at Tarapur in 1982. The nominal daily reprocessing capacity of this facility is 0.5 tonnes of uranium yielding an annual capacity of 100 tonnes of CANDU spent fuel.[66] The Chairman of the Indian Atomic Energy Commission said in late 1990 that the capacity of the facility had been increased to a maximum of 150 tonnes per year.[67] Assuming spent fuel from

---

[65] Rochlin (note 11), p. 72.

[66] 'India's significant efforts on reprocessing and vitrification', *Nuclear Europe,* Jan. 1983, pp. 45–46.

[67] Hibbs, M., 'Indian reprocessing program grows, increasing stock of unsafeguarded Pu', *Nuclear Fuel,* 15 Oct. 1990.

Indian CANDU power reactors has nominal burnups of 6700 MWd/t, it would contain about 3.5 kg of plutonium per tonne of fuel. Therefore, at maximum design fuel throughput, PREFRE could separate about 500 kg of reactor-grade plutonium a year.

However, PREFRE has never operated at this rate. For a variety of technical, logistical and political reasons production has fallen well short of design capacity. The facility was built to handle fuel primarily from the Tarapur Atomic Power Station (TAPS, BWR) and the Rajasthan Atomic Power Station (RAPS, CANDU).[68] No fuel from the TAPS reactor has been processed as this fuel is under IAEA safeguards and prior consent obligations with the United States apply to handling of its fuel. Since the mid-1970s the United States has not been willing to grant this consent. Small quantities of fuel from the RAPS 1 reactor were processed at PREFRE.[69] By late 1983, about 20 tonnes of RAPS spent fuel had been processed, much of which appeared to have had a low burn-up since it was reported to have contained only 25 kg of plutonium.[70] According to the IAEA, no additional RAPS spent fuel has been processed at PREFRE since the early 1980s.

In late 1985 or early 1986, the PREFRE facility began reprocessing spent fuel from the unsafeguarded Madras Atomic Power Station (MAPS, CANDU).[71] Processing of this fuel has been constrained by continuing technical problems at PREFRE, and the difficulty of transporting fuel from the reactor to Tarapur. Spent fuel is transported by rail but, because of a change in the railway gauge, fuel casks have to be transferred between trains *en route*. Fuel transports have therefore been kept to a minimum.

Production at PREFRE has also been limited by the amount of fuel discharged by the MAPS reactors. MAPS 1 and 2 began operating in 1984 and 1986 respectively and by June 1994 they had a cumulative load factor of 46 per cent.[72] Assuming an optimal fuel management regime, it is estimated that these two reactors together had discharged a total of 325 tonnes of spent fuel to the end of 1993. Allowing two years to cool the fuel sufficiently for handling and transportation, PREFRE could have processed no more than 240 tonnes of fuel by then. About half of this fuel (112 t) would have been the first cores containing relatively low burnup fuel with lower inventories of plutonium (see appendix B). Overall, it is estimated that, by the end of 1993, a maximum of 710 kg of plutonium could have been separated from MAPS fuel at PREFRE. At a fuel throughput of 25 tonnes per year a further 90 kg of plutonium could be separated annually, bringing the maximum up to 890 kg by the end of 1995.

---

[68] Sood, D. D., 'The role of plutonium in nuclear power programme of India', Paper presented at the IAEA Advisory Group Meeting on Problems Concerning Accumulation of Isolated Plutonium, Vienna, 26–29 Apr. 1993.

[69] 'India reprocesses RAPS fuel under IAEA eyes', *Nuclear Fuel*, 14 Feb. 1983, p. 12.

[70] Abraham, A., 'Plutonium missing in Tarapur plant', *Sunday Observer* (Bombay), 16–22 Oct. 1983.

[71] 'India's supply of unsafeguarded Pu grows as reprocessing of MAPS fuel begins', *Nuclear Fuel*, 11 Aug. 1986.

[72] Nuclear Engineering International, *World Nuclear Industry Handbook 1995* (Reed International: London, 1994), p 14.

Actual plutonium separation is thought to be considerably lower. Historically, Indian policy has been to synchronize the separation of plutonium with its use and to avoid the stockpiling of civil plutonium.[73] According to Dr A. N. Prasad, Director of the Fuel Reprocessing and Waste Management Group, Department of Atomic Energy, PREFRE has only separated plutonium for research purposes, which include a reload of the 40 MWth Fast Breeder Test Reactor (FBTR) at Kalpakkam. The FBTR has been in operation since 1987. Its core contains about 80 kg of plutonium and its annual refuelling requirements are estimated at about 30 kg plutonium per year. Maximum total requirements to the end of 1995 would have been about 320 kg of plutonium.

Since then, and in response to the cessation of French low-enriched uranium supplies for the TAPS reactor at the end of 1992, India is believed to have increased the rate of plutonium separation at PREFRE to service a small MOX fuel programme at TAPS as a way of partially making up the deficit. This policy was confirmed in late 1994 when the People's Republic of China agreed to supply LEU to keep the Tarapur reactors in operation.[74] The first shipment of LEU arrived in May 1995 at the Tarapur MOX fuel fabrication plant. No further details exist about future fuelling strategy at TAPS.

Assuming a one-third core reload, and a one-third core loading of MOX enriched with about 3.5 per cent total plutonium, annual plutonium requirements for MOX fuel for TAPS 1 and 2 would be about 300 kg. It is clear that PREFRE alone could not meet this level of demand. If fuel throughputs had been maximized at PREFRE from 1992 onwards, following a period of production to satisfy FBTR requirements alone, we estimate that by the end of 1993, some 350 kg plutonium would have been separated at PREFRE. This would have increased to 525 kg by the end of 1995.[75]

In March 1996 cold commissioning began at the Kalpakkam Reprocessing Plant (KARP), located at the Indira Gandhi Centre for Atomic Research (IGCAR) near Madras. 'Hot commissioning' was planned for the end of 1996. This facility is designed to process spent fuel from the MAPS reactors.[76] The design capacity of this facility is 100 tonnes of spent CANDU fuel a year, for an output of about 350 kg of plutonium per year.[77]

Estimating the future growth in the civilian plutonium inventory is difficult, primarily because of the lack of transparency about Indian plutonium policy, but also because of technical, political and programmatic uncertainties inherent in Indian policy. Rates of separation will be primarily determined by operating performance at the Kalpakkam reprocessing plant, progress in the construction and operation of the new Prototype Fast Breeder Reactor (PFBR) and whether secure sources of enriched uranium for TAPS can be found. At minimum, PREFRE could be expected to separate about 50 kg of plutonium a year, giving

[73] Sood (note 68).
[74] Hibbs, M., 'China will supply U, SWU to India; 70 kg MOX loaded at Tarapur BWRs', *Nuclear Fuel*, 24 Oct. 1994, p. 6.
[75] Assuming a 25 t/y fuel throughput and a 3.5 kg Pu_{tot} concentration.
[76] 'N-reprocessing plant commissioned', *The Hindu*, 28 Mar. 1996.
[77] Interview with Dr P. K. Iyengar, Bombay, 5 Sep. 1990.

a further 250 kg separated by the end of 2000. If the Kalpakkam plant is commissioned smoothly beginning in 1997 and reaches a capacity of 100 tonnes of CANDU fuel per year by 2000, it could have separated a further 980 kg or so. We therefore estimate that total Indian civil plutonium separation could be 1700–1800 kg by the end of 2000.

India has planned to accumulate a stockpile of at least 2000 kg of plutonium for the initial core-load of the PFBR which it planned to bring into operation by 2005.[78] In addition to the first core-load of plutonium, Indian officials estimate a need for another 3000 kg to refuel the reactor during the first five years of operation. After that, they plan to recover plutonium from the breeder spent fuel.

A new requirement for plutonium will arise if the decision is taken to increase CANDU fuel burnups by incorporating small amounts of plutonium in the fuel. The implementation of this plan and the requirements it imposes will depend on the future evolution of supply and demand for plutonium in India.

Argentina

In 1978, Argentina announced its decision to build a small reprocessing facility at the Ezeiza Research Complex near Buenos Aires. The plan expected to separate about 15 kg of plutonium a year from about 5 tonnes of spent fuel discharged from its Atucha 1 CANDU reactor.[79] The Ezeiza plant suffered many delays and was never completed. Argentina also planned to fabricate plutonium fuel for the Atucha 1 reactor and was building a small plutonium fuel fabrication plant at Ezeiza.

The National Atomic Energy Commission (CNEA) had plans for a larger reprocessing plant, with a design capacity of some 80–90 tonnes of spent fuel per year and recovering about 240–70 kg of plutonium annually,[80] but these plans have been cancelled (see also chapter 13).

VI. Summary of power- and fast-reactor fuel reprocessing, 1960–2000

By the end of 1993 about 158 tonnes of plutonium had been separated from power-reactor and research-reactor fuel at plants around the world (see table 6.8). At the end of 1995 this had increased to 191 tonnes of plutonium, primarily because of the rapid increase in LWR fuel reprocessing at La Hague and Sellafield. Of the total for the end of 1995, about 180 tonnes of plutonium had been separated from power-reactor fuel. About 18 per cent of the total plutonium discharged from power reactors (990 t) had therefore been separated by the end of 1995. Summary information on cumulative plutonium separation

[78] Interview with Dr P. K. Iyengar, Bombay, 18 Mar. 1992.
[79] 'Argentina's CNEA decides to reduce scope of near-term fuel reprocessing program', *Nuclear Fuel*, 12 Aug. 1985.
[80] Interview in Buenos Aires, Aug. 1988.

Table 6.8. Cumulative separation of plutonium from power-reactor and research-reactor fuel, at the end of 1970, 1980, 1990, 1993 and projected to 2000

Figures are in kilograms of total plutonium. Totals are rounded to the nearest 100.

Plant/fuel	31 Dec. 1970	31 Dec. 1980	31 Dec. 1990	31 Dec. 1993	31 Dec. 2000
France					
UP1 (Magnox)	–	1 500	7 900	10 300	12 900
UP2 (Magnox)	600	6 400	8 700	8 700	8 700
UP2 (LWR)	–	1 900	23 400	29 900	74 700
UP2 (FBR)	–	–	2 000	2 000	2 000
UP3 (LWR)	–	–	1 600	11 500	54 200
TOP/APM (FBR)	–	1 700	2 400	2 500	2 800
AT1 (FBR)	–	100	200	200	200
United Kingdom					
B205 (Magnox)	4 900	24 100	47 900	54 300	68 900
B204/205 (LWR)	200	400	400	400	400
D1206 (FBR)	–	–	3 000	3 700	5 500
THORP	–	–	–	–	17 900
Russia					
RT-1 (LWR)	–	6 500	22 000	26 000	32 000
Japan					
Tokai (LWR)	–	500	3 700	5 200	9 600
India					
PREFRE (LWR/CANDU)	–	–	100–200	300–400	600–700
KARP (CANDU)	–	–	–	–	200–1 000
United States					
West Valley (LWR)	1 200	1 505	1 505	1 505	1 505
Germany					
WAK (LWR)	–	550	1 180	1 180	1 180
Belgium					
Eurochemic (LWR/FBR/Magnox)	300	680	680	680	680
Total (thermal reactors)	7 200	44 100	119 100	149 800	283 000
Total (fast reactors)	–	1 900	7 600	8 400	10 500
Total	7 200	46 000	126 700	158 200	294 300

Source: SPRU Spent Fuel and Plutonium Database, 1994.

at each industrial-scale reprocessing plant is given in table 6.8. As in the rest of Part III of this book, the end of 1993 is used as the baseline.

Table 6.8 contains data of variable quality, the best being from facilities which have ceased operating and where separation figures have been reported in the open literature, such as Eurochemic, WAK and West Valley, the worst being for the RT-1 and Tarapur facilities for which little no published operating information is currently available. Information for the UK, France and Japan is from a variety of sources, and includes some modelling assumptions (see the relevant sections in this chapter for more detail). Plutonium separation data for all these plants are presented in appendix C. The figures for RT-1 and PREFRE should be regarded as tentative. The RT-1 figure does not include material separated from non-power-reactor fuel. All of these totals should be regarded as central estimates, with error margins of about 20 per cent for RT-1 and PREFRE, and about 5 per cent for the rest.

Estimates to the year 2000 are based on plant capacities, and on knowledge of existing contractual arrangements. We have assumed that the large European oxide-fuel reprocessing plants will operate at more or less full capacity as they process fuel under contracts covering the first 10 years of their operation. In the case of UP2 and UP3 this would take operation to the turn of the century, while THORP will continue operating until at least 2004 (and probably beyond, given its slow start-up) on this basis. B205 would continue operating at more or less full capacity to beyond 2000 even if all the British Magnox reactors were immediately shut down. Some doubt surrounds the operation of RT-1, but it can continue operating at current levels servicing just Russian VVER-440 requirements. In other words, a forecast of plutonium separation to the year 2000 is relatively robust, assuming that there are no major accidents, technical hitches or unforeseen changes in policy. By the end of the year 2000 we forecast that some 280 tonnes of plutonium will have been separated from power-reactor fuel.

The estimates above represent a slight downward revision of the separation estimates provided in *World Inventory 1992*. The primary reason is our better understanding of the operating history of the RT-1 plant at Chelyabinsk in Russia. The delay in commissioning the THORP facility in the UK has also had an effect.

The historical trend of plutonium separation in reprocessing is presented in figure 6.7. This shows that a steady buildup in world plutonium separation rates occurred from the mid-1960s to the mid-1980s, when there was a levelling off at a rate of 7–9 tonnes of total plutonium per year. From 1993 onwards however, with the commissioning of the major French and later British reprocessing plants, a rapid new increase began, leading to a doubling of plutonium separation rates to nearly 23 tonnes per year by the turn of the century. It seems likely that this will mark the apogee of civil reprocessing in the current period, the prospects being poor for a continuation of reprocessing at this level.

Plutonium separation has been a highly concentrated industrial activity. The great majority of commercial reprocessing has occurred in nuclear weapon states and at only a few sites. Four plants in three countries—Windscale/Sellafield (UK), La Hague and Marcoule (France) and Chelyabinsk (Russia)—account for over 93 per cent of the plutonium separated from power-reactor fuel to date. This concentration will increase over the coming decade. Over 95 per cent of cumulative plutonium separation will have taken place in these locations by the end of 2000.

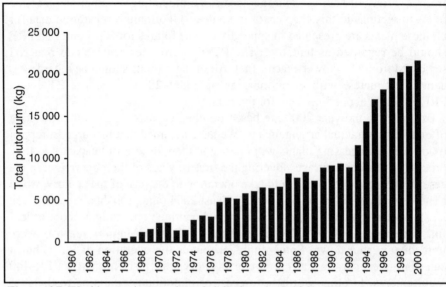

Figure 6.7. World annual separation of civil plutonium, 1960–2000

Only about 4.5 per cent of plutonium separation has so far taken place in non-nuclear weapon states (in Japan, Germany and Belgium), and the civil separation of plutonium in the threshold and *de facto* nuclear weapon states has been negligible in comparison. To a great extent, nuclear weapon states have maintained their domination over the production of plutonium. This has been due more to the poor econimics of reprocessing and difficulties in licensing reprocessing plants in NNWS, than to active policies of dissuasion by NWS.

Civil plutonium separation in states not party to the NPT is small and concentrated in one country—India. Even with a rapidly growing programme, India will have separated less than 2 tonnes of civil plutonium by the year 2000.

The early years of the next century could see a sharp decline in the rate of reprocessing, down to a level of about 9–12 tonnes of plutonium separated each year. This is mainly the result of falling throughputs at the large French, British and Russian plants.

Figure 6.8 shows the historical production of plutonium in each of the three major reprocessor countries. It shows clearly the relative stability of the British reprocessing programme between the early 1970s and the mid-1990s, compared with the steady and then accelerated expansion in the early 1990s of the French reprocessing programme. It also shows that French reprocessing has now reached a plateau. Figure 6.8 also shows that Russian plutonium separation was on a par with British and French production between the late 1970s and late 1980s, but has since fallen behind. Today the Russian reprocessing programme is at the same scale as the Japanese demonstration programme. These shifts in the importance of the three key reprocessing programmes

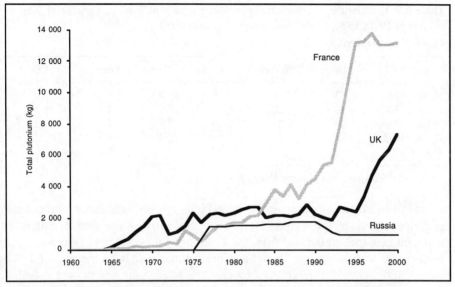

Figure 6.8. Rates of civil plutonium separation at industrial-scale reprocessing facilities in the UK, France and Russia, 1960–2000

are shown in table 6.9, which shows the proportion of cumulative plutonium separation that had taken place in different countries at specific points over the past three decades. Up to the early 1980s over half of world cumulative separation from power reactor fuel had taken place at Windscale/Sellafield. With the start of large-scale oxide fuel reprocessing at La Hague, the relative amount separated in France has risen, mainly at the expense of the UK. This trend is set to continue. By 1991–92 more power-reactor plutonium had been separated in France than in the UK. On current plans, slightly over 50 per cent of cumulative separation will have taken place in France by 2000, while the proportion separated in the UK will have fallen to one-third.

Ownership of plutonium

While plutonium separation has been carried out mainly in three nuclear weapon states, the ownership of this material is more dispersed. Much reprocessing in the NWS is of fuel from reactors in NNWS: about 70 per cent at THORP; some 50 per cent at La Hague; and, to date, over 50 per cent at Chelyabinsk.

In the European–Japanese reprocessing regime, plutonium generally remains the property of the utility in whose reactor it was produced.[81] Both the UK and France have already reprocessed substantial amounts of fuel from other countries and this will continue in the future (see table 6.10).

[81] The exception is plutonium separated from Vandellos 1 (UNGG, Spain) fuel at Marcoule (see the discussion on France above).

Table 6.9. Distribution of cumulative plutonium separation from power-reactor fuel, at the end of 1970, 1980, 1990, 1993 and projected to 2000

Figures are percentages.

Date	UK	France	Russia	Japan	Germany	Other[a]
31 Dec. 1970	73	8	0	0	0	19
31 Dec. 1980	56	22	15	1	1	5
31 Dec. 1990	40	35	19	3	1	2
31 Dec. 1993	36	40	18	4	1	2
31 Dec. 2000	33	51	12	3	0.4	1

[a] Includes Belgium, India and the United States.

Fuel from Magnox reactors in Japan and Italy has been routinely reprocessed at Sellafield since the 1960s. Details on this activity are sparse, but it is assumed here that all the fuel from the Tokai and Latina reactors is reprocessed three years after discharge. It is therefore estimated that by the end of 1993 some 2400 tonnes of foreign magnox spent fuel had been reprocessed at Sellafield,[82] and that this will rise to almost 3200 tonnes by 2002/3, when it is expected to end following the closure of the Tokai reactor. Of the 4.6 tonnes of plutonium so far separated, some 1.5 tonnes remained in store at Sellafield at the end of 1993.[83] LWR fuel has also been reprocessed from foreign reactors at both La Hague and Chelyabinsk (a small amount of LWR fuel was also processed at Windscale in the early 1970s). In all, some 28 tonnes of plutonium had been separated from foreign fuels at La Hague by the end of 1993 (compared to 12 t from French fuel).

In table 6.10 plutonium separated from foreign fuel in the UK, France and Russia is allocated to their customers' countries. By the end of 1993, it is estimated that about 46 tonnes of plutonium had been separated in the UK, France and Russia from spent fuel discharged by foreign reactors (about 30 per cent of the total plutonium separated in these three countries, see table 6.8). By the end of the year 2000 nearly 100 tonnes of foreign-origin plutonium will have been separated in the three nuclear weapon states (over one-third of the total separated). Some 42 tonnes of German-origin plutonium and 27 tonnes of Japanese-origin plutonium will have been separated in Britain and France.

We estimate that about 13 tonnes of the 26 tonnes of plutonium separated from VVER fuel at RT-1 by the end of 1993 came from non-Russian reactors.[84] None of this material will be returned under the fuel service arrangements currently in force. Under the intergovernmental agreements covering the fuel cycle of Soviet-supplied reactors, fuel remained the property of the USSR.

---

[82] Of this, 1100 t was from the Japanese Tokai 1 reactor and 1300 t was from the Italian Latina reactor.

[83] British Department of Trade and Industry, *Press Notice*, P/94/439, 19 July 1994. This had increased to 2 t by 31 Mar. 1995. DTI Press Notice, P/95/455, 13 July 1995.

[84] About 1500 t of the 3200 t handled by RT-1 is estimated to have been discharged from non-Russian reactors. We assume a mean plutonium concentration of 8 kg per tonne of irradiated fuel.

Table 6.10. Cumulative plutonium separated in the UK, France and Russia from fuel from non-nuclear weapon states, at the end of 1980, 1990, 1993 and projected to 2000
Figures are in kilograms of total plutonium.

	Past separation			Projected separation
NNWS	31 Dec. 1980	31 Dec. 1990	31 Dec. 1993	31 Dec. 2000[a]
Plutonium separated from NNWS fuel in the UK				
Italy	1 400	2 500	2 500	3 400
Japan	800	1 600	2 100	8 400
Germany	–	–	–	2 200
Netherlands	–	–	–	200
Spain	–	–	–	400
Switzerland	–	–	–	1 000
Total	**2 200**	**4 100**	**4 600**	**15 600**
Plutonium separated from NNWS fuel in France				
Japan	–	1 100	6 600	19 300
Germany	1 100	13 500	16 900	40 700
Switzerland	–	300	1 500	4 100
Belgium	–	800	2 200	5 000
Netherlands	–	–	1 200	1 900
Total	**1 100**	**15 700**	**28 400**	**71 100**
Plutonium separated from NNWS fuel in Russia[b]				
Armenia	–	–	1 800	..
Bulgaria	–	–	4 400	..
Czech Republic	–	–	1 000	..
Finland	–	–	2 100	..
GDR	–	–	1 000	..
Hungary	–	–	550	..
Slovakia	–	–	550	..
Ukraine	–	–	1 500	..
Total	–	–	**12 900**	**12 900**
Total (UK, France and Russia)	**3 300**	**19 800**	**45 900**	**97 800**[c]

[a] For fuel throughput data see tables 6.4, 6.5 and 6.7.

[b] Plutonium separation estimates for VVER fuel are given only up to the end of 1993 since only very incomplete information exists for the previous periods.

[c] Grand total assumes there is no further processing of foreign fuel at RT-1. This represents an underestimate since some foreign fuel transits to RT-1 resumed in 1995.

Final responsibility for dealing with the separated materials therefore lay with Soviet authorities. Plutonium separated from fuel reprocessed once it had been sent back to Mayak therefore belonged to Minatom. Today the situation is less clear-cut. Fuel management and reprocessing agreements have been put on a commercial basis, and this leaves open the possibility that separated plutonium may be returned to customers in Eastern Europe and former Soviet republics. For economic, environmental and security reasons, this does not appear likely to happen in the near future.

Table 6.11. Cumulative discharged power-reactor plutonium which has been separated, at the end of 1970, 1980, 1990 and 1993 and projected to 2000
Figures are in kilograms of total plutonium.

	31 Dec. 1970	31 Dec. 1980	31 Dec. 1990	31 Dec. 1993	31 Dec. 2000
Plutonium separated	7 200	44 100	119 100	149 800	283 000
Plutonium discharged	12 700	146 000	637 900	846 200	1 384 000
Percentage separated	*51*	*29*	*19*	*17*	*20*

The global significance of reprocessing as a plutonium management strategy

As a proportion of the total plutonium discharged from power reactors, plutonium separation has fallen over the past two decades, although it will probably climb again slightly during the late 1990s. This is mainly the result of the low level of oxide-fuel reprocessing in the past and the expansion of oxide reprocessing during the 1990s. Gross production and separation figures are presented in table 6.11.

Despite the great increase in reprocessing capacity, most plutonium produced in power-reactor cores (around 80 per cent) is set to remain in spent fuel. Through choice, circumstance and expediency, storage is the preferred spent-fuel management strategy around the world, although utilities face difficulties in securing additional storage capacity for mounting spent-fuel inventories in many countries. Reprocessing will remain a niche option, probably of declining importance beyond the turn of the century. Most plutonium produced in the world's power reactors will be stored for lengthy periods at the surface, before planned disposal to deep geologic repositories. Plutonium held in spent fuel is likely to remain the single largest inventory, rising from about 800 tonnes (990 t less 190 t) at the end of 1995 to over 1100 tonnes by the year 2000.

VII. Projections of plutonium separation to 2010

In this section three scenarios (maximum, current commitments and phase-out) of plutonium separation from civil fuels are presented. Under the 'maximum' scenario, all reprocessing plants operate at maximum capacity for the whole period up to 2010 and new plants are brought on-stream as planned. However, maximum scenarios have a habit of not being realized. We have therefore constructed two further scenarios which look at cases of reduced plutonium separation rates. Under the 'current commitments' scenario, only existing contracts and firm national commitments to reprocessing (in Britain, France, Japan and Russia) are assumed to be fulfilled. Under the 'phase-out' scenario, all reprocessing plants are closed down, bar one. B205 is assumed to operate until 2010 as an operational necessity.

Table 6.12. Projected annual spent-fuel throughputs and cumulative plutonium separation at industrial-scale reprocessing plants, three scenarios for 2001–10
Throughput is given in average tonnes per year; separation in kilograms of total plutonium.

Plant	Spent-fuel throughput (t/y)			Plutonium separation (kg, Pu_{tot})		
	Phase-out	Current	Maximum	Phase-out	Current	Maximum
B205	250	470	470	4 900	17 700	17 700
THORP	–	640	700	–	38 700	63 000
UP2-800	–	660	750	–	59 500	67 500
UP3	–	230	800	–	19 800	72 000
Tokai	–	20	90	–	1 500	7 200
Rokkasho-mura	–	140	370	–	11 600	30 400
RT-1	–	100	200	–	7 800	18 000
RT-2	–	–	250	–	–	22 500
PREFRE	–	–	50	–	–	1 800
Kalpakkam	–	100	100	–	3 300	3 500
Total	**250**	**2 360**	**3 780**	**4 900**	**159 900**	**303 600**

Describing a maximum forecast of plutonium separation for the period 2001–10 is relatively straightforward. French and British plants are assumed to continue operating at full capacity, although account is taken of the falling requirements for metallic fuel reprocessing towards the end of the decade as Magnox and UNGG reactors are shut down. The maximum scenario assumes that both BNFL and Cogema will be able to secure new contracts to cover all of their existing oxide reprocessing capacity. In addition, new plants at Rokkasho-mura and Krasnoyarsk are assumed to be commissioned as planned during the middle of the decade. There appears to be little likelihood that any facilities not already planned will be built.

In the 'current commitments' scenario, political and commercial pressures bring a gradual end to reprocessing. Existing baseload, options and post-2000 reprocessing contracts at THORP and UP3 are assumed to be worked out, but no new contracts are anticipated. However, it is assumed that strong national reprocessing programmes will survive. Thus in France it is assumed that EdF will renew its reprocessing arrangements with Cogema for the post-2000 period, that RT-1 will continue handling Russian spent fuel and that the Rokkasho-mura facility will be brought into operation, but at a relatively low level at first. The B205 plant also continues operating as in the maximum scenario. Under this scenario, plutonium separation at the four large reprocessing plants in the UK, France and Japan would be approximately halved in comparison with the maximum scenario.

Under the phase-out scenario an abandonment of oxide-fuel reprocessing occurs. Reprocessing contracts for the period beyond 2000 are abrogated, and all existing reprocessing policies are wound up. Scenario results are presented in table 6.12.

With such a variety of assumptions, it is to be expected that the range of possible outcomes will be wide. If the nuclear industry's present plans are carried through to the fullest extent (the maximum scenario), around 300 tonnes of plutonium would be separated from power-reactor fuel between 2001 and 2010. If only current commitments are met, that is, without substantial new ordering for reprocessing services, this will be halved. Were there to be a decisive move away from reprocessing, the quantities of plutonium produced over this timescale would obviously be very much smaller (as reflected in the phase-out scenario). The most plausible forecast probably lies somewhat below the current commitments scenario, with some of the existing contractual arrangements (such as the 'new' contracts between German utilities and BNFL and Cogema) not being fulfilled, but with strong national programmes such as those in France, the UK, Russia and Japan continuing.

VIII. Conclusions

We estimate that during 1994–2010 some 200–300 tonnes of plutonium will be separated from power-reactor fuel. This compares with the 190 tonnes separated by the end of 1995. Reprocessing is due to be carried out at 11 plants in 5 countries, 4 of which are parties to the NPT. (Reprocessing in one of these, Russia, is unsafeguarded.) The reprocessing of power-reactor fuel is a large-scale and complex industrial undertaking which today is geographically highly concentrated. To date, over nine-tenths of all civil plutonium separation has taken place in NWS. Until Japan begins operating its planned reprocessing facility at Rokkasho-mura, this geographical concentration is likely to become even more marked. However, the ownership of the plutonium separated by reprocessing is much more diffuse. As a result, substantial international flows of separated plutonium and of plutonium fuel will develop if recycling and disposition plans are implemented. Nevertheless, the number of states involved in plutonium separation and use should remain limited to those currently involved. Indeed the coming decade should see a continued waning of interest in reprocesing as more countries opt for interim storage as the preferred fuel management route. Reprocessing plans may also be affected if priority is given to the recycling of weapon-grade plutonium from dismantled Russian and US warheads (see chapter 15). The European reprocessing industry in particular seems unlikely to be able to sustain into the longer term the production levels it will experience in the short period before the century ends.

The above estimates of how much plutonium has been produced and separated are followed in the next chapter by a consideration of how separated plutonium has been used, where stocks are to be found and how widely plutonium fuels are likely to be used in the future. This allows some broad conclusions to be drawn about the future balance of the supply and demand for plutonium.

7. Commercial and research and development uses of plutonium

I. Introduction

By the end of 1995 some 190 tonnes of plutonium had been separated from power-reactor fuel around the world. This was some two-thirds of the amount separated in weapon programmes. The great majority of civil reprocessing has been carried out in the nuclear weapon states, and specifically in the UK, France and Russia. In the future a growing proportion of the plutonium separated will belong to a small number of non-nuclear weapon states. The final part of this assessment of civil plutonium considers how much separated material has been used to fuel nuclear reactors and estimates the present day and future balance between supply and demand at national and global levels.

Plutonium was seen, from the beginning of the nuclear enterprise, as both a weapon material and a valuable fuel. It was also considered that the most effective way of using the material would be in fast reactors, rather than in thermal reactors. In some of the more ambitious plans laid for nuclear power in the 1970s, large programmes for plutonium-fuelled fast reactors would replace thermal reactor programmes as uranium scarcity began to limit the growth of nuclear power.[1] A more or less self-sustaining 'plutonium economy' would then be established, plutonium being both created and burned in 'breeding' fast reactors. Continuous reprocessing of nuclear fuels would be an integral part of the fuel cycle. Fast-reactor research and development programmes were therefore launched in a number of countries, often involving international collaboration.

Different models of the transition from thermal to fast reactors were proposed. In the simplest, the first generation of thermal reactors would be directly replaced by fast reactors. Other models held that the transition would not necessarily be smooth, and that, as fast-reactor programmes were being established, plutonium would also be used to fuel thermal reactors. Research and development into thermal-reactor plutonium fuel was therefore carried out at Belgian, Dutch, German, Italian and US reactors.[2]

As fast-reactor commercialization became a more distant prospect, plutonium recycling in thermal reactors gradually came to be regarded as the only means of large-scale burning of the plutonium being separated at commercial repro-

---

[1] An influential report came from the World Energy Conference, *Nuclear Resources: The Full Reports to the Conservation Commission of the World Energy Conference* (IPC Science and Technology Press: Guildford/New York, 1978).

[2] Bairiot, H., 'Laying the foundations for plutonium recycle in light water reactors', *Nuclear Engineering International*, Jan. 1984, pp. 27–33.

cessing plants. By the early 1980s many of the countries committed to reprocessing had begun to develop programmes for plutonium recycling in conventional thermal reactors. Although energetically a less effective way of using plutonium (see chapter 2, section III), this policy allowed nuclear operators to dispose of a material which for political and economic reasons they did not want to store. In coming years, thermal recycle is likely to be the main disposal route for separated civil plutonium.

This chapter is concerned with the use of plutonium separated from power-reactor fuels in civil programmes. Sections II–IV deal with fast-reactor plutonium fuel. Sections V and VI are concerned with the recycling of plutonium in thermal reactors. There has been considerable speculation recently that plutonium extracted from dismantled nuclear weapons might be used to fuel power reactors. This issue is considered in chapter 15.

II. Fast-reactor fuel cycles

Fast reactors may be fuelled with either highly enriched uranium or plutonium. Prototypes have often been fuelled with HEU, although plutonium has always been the preferred fuel. Because fission cross-sections are much lower in fast than in thermal reactors (see chapter 2), fuel must contain higher concentrations of fissile material. The proportions of fissile uranium or plutonium contained in fast-reactor fuel therefore range from 15 to 30 per cent as compared with 3 to 5 per cent in thermal reactors. Fast-reactor cores are also less uniform than thermal-reactor cores. Fuel cores loaded with 'driver assemblies' are typically surrounded by arrays of depleted or fresh uranium fuel assemblies (blankets) in which plutonium is produced by neutron capture. This material could then be recovered by reprocessing and recycled as fuel. Today, given that the priority is to dispose of plutonium, fast reactors are typically no longer set up to 'breed', and blanket assemblies have been removed or substituted.

In a fast reactor operating to design capacities, between 30 and 40 per cent of the fuel core is replaced each year. Once discharged, the fuel is stored and allowed to cool, either under water or (as at Superphénix) in sodium. The fuel may then be reprocessed. Fast-reactor fuel cycles were originally planned to be independent and self-sustaining, although nowhere has this been achieved. Plutonium-fuelled fast reactors have in most cases been fuelled with material separated from thermal fuels. Only in France and the UK has fast-reactor fuel been consistently reprocessed and separated plutonium recycled, although a fast-reactor fuel reprocessing facility is currently being built in Japan.

The reprocessing of fast-reactor fuels involves the same technology as thermal-reactor fuel reprocessing, with some modifications to take account of the greater plutonium inventories and the more intense radioactivity produced at higher levels of irradiation. Fast-reactor fuel reprocessing capacities are reviewed in chapter 6.

When used as a fuel, plutonium is blended with natural or depleted uranium to form a mixed oxide (MOX). The greater part of the fissile worth of the fuel is then provided by the plutonium. This mixture can be sintered into fuel pellets which are inserted into a steel or zircalloy cladding. Handling plutonium is more hazardous and more complex than handling uranium because of its radioactivity and toxicity and because of the need to guard against criticality incidents. Inhalation must be avoided, and extensive radiation shielding (not necessary in uranium fuel fabrication) is also usually required. These safety precautions lead to a more capital-intensive and costly fabrication process for MOX fuel (see section V below for more details).

III. Plutonium use in fast reactors

By the end of 1995, 19 fast reactors had been built in eight countries of which 8 remained operational. They were all either prototype or demonstration plants. The only commercial-scale reactor, Superphénix at Creys-Malville in France, has now been converted into a research facility. The full list of fast reactors, together with basic fuelling information, is given in table 7.1.

The quantities given in table 7.1 refer to fissile rather than total plutonium. This is because data on reactor cores are usually published in this form, without details of the origin of the material. Plutonium used to fuel fast reactors in Europe and the USA included low-burnup material discharged from Magnox reactors (about 80 per cent fissile) as well as material separated from LWR fuel (about 70 per cent fissile). In addition, in France and the UK some fast-reactor core plutonium has been recycled. To avoid the complications introduced by estimating the fissile content, fissile amounts (Pu_{fiss}) are used in calculating plutonium in initial cores. Estimates of total plutonium (Pu_{tot}) amounts are given where necessary. For these estimates a general assumption is made that plutonium used to fuel fast reactors is 75 per cent fissile.

According to the data in table 7.1, about 9.5 tonnes of fissile plutonium were needed to fabricate the first cores for the eight reactors fuelled initially with plutonium. Maximum demand for plutonium from currently operational fast reactors is about 3 tonnes per year. In practice, far less plutonium has been loaded into these reactors. Fast reactors have, on the whole, operated erratically. Most have been experimental facilities not intended to operate at high capacity factors, and several have encountered technical problems which have led to regular and prolonged shut-down. Current annual plutonium consumption in fast reactors is less than 500 kg, almost all of this in Japan.

Operating, shut down and planned commercial plutonium fuel fabrication plants are listed in table 7.2. These include plants producing fuel for fast reactors, thermal reactors and dual-purpose facilities. The major fast-reactor MOX fuel fabrication plants operating today are at Cadarache in France and Tokai-mura in Japan.

Table 7.1. Fast reactors: retired and operating in 1995

Status/ Country	Reactor	Period of Operation	Power (MWth)	Fuel	Initial core (THM)	Initial core (kg Pu$_{fiss}$)	Maximum reload (kg Pu$_{fiss}$)
Shut down or abandoned							
USA	EBR1	1951–63	1–2	HEU	0.1	0	0
Russia	Obinsk BR5	1954–59	5–10	Pu + U	0.15	~20	0
UK	Dounreay DFR	1962–77	60	HEU	0.34	0	0
USA	EBR2/IFR	1963–94	62.5	HEU	0.46	0	0
USA	Fermi 1	1966–72	300	HEU	2.6	0	0
France	Rapsodie	1967–83	20	HEU/Pu	0.040	0[a]	0
UK	Dounreay PFR	1974–94	600	Pu + U	4.1	1 100	400–500
Germany	KNK-2	1977–91	~100	HEU/Pu	0.73	430[a]	0[b]
USA	FFTF	1980–94	400	Pu + U	1.87	640	200
USA	ZPPR	1969–94	0	Pu metal	..	..	Variable
Germany	Kalkar SNR300	—	762	Pu + U	5.1	1 400	350–400
Operational							
Russia	BOR60	1969–	60	HEU	0.176	0	Experimented with Pu
Kazakhstan	BN350	1972–	1 000	HEU/Pu	1.22	0[a]	Experimented with Pu
France	Phénix	1973–	560	HEU/Pu	4.3	0[a]	400–500
Japan	Joyo	1977–	100	Pu + U	0.76	0[a]	60–120
Russia	BN600	1980–	1 470	HEU/Pu	8.5	0[a]	Experimented with Pu
France	Superphénix	1985–	2 900	Pu + U	31.5	4 800	1 600
India	FBTR	1987–	40	Pu + U	0.19	30	30
Japan	Monju	1994–	714	Pu + U	5.9	1 100	540

[a] The initial core was enriched uranium.
[b] Only two full cores were fabricated and inserted.

Note: THM = tonnes of heavy metal.

Table 7.2. Plutonium fuel fabrication facilities
Capacity is in tonnes of MOX fuel per year.

Status/Country	Facility	Operator[a]	Period of operation	Fuel	Capacity
Shut down or cancelled					
UK	Sellafield	UKAEA/BNFL	1970–89	FBR	4
Germany	Hanau BEW1	Alkem/Siemens	1972–92	FBR/LWR	25
Germany	Hanau BEW2	Siemens	–	LWR	120
India	Trombay MOXFFP	DAE	?	LWR	10
Operating					
France	Cadarache ATPu	CEA	1970–89	FBR	15
France	Cadarache CFCa	Cogema	1989–	FBR	10
				LWR	15
France	Marcoule Melox	Cogema	1995–	LWR	120–160
Japan	Tokai PFFF	PNC	1972–	FBR	1
				ATR	9
Japan	Tokai PFPF	PNC	1988–	FBR	4
Belgium	Dessel P0	BN	1973–	FBR/LWR[b]	35
India	Trombay	DAE/BARC	1982–	FBR	0.2
India	Tarapur	DAE	1992?–	LWR	20
Russia	Chelyabinsk-65 Granat	Minatom	1988–	FBR	~1
Russia	Chelyabinsk-65 Paket	Minatom	1988–	FBR	~1
UK	Sellafield MDF	BNFL	1993–	LWR	8
Planned					
UK	Sellafield SMP	BNFL	1997–	LWR	120
Russia	Chelyabinsk-65 Complex 300	Minatom	2000?–	FBR	40?
Japan	Rokkasho-mura	Not decided	2005?–	LWR	100?
Russia	Krasnoyarsk-26	Minatom	2005?–	LWR	150?
Belgium	Dessel P1	BN	?	LWR	40

[a] BARC = Bhabha Atomic Research Centre; BN = Belgonucléaire; BNFL = British Nuclear Fuels; CEA = Commissariat à l'Énergie Atomique; Cogema = Compagnie Générale des Matières Nucléaires; DAE = Department of Atomic Energy; JNFS = Japan Nuclear Fuel Services Company; MAPI = Ministry of Atomic Power and Industry; MOXFFP = MOX Fuel Fabrication Plant; PNC = Power Reactor and Nuclear Fuel Development Corporation.

[b] FBR–MOX production ended at Dessel in 1985.

IV. Past and projected plutonium use in fast reactors

Published information on fast-reactor fuel cycles is rather scarce. However, there are relatively few operating reactors and their basic characteristics are published in open literature. It is therefore possible to go some way towards reconstructing fuelling histories by taking account of their operating performance and by balancing this against what is known about production at MOX fuel fabrication plants.

The United Kingdom

Two fast reactors were built and operated in the UK—the Demonstration Fast Reactor (DFR) and the Prototype Fast Reactor (PFR)—both located at Dounreay in the north of Scotland. Dedicated fuel reprocessing facilities were also constructed at Dounreay. Fuel fabrication for both reactors was carried out at Windscale/Sellafield, until in 1989 PFR fuel production was moved to the CFCa plant at Cadarache in France. Fuel-cycle strategy at British fast reactors included reprocessing and plutonium recycling. The UKAEA and the CEA in France are unique in having fuel-cycle infrastructures in which fast-reactor fuel has been recycled.

The DFR reactor, which operated from 1962 until 1977, was fuelled exclusively with HEU fuel. In total a little over 10 tonnes of fuel containing some 7.5 tonnes of ^{235}U were loaded into the reactor. However, the total amount of HEU dedicated to the DFR fuel cycle was much smaller since fuel was reprocessed and HEU rapidly recycled. At any one time no more than about 1 tonne of HEU was therefore committed to the DFR fuel cycle.

By the time the second fast reactor, PFR, began operating in 1974 a net surplus of civil plutonium existed in the UK. For the first six years of operation discharged fuel was simply stored. Nevertheless, PFR driver fuel has been steadily reprocessed and recycled since 1980. In total some 25 tonnes of PFR driver fuel were produced and loaded into the reactor by 1992 (see chapter 6, section V). The reactor was shut down in March 1994 and reprocessing of the final core plus an estimated 7 tonnes of blanket fuel is due to be completed by 1997/98. It is estimated that no more than 6 tonnes of PFR driver fuel were produced at Cadarache. The balance—19 tonnes—is therefore taken to have been produced at Sellafield. Assuming a typical fissile plutonium enrichment in fresh PFR MOX driver fuel of 25 per cent, it is further estimated that a total of some 6.8 tonnes of total plutonium were fabricated into MOX at Sellafield. Outer core fuel would have been fabricated using depleted uranium.

As with HEU loaded at the DFR, however, the total amount of plutonium devoted to the PFR cycle was rather less than this because material was routinely recycled after 1980. The published amount of total plutonium in the fast-reactor fuel cycle in the UK—including that in reactor cores, in spent fuel held in storage ponds, in the process of extraction at the Dounreay reprocessing plant, being fabricated into MOX fuel, and in fresh fuel—is about 5 tonnes (to the nearest half tonne). This has changed little since figures were first published in 1982.[3] We would regard this as close to the minimum required to keep the PFR reactor regularly fuelled, considering the delays and backlogs inherent within a fuel cycle. It is therefore likely that some plutonium separated at

---

[3] British Department of Energy, Press Release no. 178, 17 Oct. 1991, states that 5 t of plutonium have been sold by utilities to the UKAEA for fast-reactor R&D since 1969. This includes a small quantity of material used for other research purposes. *Parliamentary Debates, House of Commons [Hansard], Official Report*, 1 Apr. 1982, col. 169, states that 5.5 t of plutonium were sold or leased to the UKAEA between 1969 and 1981.

Windscale/Sellafield before 1969, and thus belonging to the UKAEA, was also used in fabricating the PFR fuel.[4]

France

Three fast reactors have been operated in France: Rapsodie, Phénix and Superphénix. All have been fuelled predominantly with plutonium. The main information about fuelling regimes for each of these reactors comes from reports about how much fast-reactor fuel has been reprocessed in France. Only sketchy information exists on fast-reactor MOX fuel fabrication and the amount of plutonium incorporated into this fuel. All fast-reactor MOX fuel for French reactors has been produced at the CFCa plant at Cadarache. The only hard data about the operating history of this plant is that between 1964 and the end of 1993, a total of 103 tonnes of fast-reactor fuel were produced at CFCa.[5]

The Rapsodie reactor operated between 1967 and 1983, and had a small fuel core weighing just 40 kg. 960 kg of Rapsodie MOX fuel are reported to have been reprocessed at the AT1 plant at La Hague and the APM plant at Marcoule.[6] We take this to be the total amount of MOX fuel inserted into the reactor. Assuming a mean plutonium inventory of 25 per cent, it is estimated that about 240 kg of fissile plutonium (about 300 kg of Pu_{tot}) were inserted into the reactor.

Few data have been published about fuelling strategy at the Phénix reactor, largely because it has been operated by the CEA. The first core at Phénix was composed of HEU fuel. Since then it has been fuelled with MOX, primarily with plutonium belonging to the CEA. A proportion of this material appears to have come from the production reactors at Marcoule. In addition, it is believed that Phénix was itself used as a source of weapon-grade plutonium (see chapter 3, section VII).

It is only possible to draw certain upper bounds on how much plutonium may have been loaded into Phénix. One way is to look at the reactor's operation, the other is to look at fuel-fabrication capacity. At the end of 1990, shortly before it was shut down for safety reasons, Phénix had a cumulative lifetime load-factor of about 56 per cent.[7] This is a remarkably high capacity factor for a prototype reactor. If we assume that on average 35 per cent of its core was replaced each year,[8] and that reloads were enriched to a mean of 24 per cent with plutonium,[9]

---

[4] Before 1969, the UKAEA owned all nuclear fuel in the UK. Ownership rights were transferred to utilities in 1969 and 1971 (see chapter 3).

[5] Laurent, J.-P. et al., 'Reprocessing and plutonium recycling: the French view', Paper presented at the International Conference on the Nuclear Power Option, IAEA-CN-59/71, Vienna, 5–8 Sep. 1994.

[6] Mégy, J. et al., 'The fast breeder reactor fuel cycle in Europe—present status and prospects', *Nuclear Technology*, vol. 88 (Dec. 1989), pp. 283–89.

[7] Nuclear Engineering International, *World Nuclear Industry Handbook 1992* (Reed Publishing: London, 1991), pp. 20–21.

[8] Superphénix was designed to have an operating cycle of 14 months, i.e., on average some 40% of the core would be replaced each year. Nuclear Engineering International (note 7), p. 65.

[9] The Phénix fuel core weighs 4.3 t (35% is 1.5 t). Fuel enrichments range from 19.2% to 27.2% fissile plutonium. NERSA, *The Creys-Malville Plant*, Paris, 1987, p. 8.

then over the 17 years of operation to 1990 some 6.9 tonnes of fissile plutonium could have been loaded into Phénix. Adding the core (830 kg Pu_{fiss}) and assuming one reload of fresh fuel had been fabricated (about 460 kg Pu_{fiss}) gives a total consumption of 8.2 tonnes of Pu_{fiss} (about 11 t of Pu_{tot}).[10] It is estimated that this plutonium was fabricated into a total of about 34 tonnes of MOX at Cadarache.

As with the PFR in the UK, the amount of plutonium committed to the Phénix fuel cycle was substantially smaller than this. A total of about 23 tonnes of Phénix fuel had been reprocessed by the end of 1993 (see chapter 6, section V). It is estimated that some 6.7 tonnes of total plutonium had been separated from this fuel. For logistical reasons—the delays inherent in the MOX fuel fabrication process—no more than about 5 tonnes of this could have been recycled into Phénix fuel. It can therefore be inferred that some 6 tonnes of total plutonium (4.3 t of fissile plutonium) have been devoted to the Phénix fuel cycle—an amount comparable with the PFR cycle. Phénix was brought back into operation primarily as a fuel-testing reactor within the CEA's SPIN programme at the end of 1994. Plutonium consumption at the reactor is likely to be at a markedly lower rate in the future.[11]

The Superphénix (1240 MWe) reactor is by some distance the largest fast reactor in the world. It has been entirely fuelled with MOX, the initial core weighing 38 tonnes and containing about 4300 kg of fissile plutonium (about 5700 kg Pu_{tot}).[12] The reactor reached full power at the end of 1986 and operated intermittently until July 1990 when it was shut down for safety reasons. The reactor still contains its first core of fuel, although two reloads (one full fuel core) have been fabricated. The two fabricated reloads (one-half core each) contain about 4400 kg of fissile plutonium (about 5800 kg of Pu_{tot}).[13] In 1987, 13 fuel assemblies were removed from the reactor, but they were put back into the core owing to a leakage in a fuel storage vessel. Thereafter, the reactor operated under a licence not requiring operation of the fuel store.

Not all of the plutonium loaded at Superphénix is French material. Électricité de France owns only 51 per cent of the reactor, the rest being divided between the Italian utility ENEL (33 per cent) and SBK, a consortium of other European utilities (16 per cent). The foreign component of the fuel is therefore about 4.3 tonnes.

Two years after the reactor last operated, in July 1992, the French Government allowed the existing operating licence for Superphénix to lapse by refus-

[10] See van Dievoet, J. et al., 'MOX fuel and its fabrication in Europe', Paper presented at the 7th Pacific Basin Nuclear Conference, San Diego, 4–8 Mar. 1990.
[11] Prunier, C. et al., 'The CEA SPIN programme: minor actinide fuel and target aspects', Paper presented at Global '95, Versailles, 11–14 Sep. 1995.
[12] Lacroix, A. and Gloaguen, A., 'Use of the Creys-Malville reactor as a plutonium and/or actinide burner', Paper presented at the Uranium Institute Symposium 1993, London, Sep. 1993.
[13] A total of 11.5 t of plutonium have been fabricated into MOX for Superphénix. M. A. Gloaguen, EdF, personal communication, May 1994.

ing permission for the reactor to restart.[14] Instead, a report into the future of the reactor was commissioned from Hubert Curien, a former Minister for Research and Space. The Curien Report, published in December 1992, argued that an 'adjustment' should be made to allow Superphénix to be converted into a test bed for plutonium and minor actinide 'consumption'.[15] Assent to this research programme, so clearing the way for the reactor's restart, was given in July 1994. Under the CARPA programme the whole of the first core would be replaced in 1996–97 with the already fabricated core reload. A third core—not yet fabricated—would be inserted 'around 2001'.[16] This core would be similar to previous reloads, except for insertion of some 20 'high plutonium consumption' sub-assemblies.[17] Beyond this, future fuelling strategy will depend on the success of the research and development programme and the level of interest in plutonium and actinide transmutation in a decade's time.

Taking Phénix and Superphénix together, it can be estimated that up to the end of 1993 about 8.7 tonnes of French fissile plutonium (4.3 t plus 4.4 t) and 4.3 tonnes of foreign-owned plutonium (13 t in total) were dedicated to fast reactors in France (about 11.8 t of French and about 5.7 t of foreign-owned total plutonium). These figures were valid at the end of 1995.

Japan

Plutonium has been used in six separate R&D applications in Japan: the Japan Atomic Energy Research Institute's fast critical-fuel assembly and Light Water Critical Assembly; the deuterium critical assembly of the Power Reactor and Nuclear Fuel Development Corporation (PNC); the Joyo fast reactor; the Fugen advanced thermal reactor; the Monju fast reactor; and MOX assemblies in operating LWRs. The three critical assemblies are experimental rigs together containing some 425 kg of total plutonium.[18] It is not clear which of these assemblies were fabricated with plutonium separated from Japanese fuel and which were purchased from the United States or other suppliers.

The Joyo reactor is an experimental facility commissioned in 1977. The initial Joyo I core was fuelled with enriched uranium, but this was replaced in 1979 with a new core (Joyo II) containing MOX. Since then it has been fuelled with plutonium fuel. No fuelling information has been published for the reactor, but it is known that the initial core contained 12 per cent enriched uranium.

---

[14] MacLachlan, A., 'Prime Minister orders more work, public inquiry for Superphenix', *Nucleonics Week*, 2 July 1992.

[15] Ministère de la Recherche et de l'Espace, *Le traitement des produits de la fin du cycle electronucléaire et la contribution possible de Superphénix*, Rapport du Ministre de la Recherche et de l'Espace à Monsieur le Premier Ministre, Paris, 17 Dec. 1992.

[16] Anzieu, P. and Del Beccaro, R., 'Plutonium burning and actinide transmutation in Superphenix', Paper presented at *Global '95*, Versailles, 11–14 Sep. 1995.

[17] Lacroix and Gloaguen (note 12); and MacLachlan, A., 'Test program for Superphenix set, meeting last restart condition', *Nucleonics Week*, 28 July 1994, pp. 8–10.

[18] 'Atomic Energy White Paper unveils conditions of Japan's plutonium inventory', *Atoms in Japan*, Nov. 1994, p. 5; and Dr J. Takagi, Citizens' Nuclear Information Center, Tokyo, personal communication, Dec. 1994.

Initial MOX reloads were fabricated at the Tokai Plutonium Fuel Fabrication Facility (PFFF), starting in 1978. Since October 1988 Joyo fuel has been produced at the new Plutonium Fuel Production Facility (PFPF) plant.[19] Annual reloads at Joyo have reportedly ranged between 60 and 120 kg of total plutonium. Assuming that the first MOX reload took place in 1979, a maximum of 1700 kg could have been loaded into the reactor by the end of 1993.[20] Takagi suggests that the true figure is slightly lower, estimating that by the end of 1992 some 1100 kg Pu_{fiss} (1550 kg Pu_{tot}) had been inserted into Joyo.[21] Considering the production difficulties which have been experienced at the PFPF plant in producing fuel for the Monju reactor (see below), it is assumed that no further Joyo fuel was produced during 1993. Future plutonium consumption at Joyo is put at 100–140 kg Pu_{tot} per year in Japan Atomic Energy Commission projections.

The 165-MWe Fugen ATR, which began operating in 1978, has been fuelled with both enriched uranium and plutonium fuel (about 55 per cent of the total).[22] MOX fuel for the reactor has been produced since 1977 at the ATR line of Tokai's PFFF MOX fuel fabrication plant. By the end of March 1993, about 84.2 tonnes of MOX fuel had been loaded into Fugen.[23] Assuming that this fuel had a mean plutonium enrichment of 1.9 per cent, this represents some 1600 kg of total plutonium. MOX fuel reloads have averaged at about 4.6 tonnes per year, containing nearly 90 kg of total plutonium. There are plans to maximize plutonium consumption at Fugen by converting the entire core to MOX fuel. The reactor would then be able to consume a maximum of about 200 kg Pu_{tot} each year.

A new prototype fast reactor, Monju (280 MWe) went critical on 5 April 1994. The reactor is entirely fuelled with MOX, and the initial core contained 920 kg of fissile plutonium (about 1200 kg Pu_{tot}).[24] Fuel production began at the PFPF plant at Tokai in 1990. Owing to a number of manufacturing problems beginning in 1991, fabrication of the entire first core was not completed

---

[19] Nakano, H. et al., 'Automation improves safety at Japan's Tokai MOX plant', *Nuclear Engineering International*, Dec. 1989, pp. 27–29. The PFPF began operating in 1965. 1.5 t of MOX fuel were produced for irradiation test studies. The PFFF began operating in 1972 and produced fuel for Fugen and Joyo. The FBR line at the PFPF began operating in 1988 on Joyo and Monju fuel. An ATR line is due to be added to PFPF to produce fuel for the planned Ohma DATR.

[20] Plutonium enrichments range from 20% to 40%. Berkhout, F., Suzuki, T. and Walker, W., 'Surplus plutonium in Japan and Europe: an avoidable predicament', MIT Working Paper, MITJP 90-10, Cambridge, Mass., Sep. 1990, p. 9.

[21] Japan's cumulative plutonium inventory (as of 31 Dec. 1992), CNIC, Tokyo, mimeo, Nov. 1993.

[22] About 1000 fuel assemblies have been loaded at Fugen, of which some 550 were MOX assemblies. 'ATR "Fugen" marks 15th year of operation since the first criticality', *Atoms in Japan*, Mar. 1993, pp. 9–13.

[23] The total weight of the Fugen fuel core is 34.3 t, held in 224 calandria pressure tubes each with one fuel assembly.

[24] Takahashi, T., Yamaguchi, O. and Kobori, T., 'Construction of the Monju prototype fast breeder reactor', *Nuclear Technology*, vol. 89 (Feb. 1990), p. 164; and Akebi, M. et al., 'Building Monju—Japan's prototype FBR', *Nuclear Engineering International*, Oct. 1991, pp. 37–44. Fuel assembly weight: 0.0298 t, 108 inner core assemblies (15% enriched), 90 outer-core assemblies (20% enriched). Reload enrichments will be 16% and 21%, respectively.

until January 1994.[25] The reactor was scheduled to operate on 14-month cycles. Annual plutonium requirements for Monju were projected to be 550 kg Pu_{tot} (400 kg Pu_{fiss}). However, following a sodium leak at Monju in December 1995, the reactor is expected to be shut down for at least two years.[26]

We have estimated that Japan's three experimental reactors had consumed a total of about 4400 kg of total plutonium by the end of 1993. An additional 400 kg of plutonium was embedded in critical assemblies. Future consumption in these reactors will depend principally on when Monju is brought back into operation. With all three reactors operating, and with plutonium throughput at Fugen maximized, total annual consumption could reach about 900 kg Pu_{tot} (650 kg Pu_{fiss}).

One further plutonium-fuelled demonstration reactor is planned in Japan over the next decade: a Demonstration Fast Breeder Reactor (DFBR—670 MWe, construction to begin in 2000–2005) for which a site has not yet been chosen. The size and fuelling requirements of the DFBR have not yet been published. However, scaling up from the Monju core it is estimated that the DFBR initial core would contain about 2200 kg Pu_{fiss} (3100 kg Pu_{tot}). Annual plutonium requirements for the reactor would be about 1300 kg Pu_{tot}.

Another planned reactor, the Demonstration Advanced Thermal Reactor (DATR), was cancelled in August 1995.[27] Instead, three advanced boiling water reactors (ABWRs, 1100 MWe each) with full MOX core loading are now planned, each consuming about 1200 kg Pu_{tot} annually.

Germany

Plutonium was used in the production of fuel for two German fast reactors, although only one of these was operated.[28] The KNKII research reactor (17 MWe) at Karlsruhe operated with a fast core between 1977 and 1991. Two separate fuel cores were loaded at the reactor, the first inserted in 1977 and discharged in 1982, the second in 1982 and discharged in 1991.[29] The first core was composed mostly of HEU, although it also contained 110 kg of US-origin plutonium (see chapter 6, section V).[30] The second core contained plutonium fuel. Assuming a mean enrichment of 59 per cent,[31] the total use of plutonium

---

[25] 'Fabrication of MOX fuel for Monju completed at PNC's Tokai', *Atoms in Japan*, Jan. 1994, p. 29.

[26] Hibbs, M., 'Monju leaks secondary sodium; fast breeder's future questioned', *Nucleonics Week*, 14 Dec. 1995, pp. 1–7.

[27] 'ABWR at Ohma may upset Japan's plutonium project', *Nuclear Engineering International*, Sep. 1995, p. 2.

[28] During the 1960s, Euratom bought small quantities of plutonium for experimental use at the Masurca (France) and SNEAK (FR Germany) reactors. Under the first agreement, 350 kg of plutonium for critical assemblies were sold to Euratom. A second agreement made provision for an extension of this amount to 500 kg. However, this was not taken up, the total amount purchased coming to 460 kg. About one-third of this material was used in Germany. Personal communication from W. Stoll, 25 Mar. 1992.

[29] *Nucleonics Week*, 5 Sep. 1991.

[30] 'KNK II: An experimental power station equipped with a fast core', *Nuclear Engineering International*, Jan. 1979, pp. 41–44.

[31] Nuclear Engineering International (note 7), p. 77.

by the reactor was 430 kg of fissile plutonium (about 550 kg Pu_{tot}). A third core was fabricated, but never loaded. Assuming that this was equivalent to the second, the total plutonium use in KNK II fuel was 1.2 tonnes Pu_{tot}.

Better information is available about the first core of the ill-fated Kalkar (SNR-300, 295 MWe) reactor. Although fully constructed, and provided with a complete first core of fuel, this reactor did not operate before the project was abandoned in 1991.[32] The fuel assemblies for the initial core were fabricated at the Belgonucléaire facility at Dessel (40 per cent) and the Siemens plant at Hanau (60 per cent).[33] They contained 1385 kg of fissile plutonium (about 1800 kg Pu_{tot}), giving an enrichment of about 27 per cent.[34] This fuel is assumed still to be stored at Dessel and Hanau.

In sum, about 3.25 tonnes of total plutonium have been used in fuel assemblies for two German fast reactors. Since neither reactor will operate again, no further fast-reactor demand for plutonium is anticipated.

Russia and Kazakhstan

Plutonium use in Russian and Kazakh fast reactors is still at the research and development stage. Tests on plutonium fuel have been carried out at various sites since the late 1950s. A plutonium alloy core was produced for the IBR-20 pulsed reactor at Dubna in 1959, while the BR-5 fast reactor at Obninsk was based on a plutonium dioxide fuel core. The same fuel type was used to fuel the IBR-2 pulsed reactor in 1965.[35]

Systematic research did not start until 1970 with fuel tests being carried out at the BOR-60 and SM-2 reactors at the Scientific Research Institute for Nuclear Reactors (NIIAR), Dimitrovgrad. This work led to a more extensive programme of tests carried out at the BN-350 (Aktau, Kazakhstan), starting in 1980, and at BN-600 (Beloyarsk 3, Russia) in 1990. These reactors were designed around a uranium dioxide core (enriched to 20–25 per cent ^{235}U), and cannot be completely converted to MOX use. Each reactor can be used to test up to 30 MOX fuel assemblies at a time, that is, about 700 kg MOX at BN-600 and 160 kg MOX at BN-350.[36] MOX fuel for both reactors is produced at the 'Paket' pilot MOX fabrication facility at the Mayak complex in

---

[32] Hibbs, M., 'BMFT confirms scuttling of SNR-300, Siemens to take INTERATOM in hand', *Nucleonics Week*, 28 Mar. 1991.

[33] Hibbs, M., 'Bonn will retain Pu custody but may transfer inventory', *Nucleonics Week*, 30 Apr. 1990.

[34] This plutonium came from a variety of sources. About 550 kg originated from LWR fuel reprocessing of Rheinische-Westfälische Elektrizitätswerk AG (RWE) fuel at La Hague; about 500 kg was plutonium from French Magnox reactors, about 170 kg came from the Netherlands reactor, Borselle; and most of the remainder was separated at the Weideraufarbeitungsanlage Karlsruhe (WAK). Because of the different isotopic compositions of this material, it is difficult to estimate precisely what this fissile quantity is equivalent to in total plutonium.

[35] Bibilashvili, Yu. K. and Reshetnikov, F. G., 'Russia's nuclear fuel cycle: An industrial perspective', *IAEA Bulletin*, no. 3 (1993), pp. 28–33.

[36] The BN600 core contains 8.5 t of fuel distributed between 370 fuel assemblies. The BN350 core weighs 1.22 t and contains 222 fuel assemblies. Nuclear Engineering International: *World Nuclear Industry Handbook 1995* (Reed International: London, 1994), pp. 40–42.

Table 7.3. Estimated plutonium consumption in nuclear R&D reactors in Russia and Kazakhstan, 31 December 1993
Figures are in kilograms Pu_{tot}.

Reactor	Fuel production facility[a]	Total plutonium consumed
BR-2	SRIIM and Mayak (pilot bay)	20
BR-5 (BR-10)	SRIIM and Mayak (pilot bay)	30
IBR-2	Mayak (pilot bay)	25
IBR-30	Mayak (pilot bay)	100
BOR-60	SRIAR, Paket and Granat	350
BN-350	Paket, Zmechug	70
BN-600	Paket, Granat	60
MIR	Granat	?
Total		**655**

[a] SRIIM, Scientific Research Institute for Inorganic Materials, Moscow; and SRIAR, Scientific Research Institute for Atomic Reactors, Dimitrovgrad.

Sources: Solonin, V. N., 'Utilization of nuclear materials released as a result of nuclear disarmament', Paper presented at the International Symposium on the Conversion of Nuclear Warheads for Peaceful Purposes, Rome, 15–17 June 1992; Bibilashvili, Yu. K. and Reshetnikov, F. G., 'Russia's nuclear fuel cycle: an industrial perspective', *IAEA Bulletin*, no. 3 (1993), pp. 28–33; and Mogorov, V. M., 'Utilization of plutonium in Russia's nuclear power industry', Paper presented at the International Policy Forum: Management and Disposition of Nuclear Weapon Materials, Leesburg, Va., 8–11 Mar. 1994.

Chelyabinsk-65. Production capacity at this plant is 10 fuel assemblies per year—a maximum of about 200 kg MOX containing about 50 kg Pu_{tot}.

Three manufacturing processes for MOX fuel have been developed at different institutes and industrial facilities in Russia: the 'sol-gel' approach, which was developed at the Zhemchug facility in the Mayak complex (operated 1986–87); the 'ammonia co-precipitation' process, developed for the Granat plant, also at Mayak (start of operation 1988); and the 'vibro-compaction' process, for which a pilot facility was constructed at NIIAR, Dimitrovgrad.

Commercial fast reactors (BN800-type) planned for the South Urals project at Yuzhno-Uralskaya and at Beloyarsk have been designed with MOX fuel cores. Although construction began at the South Urals site in 1984, work was abandoned in 1987 with only concrete footings in place for the first two reactors.[37] Initial cores for these reactors are designed to be about 2.3 tonnes of total plutonium, while annual reloads would be about 1.6 tonnes.[38] To support the fuelling of the BN800 reactors, work was begun in 1984 on the Complex 300 (also

---

[37] Cochran, T. B. and Norris, R. S., *Russian/Soviet Nuclear Warhead Production*, Nuclear Weapons Databook Working Paper NWD 93-1 (Natural Resources Defense Council: Washington, DC, 8 Sep. 1993), pp. 60–61.

[38] Mogorov, V. M., 'Utilization of plutonium in Russia's nuclear power industry', Paper presented at the International Policy Forum: Management and Disposition of Nuclear Weapon Materials, Leesburg, Va., 8–11 Mar. 1994.

known as A-300) MOX fuel fabrication facility at Mayak. Work on this plant was also suspended in 1987 when it was 50 per cent completed. If the BN800 projects are revived, the Complex 300 plant, with a capacity of about 40 tonnes MOX per year, could be completed by about 2000. A Franco-German-Russian plan to build a TOMOX fuel fabrication plant capable of processing 1.3 tonnes of plutonium per year (about 35 t MOX, enough for the BN-600 reactor and four VVER-1000s at Balakovo) is also under consideration. This plant, which could be coupled to A-300, is still in an early planning phase.[39]

Drawing together a complete picture of plutonium use in Russian and Kazakh R&D activities is complicated by the large number of pilot and 'semi-commercial' plutonium fuel fabrication facilities which have operated in Russia. Solonin lists eight separate facilities producing fuel for eight different reactors and a critical assembly.[40] In addition, a variety of Russian authors have published conflicting figures. The authors' own consolidated interpretation of these figures is presented in table 7.3, showing that a total of about 650 kg of total plutonium had been consumed up to the end of 1993 in Russian and Kazakh MOX fuel R&D. Without progress in the BN-800 series of reactors, plutonium consumption by fast reactors in Russia will remain at a very low level.

India

Indian MOX fast-reactor fuel fabrication capacity is small-scale and dedicated to supplying fuel for its critical assembly facility Purnima II (5 kWe, commissioned 1984) and for its Fast Breeder Test Reactor (FBTR, commissioned 1987). A small plutonium fuel-fabrication facility was set up at the Bhabha Atomic Research Centre, Trombay to produce the Purnima assembly and this was subsequently expanded to produce fuel for the FBTR. The design capacity of the facility is 200 kg of MOX per year. However, maximum annual reloads at the FBTR are about 90 kg MOX, and it seems likely that the plant has therefore operated at lower capacity.[41]

We estimate that the Purnima assembly has consumed no more than about 15 kg Pu_{tot}.[42] Estimating plutonium consumption at the FBTR is more difficult because no operating information has been published for the reactor. The fuel core at the FBTR weighs 190 kg. Assuming a 30 per cent plutonium enrichment, we can estimate that the first core consumed about 60 kg Pu_{fiss} (about 80 kg Pu_{tot}). Assuming that the reactor has operated with a 50 per cent capacity

[39] MacLachlan, A., 'French and Russians study MOX plant for fast reactors, VVERs', *Nuclear Fuel*, 9 Oct. 1995, pp. 6–8.

[40] Solonin, V. N., 'Utilization of nuclear materials released as a result of nuclear disarmament', Paper presented at the International Symposium on the Conversion of Nuclear Warheads for Peaceful Purposes, Rome, 15–17 June 1992.

[41] Sood, D. D., 'The role of plutonium in the nuclear power programme of India', Paper presented at the Advisory Group Meeting on Problems Concerning the Accumulation of Separated Plutonium, IAEA, Vienna, 26–29 Apr. 1993.

[42] The Japanese critical assemblies, DCA and FCA have power outputs of 1 kW and 2 kW, respectively. The plutonium inventory of DCA is 180 kg Pu_{fiss}, and of FCA is 300 kg Pu_{fiss}. The Purnima assembly has a power output of one-twentieth of the DCA assembly.

factor since 1987, that fuel burnups have been optimized and that plutonium enrichments for reloads are 30 per cent, we estimate that the reactor consumes around 15 kg Pu_{tot} per year. Between 1987 and the end of 1993, we estimate that about 90 kg Pu_{tot} were consumed in fuel reloads. Under these conditions, total plutonium consumption at the FBTR would have been about 170 kg Pu_{tot}.

Our upper estimate for total plutonium consumption at research facilities and fast reactors in India therefore comes to 190 kg. However, this figure must be treated with caution. In particular, there is a good chance that plutonium requirements at the FBTR reactor have been lower owing to poor operating performance, as has been the case at many other fast reactors around the world. A lower bound estimate would be one in which no fresh fuel has been loaded at the FBTR. In this case, maximum plutonium consumption in experimental and fast-reactor MOX in India would be no more than 90 kg Pu_{tot}.

Operation of the follow-on to the FBTR, the Prototype Fast Breeder Reactor (PFBR, 500 MWe) is currently planned for around 2005. The design of the reactor has been finalized and government approval for the project has been given. In the first phase this reactor would be fuelled with standard MOX (initial core, 2000 Pu_{tot}), although work is also being carried out to develop advanced fuels. To meet the demands of the PFBR a new plutonium fuel fabrication plant would need to be built, although this is still at an early planning stage. Cut-backs in the budget for nuclear programmes in India have meant that, as in Russia, although a clear government policy exists to promote fast reactors, capital scarcity is a major constraint. Further delays are likely, and forecasts of plutonium consumption are extremely difficult to make.

The United States

Three small fast reactors—EBR1, EBR2 and Fermi 1—were built and operated in the United States. All three were uranium-fuelled and have been shut down, the last, EBR2, in September 1994.

Fast-reactor research was conducted at a number of other sites under the auspices of the US Department of Energy. Two facilities, the Fast Flux Test Facility (FFTF) at Hanford, and the Zero Power Plutonium Reactor (ZPPR) at the Idaho National Engineering Laboratory (INEL), were fuelled with plutonium. Much of this material is believed to have come from the UK (see chapter 3).

The FFTF reactor (400 MWth) operated between 1980 and 1994. Four core loadings of fuel (7.5 t of MOX) containing just over 2.9 tonnes of fuel-grade plutonium (nominal 12 per cent ^{240}Pu) were fabricated for the reactor by the Kerr-McGee Corporation during the 1970s.[43] Although there were plans to produce additional fuel at Los Alamos during the late 1980s, no further reloads were produced. Cochran *et al.* report that the ZPPR Project consumed

---

[43] Cochran, T. B., *et al.*, NRDC, *Nuclear Weapons Databook, Vol. II: US Nuclear Warhead Production* (Ballinger: Cambridge, Mass., 1987), p. 76.

3.8 tonnes of plutonium (3.4 t fuel-grade, 0.2 t weapon-grade and 0.2 t reactor-grade).[44] The ZPPR reactor was also shut down in 1994.

Total plutonium use in US fast-reactor research has therefore been about 6.7 tonnes.

Summary of plutonium use in fast reactors: past and projected

Taking all these figures together, it can be estimated that by the end of 1993 fast reactors and critical assemblies had used nearly 38 tonnes of plutonium. A summary of past consumption figures, together with estimates for projected demand from operating and planned reactors, is shown in table 7.4. What is striking about these estimates as compared with those presented in *World Inventory 1992* is that very little additional plutonium has been consumed in fast reactors in the intervening years, and projections for future use have come down sharply. This is especially so for the current decade in which the French and Russian fast-reactor programmes now seem set to consume small amounts of plutonium.

In drawing up the future scenarios in table 7.4, the following assumptions have been made:

1. *High scenario*: In France, Superphénix and Phénix are brought back into operation under the CAPRA and SPIN programmes, Phénix operates until 2005, Superphénix operates to 2010; all the Japanese reactors are constructed and operate as planned (Joyo, Fugen, Monju and DFBR); FBTR continues operating and PFBR is commissioned on line in India; and a single BN800 reactor is commissioned in Russia in 2005.

2. *Low scenario*: Superphénix requires no further fuel reloads beyond those already fabricated; Phénix operates between 1996 and 2001; Joyo and Fugen operate until 2000 and Monju operates from 1998 at 50 per cent capacity in Japan; no BN800 reactors are brought into operation.

As with the projections for plutonium separation, the results show a wide range of possibilities, although as time goes by the range does appear to be narrowing and is tending to fall. This is as a consequence of the abandonment of several important fast-reactor programmes in the USA, Germany and the UK, the recasting of the French programme and the serious problems affecting the Japanese programme.

Historically, fast and experimental reactors have consumed about one-quarter of the plutonium separated from power-reactor fuel (38 t Pu_{tot} out of 157 t Pu_{tot} by the end of 1993). Over the next two decades, the proportion of separated plutonium loaded at fast reactors will fall. Because of stalled programmes elsewhere, future fast-reactor consumption of plutonium will be determined by new reactors being built in Japan, India and Russia and on the progress made

[44] Cochran *et al.* (note 43), p. 76.

Table 7.4. Plutonium consumed in fast and experimental reactor fuel, 31 December 1993, and high and low scenarios for 1994–2000 and 2001–10

Figures are in tonnes of Pu_{tot}.

Country	31 Dec. 1993	1994–2000[a] High	Low	2001–10[a] High	Low
France	17.5	2.0	0.9	15?	–
Germany	3.3	–	–	–	–
India	0.1	0.1	–	3	–
Japan	4.4	3.0	0.5	15	1.5
Russia/Kazakhstan	0.7	0.5	–	7	–
United Kingdom	5	–	–	–	–
United States	6.7	–	–	–	–
Total	**37.7**	**5.6**	**1.4**	**40**	**1.5**

[a] The projected plutonium consumption figures do not take into account recycling of plutonium separated from fast-reactor fuel. Assuming that fast-reactor fuel reprocessing continues and that separated plutonium is recycled, the projections above are over-estimates.

towards plutonium 'burning' in the French programme. The French programme alone represents about one-half of maximum consumption in the post-2000 period.

Given the technical, economic and political problems facing fast reactors in recent years, it seems extremely unlikely that the Japanese, Indian and Russian policies represented in the high scenario will be realized. Taking account of the past record of performance of Phénix and Superphénix, French fast-reactor plutonium consumption is also likely to fall far below the 'high' scenario. As a result, plutonium consumption in fast and experimental reactors may be closer to the low than the high scenario presented here. In our view, less than 10 tonnes of plutonium will be consumed in fast reactors between 1994 and 2010 compared to total separation of perhaps 270 tonnes.

As R&D reactors can only consume a small proportion of plutonium arisings under any of the scenarios presented above, the pressure on other methods of disposing of plutonium will continue to grow if reprocessing goes ahead as planned and if disposition of surplus weapon plutonium is actively pursued. Recycling in thermal reactors is currently the only available alternative.

V. Plutonium use in thermal reactors

The recycling of plutonium in thermal reactors has a considerable history. During the 1950s there was a body of technical opinion in Europe and the USA that while there was likely to be a transition to fast reactors, this transition would not necessarily be a smooth one. Imbalances between the separation of plutonium in civil reprocessing and requirements for the fuelling of fast reactors would be likely during this period. To avoid the accumulation of plutonium surpluses, nuclear planners in some countries therefore began to study the feasibility of using plutonium to fuel thermal reactors.

Research and development work sponsored by the OECD Nuclear Energy Agency and fuel suppliers such as Westinghouse continued through the 1960s and early 1970s, during which time MOX elements were loaded at 19 LWRs in 10 countries.[45] For the most part, small amounts of plutonium were involved, priority still being given to fast-reactor research. The largest of these thermal-reactor MOX programmes developed in Belgium, centred around the BR3 PWR, and in Germany between 1966 and 1977 where utilities collaborated in the development of plutonium fuel technology.[46]

During the 1970s doubts increased about future plutonium supply. Difficulties were experienced in both Europe and the USA in bringing large reprocessing plants into operation, and by 1977 President Carter had put forward his policy aimed at restricting plutonium production and use. In Europe and Japan early commercialization of fast reactors remained a policy objective, and the restriction on plutonium supply meant that interest in thermal recycling faded in most countries, even in Belgium and Germany.

In about 1980, interest in thermal MOX fuelling was revived in Western Europe and Japan. Prospects for commercial reprocessing revived while fast-reactor programmes fell into decline, so generating surpluses of separated plutonium. New thermal MOX programmes were initiated first in Germany, and a little later in Switzerland, France and Belgium. Towards the end of the decade, major plans for thermal MOX were also revealed in Japan.[47] Russia is also considering thermal recycling. Disposition in thermal reactors is also an option for rendering secure excess stocks of US and Russian weapons plutonium. Preliminary work has been completed on the technical basis for disposing of plutonium in Russian and US reactors.[48]

The MOX production process

Unlike the production of uranium oxide fuel, the fabrication of plutonium–uranium fuels must be handled remotely. Additional shielding and remote operation are necessary when handling plutonium, especially when it has been extracted from higher burnup fuels. These requirements have hampered the rapid installation of MOX fuel fabrication capacity. Up until 1995, most MOX

[45] Bairiot, H., 'Laying the foundations for plutonium recycle in light water reactors', *Nuclear Engineering International*, Jan. 1984, pp. 27–33.
[46] About 214 MOX fuel assemblies were loaded at 5 LWRs. Schlosser, G. J. and Winnik, S., 'Thermal recycling of plutonium and uranium in the Federal Republic of Germany: strategy and current status', IAEA-SM-294/33, in *The Back-End of the Nuclear Fuel Cycle: Strategies and Options* (IAEA: Vienna, 1987), pp. 541–49.
[47] 'Japan's fuel recycling policy: AEC formulates plan to use plutonium mainly for LWRs', *Atoms in Japan*, Aug. 1991, pp. 4–10.
[48] Novikov, A. N. *et al.*, 'Use of MOX fuel in VVER-1000', Paper presented at the Workshop on Managing the Plutonium Surplus: Applications and Options, Royal Institute of International Affairs, London, 24–25 Jan. 1994.

fuel fabrication remained at a sub-commercial scale. Two large new commercial-scale LWR–MOX fuel fabrication plants are being commissioned in Europe (Melox at Marcoule in France, and SMP at Sellafield in the UK). These plants are designed to absorb civil plutonium separated at La Hague and Sellafield. A third plant, SBH at Hanau in Germany, was almost completed but will not be commissioned.

The two most problematic contaminants of plutonium are americium-241 and plutonium-238, and their prevalence will, to a large extent, be determined by the level of irradiation (burnup) of the fuel from which any given batch of plutonium was separated. Americium-241 is a strong X-ray and gamma-ray emitter which results from the decay of ^{241}Pu.[49] Americium ingrowth approximately doubles the initial radiation hazard of reactor-grade plutonium within two years. Older, first-generation MOX fabrication facilities cannot handle plutonium containing more than 1.3 to 1.5 per cent of ^{241}Am (equivalent to reactor-grade plutonium stored for 3–4 years after reprocessing). New commercial facilities will be able to handle plutonium containing up to about 2.5 per cent ^{241}Am, thereby extending storage times by a further two or three years, assuming standard burnup and fuel cooling-times.

However, while ^{241}Am can be chemically separated from a sample of aged plutonium, ^{238}Pu cannot. Plutonium-238 is a spontaneous neutron-emitting isotope and an alpha emitter (causing heat production).[50] The neutrons need to be shielded against and the heat production must be taken into account during manufacture and transit. Although there are no technological limits to handling material containing high levels of ^{238}Pu (very pure samples of ^{238}Pu have routinely been used in the production of heat sources for spacecraft), licensing requirements which apply even to the new, more heavily shielded fabrication facilities currently limit the isotopic composition of plutonium which can be used in the production of MOX.[51]

Securing low operator radiation doses in the face of these hazards greatly increases the investment cost of MOX fabrication plants, and pushes up the unit cost of MOX fuel. Plutonium fuel fabrication is therefore considerably more expensive than enriched-uranium fuel production even in larger plants. Successive studies by the OECD Nuclear Energy Agency have found that MOX fabrication prices are between three and six times higher than the prices for uranium fuel fabrication.[52] Practical commercial experience, especially in Germany, suggests that the difference may at times be even greater. The expectation is that production costs will in time fall with increased volume of output. Such economies of scale and standardization will depend largely on whether large MOX plants operate at high capacity factors.

---

[49] Half-life 14.4 years, making up about 9% of LWR plutonium with a burnup of 33 GWd/t.

[50] Half-life 87.7 years, making up about 1.3% of LWR plutonium with a 33 GWd/t burnup.

[51] The MOX refabrication line at the Siemens Hanau plant held licences valid for plutonium with a 2% ^{238}Pu content (equivalent to a PWR burnup of 40 GWd/t). Kessler, G., 'Direct disposal versus multiple recycling of plutonium', Paper presented at the German RSK/Japanese NSC Meeting, Tokyo, Nov. 1992.

[52] OECD/NEA, *Plutonium Fuel: An Assessment* (OECD: Paris, 1989), pp. 68–73; and OECD/NEA, *The Economics of the Nuclear Fuel Cycle* (OECD: Paris, 1994), p. 41.

Even with high throughputs and rapid fuel fabrication at existing and planned facilities, the rates of reprocessing and MOX fuel fabrication are unlikely to be balanced. One of the key logistical bottlenecks in Europe and Japan is the lag between MOX fabrication and reprocessing for much of this decade and beyond, despite MOX programmes being brought forward in a number of countries to ensure increased plutonium consumption. Reprocessing schedules are generally set by the reprocessors, who seek to maximize fuel throughput at their plants. Unless these schedules are adjusted downwards, substantial stocks of plutonium are likely to accumulate at reprocessing sites (see section VIII of this chapter and chapter 14). These stocks would quite rapidly become unusable due to the buildup of ^{241}Am. It remains to be seen whether reprocessors will be prepared to reduce the rate of production to avoid plutonium stockpiling. The dilemma they face is that by slowing the rate of reprocessing, they will increase their costs.

Under current 'baseload' reprocessing contracts, utilities are left with little room for manoeuvre. Plutonium separation rates are decided by the reprocessor, and the rate at which this material is fabricated into fuel is determined mainly by existing capacities at MOX fuel plants which are also predominantly in the hands of reprocessors. For individual utility companies it is therefore difficult to match supply with demand so as to avoid the costly consequences of long-term plutonium stockpiling.

In normal circumstances, orders for MOX fuel must be placed by a utility with the fabricator some 18 months before plutonium is dispatched to the fabrication plant. On arrival, the material is typically stored for up to 12 months, followed by a chemical analysis. Fabrication may take about two months, with a further two-month delay before the fuel is delivered to the reactor. In all, the time between initial order and delivery, assuming all goes to plan, is between 30 and 36 months.

VI. National programmes for thermal plutonium recycling

Accounts of thermal plutonium recycling activities around the world are generally sketchy. Since plutonium fuel use has not been concentrated in one particular reactor type, and because it has until recently been on a pre-industrial scale, estimates of historical plutonium consumption must be based almost entirely on data about amounts of MOX fuel fabricated and on assumptions about plutonium enrichments in the fuel. Projections of plutonium consumption in commercial-scale recycling programmes can be based on plans published by utilities.

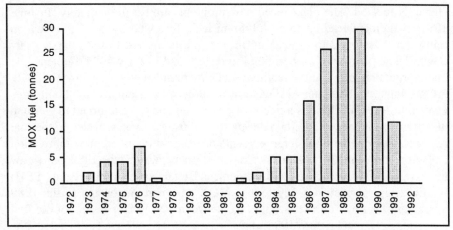

Figure 7.1. LWR–MOX production at the German Hanau fuel fabrication plant, 1972–92

Germany

The first German MOX R&D programme ran from 1966 until 1977. It was funded by the federal government and was concentrated at three reactors: the VAK reactor at Kahl (MOX first loaded in 1968); the Gundremmingen A BWR; and the Obrigheim PWR. Altogether some 21 tonnes of MOX fuel containing about 600 kg of fissile plutonium were produced for this programme at the Alkem plant at Hanau.

A second 'pilot' MOX agreement was concluded between Siemens and German nuclear utilities in 1980. This covered the period 1982–88 and involved the loading of 201 fuel assemblies (84 t MOX, about 2.3 t fissile plutonium) at seven PWRs and two BWRs.[53] As with the first programme, all of this fuel was manufactured at Hanau. Figure 7.1 shows the production record at Hanau between 1972 and 1992. A total of 158 tonnes of LWR–MOX fuel were produced containing about 4400 kg fissile plutonium (6.2 t of Pu_{tot}).

From the mid-1980s onwards, plutonium fuel fabrication activities at Hanau were subject to political and legal challenge as the plant became the focus of anti-nuclear campaigning in Germany. MOX fabrication continued, however, to be relatively undisturbed during the 1980s. In 1991 a Social Democratic Party (SPD)–Green coalition government was re-elected in the state of Hesse, where

---

[53] Participating reactors were: Obrigheim, Neckarwestheim 1, Grafenrheinfeld, Unterweser, Grohnde, Brokdorf, Phillipsburg 3 (PWRs), and Gundremmingen and Kruemmel (BWRs). Schmiedel, P., 'Experience with plutonium recycling in the Federal Republic of Germany', Paper presented at the Uranium Institute Thirteenth Annual Symposium, London, 7–9 Sep. 1988; Schlosser and Winnik (note 46); Hibbs, M., 'German utilities bracing for MOX fuel cost increases', *Nuclear Fuel*, 6 Jan. 1992, pp. 10–11; Brandsetter, A., Schmiedel, P. and Stoll, W., 'Einsatz von Plutonium in LWR und SBR' [The use of plutonium in LWR and SBR], *Atomwirtschaft*, Aug./Sep. 1984, pp. 453–58; and Schmiedel, P., 'Recycling and its implications for uranium demand: the German perspective', Paper presented at the Uranium Institute Fifteenth Annual Symposium, London, 5–7 Sep. 1990. The latter paper states that by Sep. 1990 Siemens had manufactured almost 70 000 fuel rods weighing 133 t and containing 3.7 t of fissile plutonium (a mean plutonium enrichment of 2.78%).

Hanau is located. The plant faced new problems almost immediately. In June 1991 an incident leading to a plutonium leak, followed by another in which rainwater was found to be leaking into the building, led to the plant's shutdown.[54] The plant was not permitted to restart, and in early 1994 Siemens, the operator, announced that the facility was being permanently shut down.

On 1 January 1989 a new MOX agreement between Siemens and the utilities came into force. Under this agreement plutonium due to be returned to German utilities from France and Britain during the 1990s was to be recycled in thermal reactors. To cope with the large amounts of material involved, new fabrication capacity was planned at Hanau. Construction began in late 1987 on a new bunkered MOX fabrication facility (Siemens Brennelementwerke Hanau, SBH) with a design capacity of 120 tonnes MOX per year. According to the 1989 agreement, German utilities would take between 70 and 80 per cent of the fuel produced at Hanau up until 1998.

Construction of the new Hanau plant suffered from opposition by the Hesse state government. Despite sustained efforts to keep the project alive, at the end of June 1995 German nuclear operators and Siemens decided to abandon the SBH facility for civil plutonium fuel fabrication.[55] Alternative uses for SBH, such as conversion to a 'disarmament' facility processing Russian weapon plutonium into MOX, were considered during 1995, but these discussions came to nothing.

With the closure of the old Alkem plant and the abandonment of the new SBH plant, German utility MOX strategy will depend on fuel fabrication in Belgium, France and possibly Britain. Since 1992 all German MOX has been fabricated at Dessel. This has led to a sharp slow-down in MOX fuel insertion at German reactors. Between the end of the second Siemens agreement in 1988 and the end of 1993, just 32 tonnes of MOX were loaded. A total of 158 tonnes MOX had therefore been irradiated, containing some 6.4 tonnes Pu_{tot}. A further 0.5 tonnes of total plutonium had been used to fabricate fuel not yet inserted into reactors.[56]

At the end of 1995, German utilities held contracts for the fabrication of 323 tonnes MOX, to be produced at Dessel and Melox in the period 1995 to 2003. This could absorb between 16 and 20 tonnes of plutonium, depending on the matrix material used and burnup assumptions. This compares with a total of about 42 tonnes of plutonium which will be available to German utilities between 1995 and 2003.

As well as securing new MOX fabrication capacity, German utilities have sought to extend the capacity to load MOX fuel at operating reactors. Applications were made in the late 1980s for licences to load MOX at all but three power reactors (making a total of 18). In 1995, 12 reactors held MOX licences,

---

[54] 'Incidents close Hanau plant', *Nuclear Engineering International,* Aug. 1991, p. 3.

[55] 'Hanau MOX plant abandoned', *Nuclear Engineering International,* Sep. 1995, p. 7.

[56] Thomas, W., 'Use of mixed oxide fuel in existing light water reactors in Germany', eds E. R. Merz *et al.*, *Mixed Oxide Fuel Exploitation and Destruction in Power Reactors* (Kluwer: Dordrecht, 1995), pp. 113–22.

but difficulties had been encountered negotiating the remainder. Licences have been awarded for the loading of up to 172 PWR assemblies and 138 BWR assemblies per year, together containing about 4 tonnes of fissile plutonium (5.6 t Pu_{tot}). Existing MOX fabrication contracts would consume plutonium at a rate of about 2 tonnes total plutonium per year.

Belgium

Belgonucléaire, together with Siemens, has been a pioneer of LWR–MOX fuel fabrication, beginning with pilot-scale production at Mol of plutonium fuel for the BR3 reactor in 1963. New production capacity, capable of fabricating FBR and LWR–MOX fuel, was established at Dessel in 1973.[57] Production remained at a low level until 1983/84, with some 240 fuel assemblies being produced for two BWRs (Garigliano and Dodewaard) and one PWR (Chooz), besides BR3 fuel.[58] During this period somewhat less than 10 tonnes of MOX were manufactured, containing about 550 kg total plutonium.[59]

In 1984, under the joint COMMOX initiative between Cogema and Belgonucléaire, the plant (renamed P0) was backfitted to give it a rated capacity of 35 tonnes of LWR–MOX per year. It was gradually brought to full capacity in 1989 (see figure 7.2). Over 80 per cent of this capacity has been taken up with orders for French reactors, the rest being produced for BR3, Swiss and German reactors. By the end of 1993, 217 tonnes of fuel containing 10.5 tonnes of total plutonium had been produced in this new phase of production. Total plutonium consumption in LWR–MOX production at Dessel until the end of 1993 was therefore about 11 tonnes. In the late 1980s, Belgonucléaire began planning to build a second MOX fabrication module at Dessel (P1). These plans were suspended following legal challenge.

COMMOX held contracts with the French utility EdF to supply about 25 tonnes of MOX annually up to 1996. These contracts, together with further demand from Swiss, Belgian and German utilities, assure production at Dessel P0 until the turn of the century.

Apart from being a producer of MOX, Belgium is also a consumer of plutonium fuel. Belgian policy is to recycle rapidly all the plutonium separated in reprocessing in two operating PWRs. A total of about 4.7 tonnes of plutonium separated at La Hague is due to be recycled in Belgian reactors between 1995 and 2005. The fuel will be produced by Fragema, subcontracting to Belgonucléaire the production of MOX pellets at Dessel. Some 144 MOX assemblies

---

[57] Le Bastard, G., 'MOX fuel fabrication: present and future', Paper presented at the Uranium Institute Fourteenth Annual Symposium, London, 6–8 Sep. 1989.

[58] A total of some 220 BR3 assemblies containing about 300 kg total plutonium were produced at Dessel. See van Dievoet, J. et al. (note 10); and Nuclear Engineering International (note 7), p. 85.

[59] Bairiot, H., et al., 'Foundations for the definition of MOX fuel quality requirements', Paper presented at the Characterization and Quality Control of Nuclear Fuel Conference, Karlsruhe, 19–21 June 1990.

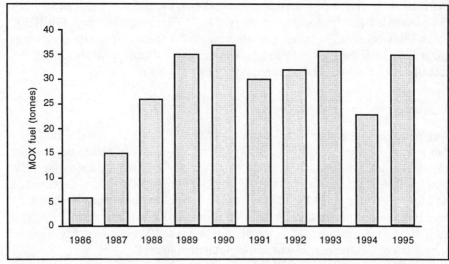

Figure 7.2. LWR–MOX production at the Belgian Dessel P0 fuel fabrication plant, 1986–95

(65 tonnes MOX) will be produced and these will be inserted at the Tihange-2 and Doel-3 PWRs.[60]

MOX licences for both reactors were awarded in June 1994, following a public and parliamentary debate about nuclear fuel policy and a difficult licensing process.[61] The first MOX reloads were inserted at Tihange and Doel in 1995. Under current plans fuel will be delivered at a rate of about 8 tonnes MOX per year between 1995 and 2002, all of which will be produced at Dessel. Higher plutonium enrichments will be achieved through the use of depleted uranium as the fuel matrix and by driving up average fuel burnups to 45 GWd per tonne.

France

Plutonium recycling in LWRs was not considered by EdF until the early 1980s when, because of reduced forecasts of fast-reactor plutonium consumption, a series of feasibility studies was commissioned.[62] In 1985 agreements were signed by EdF and a number of French fuel-cycle companies which supported a French MOX programme based on the 900 MWe class of PWRs. Under this programme a new plutonium fuel production facility has been built at Marcoule and a MOX recycling programme launched by EdF. The Melox plant, commissioned in 1995 by COMMOX, has a design capacity of 120 tonnes MOX per annum (6–8 t of Pu_{tot}, depending on the plutonium enrichment), although

---

[60] Resteigne, L., 'Utilisation of MOX fuel assemblies in Belgian PWRs', Paper presented at *Global '95*, Versailles, France, 11–14 Sep. 1995.
[61] MacLachlan, A., 'Belgian MPs approve use of MOX but put moratorium on reprocessing', *Nuclear Fuel*, 3 Jan. 1994, pp. 1–4.
[62] EdF, Service des Combustibles 'Retraitement recyclage', Paris, 6 Mar. 1990, mimeo.

capacity is due to be expanded to 160 tonnes. In 1995, 25 tonnes of MOX pellets were produced at Melox and the plant will reach full capacity in 1997.[63] Since 1988, most EdF MOX requirements have been met with production at Dessel. The remainder has been produced at CFCa at Cadarache, refurbished to produce LWR–MOX in 1989. By the end of 1993, 32 tonnes MOX (containing about 1500 kg Pu_{tot}) had been produced at Cadarache.[64]

MOX was first loaded into a French PWR in September 1987, when 16 assemblies were inserted into the St Laurent B1 reactor.[65] Since then MOX has been loaded at six further 900-MWe class PWRs (St Laurent B2; Gravelines 3 and 4; Dampierre 1 and 2; and Blayais 2). By the end of 1993, 153 tonnes of MOX (containing about 7.8 t Pu_{tot}) had been inserted at French reactors, most of it produced at Dessel.[66] By the end of 1995 some 245 tonnes of MOX had been inserted, containing 11.7 tonnes PU_{tot}.[67]

EdF currently holds licences to fuel 16 of its reactors (all in the 900-MWe class) with one-third core of MOX, and plans to seek changes to the licences of a further 12. In the first '16 reactor' phase of the programme, a 'hybrid' fuel management scheme has been implemented in which enriched uranium fuel operates for four fuelling cycles to a higher burnup while MOX operates at 33 GWd/t over three cycles. EdF's historical MOX fuel loadings are shown in figure 7.3. Over the next few years EdF expects to be able to increase MOX burnups, thereby increasing the plutonium enrichment of the fuel, maximizing the amount of plutonium consumed in these reactors and avoiding the economic penalties of the hybrid scheme. Under the current plan, some 22 reactors will be fuelled with MOX, so raising EdF's MOX fuel requirement from a current 45 tonnes per year (consuming about 2.3 t Pu_{tot}), to about 140 tonnes (consuming up to 10 t of Pu_{tot}) in 2000 and beyond.[68] There are no plans to fuel the larger 1200-MWe class of PWRs with MOX.

Under the 'equality of flows' strategy, EdF aims to bring into equilibrium the separation of plutonium from its spent fuel, the fabrication of this material into MOX and the insertion of MOX fuel into reactors. EdF is therefore seeking to increase the number of reactors where MOX can be loaded, and to maximize the amount of plutonium that can be consumed by MOX reloads by driving up MOX fuel burnups and by using a depleted uranium fuel matrix. A key goal of

[63] Gloaguen, A., EdF, personal communication, 21 Sep. 1995.
[64] Laurent, J. P. et al., 'Reprocessing and plutonium recycling: the French view', Paper presented at the International Conference on the Nuclear Power Option, IAEA, Vienna, 5–8 Sep. 1994.
[65] 'Saint Laurent inaugure les "MOX"', Revue Générale Nucléaire, no. 5 (Sep./Oct. 1987), p. 490.
[66] Simon, M. A., 'Recycling developments in France', Paper presented at the Uranium Institute Fourteenth Annual Symposium, London, 5–7 Sep. 1990; and MacLachlan, A., 'EdF firms up Pu use strategy, will licence 12 more PWRs for MOX', Nuclear Fuel, 20 June 1994, pp. 12–15.
[67] Fournier, W. and Dalverny, G., 'Experience of Melox start-up', Paper presented at Global '95 (note 60).
[68] French 900 MWe reactors operating with one-third core MOX can each accept about 8 t of MOX per year with a plutonium enrichment of about 5% Pu_{tot}, assuming a depleted uranium fuel matrix. Under a quarter-core refuelling regime, about 4.5 t of MOX could be inserted annually into these reactors with a plutonium enrichment of about 7% Pu_{tot}. Nigon, J.-L. and Golinelli, C., 'MOX in France: Domestic programme and MELOX plant', eds E. R. Merz et al. (note 56), pp. 235–40.

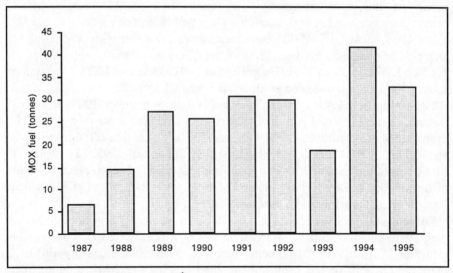

Figure 7.3. MOX fuel reloads at Électricité de France reactors, 1987–95

this policy is to avoid the unnecessary storage of separated plutonium. The EdF strategy calls for 'mono-recycling' of plutonium and reprocessed uranium. In meeting all these objectives, EdF enjoys a considerable amount of flexibility over reprocessing schedules which, until now, has been denied to other utility customers of Cogema and BNFL.

Switzerland

Details of thermal recycling in Switzerland are relatively scarce. No official government policy exists on the use of plutonium,[69] and the utilities themselves have published little about their activities. As they are not members of fast-reactor consortia, their only use for separated plutonium is as LWR–MOX. The first demonstration loading of MOX assemblies took place at the Beznau 1 reactor in 1978.[70] This fuel was fabricated by Siemens at the Hanau plant.

In pursuing a recycling policy since the late 1970s, the utilities have been constrained by US nuclear export policy. Under the US 1978 Nuclear Non-Proliferation Act, 're-transfers' (as from Cogema to a MOX fuel fabricator) of plutonium produced from US-origin uranium must receive 'prior consent' from the US Congress and regulatory bodies. Since Switzerland is not a member of the European Union, retransfer arrangements had to be negotiated through the US–Swiss nuclear cooperation agreement. Because of concerns in the USA about Swiss nuclear export policies, the US Congress insisted on approvals

---

[69] OECD/NEA, 1989 (note 52), Annex F, p. 114.
[70] van Dievoet et al. (note 10).

being made case-by-case, which has been time-consuming for the Swiss to arrange. Approvals from the USA have taken up to four years to be awarded.[71]

Four approvals were granted during the 1980s—for 200 kg in 1985; 108 kg in 1988; and 150 kg and 132 kg in 1990,[72] making a total of 590 kg fissile plutonium (about 730 kg of Pu_{tot}). It is also known that 612 kg of total plutonium were fabricated into MOX fuel at Dessel during 1990 for insertion into the Beznau 2 reactor.

To date, therefore, it is estimated that a total of about 1300 kg Pu_{tot} have been inserted into Swiss reactors, mainly Beznau 1 and 2. However, of this material, only some 500 kg had been separated from Swiss fuel. The rest was borrowed from other utilities under 'swap' arrangements. In late 1994, a total of about 5.8 tonnes of fissile plutonium were still to be separated from Swiss fuel at La Hague and Sellafield. About 700 kg of this will be swapped back to non-Swiss utilities, leaving a total of about 5.2 tonnes (7.1 t Pu_{tot}, about 140 t of MOX fuel) to be recycled in Switzerland. This fuel is expected to be produced at Dessel, Melox and the MDF plant at Sellafield. Swiss utilities currently hold contracts covering about 3.4 tonnes of total plutonium.

Swiss utility policy is that each utility will recycle its own plutonium, and that no further reprocessing contracts will be entered into.

Japan

As in France, the policy of recycling plutonium in LWRs was slow to be adopted in Japan. Until the mid-1980s the government and utilities still believed that plutonium demand from fast reactors and advanced thermal reactors would be sufficient to absorb the material separated in Europe and Japan during the 1990s.

Adjustments in Japanese reprocessing–recycling policy began in earnest with the creation in 1984 of the 'Study Group on Long Term Nuclear Development Strategy', by the Japan Atomic Energy Commission (JAEC). This group was concerned with the problem of 'surplus plutonium' and recommended that in addition to planned fast reactors, 5 ATRs and 10 LWR units would be required to absorb some 100 tonnes of plutonium arising from reprocessing. An apparently unrelated proposal from another group within the JAEC recommended in 1985 that 7 ATRs and 17 LWRs would be required by 2010 to recycle available plutonium.[73]

These recommendations were confirmed as official policy in the 1986 Nuclear Energy Vision published by the Ministry of International Trade and

---

[71] Bay, H., 'Problems concerning the accumulation of isolated plutonium—the Swiss situation', Paper presented at the Advisory Group Meeting on Problems Concerning the Accumulation of Separated Plutonium, IAEA, Vienna, 26–29 Apr. 1993.

[72] The last transfer was from La Hague to Dessel. Knapik, M., 'US appears set to bless commercial use of plutonium in Swiss nuclear plants', *Nuclear Fuel*, 3 Sep. 1990.

[73] Skolnikoff, E., Suzuki, T. and Oye, K., *International Responses to Japanese Plutonium Programs*, Report C/95-5, Center for International Studies, MIT, Cambridge, Mass, Aug. 1995, p. 6.

Industry (MITI).[74] In the same year a small-scale MOX demonstration project was launched with the insertion of two fuel assemblies at the Tsuruga 1 BWR.[75] Since then, plutonium and recycling policy has been the subject of more or less continual review by a number of government and industry agencies. This has led to significant evolution in policy, characterized mainly by postponements and drift. Today there is still great uncertainty surrounding the eventual shape of Japanese plutonium-use programmes.

The 1987 JAEC Long Term Programme proposed yet another programme including one Demonstration ATR (DATR) and 10 LWRs being fuelled with plutonium by the late 1990s. The most recent policy statement, the Long Term Programme of 1994, stated that 'a few' PWRs and BWRs would be fuelled with plutonium 'in the latter half of the 1990s'.[76] The objective was to fuel 'over 10' LWRs with MOX by 2010. By then, the current plan states that about 3 tonnes of fissile plutonium (a little over 4 tonnes Pu_{tot}) will be recycled in LWRs per year. With the cancellation of the DATR reactor in 1995 there are plans to construct a further three ABWRs, also to be fuelled with MOX.

Some of the confusion stems from the principle that each utility should initially recycle its own plutonium, although there may be more cooperation in the future. It is generally expected that the large utilities—Tepco and Kansai—will take the lead and that MOX will be produced in Europe. No utility has yet come forward publicly with a definite recycling programme. There have been logistical, bureaucratic and political obstacles to the creation of a practicable LWR–MOX programme. Utilities have had to reach agreement among themselves and with the regulator on practical steps to fabricating and loading plutonium fuel into reactors. They have also had to contend with uncertainties associated with MOX fuel fabrication in Europe. Another obstacle is potentially hostile public reaction in areas around reactors fuelled with plutonium fuel. Consent from local governors is required before plutonium recycling can commence. Following the December 1995 accident at Monju, governors of three leading prefectures submitted an official letter to the Prime Minister asking for a delay to the start of plutonium recycling until the problems at Monju are resolved. Monju is unlikely to operate again before 1998.

Lastly, utility companies have sought to avoid the controversy which surrounded the 1992 shipment of plutonium from France to Japan.[77] Future shipments will be in the form of fabricated MOX fuel, rather than plutonium oxide. Before placing MOX fabrication contracts with European suppliers however, the utilities are seeking assurances that regular shipments of MOX fuel can be

---

[74] MITI, *Nuclear Energy Vision*, Tokyo, Sep. 1986, pp. 39–42.

[75] This was followed in 1988 by 4 assemblies (containing just 15 kg of total plutonium) being loaded at the Mihama 1 PWR. Matsuoka, Y. and Abeta, S., 'Mihama MOX trials meet with success', *Nuclear Engineering International*, Dec. 1989, pp. 24–25. If the Tsuruga assemblies contain a similar percentage of plutonium then the total plutonium loaded as MOX in Japan is about 20 kg.

[76] 'Fuel recycling reaffirmed by new long-term program as summary comes out', *Atoms in Japan*, May 1994, pp. 4–9.

[77] Separated plutonium was returned to Japan from Britain on several occasions during the 1970s and 1980s.

made without international and domestic rancour. There is as yet no sign that this attempt at finding national and international consensus has succeeded, and without it, no firm plans can be put in place. In total, about 43 tonnes of total plutonium are due to be returned to Japan over the next 10 years or so. In low burnup LWR fuel this would be enough to produce somewhat over 1000 tonnes of MOX fuel.[78]

The United Kingdom

While plutonium recycling in British gas-cooled reactors was considered during the 1950s, no economic case was demonstrated. Instead, interest in plutonium as a fuel focused on fast reactors. When surpluses of civil plutonium first became apparent in the early 1970s, a policy of plutonium storage was adopted and this has continued to the present day. More recently, new studies into plutonium recycling in AGRs have been launched. A recent Uranium Institute report makes clear that one-third core MOX fuelling is planned to begin at the Sizewell 'B' PWR in 2000/2001.[79]

By the late-1980s, as large-scale reprocessing capacity began to come into operation in Europe, BNFL decided to establish a MOX fabrication capacity at Sellafield, primarily with an eye on the potential for business in fabricating Japanese plutonium separated at B205 and THORP into MOX. The first step was achieved in spring 1993 when the MOX Demonstration Facility (MDF) was commissioned. This plant has a design capacity of 8 tonnes MOX per year and results from the upgrading of the mothballed UKAEA Fuel Services plant at Windscale, which up until 1989 produced fuel for the Dounreay PFR. The first order for about 10 tonnes of MOX was signed with the Swiss utility Nordostschweizerische Kraftwerke AG in early 1992, and BNFL has stated that 'more than half the capacity of MDF' has been reserved.[80]

The next phase of expansion for BNFL involved the rapid construction of a new facility, the Sellafield MOX Plant (SMP). SMP has a planned capacity of 120 tonnes MOX per year (5–6 t Pu_{tot} per year) and shares some technology of the SBH facility at Hanau. The plant received planning approval in early 1994 and BNFL expects to produce its first MOX fuel at SMP in 1997.[81]

India

India is committed to establishing a fast-reactor programme based on plutonium fuel. Since the late 1970s a small plutonium fuel programme has also

---

[78] Assuming natural uranium matrix and 33 GWd/t burnup, plutonium enrichments would be about 4.1% Pu_{tot}.

[79] Uranium Institute, *The Recycling of Fissile Materials*, Report of the Recycling Group, London, Aug. 1996.

[80] Marshall, P., 'BNFL wins order to supply MOX fuel for Switzerland's Beznau-1', *Nuclear Fuel*, 20 Jan. 1992, p. 13; and Marshall, P., 'BNFL ships MOX to Switzerland's NOK', *Nuclear Fuel*, 4 July 1994, p. 13.

[81] Marshall, P., 'BNFL given local approval to build commercial MOX fabrication plant', *Nuclear Fuel*, 28 Feb. 1994, pp. 8–9.

been put in place for the Tarapur BWR station (2 x 160 MWe, commissioned 1969). The Tarapur reactors are fuelled with enriched uranium. Since India has no domestic commercial-scale enrichment capacity, foreign supplies have needed to be secured. Initially, enriched uranium was provided by the United States, but in 1976 these supplies were restricted as a way of bringing pressure on India to curtail its suspected nuclear weapon programme. US supplies were suspended in September 1979. It was not until late 1982 that a new supply of uranium was arranged with France.

In the interim period it was decided to set up a Mixed Oxide Fuel Fabrication Plant at the Trombay nuclear research centre.[82] The plant had a design capacity of 10 tonnes MOX per year. Following the agreement with France, this plant was not operated. By the time the French supply agreement expired in 1992, a new plutonium fuel plant had been completed at Tarapur. The plant has a published design capacity of 20 tonnes MOX per year. In late 1994 it was reported that India had inserted 70–80 kg MOX (containing perhaps 3 kg Pu_{tot}) into the Tarapur reactor.[83]

India is expected to continue loading MOX at the reactor in future years. The original plan to go for a full core of MOX at both Tarapur units need not now be implemented in the short term, since China has agreed to supply enriched uranium. Indeed, implementing a full-core MOX scheme is currently beyond India's means. To produce a one-third core reload for both reactors would require some 300 kg of total plutonium per year. It is estimated that the total Indian plutonium inventory stood at about 150 kg at the end of 1993 (350 kg separated minus about 200 kg consumed in fast reactor and critical assembly MOX fuel) with annual separation at a rate of 50–100 kg total plutonium. Future MOX production for Tarapur will depend on several factors, including the availability of long-term supplies of enriched uranium, the rate of reprocessing at PREFRE and the new plant at Kalpakkam, and plutonium requirements in the Indian fast reactor programme.

Russia

Very little plutonium has been put to civil use in Russia. With the large and growing stockpile of civil plutonium at Mayak, and the anticipated commissioning of the RT-2 reprocessing facility next century, efforts are underway to develop a concerted policy of civil plutonium disposition, including in commercial reactors. There will also be a need in the longer term to develop ways of disposing of the large amounts of plutonium released from the dismantlement of nuclear warheads.

For Russian planners the preferred means of plutonium disposition is still as fuel in fast reactors (see section IV above), although light water reactors are

[82] Sood (note 41), p. 5.
[83] Hibbs, M., 'China will supply U, SWU to India: 70 kg MOX loaded at Tarapur BWRs', *Nuclear Fuel*, 24 Oct. 1994, p. 6.

Table 7.5. Plutonium consumption in LWR–MOX fuel fabrication, up to 31 December 1993

Figures for MOX fuel are in tonnes of heavy metal; figures for plutonium are in kilograms. Totals are rounded.

Country	MOX fuel	Fissile plutonium (Pu_{fiss})	Total plutonium (Pu_{tot})
Belgium	227	7 900	11 000
France	32	1 100	1 500
Germany	158	4 400	6 400
Japan	> 1	40	50
India	> 1	3	3
United Kingdom	3	150	200
Total	**420**	**13 600**	**19 150**

now also being considered.[84] Plans for French and German MOX fabrication technology to be transferred to Russia have been announced, but a VVER–MOX programme is still a long way from being launched. If it is, this will be seen as a reversal by the Russian fast-reactor lobby.[85]

VII. Summary of plutonium use in thermal reactors: past and projected

Estimates of historical MOX fuel fabrication for thermal reactors are summarized in table 7.5. About 19 tonnes of total plutonium had been used to produce MOX fuel to the end of 1993 in the production of about 420 tonnes of MOX fuel. Annual MOX fuel production in 1993 stood at a little over 50 tonnes per year, consuming about 2500 kg of total plutonium. Significant amounts of plutonium have been inserted into 17 thermal reactors in just three countries: France, Germany and Switzerland, for a total of 17 tonnes of total plutonium. By 1995 a major increase in MOX fabrication was well underway. Total MOX production in 1995 was almost 90 tonnes, containing about 4.8 tonnes total plutonium. Germany has been overtaken since 1990 as the major producer of LWR–MOX by Belgium, while in terms of MOX fuel inserted into reactors France overtook Germany. By the end of 1995 some 27.5 tonnes of plutonium had been consumed in LWR–MOX fabrication.

Since the publication of *World Inventory 1992* the future of MOX fuel production in Europe has become clearer: the operating record of P0 at Dessel is well established; the Melox facility at Marcoule has been commissioned and

---

[84] Kudriavtsev, E. G. and Mikerin, E. I., 'Russian prospects for plutonium utilization', Proceedings of the *1993 International Conference on Nuclear Waste Management and Environmental Remediation*, Prague, 5–11 Sep. 1993.

[85] Mikhailov, V. N., *et al.*, 'Prospects for using plutonium as a fuel in Russia', Paper presented at the Disposal of Weapons Plutonium—Approaches and Prospects, NATO Advanced Research Workshop, St Petersburg, 14–17 May 1995.

Table 7.6. Scenario for plutonium consumption in LWR–MOX fuel, 1994–2000

Figures for MOX fuel burnup are given in MWd/t heavy metal, figures for fuel enrichment are given as a percentage of total plutonium, figures for MOX fuel loaded are in tonnes of MOX heavy metal and figures for plutonium are in kg Pu$_{tot}$.

Country	MOX fuel matrix	MOX fuel burnup	MOX fuel enrichment	No. of reactors fuelled with MOX	Total MOX fuel loaded into reactors	Total plutonium loaded into reactors as MOX
Belgium	Depleted U[a]	43 000	7	2 (1996–2000)	48	3 400
France	Depleted U	33 000	4.8	7 (1994–96)	110	5 400
		43 000	7	12 (1997–2000)	220	15 100
Germany	Natural U	43 000	5	5 (1994–2000)	150	7 500
Japan	Natural U	33 000	4.1	4 (1998–2000)	65	2 600
Switzerland	Natural U	43 000	5	2	20	1 000
Total				Up to 25	613	**35 000**

[a] Assume 0.225% ^{235}U.

Source: Adapted from table A-4 in Berkhout, F., *et al.*, 'Disposition of separated plutonium', *Science and Global Security*, vol. 3 (1993), p. 197.

will be expanded, while government and utility support for MOX has remained strong in France; MOX production has commenced at Sellafield and the SMP plant is nearly complete. On the other hand, it is now clear that MOX production in Germany has ended, and that the SBH facility will not operate. The planned expansion of MOX production at Dessel in Belgium has also been much delayed. Over the longer term, the main uncertainty is over the future of Japanese MOX policy and production.

Strategies for fuelling reactors with MOX have also been clarified in many countries, so that it is now easier to judge what the demand for MOX fuel will be from utilities. This is useful because the true demand for plutonium is finally generated by the implementation of MOX policies by utilities, not by the availability of MOX fuel fabrication capacity. Most projections of plutonium consumption take MOX fuel fabrication capacity as their baseline. For a utility implementing a plutonium recycling strategy, there are at least two separate hurdles which need to be crossed. The first is to contract for MOX fabrication capacity and the second is to relicense reactors so that they can be loaded with plutonium fuel. Very few currently operating reactors were licensed to accept MOX when they began operating. In addition to these two hurdles, Japanese utilities have the additional obstacle of securing a trouble-free, routine transfer of plutonium from Europe to Japan (the transfer of MOX fuel from Sellafield to German reactors may also prove troublesome). All of these factors need to be taken into account in making forecasts about future plutonium consumption in thermal reactors.

1994–2000

As always when making forecasts, there is a need for caution. Executing a MOX programme is logistically complicated and in most countries politically controversial. While the intentions of the many actors involved, as articulated in policy, may now be clearer, there are still many uncertainties about their capacity to implement these policies. Knowledge about MOX policies can be used to build up a picture of near-term plutonium consumption in these programmes. Table 7.6 provides one scenario, which we take to be a reasonable central case, for plutonium consumption in LWR–MOX in the period 1994–2000. This shows that France and Germany will continue to be the main consumers of plutonium in the near-term, and estimates that some 35 tonnes of plutonium will be inserted into reactors as LWR–MOX in 1994–2000. This is more than double the amount of plutonium which had been loaded at thermal reactors by the end of 1993.

The rapid growth of thermal recycling is founded primarily on the expansion of MOX fuel fabrication capacity in France and Britain. LWR–MOX fuel fabrication capacity is planned to rise from around 50 tonnes MOX per year in the early 1990s to a little over 300 tonnes by 2001/2002 (see figure 7.4). This represents 'committed' facilities which are either already operating, in con-

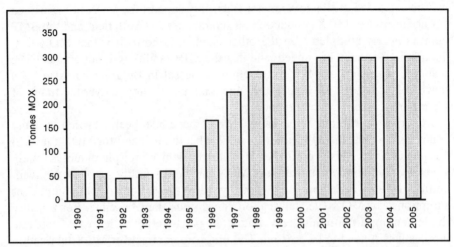

Figure 7.4. Projected world LWR–MOX fuel fabrication capacity: committed facilities, 1990–2005

struction or have planning approval—P0, CFCa, Melox, MDF and SMP. The projection does not include the P1 plant at Dessel. Beyond the year 2005, there are plans for new capacity to come on line in Japan and Russia.

The first question to ask is whether there will be sufficient MOX fabrication capacity to meet the demand revealed in utility MOX plans. Table 7.7 sets out estimates for the maximum production of MOX fuel and for plutonium consumption in the 'committed' MOX fabrication facilities. This shows that, if all goes to plan, there will be almost twice as much MOX fabrication capacity as is required to meet declared MOX programmes (35 tonnes). Some 66 tonnes of plutonium could be fabricated into MOX by committed MOX fabrication plants. There is therefore likely to be a MOX fabrication over-capacity. It is estimated that in the period 1994–2000 some 600 tonnes of LWR–MOX will be inserted in reactors, against a fabrication capacity of about 1200 tonnes MOX in committed plants. This may explain why, in the face of mounting plutonium surpluses, there has been hesitation among fuel-cycle companies to set up new MOX production capacity. It also suggests that, as with reprocessing, MOX suppliers will need to cut prices.

The mismatch between MOX supply and demand shows that forecasts of plutonium consumption based on MOX fuel production capacity risk overestimating the true rates of recycling in thermal reactors. In forecasting national or world surpluses of plutonium there is still a tendency to assume that the key bottleneck is MOX fuel fabrication. In order to derive plutonium balances, analysts therefore have tended to compare plutonium arisings at reprocessing plants with plutonium consumption at MOX fuel fabrication plants, generally assuming that these MOX plants would operate at full capacity and that all the

Table 7.7. LWR–MOX fabrication capacity, 1993–99[a]

Figures for MOX fuel production are in tonnes of MOX heavy metal; figures for plutonium consumption are in kg Pu_{tot}.

Country	Plant	Maximum MOX fuel production[b]	Maximum plutonium consumption
Belgium	P0	235	15 000
France	CFCa	165	9 200
	Melox	510	29 300
United Kingdom	MDF	50	2 300
	SMP	220	9 900
Total		**1 180**	**65 700**

[a] A one-year delay is assumed between MOX fuel fabrication and insertion into the reactor.
[b] Assumed capacities:
P0:   1993–95, 39 t MOX; 1996–99, 44 t MOX.
CFCa: 1993, 15 t MOX; 1994–96, 20 t MOX; 1997–99, 30 t MOX.
Melox: 1995, 50 t MOX; 1996, 100 t MOX; 1997–99, 120 t MOX.
MDF: 1993–99, 8 t MOX.
SMP: 1998, 40 t MOX; 1999, 80 t MOX.

Source: SPRU Spent Fuel and Plutonium Database, 1995.

fuel they produced would be taken up by utilities.[86] This was also the approach used in *World Inventory 1992*. The analysis here shows that such an approach will overstate plutonium consumption, and therefore understate plutonium stockpiles. Excess MOX fabrication capacity seems unlikely to be filled by advance orders for MOX fuel given continued uncertainty in key MOX programmes in Japan and Germany. Moreover, at the end of 1993 about 2.3 tonnes of plutonium were held in fresh MOX fuel not yet inserted into a reactor (see section VIII).

Beyond 2000

Beyond the year 2000 it is more difficult to predict utility recycling policy, partly because it will to a large degree reflect the success or failure of programmes which develop over the next five years or so. Utilities in the 'core' plutonium-using countries—France, Germany, Japan, Switzerland and Belgium—currently aim for legal, economic and political reasons to consume all of their separated plutonium in MOX, most of it in thermal reactors. Utilities in three of these countries—Germany, Switzerland and Belgium—are in the process of retreating from a reprocessing strategy, so that they will have a finite amount of plutonium to dispose of. Altogether, some 64 tonnes of plutonium will be separated from fuel originating in these countries.[87] To dispose of this

---

[86] For example: Chantoin, P. M. and Finucane, J., 'Plutonium as an energy source: quantifying the commercial picture', *IAEA Bulletin*, no. 3 (1993), pp. 38–43.

[87] Baseload and follow-on contracts will generate the following amounts of total plutonium: Germany, 51 t; Switzerland, 8 t; Belgium, 4.2 t. Schmidt, U., 'Problems concerning accumulation of isolated pluto-

material these three countries would need to contract for about 1100 tonnes MOX, representing about four years' worth of capacity at committed fabrication plants.[88] Only France and Japan have long-term open-ended commitments to both reprocessing and to thermal plutonium recycling. Each has or plans to establish independent LWR–MOX fabrication capacities which are designed to absorb plutonium separated from domestic reprocessing plants. Utilities in these countries aim to match the separation of plutonium closely with its disposal in MOX fuel.

What is emerging, therefore, is a retrenchment of the global reprocessing and plutonium fuel services market created in the 1980s. For an interim period lasting to about 2010, while baseload contracts and post-baseload contracts are worked out, there will be a finite demand for MOX fabrication capacity from utilities in Belgium, Germany and Switzerland. There will also be a large and sustained demand from utilities in France and Japan. From 1995 most French requirements will be met in France itself. Japanese utilities also eventually plan to meet their demand for MOX fuel at home. Outside this core group will be utilities in the other reprocessor customer countries—Italy, the Netherlands, Spain—as well as Britain and Russia, whose plutonium inventories will probably be stored, possibly against future use.

Converting these observations into estimates for plutonium consumption in thermal MOX is problematic. We argue that demand, if it materializes, will be composed of sustained MOX policies in France and Japan (which together may require 150–170 tonnes MOX per year by early next century) and other European demand which will amount to no more than 1200 tonnes MOX before reprocessing is abandoned in those countries. Assuming that the only new MOX fabrication plant to be commissioned is at Rokkasho-mura, world MOX fabrication capacity could rise to 380 tonnes MOX per year.[89] A surplus in MOX fuel fabrication is therefore likely to persist beyond the year 2000. Since French and Japanese capacities are intended to satisfy domestic demands, the likely loser in the longer term will be BNFL, with no domestic demand to meet.

Given this broad demand–supply picture, plutonium consumption in the post-2000 period has been modelled. In table 7.8, two scenarios for MOX fuel fabrication are given for the period 2001–10. As discussed above, these estimates need to be treated with great caution. In the low scenario, German, Swiss and Belgian requirements are worked out and the Japanese LWR–MOX programme is cut back. Production is therefore heavily weighted towards the Melox plant in France. If there is to be a rationalization of European MOX capacity, it is likely to lead to greater concentration, not to more plants operating at lower capacities. The Melox plant, assuming it operates efficiently, seems the most

---

nium: situation in Germany', Paper presented at the Advisory Group Meeting on Problems Concerning the Accumulation of Separated Plutonium, IAEA, Vienna, Apr. 1993; and Bay (note 71).

[88] Assuming an average total plutonium enrichment of about 6%.

[89] This compares with an estimate of 350 t MOX/y industry estimate presented in 1993. Cornet, G. *et al.*, 'European experience and potential in use of plutonium in LWRs', Paper presented at TOPNUX'93, The Hague, Apr. 1993.

Table 7.8. Projected MOX fuel fabrication capacity, high and low scenarios for 2001–10

Figures for MOX fuel production are in tonnes of MOX heavy metal; figures for plutonium consumption are in kg Pu_{tot}.

	Plant	MOX fuel production		Plutonium consumption	
		High[a]	Low	High	Low
Belgium	P0	440	–	30	–
France	CFCa	300	150	20	10
	Melox	1 500	1 000	110	75
India	Tarapur	20	–	< 1	–
Japan	Rokkasho[a]	300	150	15	8
UK	MDF	80	–	4	–
	SMP	1 200	–	63	–
Total		**3 840**	**1 300**	**242**	**93**

[a] The Rokkasho MOX fabrication plant is assumed to begin operation in 2005 with a design capacity of less than 100 t MOX per year.

likely to survive in such circumstances. The high scenario assumes that all plants consistently operate at maximum capacities, possibly to meet demand generated by the need to build down civil and weapon plutonium surpluses.

Table 7.8 shows that committed MOX plants will need to operate at near full capacity to fabricate Belgian, French, German, Japanese and Swiss plutonium into fuel in the first decade of the next century. Assuming that the Rokkasho-mura reprocessing plant is commissioned as planned, and that French reprocessing continues at current levels (850 t fuel per year), some 250 tonnes of plutonium would need to be disposed of in these programmes.

VIII. Commercial and R&D plutonium use compared with quantities separated

In chapter 6 it is estimated that by the end of 1993 about 150 tonnes of total plutonium had been separated from power reactor fuel at reprocessing plants (this does not include material separated from fast reactor fuel). In this chapter it is estimated that about 57 tonnes of plutonium have been fabricated into plutonium fuel for fast and thermal reactors (37.6 t for fast reactors, 19 t for thermal reactors). Using this approach, we estimate that a residual amount of about 92 tonnes was being stored at the end of 1993. This represents an increase in the world inventory of civil plutonium of about 20 tonnes in the period 1991–93.[90] More than 60 per cent of the plutonium separated from power-reactor fuel by the end of 1993 was therefore held in store.

[90] See *World Inventory 1992*, p. 142.

Table 7.9. Power-reactor plutonium separation and use, to 31 December 1993
All figures are in kilograms of total plutonium.

Country	Plutonium separated	Plutonium use Fast reactors	Plutonium use Thermal reactors	Plutonium balance[a] By ownership	Plutonium balance[a] By location
Belgium	2 200	0	600	1 600	3 500[b]
France	32 300	8 900[c]	7 800	15 600	27 300[d]
Germany	18 900[e]	2 100[f]	6 900	9 900	2 400
India	400	100	10	300	300
Italy	2 500[g]	1 900[h]	50	550	0
Japan	14 300[i]	3 600[j]	50	10 650[k]	4 700
Netherlands	1 200	0[l]	100	1 100	0
Russia	26 500	0[m]	0	26 500	26 500
Switzerland	1 500	0	1 300	200	0
UK	43 900[n]	4 700[o]	0	39 200[p]	40 900[q]
USA	1 500	0[r]	0	1 500	1 500
Total	**145 200[s]**	**21 300[t]**	**16 810[u]**	**107 100[v]**	**107 100**

[a] The surplus includes material which has been fabricated into MOX fuel, but not yet inserted into a reactor. This is the approach used by the Japanese Atomic Energy Commission in the publication of its plutonium inventory figures in late 1994. See 'Atomic Energy White Paper unveils conditions of Japan's plutonium inventory', *Atoms in Japan*, Nov. 1994, pp. 4–7.

[b] Material in store at Dessel in the form of plutonium oxide powder, fresh LWR–MOX fuel and Kalkar fuel.

[c] This does not include 6.2 t of total plutonium embedded in fabricated MOX fuel for Phénix and Superphénix which has not been inserted into those reactors.

[d] Includes material separated from Belgian, German, Japanese and Swiss fuel not yet repatriated.

[e] This includes some 750 kg of plutonium imported from the USA for fast-reactor use.

[f] Includes 1200 kg used in KNK11 fuel and 900 kg as SBK component of Superphénix first core, but excludes plutonium used to produce first core for Kalkar reactor.

[g] Includes some 130 kg of plutonium imported from the USA for fast-reactor use.

[h] Represents the 33% stake held by the Italian utility ENEL in the Superphénix reactor.

[i] Includes some 160 kg of plutonium imported from the USA as a critical assembly.

[j] Includes plutonium used to fuel the Fugen ATR, but not material used to fuel the Monju fast reactor (about 1200 kg Pu_{tot}) since this reactor did not start operation until early 1994.

[k] The published Japanese plutonium inventory at the end of 1993 stood at 10 880 kg Pu_{tot}. However, this included an amount of material allocated to Japanese utilities by Cogema at La Hague which had not been separated from Japanese fuel. The total amount of plutonium stored at La Hague and Sellafield was 6197 kg Pu_{tot}.

[l] The Dutch stake in Kalkar FBR fuel is 170 kg Pu_{tot}. See: Tweede Kamer der Staten-Generaal, Vergaderjaar 1990–1991, 21 800 V, The Hague, no. 80, pp. 1–2.

[m] Most of the plutonium used in Russian fast and experimental reactor MOX tests has had a military origin.

[n] This includes material separated at Windscale/Sellafield since Mar. 1971 from Magnox power reactors and from UKAEA reactor fuel. Material separated before 1971 was not formally designated as 'civil' and has not been included by the British Government in its annual inventory declarations.

COMMERCIAL AND R&D USES OF PLUTONIUM   231

^{o} Assumes that c. 800 kg of the 5 t currently in the PFR fuel cycle came from the UKAEA stock composed of plutonium separated from Magnox power-reactor fuel between 1969 and Mar. 1971. We estimate that a further 300 kg of this inventory is currently in process at Sellafield.

^{p} At the end of Mar. 1994 the British inventory was 40 t. We estimate that in the period Jan.–Mar. 1994 about 800 kg Pu_{tot} were separated at B205. British Department of Trade and Industry, Press Notice P/94/439, 19 July 1994.

^{q} Includes 1.7 t of plutonium separated from Japanese and Italian magnox fuel, but not repatriated.

^{r} A zero is inserted here because the 6.7 t of plutonium used to fuel the ZPPR and FFTF reactors came mainly from British Magnox reactors (about 4 t), traded for HEU under the US–UK Mutual Defence Agreement, or was produced at the 'N' military reactor at Hanford. See chapter 3, section VI.

^{s} The discrepancy with table 6.8 results principally from a different counting convention used in estimating historical British plutonium separation. The figure in table 6.8 (149.8 t) includes plutonium separated from power-reactor fuel at the B205 reprocessing plant at Windscale/Sellafield between 1964 and 1971 (about 4.2 t). The figure in this table refers only to material separated since Mar. 1971.

^{t} The discrepancy with table 7.4 results because plutonium embedded in fresh, unirradiated MOX fuel is not included here. It also includes: 6.7 t of plutonium used to fuel US fast reactors; 900 kg embedded in Superphénix reload owned by the SBK consortium of European utilities; 700 kg of military plutonium used in Russian fast-reactor research; material used to fabricate fuel for the first core of the Monju fast reactor (1.2 t); and about 300 kg of British PFR plutonium currently being processed at Sellafield (total 9.8 t Pu_{tot}).

^{u} The discrepancy with table 7.5 results from a different counting convention. The figure in table 7.5 includes plutonium fabricated into MOX but not inserted into a reactor. The difference of 2340 kg Pu_{tot} represents the amount held in fresh MOX fuel at fabrication plants and reactor sites.

^{v} Taking account of plutonium fabricated into MOX fuel which is set to be irradiated but not allocated here—one reload at Superphénix, the first fuel core for Monju, material in fresh LWR–MOX fuel and material 'in process' (total: 9200 kg Pu_{tot})—the surplus can also be given as 97 200 kg Pu_{tot}.

National balances

By comparing the estimates of plutonium production and use in chapters 6 and 7, national and plutonium balance estimates can be derived. Table 7.9 shows the amount of plutonium separated, the amount consumed in MOX and the remaining balance for each of the countries where significant amounts of spent fuel had been reprocessed at the end of 1993. The total balance of separated plutonium and plutonium in unirradiated fuel is estimated to be 107 tonnes at the end of 1993. The national plutonium balance figures are stated by ownership (nationality of the utility owning the plutonium) and location (the physical location of the plutonium). The table indicates the scale of plutonium transfers that have taken place for fuel reprocessing and MOX fuel fabrication.

The counting conventions used in table 7.9 are different from those used in earlier sections. This is why the world surplus is given here as being about 107 tonnes rather than 92 tonnes. In particular, fresh MOX fuel which has not

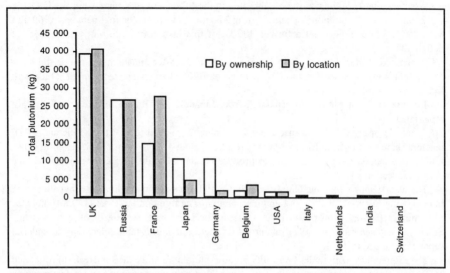

Figure 7.5. Estimated plutonium balances by ownership and location, 31 December 1993

yet been inserted into a reactor is included in the balance of separated plutonium, rather than in the figures relating to recycled material. This is the convention used in the November 1994 Japanese declaration of plutonium stocks, which seems likely to be a forerunner of future national declarations.[91] We estimate that about 7.0 tonnes of total plutonium were held in fresh fast reactor fuel and 2.3 tonnes were held in fresh LWR–MOX fuel at the end of 1993. In addition, the figure given for plutonium separation in the UK is lower in table 7.9 than in table 6.8 because some 4.2 tonnes of power-reactor plutonium separated at B205 at Sellafield before March 1971 are not included. This material is not included in the British Government's own declarations of plutonium inventories, and it is excluded here.

The aggregate result of these differences in counting convention is that, relative to tables 6.8, 7.4 and 7.5, the world plutonium balance shown in table 7.9 is increased by some 14 tonnes. The two estimates of world surplus can therefore be reconciled. Establishing common ground rules for plutonium declarations will clearly be an important task in any attempt to encourage universal reporting of inventories. This issue is further discussed in chapter 14.

The great variation in national inventories is testament to the variety of reprocessing and plutonium use policies which have been pursued in different countries. In general, countries with large reprocessing commitments have not been able to curtail the buildup of significant plutonium stockpiles, even where that has been the expressed policy. The largest inventories are in the UK (39.2 t) and in Russia (26.5 t). About 65 per cent of the world surplus of pluto-

[91] 'Atomic energy White Paper unveils conditions of Japan's plutonium inventory', *Atoms in Japan*, Nov. 1994, pp. 4–7.

Table 7.10. Civil plutonium balances as declared by states, as of 31 December 1995
Figures are in kilograms of total plutonium.

Country	Pu balance, by location
Belgium	2 000[a]
France	55 300[b]
Germany	2 400
Japan	4 700
United Kingdom	48 500
United States	100

[a] The running inventory at the Dessel P0 plant is now put at 1300 kg. We include c. 700 kg contained in Kalkar fuel still stored at Dessel.

[b] The French Ministry of Industry, Post and Telecommunications published figures showing that on 31 Dec. 1995 there were: 36.1 t of separated plutonium at reprocessing plants; 5.5 t of separated plutonium at other sites; 10.1 t in irradiated MOX or in fabrication; and 3.6 t in French MOX at reactors. Of the total, 25.7 t belonged to foreign utilities, the balance, 29.6 t, belonging to French organizations.

Sources: PPPN Newsbrief (University of Southampton), no. 34 (second quarter, 1996), p. 9; Hibbs, M., 'Bonn expects to abandon Hanau plutonium custody', *Nuclear Fuel*, 12 Aug. 1996, pp. 4–5; British Department of Trade and Industry, Press Notice P/96/563, 18 July 1996; US Department of Energy, *Plutonium: The First 50 Years: United States Plutonium Production, Acquisition and Utilization from 1944 to 1994* (Department of Energy: Washington, DC, Feb. 1996), p. 37.

nium is therefore stored in just two sites, Sellafield and Chelyabinsk-65. Even in France, however, a quite significant inventory of plutonium has developed, especially when the quantities embedded in fresh fast-reactor fuel are included. Only smaller countries, such as Switzerland and Belgium, have so far succeeded in managing their plutonium balances to maintain small stocks. National inventories are shown in graphical form in figure 7.5.

World Inventory 1992 recommended that national balances should be routinely published by all countries holding or owning plutonium. Since 1995, national plutonium balances have been published by a number of countries, partly as a result of discussions between the International Plutonium Management group of countries in Vienna, and partly as a result of 'openness' policies. Although these efforts are to be welcomed, they still present problems for the analyst. No unified definition of inventories is yet being applied. These definitional problems are especially acute in nuclear weapon states where the scope of 'civil' and 'defence-related' materials varies from country to country. In table 7.10 we present a summary of inventory declarations as at the end of 1995. These numbers are not strictly comparable with those presented in table 7.9. The apparently very large increase in the French plutonium stock can be explained by the near full-scale production of UP2 and UP3, and by the inclusion of previously unknown stocks into the civil inventory (see note *b*).

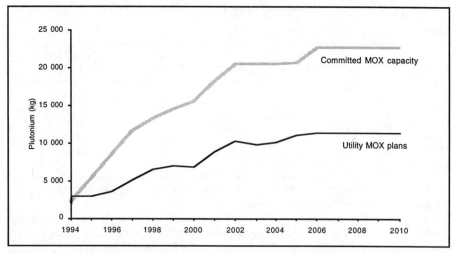

Figure 7.6. Plutonium consumption forecasts: committed MOX fabrication capacity and utility MOX plans, 1994–2010

Forecasting plutonium balances

Plotting future plutonium balances is more difficult. As argued above, three different processes need to be modelled: the rate of plutonium separation, subject to change for a variety of reasons; the rate of plutonium consumption in MOX fabrication; and the rate at which MOX is loaded into reactors. Each of these rates is variable and may lead utilities and reprocessing companies to change their behaviour. In particular, plutonium policies may be changed in direct response to the state of national surpluses.

Rather than repeating the scenario approach used in *World Inventory 1992*, two different approaches to the problem of plutonium balance forecasting are presented here. First we present an analysis of aggregate balances based on an assessment of the plutonium set to be consumed in LWR–MOX programmes announced by utilities. As an alternative, we also present an assessment of world balances, which is based on the simpler comparison of plutonium separation rates and forecast MOX fuel fabrication rates. Annual plutonium consumption under both approaches over the period 1994–2010 is set out in figure 7.6.

The derivation of plutonium balance forecasts is possible by comparing estimates of separation (we assume the 'current commitments' scenario developed in chapter 6) with estimates of consumption. Using the first approach for forecasting plutonium consumption, based on declared utility MOX plans, we predict that national surpluses will grow in all cases, unless there is a slow-down in separation rates. Table 7.11 shows that national plutonium balances are likely to grow in the period up to the year 2000, even in countries with active MOX programmes (Belgium, France, Germany, Japan and Switzerland, BFGJS). In the period 1994–2000 some 95 tonnes of

Table 7.11. National plutonium balances assuming utility MOX policies are implemented, 1994–2000

Figures are in kilograms of total plutonium.

Country	Balance 1993	Plutonium Separation 1994–2000	Plutonium use 1994–2000		Balance 2000
			Fast	Thermal[a]	
Belgium	1 600	2 700	–	3 400	900
France	18 200	40 800	6 400[b]	20 500	32 100
Germany	10 500	25 000	–	7 500	28 000
Japan	10 700	22 900	2 000	2 600	29 000
Switzerland	200	3 500	–	1 000	2 700
Total	41 200	94 900	8 400	35 000	92 700

[a] See table 7.6 for derivation
[b] This figure assumes that the initial core at Superphénix is replaced before 2000 under the CAPRA programme and that one reload is inserted at Phénix.

plutonium are expected to be separated from fuel originating in these countries at La Hague, Sellafield and Tokai-mura, while only about 43 tonnes will be consumed in fast reactor and thermal reactor MOX programmes. We predict that plutonium balances for the BFGJS group of countries (41 t total plutonium in 1993) will more than double to over 90 tonnes by the year 2000.

Projections of utility policy beyond the year 2000 are inevitably much less certain, partly because utility companies have, in many cases, still to set policy much beyond the turn of the century. To take account of this uncertainty, and in order to compare plutonium balance projections based on utility plans with those based on MOX fabrication capacity, the remainder of this analysis of plutonium balances is presented in more aggregated form.

Results of this analysis of cumulative balances are presented graphically in figure 7.7. Two sets of lines are presented. The thicker, grey lines represent total world civil plutonium balances, while the thinner, black lines show plutonium balances in countries with active plutonium recycling policies (Belgium, France, Germany, Japan and Switzerland, BFGJS).[92] The difference between the sets of lines is made up of balances in Britain, Russia and India. For both balances (world and BFGJS) two scenarios are presented for the period 1990–2010: the 'committed MOX capacity' and the 'utility MOX plans' scenarios.

The 'committed MOX capacity' scenario assumes that all committed MOX fabrication facilities will operate smoothly at full capacity, absorbing plutonium from national balances. Figure 7.7 shows that under this assumption aggregate world plutonium balances (represented by the lower grey line) will grow to a maximum of some 150 tonnes by the year 2000, and thereafter shrink as European and Japanese MOX programmes begin to consume plutonium at a rate

---

[92] For ease of presentation, the BFGJS totals include the small balances of The Netherlands and Italy.

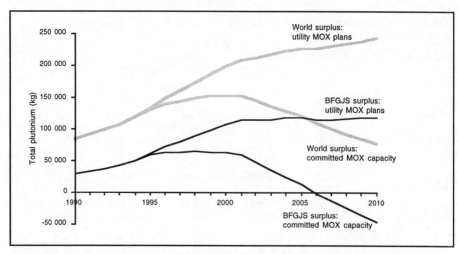

Figure 7.7. Projected world and European/Japanese plutonium surpluses, 1990–2010
Note: BFGJS: Belgium, France, Germany, Japan and Switzerland.

faster than that at which it is separated. Under this scenario, world plutonium balances will stand at about 75 tonnes by 2010.

Under the 'committed MOX capacity' scenario the aggregate balance of the BFGJS group of countries will increase to about 60 tonnes by 2000, before declining to zero in about 2006. Thereafter, this scenario suggests a net deficit of plutonium in these countries. Such a deficit could only be maintained by transfers of plutonium from existing civil stocks in Britain and Russia into MOX recycling programmes in the BFGJS countries. At present such transfers seem unlikely, especially since disposition of weapon plutonium may be seen as a greater priority than disposition of civil surpluses. Assuming that MOX strategies in the BFGJS countries are to achieve zero national plutonium balances, rather than to consume the surpluses of others, figure 7.7 suggests that beyond 2005 demand for MOX fuel will fall dramatically. Committed MOX facilities would have the capacity to fabricate annually about 23 tonnes of total plutonium into MOX, while the demand in active MOX programmes would fall to between 10 and 13 tonnes post-2005. This suggests that a substantial proportion of new MOX fabrication capacity will have an operating life of no more than about 10 years.

By contrast, the 'utility MOX plans' scenario assumes that plutonium will be consumed at a rate set by utility MOX policy (as represented in tables 7.6 and 7.11), rather than by MOX fabrication capacity. Using this approach, figure 7.7 shows that world plutonium balances (represented by the upper grey line) will grow rapidly to stand at about 200 tonnes in the year 2000, and continue to grow at a slower rate to about 245 tonnes in 2010. Under this scenario, which we argue above is a more realistic view of the future, the rate of separation of plutonium will therefore continue to outpace the rate at which plutonium is inserted as MOX into reactors.

Among the BFGJS group of countries a stabilization of aggregate plutonium balances is forecast under the 'utility MOX plans' scenario. Balances will continue to grow to a level of about 110 tonnes in the year 2000, and grow slowly thereafter. An approximate equilibrium will then have been established between rates of plutonium separation and use in this group of countries. Growth in the world balance under this scenario will therefore be entirely the result of growth in British, Indian and Russian surpluses.

IX. Conclusions

In this chapter we seek to provide a comprehensive analysis of historical and future trends in the use of plutonium as a fuel. We show that up until the mid-1980s plutonium had predominantly been used to fuel prototype fast reactors. With the advent of major expansions in reprocessing during the 1980s, utilities were forced to turn to thermal recycling as the only available way of disposing of plutonium. Major new investments were therefore made in the construction of new MOX fuel fabrication capacity. As with reprocessing, a German project was abandoned in the face of political opposition, and MOX fabrication capacity has come to be concentrated with French and British reprocessing companies. At the same time, utilities have sought to gain licences to load plutonium fuel into operating light water reactors.

However, although the transition to large-scale thermal MOX recycling has been achieved efficiently in some countries (France is the outstanding example), it has proven to be problematic in many others. This is primarily due to declining economic incentives, a lack of widespread political approval for the use of plutonium and effective campaigning by environmental organizations. Under these circumstances, utilities have found it difficult to make the demanding contractual and licensing arrangements required to develop an effective MOX programme. The result is that many utilities will not be able to use all of the plutonium becoming available to them at reprocessing plants.

We estimate that the world balance of civil plutonium stood at about 107 tonnes at the end of 1993 (including plutonium fabricated into fresh MOX but not yet inserted into a reactor). This balance had grown to some 132 tonnes by the end of 1995, and under the more pessimistic of the plutonium use scenarios developed in this chapter will grow to about 200 tonnes by the year 2000. Under the more optimistic scenario, the world plutonium balance will grow to about 150 tonnes by the year 2000.

This chapter has sought to characterize national plutonium balances. This is in anticipation of universal declarations of national plutonium inventories, both civil and military. It should be possible to corroborate declarations by national authorities with estimates made on the basis of existing public knowledge.

8. Civil highly enriched uranium inventories

I. Introduction

Unlike plutonium, highly enriched uranium is not a by-product of civil nuclear power programmes. As a result, the international community has had an easier time in its attempts to reach a consensus that civil HEU fuels should be eliminated or replaced by low-enriched uranium fuels. Currently, HEU plays a diminishing role in civil research programmes and almost no role in nuclear power programmes worldwide. Nevertheless, the complete elimination of civil HEU fuels in research programmes remains a long-term goal, and the attempt to further this goal continues to generate controversy.

Out of a total of more than 300 research and test reactors operating currently in over 50 countries, about 175 of them in 39 countries have used HEU fuels.[1] Almost all of this HEU was produced in military enrichment plants in the nuclear weapon states.

Much of the spent HEU fuel in non-nuclear weapon states has been or will be returned to the NWS for storage or reprocessing, with the result that civil HEU inventories have stayed relatively low. In the USA, the returned HEU was mixed with military HEU stocks until 1988 when it was decided not to accept any more foreign spent fuel, pending the outcome of environmental impact analyses. In early 1996, the USA announced its decision once again to accept and manage foreign research reactor fuel originally supplied by the USA. In Russia, the precise fate of civil spent HEU fuel is unknown, although it has reportedly been reprocessed and the recovered HEU recycled in both military and civil reactors. So far, Russia has not taken back much of the spent fuel from the foreign reactors it supplied.

Although civil research reactor inventories are now expected to decrease, in the 1970s relatively large inventories of civil HEU, mostly containing over 90 per cent ^{235}U, and thus weapon-grade material, were accumulating in Europe and elsewhere. Despite all this material outside the NWS being under IAEA or Euratom inspection, concern grew that terrorist groups or clandestine national efforts would divert some of it and use it to make nuclear weapons. This concern turned out to be justified when, in August 1995, Iraq admitted that in 1990 Saddam Hussein had ordered the production of a nuclear weapon using a stock of about 35–40 kg of safeguarded HEU (over 80 per cent enriched), supplied by France and Russia for Iraq's civil research reactors (see chapter 11). Iraq planned to divert both fresh and irradiated fuel.

---

[1] Travelli, A., 'Overview of the U.S. Reduced Enrichment Research and Test Reactor (RERTR) Program', viewgraphs presented at the Institute for Science and International Security (ISIS) Non-Proliferation Workshop, Alexandria, Va., 6 Jan. 1994.

Because of the danger posed by civil HEU, the USA and other Western countries have cooperated since the late 1970s to reduce the amount of HEU in civil nuclear programmes. The main technical focus of these efforts has been the development of new LEU fuels for thermal research reactors that can replace the HEU fuels without affecting the performance of the reactor.

II. Civil suppliers of highly enriched uranium

The USA

By far the largest supplier of civil HEU to research reactors has been the USA, which over the years has exported about 26 000 kg of HEU, containing over 18 000 kg of ^{235}U.[2] A significant quantity of this material still remains overseas. The USA has also supplied about 30 000–35 000 kg of weapon-grade uranium to its own thermal, research and early power reactors.[3]

The average amount of HEU exported for research reactors by the USA declined dramatically during the 1970s and 1980s. The amount of ^{235}U in HEU exported during 1970–77 averaged 660 kg per year.[4] This amount decreased to an average of 380 kg per year during 1978–82, then further to an average of 160 kg per year during 1983–88 and reached zero by 1993. The USA stopped exporting HEU for civil purposes in the early 1990s as a result of congressional action motivated by continuing concerns about HEU diversions. The Clinton Administration has no plans to try to reverse this statutory limitation which, in effect, bars exports.[5]

---

[2] The United States Nuclear Regulatory Commission's Report to Congress on the Disposition of Highly Enriched Uranium Previously Exported from the United States, Nuclear Regulatory Commission, Washington, DC, Jan. 1993 (hereafter NRC Report to Congress); Albright, D., 'Civilian inventories of plutonium and highly enriched uranium', eds P. Leventhal and Y. Alexander, *Preventing Nuclear Terrorism* (Lexington Books: Lexington, Mass., 1987), pp. 265–91; and Letter from Armando Travelli, Reduced Enrichment for Research and Test Reactors (RERTR) Program Manager, to Leonard Weiss, Staff Director, US Senate Committee on Governmental Affairs, 9 June 1989. Most of the information about HEU exports and returns is found in the Department of Energy's Nuclear Materials Management and Safeguards System and this information has been obtained by one of the authors through Freedom of Information requests and from Congressional oversight committees.

[3] Albright (note 2); Cochran, T. B. *et al.*, NRDC, *Nuclear Weapons Databook, Vol. II: US Nuclear Warhead Production* (Ballinger: Cambridge, Mass., 1987), appendix D.

[4] Travelli to Weiss (note 2).

[5] The 'Schumer Amendment' to the Energy Policy Act, which was passed in late 1992, stipulates that HEU exports are barred unless 3 conditions are met: (*a*) no alternative fuel exists; (*b*) the recipient promises to use an alternative fuel if possible; and (*c*) the US Government works to develop an alternative fuel. A loophole in this law was an export of almost 300 kg of WGU in unused fuel originally intended for use in the Fort St Vrain reactor in Colorado. The reactor was shut down because it operated so poorly, and the reactor owners in 1993 applied for an export permit to send the material to France which would recover the HEU and return it to its owner in the United States. Hiruo, E., 'Transnuclear seeks NRC okay to ship HTGR fuel to France for processing', *Nuclear Fuel*, 7 June 1993, p. 10–11. The Washington group Nuclear Control Institute (NCI) intervened with the Nuclear Regulatory Commission to block the shipment. Dizard, W., 'Nuclear Control Institute seeks to block export of HEU to France', *Nuclear Fuel*, 5 July 1993, p. 16. An NCI member and a State Department official said that a settlement was reached whereby the HEU would be extracted in France but the HEU would be blended down to 19% enriched material.

Russia

The other major supplier of civil HEU has been Russia, although it has mainly supplied HEU to research and fast reactors in the former Soviet Union and Eastern Europe. An estimated 10 000 to 20 000 kg of HEU went to Soviet research reactors, and about 2000–3000 kg were exported to Eastern Europe, Iraq, Libya, North Korea and Viet Nam.

Russia has supplied a much larger quantity of HEU to two, large fast breeder reactors (one in Russia and the other in Kazakhstan), although this material has initial enrichments of only 20–25 per cent. In total, Russia has provided about 150 000 to 200 000 kg of 20–25 per cent enriched fuel to these reactors.[6] The spent fuel from these reactors is reprocessed and recycled.[7] A fraction of the recovered enriched uranium is LEU because a significant fraction of the ^{235}U was consumed while in the reactors.

In June 1996, Russia agreed to supply France with 625 kg of HEU (over 90 per cent enriched) over nine years to fuel two research reactors, the RHF and Orphee reactors.[8] Reportedly, the HEU cannot be transferred to another European country without Russian consent.[9] This sale, however, is expected to release HEU for other reactors in Euratom. The USA has stopped civil HEU exports to Euratom and views this Russian HEU sale as undercutting its policy to eliminate the use of HEU in civil research programmes.

This contract caps a long period of negotiation between Russia and France. In 1993, Russia reportedly first offered to sell France 50 kg of HEU for one of its research reactors.[10] Whether Russia will try to sell HEU to other European reactor owners remains unknown and highly controversial.

Other suppliers

China, France and the UK have supplied small amounts of HEU fuel to a few countries. France was to have supplied Iraq with roughly 30–40 kg of weapon-grade uranium a year to fuel the 40 MWth Osirak research reactor. After the Israeli bombing of this reactor, this contract was stopped, but not before about 12 kg of weapon-grade material had been delivered. China supplied small amounts of HEU fuel to Iran, Pakistan and Ghana in the 1990s.

---

[6] Bukharin, O., 'Analysis of the size and quality of uranium inventories in Russia', Paper presented at the Nuclear Energy Institute's International Uranium Seminar, 8–11 Oct. 1995, Williamsburg, Va.

[7] Bukharin (note 6).

[8] Hibbs, M., 'French–Russian HEU accord signed. EU agreed to Russian prior consent', *Nuclear Fuel*, 17 June 1996; and Hibbs, M., 'Euratom could provide Russian HEU to Germany's FRM-2', *Nuclear Fuel*, 6 Nov. 1995, pp. 11–12.

[9] Hibbs 1996 (note 8).

[10] MacLachlan, A., 'MINATOM proposes to supply HEU for France's Grenoble reactor', *Nuclear Fuel*, 7 June 1993, p. 5.

III. Civil reactors using HEU fuels, 1995

Over half the research reactors that currently use HEU fuel have lifetime cores and require no additional HEU fuel. The others require roughly 1050 kg of ^{235}U in HEU fuels per year, and about 60 per cent of this material is used in US and Russian research reactors (see appendix D). Most of this material is weapon-grade.

In addition to the amount of HEU used as fuel in these reactors each year, a larger amount of HEU is in the 'pipeline'. This includes all HEU that is held by fuel fabricators, stored at reactor sites as fresh or irradiated fuel, reprocessed or transported between any of these facilities. If it takes four years for the HEU fuel to go through the entire cycle from enrichment to reprocessing, up to four times the amount required by reactors each year could be in circulation. In addition, European fuel fabricators and brokers have stockpiled an inventory of fresh HEU. Both operating and closed reactors throughout the world have built up inventories of spent fuel awaiting shipment to reprocessors or storage sites (see below).

Several civil reactors are not included in the above estimate. The two largest FBRs in the former USSR, with a combined electrical generating capacity of about 950 MWe, require roughly 10 000–15 000 kg of 20–25 per cent enriched uranium annually.[11]

Russia's three remaining military production reactors, which are now dedicated to civil purposes, use a few hundred kilograms of weapon-grade uranium fuel each year in combination with a much larger quantity of natural uranium fuel (see chapters 3 and 4). Russia is considering several options to convert the nuclear cores of these reactors so that they would produce little plutonium. One of these options, however, includes the use of an all weapon-grade uranium core. If implemented, this option would dramatically increase the annual civil WGU demand in Russia.

Two high-temperature power reactors in the USA and Germany would have required about 300 kg of weapon-grade uranium per year, but they were shut down. Plans to build more of them have been shelved. As a result, HEU now plays no part in Western nuclear power production programmes.

The United States was planning to build an Advanced Neutron Source reactor, with a baseline power of 330 MWth and an annual requirement of about 350 kg of 93 per cent enriched uranium. However, the USA cancelled this reactor because of cost reasons and concerns that this reactor could undermine the US policy of discouraging civil HEU use internationally.

A few countries are building or planning to build research reactors that would use a significant quantity of HEU fuel. In the West, only Germany is known to have decided in recent years to start a reactor project that would use HEU fuel. (All Western civil research reactors built since 1980 have used LEU fuels.[12])

[11] Bukharin (note 6).
[12] Travelli (note 1).

Germany has started building the FRM-2, a 20 MWth reactor at the Technical University outside Munich, which would require about 40 kg a year of 93 per cent enriched fuel and which would be finished early next century.[13] The reactor operators are reported to have told Federal German officials that the reactor could be redesigned to use 70 per cent enriched uranium, without triggering a controversial relicensing process that could lead to the reactor's cancellation as a result of growing opposition to the project.[14] Nevertheless, the operators have shown no desire to actually reduce the enrichment level of the fuel. They have also stated that they have obtained a 10-year supply of HEU fuel (over 90 per cent enriched) from the broker NUKEM.[15] However, evidence suggests that only a five-year supply is firmly in hand. The other five years are covered by an 'option to buy'. A US official said that the US Government has a written statement from the Germans that none of this HEU will come from Russia. Whether this statement is binding is doubtful. However, no information has emerged to suggest that this reactor would receive any HEU from Russia.

Russia is building a 100 MWth research reactor, called PIK, near St Petersburg that is designed to use 90 per cent enriched uranium fuel.[16] Construction started in 1976.[17] So far, however, according to a US official, the reactor is only about 30 per cent complete, and the project does not have sufficient funds to finish its construction. Russia is also planning to build the 100 MWth MPR reactor in Obninsk, near Moscow, that is being designed to use 90 per cent enriched uranium fuel.[18] Russian officials have also stated that the 200 MWth SPHINKS reactor being designed would use weapon-grade uranium fuel. The exact status of these two planned reactors is unclear, but construction is not believed to have started or to be imminent.

China's activities are unknown. However, US officials were told by Chinese officials that they are considering building a new research reactor which could use HEU fuel.

IV. Converting to low-enriched uranium fuels

For almost 20 years, countries have worked to eliminate or reduce the amount of civil HEU in commerce. The principal focus of these international efforts has

---

[13] Mo, S. C., Hanan, N. A. and Matos, J. E., 'Comparison of the FRM-II HEU design with an alternative LEU design', Argonne National Laboratory, Paper presented at the 1995 International Meeting on Reduced Enrichment for Research and Test Reactors, 18–21 Sep. 1995, Paris, France.

[14] Abbot, A. and Schiermeier, Q., '... as leak fuels German uranium debate', *Nature*, vol. 379 (25 Jan. 1996), pp. 284–85; and Hibbs, M., 'US–German meeting on FRM-2 ends with no consensus on HEU', *Nuclear Fuel*, 29 Jan. 1996, pp. 5–6.

[15] Hibbs, M., 'High FRM-2 HEU cost would exceed budget allocation', *Nuclear Fuel*, 8 Apr. 1996; Hibbs, M., 'Half of FRM-2 10 year HEU covered by expensive option', *Nuclear Fuel*, 25 Mar. 1996; and Hibbs, M., '"Nearly all" HEU for FRM-2 was obtained from European enricher', *Nuclear Fuel*, 27 Feb. 1995, p. 11.

[16] Arkhangelsky, N. V. et al. 'State-of-the-art and trends in the development of fuel elements for research nuclear reactors in Russia', Paper presented at the 1993 International Meeting on Reduced Enrichment for Research and Test Reactors, 3–7 Oct. 1993, Oarai, Ibaraki, Japan.

[17] IAEA, *Nuclear Research Reactors in the World* (IAEA: Vienna, Dec. 1994).

[18] Arkhangelsky (note 16).

been the development of alternative LEU fuels that can substitute for existing HEU fuels.

The US programme

The effort to eliminate civil HEU fuels in thermal research reactors was launched by the USA in 1978 with the formation of the Reduced Enrichment Research and Test Reactors (RERTR) programme. Since then, this programme has spawned many other national programmes; more than 27 countries have programmes today. The reduction in annual US HEU exports has depended on the success of the RERTR programmes and cooperation among these many nations.

In addition to developing new fuels, the programme assists in the development of qualified LEU fuel suppliers, encourages suppliers of research reactors to design and market only LEU-fuelled reactors, and spurs reactor operators to convert existing reactors to LEU fuels.

The US RERTR programme has targeted for conversion about 42 non-US research reactors in about 20 countries with a power of at least 1 MWth that have received HEU from the USA or other Western countries.[19] Before any conversions, these reactors together required about 445 kg of ^{235}U in HEU per year. There are another 40 or so operating reactors that have been supplied by Western countries that use HEU fuels, but these are small and probably have enough HEU fuel at their sites to last until they are shut down.

Twelve of the targeted reactors, which together used to require HEU containing about 210 kg of ^{235}U per year, have fully converted to LEU fuels. Another 26 of them, which annually require HEU with a total of 145 kg of ^{235}U, are in various stages of conversion. In addition, nine US university reactors, which are not considered target reactors of the RERTR programme, have also converted. These conversions were ordered by the US Nuclear Regulatory Commission in the 1980s.

Despite the successes of the programme, it faces several problems in accomplishing its goal of converting all research reactors to LEU fuel, including:

1. In the USA lack of funding has delayed by years the effort to develop new LEU fuels adequate for the three largest Western research reactors (one in Belgium and two in France which together require about 100 kg of ^{235}U per year), several Russian research reactors and a few US research reactors.

2. Operating reactors in Holland and South Africa which can convert have not done so. The Dutch Petten reactor has refused to convert. In 1994, the Petten operators listed a set of conditions under which it would convert, princi-

---

[19] This section is largely based on Travelli, A., 'The RERTR program: a status report', Paper presented at the 1991 International Meeting on Reduced Enrichment for Research and Test Reactors, Jakarta, 4–7 Nov. 1991; and Travelli, A., '"The RERTR Program": status and progress', Paper presented at the 1995 International Meeting on the Reduced Enrichment for Research and Test Reactors, Paris, 18–21 Sep. 1995.

pally a commitment by the USA to accept all its spent fuel under reasonable price conditions.[20] However, in early 1996, a US official, long familiar with the Petten conversion discussions, said that 'Petten is not quite committed to conversion.' In late 1995, the South African Atomic Energy Corporation (AEC) told the US Government that it had decided not to convert the SAFARI-1 reactor to LEU fuels at the present time, because of cost concerns.[21] Instead, it is going from the use of 45 per cent enriched fuel to weapon-grade uranium fuel. The AEC said that it would periodically re-evaluate this decision, and the US Government is attempting to reverse this decision.

3. The German research reactor FRM-2, whose construction has started, is designed to use fully enriched fuel, and its operators have actively resisted conversion. They typically cite cost, delays and re-licensing requirements in their refusal to use a LEU core.[22]

4. The US Government has refused to convert its own civil research reactors that use HEU fuels. Together these reactors require up to 400 kg of HEU per year (see appendix D).

5. China has not finalized a cooperation agreement with the US RERTR programme. In February 1995, the USA and China signed a statement of intent endorsing cooperative activities between the programme and Chinese laboratories involved in similar activities. However, the expected cooperation had not materialized by August 1996. US officials blamed the general political problems between the two countries and a Chinese desire to require that the broader, moribund US/Chinese nuclear cooperation agreement come into force before cooperation on new LEU fuels commences.

The Russian RERTR programme

In parallel to the USA, Russia secretly started a programme in the late 1970s to reduce the enrichment of the HEU exported to civil research reactors it supplied. By the mid- to late-1980s, the programme successfully developed and demonstrated a 36 per cent enriched uranium fuel that could replace the 80–90 per cent enriched fuel. In 1988 or 1989, it started to export the new fuel. So far, the Czech Republic, North Korea and Poland have received supplies of the 36 per cent enriched fuel. The next phase of the programme, namely to develop 20 per cent enriched fuel, was cancelled in 1988 because of a lack of funding.

In mid-1992, with no knowledge of Russia's earlier conversion efforts, the USA began discussions with the Russian Government on a joint programme to convert research reactors supplied by the former USSR. In 1993, the USA

---

[20] Knapik, M. and Hibbs, M., 'EU agrees to convert Petten reactor if US meets certain conditions', *Nuclear Fuel*, 14 Mar. 1994, p. 12.

[21] Letter from K. F. Fouche, Executive General Manager, Technology Development, AEC, to Armando Travelli, Manager US RERTR, 'Financial implications of converting SAFARI-1 to LEU silicide fuel', 30 Oct. 1995.

[22] For example, see MacLachlan, A., 'Argonne says FRM-2 could run on LEU; Germans say it would kill project', *Nuclear Fuel*, 25 Sep. 1995, p. 7.

reached a cooperation agreement with Russia to develop and demonstrate new LEU fuels for Russian-supplied research reactors. This programme, which is partially funded by the USA, hopes to develop new fuels within the next five or six years adequate for the conversion of the 13 HEU-fuelled research reactors (> 0.5 MWth) located outside Russia, most of the 13 HEU-fuelled research reactors located in Russia (2–100 MWth), and any new reactors built in Russia or abroad. Conversion of Russia's highest-power research reactors to LEU fuels is currently expected to take longer. In addition, the cooperation does not encompass the conversion of the 60 MWth BOR-60 fast reactor.

As of 1996, the US–Russian cooperation is focusing on the technical aspects of reactor conversions. The first priority is to finish the development and testing of LEU fuels that was interrupted in the late 1980s because of a lack of funding. Future areas of technical cooperation are expected to include the development of more advanced LEU fuels that could allow all Russian reactors to convert. In addition, US officials hope that the cooperation will work to establish and implement a common policy to minimize and eventually eliminate the use of HEU in civil programmes worldwide. Key aspects of such a policy would include the development of new Russian policies to limit the export of civil HEU and to take back HEU spent fuel of Russian origin from abroad.

V. Spent fuel take-back and reprocessing of HEU fuels

Because irradiated HEU fuel contains a considerable amount of residual ^{235}U, it has been reprocessed and the HEU recycled. Belgium, France, Russia, the UK, the USA and perhaps China have all reprocessed civil HEU fuels. In addition, the USA has accepted and reprocessed foreign spent fuel if it originally provided the enriched uranium.

The United States

The USA has reprocessed and recycled most of its spent domestic research reactor fuel. This includes HEU fuel from experimental breeder reactors. This reprocessing occurred in Idaho and in South Carolina at reprocessing facilities that were part of the nuclear weapon production complex. Any recovered HEU was assigned to the US military inventory.

Spent HEU fuel from the two closed high-temperature reactors has not been reprocessed. It is currently stored in Idaho pending final disposal.

For several years, the USA has not reprocessed its spent domestic research reactor fuel. Its reprocessing facilities in Idaho have been closed down and the ones in South Carolina have been renovated in order to process a stock of residual production reactor fuel and targets that remained after the production reactors were closed. Currently, it is unclear whether the domestic spent HEU research reactor fuel will be reprocessed, although shipments to the Savannah River Site are expected to increase during the next several years. The fate of domestic spent fuel is expected to match that of the foreign spent fuel.

The US spent-fuel take-back policy

Until the end of 1988, the USA routinely took back and reprocessed foreign spent HEU fuel if it was the original supplier of the HEU and if the fuel was an aluminium-based or TRIGA reactor-type fuel. The reactor owners then received a credit for any leftover ^{235}U when they purchased fresh HEU fuel, obtaining a relatively low-cost method of disposing of spent fuel and obtaining fresh fuel. By 1988, the USA had received about 3800 kg of ^{235}U in an estimated 5700 kg of HEU in foreign spent fuel.[23] The Nuclear Regulatory Commission (NRC) estimated that the total amount of HEU returned to the United States by the end of 1988 amounted to the equivalent of 8394 kg.[24] This study defines equivalent as the initial mass of the HEU, which means it includes both the amount of HEU contained in the spent fuel and the HEU consumed during irradiation. Combining the above information, about 2700 kg of ^{235}U were consumed during irradiation, or about 42 per cent of the total ^{235}U.

In 1994 and 1995, the USA accepted a small amount of foreign fuel on what it called an 'urgent relief' basis. Several reactors were running out of on-site storage space, and were being forced to consider either shutting down or sending the fuel for reprocessing in Europe. The USA viewed either option as detrimental to its policy to seek the conversion of civil reactors and eliminate civil HEU use.

Meanwhile, the Department of Energy launched a comprehensive environmental impact statement (EIS) for a programme to take back spent fuel of US origin. In May 1996, following the completion of the EIS, the DOE issued a decision to resume taking back and managing spent foreign HEU and LEU aluminium-based and TRIGA fuel containing uranium enriched in the United States.

This policy will be in effect for this type of spent fuel either in storage or generated during the next 10 years. Actual shipments could be made for a period of 13 years, allowing for cooling of fuel and unforeseen events. This period is believed by US officials to be sufficient to bring back most of these types of spent fuel, convert most overseas reactors to LEU and provide sufficient time for foreign research reactor operators to find another method of disposing of their LEU fuels and other types of US-origin irradiated HEU.

The DOE does not favour the reprocessing of this spent fuel. However, the decision outlines a management strategy which leaves open the possibility of reprocessing the spent aluminium-based fuel if suitable long-term storage and disposal technologies are not ready for implementation by the year 2000. In addition, if during storage the spent fuel develops problems that pose a health and safety risk, and no suitable storage alternative exists, the Department may

---

[23] Cochran *et al.* (note 3); US Department of Energy official, private communication with the authors, Feb. 1992. This number is estimated from Nuclear Materials Management and Safeguards System data complete through 1983, which indicate that a total of 5077 kg of HEU containing 3419 kg of ^{235}U was returned from the foreign research reactors to the USA. The average enrichment of this material was 67%.

[24] NRC Report to Congress (note 2). The quantities of HEU provided in this report were developed in part by experts in the US RERTR programme at Argonne National Laboratory.

also reprocess this spent fuel. If reprocessing occurs, the DOE would add depleted uranium during the beginning of the reprocessing operation so that only LEU would be produced as a final product.

In total, the USA expects up to 104 operating and closed research reactors in 41 countries to send back a total of about 3000 kg of HEU containing about 2000 kg of ^{235}U (average about 67 per cent enriched) and more than 12 000 kg of LEU (average enrichment about 10 per cent).[25] This amount of spent HEU fuel corresponds to an initial mass of 4600 kg of HEU and 14 600 kg of LEU, for a total of 19 200 kg of enriched uranium.[26] Only about 1000 kg of TRIGA fuel is expected to be returned; the rest is aluminium-based fuel.

The USA will pay the full cost for transporting and managing the spent fuel from developing countries. Developed countries will pay those costs themselves.

Other countries

The former USSR is believed to have reprocessed its domestic civil research reactor fuel.[27] With the exception of Iraq, whose research reactor spent fuel was sent to Russia following the Persian Gulf War under a mandate of the UN Security Council, Russia is not believed to have taken back spent HEU research reactor fuel from outside the former Soviet Union.

The UK has reprocessed HEU fuel from thermal research reactors. The UK has recovered more than 1.7 tonnes of HEU from thermal research reactor fuel, although the reference did not provide the amount of ^{235}U contained in the recovered material or its fate.[28]

The UK wants to keep its research reactor fuel reprocessing plant open at Dounreay, Scotland and is seeking customers.[29] Its goal is to attract spent fuel of British, US and perhaps Russian origin, although its prospects with regards to US-origin fuel as of early 1996 had diminished considerably following the US decision to take back its spent fuel. One customer is Australia, which agreed in late 1995 to send some of its British-origin fuel to Dounreay and may agree to send more later. However, it plans to send its US-origin fuel to the USA. Spent fuel from the German FRM-2 could also be sent to Dounreay. The UK has not announced whether it will recover HEU or blend it down to LEU, although the expectation is that recovered HEU would be returned to the research reactor owner for reuse.

---

[25] Matos, J. E., 'Foreign research reactor irradiated nuclear fuel inventories containing HEU and LEU of United States origin', ANL/RERTR/TM-22, Argonne National Laboratory, Dec. 1994, table 1.
[26] Matos (note 25).
[27] Cochran, T. B., Norris, R. S. and Bukharin, O. A., *Making the Russian Bomb: From Stalin to Yeltsin* (Westview Press: Boulder, Colo., 1995), pp. 83–84.
[28] 'The Dounreay fuel cycle facilities of AEA Fuel Services', *Atom*, Jan. 1992, pp. 18–20.
[29] Marshall, P., 'UKAEA looks forward to reprocessing MTR fuel for at least 15 years', *Nuclear Fuel*, 6 Nov. 1995, pp. 14–16.

Table 8.1. Summary of the distribution of US HEU exports
The initial mass of the HEU is expressed in kilograms.[a]

	HEU exports	Total in minus total out[b]
Euratom	21 238	13 677
Non-Euratom	4 637	3 812
Total	**25 875**	**17 489**

[a] Defined as the initial mass of uranium before insertion into a reactor.

[b] This column represents the total quantity of US-origin HEU imported by a country less the quantities retransferred out of this country or returned to the USA. The total amount of imported HEU represents shipments to this country directly from the USA and shipments via an intermediary country. The latter occurred, for example, if the USA shipped HEU to an overseas fuel fabricator which in turn shipped the fabricated HEU fuel to this country.

Source: The United States Nuclear Regulatory Commission's Report to Congress on the Disposition of Highly Enriched Uranium Previously Exported from the United States, Jan. 1993.

France has reprocessed spent HEU fuel from its research reactors at the UP1 plant at Marcoule, which is devoted mainly to gas-graphite reactor fuel (see chapters 3 and 6), France intends to mix recovered HEU with depleted uranium to get LEU.[30]

Partly because it will close the UP1 plant in 1997, France announced in early 1996 that it had decided to offer to reprocess foreign research reactor spent fuel in UP2-800 at La Hague, which is dedicated to French LWR fuel (see chapter 6).[31] Whether France will find customers or require them to take back the nuclear waste is unclear, as of the summer of 1996.

VI. US exports of highly enriched uranium

The most detailed information about civil inventories of highly enriched uranium has been developed by the USA in order to better understand the fate of the HEU it exported over the past 40 years.

The NRC study

Tables 8.1 and 8.2 are from the January 1993 Nuclear Regulatory Commission Report to Congress on the disposition of US-origin HEU.[32] In each table, the first column lists the amount of HEU originally exported from the USA, and the values in the second column represent the total quantities of US-origin HEU

---

[30] Ballagny, A., 'Status of French reactors', Paper presented at the 1995 International Meeting on the Reduced Enrichment for Research and Test Reactors, Paris, 18–21 Sep. 1995.

[31] MacLachlan, A., 'Cogema to offer reprocessing services for research reactor spent HEU fuel', *Nuclear Fuel*, 29 Jan. 1996, p. 6.

[32] NRC Report to Congress (note 2).

8.2. United States HEU exports and retransfers for non-Euratom countries
The initial mass of the HEU is expressed in kilograms.[a]

Country	HEU exports	Total in minus total out
Argentina	58	58
Australia	10	146
Austria	8	39
Brazil	8	9
Canada	2 169	1 184
Chile	0	12
Colombia	3	3
Iran	6	6
Israel	19	34
Jamaica	0	1
Japan	2 054	1 973
Mexico	13	12
Norway	0	4
Pakistan	6	16
Philippines	3	3
Romania	39	39
Slovenia	5	5
South Africa	33	10
South Korea	25	25
Sweden	148	127
Switzerland	9	82
Taiwan	10	10
Thailand	5	5
Turkey	5	8
15 other countries	1	1
Total	**4 637**	**3 812**

[a] Defined as the initial mass of uranium before insertion into a reactor.

Source: The United States Nuclear Regulatory Commission's Report to Congress on the Disposition of Highly Enriched Uranium Previously Exported from the United States, Nuclear Regulatory Commission, Washington, DC, Jan. 1993.

transferred into an individual country, less the quantities retransferred out of this country or returned to the United States. As such, the values in the second columns reflect actual inventories in terms of the initial mass of the HEU.[33] The values in the second columns, however, have not been corrected for the decrease in HEU mass due to the fissioning of ^{235}U in a reactor. Thus, the inventories in this column are calculated in terms of the initial mass of HEU that was supplied. The actual net stock of HEU left after irradiation could be as little as one-half the listed values.

[33] Defined as the initial mass of uranium before insertion into a reactor.

Table 8.3. United States HEU exports and external retransfers for Euratom countries[a]
The initial mass of the HEU is expressed in kilograms.[b]

Country	HEU exports	Total in minus total out
Belgium	187	505
Denmark	26	4
France	7 250	4 104
Germany	11 327	6 817
Greece	7	28
Ireland	0	0
Italy	354	315
Luxembourg	0	0
Netherlands	64	102
Portugal	8	8
Spain	9	1
UK	2 006	1 793
Total	**21 238**	**13 677**

[a] Exports to, and external retransfers to and from, the Euratom bloc of countries are shown. Information on retransfers includes incoming and outgoing HEU to and from Euratom countries, and to and from non-Euratom countries. This information is based on US records and calculations which have not been reconciled with Euratom records. Euratom and several member states have been requested, through US diplomatic channels, to volunteer such information.

[b] Defined as the initial mass of uranium before insertion into a reactor.

Source: The United States Nuclear Regulatory Commission's Report to Congress on the Disposition of Highly Enriched Uranium Previously Exported from the United States, Nuclear Regulatory Commission, Washington, DC, Jan. 1993.

In table 8.1, Euratom, which includes 12 European countries with US-origin HEU, is considered as a single entity because retransfers within Euratom are not reported to the US Government. Thus, in terms of inventories in individual countries, Euratom represents a 'black box', or undifferentiated entity, to the US Government. Nevertheless, the NRC attempted to estimate country-specific inventories within Euratom, as shown in table 8.3, utilizing the best information it had at hand. However, these inventories represent crude approximations at best, because Euratom refused to provide the US Government with country-specific information.

In total, the NRC estimated that 17.5 tonnes (in terms of initial mass) of HEU of US-origin remained overseas as of the end of 1992. Euratom is estimated to possess about 13.7 tonnes of this US-origin HEU.

Euratom told US officials, however, that it had only 11.3 tonnes of US-origin HEU. Euratom officials were unwilling to explain the difference in the numbers. US officials believe that the discrepancy can be mostly explained by the recovery and blending down to LEU of about 2000 kg of 35 per cent enriched uranium taken from the German SNEAK critical facility. Euratom safeguards officials, however, have refused to confirm or deny this possibility.

Table 8.4 Amount of US-origin HEU projected to remain overseas
The initial HEU mass is given in tonnes.[a]

	Initial HEU mass
US exports	25.9
Returns through 1995	– 8.4
Sub-total	**17.5**
Returns from 1996–2009	– 4.6
Total overseas	**12.9**

[a] Defined as the initial mass of uranium before insertion into a reactor.

Source: The United States Nuclear Regulatory Commission's Report to Congress on the Disposition of Highly Enriched Uranium Previously Exported from the United States, Nuclear Regulatory Commission, Washington, DC, Jan. 1993; and Matos, J. E., 'Foreign Research Reactor Irradiated Nuclear Fuel Inventories Containing HEU and LEU of United States Origin,' ANL/RERTR/TM-22, Argonne National Laboratory, Dec. 1994, table 1.

A fraction of the HEU in Euratom is fresh weapon-grade uranium, and is available for use in Euratom research reactors. US officials estimate that the amount of fresh weapon-grade uranium totalled about 1–2 tonnes as of early 1993. This material originated from many different shipments from the USA over many years. The US Government is unaware of the exact amount or location of this WGU, but as of early 1993 this stock included WGU in unused fuel originally intended for the closed German high-temperature reactor (350 kg), unused fuel for the last core of the KNK-2 reactor (60–90 kg), accumulated stocks at the German fuel fabricator NUKEM when it closed in 1988 that were transferred to CERCA, the fuel fabricator in France (> 500 kg), and additional WGU stocks accumulated over time at CERCA either directly or from other Euratom countries. Fresh weapon-grade stocks exist in other countries, such as a small amount in the UK (less than 100 kg). In addition, relatively small stocks are located at individual reactor sites.

The amount of fresh HEU in stock in 1996 at various locations in Euratom could not be determined, but this quantity is believed to be significantly less than it was in 1993. Euratom reactors require about 200 kg of HEU per year, of which about 90 per cent is weapon-grade (see appendix D). Thus the 1993 fresh HEU inventory could have been reduced by about 600 kg by now. The exact amount would depend on actual reactor performance.

Most of the remainder of the HEU is in the form of aluminium-based or TRIGA irradiated fuel, other types of irradiated fuel, recovered HEU in storage, or fuel in critical assemblies. In addition, much of the HEU has also been consumed during irradiation.

HEU returns to the USA

Under the foreign spent-fuel take-back policy recently announced by the DOE, the USA expects to receive a total of 4610 kg of HEU (initial HEU mass) during the 13-year period the policy is in effect. Table 8.4 shows that taking account of returns to the USA, about 12.9 tonnes of HEU (initial HEU mass) will remain overseas after the take-back policy expires.

There are several reasons why so much HEU will not be returned to the USA. Some of the fuel is uranium oxide or metal, and is ineligible for return to the USA. Some of the HEU has already been recovered and blended down to LEU, such as the SNEAK fuel. Roughly 4 tonnes of HEU (initial mass and initial enrichment of 20–60 per cent) are in critical facilities in France or at the Joyo fast reactor in Japan, and this material cannot be returned under the current policy.

How much HEU has been consumed?

It is possible to estimate crudely the amount of US-origin HEU consumed by reactors to date, resulting in a rough estimate of the actual amount of HEU remaining in countries supplied by the USA. From 1978 to the end of 1988, the USA issued export licences for 3289 kg of ^{235}U in HEU and accepted the return of 1629 kg of ^{235}U.[34] Another 1570 kg of ^{235}U were consumed in reactors during this period. During the 1980s, US exports of ^{235}U therefore nearly matched the return of ^{235}U in spent fuel and the amount consumed in reactors. Therefore, HEU inventories in countries supplied by the USA have not increased significantly during the period 1987–88.

For earlier periods, however, significantly more HEU was exported than returned. Only about 2150 kg of ^{235}U in spent HEU fuel had been returned by the end of 1977, out of a total of about 14 700 kg exported by then. Assuming that the returned HEU matched that returned to the USA in the 1980s (in which case about 40 per cent of the ^{235}U was consumed), then about 11 100 kg ^{235}U supplied by the USA remains in Western countries. An additional quantity of ^{235}U was shipped after 1988, which is estimated at about 600 kg ^{235}U, bringing the total to 11 700 kg.

A significant proportion of this 11 700 kg of ^{235}U was either consumed in reactors, remains in their cores, or was recovered in Europe. A fraction—up to 1000–2000 kg—remains in storage or in the fuel cycle as fresh material (see above). Assuming that on average about 40 per cent of the remainder has been consumed, about 7000–7400 kg of ^{235}U (5820–6420 kg in irradiated form and up to 1000–2000 kg in unused form) in HEU remains in these US-supplied research reactor programmes. If we include 6300 kg of ^{238}U in the original fresh HEU fuel (initial HEU enrichment of 70 per cent on average) and any ^{236}U pro-

---

[34] Letter from Armando Travelli, RERTR Program Manager, to Leonard Weiss, Staff Director, US Senate Committee on Governmental Affairs, 19 Apr. 1988; and Travelli (note 19).

duced in the fuel (< 1000 kg), the amount of HEU supplied by the USA and not returned or consumed is roughly 14–15 tonnes. This value needs to be reduced by about 2000 kg, which is thought to have been blended down in Euratom into LEU, giving a total of about 12–13 tonnes of HEU. This amount is corrected for the irradiation of the HEU and represents about half the total HEU exported by the USA.

VII. Civil inventories of highly enriched uranium

No country publishes information on its stock of civil HEU. Complicating estimates is the historical practice in the USA and the former Soviet Union of reprocessing civil spent HEU fuel and allocating the recovered HEU to military programmes. In the military programmes, however, the recovered HEU forms a small fraction of the total amount of recovered HEU and is rarely totalled (see chapter 4).

It is possible to develop crude estimates of the total amount of civil HEU assigned to all civil research reactor programmes. A lower bound on this quantity is the roughly 9 tonnes of HEU listed in the 1994 Annual Report of the IAEA that were under IAEA safeguards. Almost all of this HEU is located in non-nuclear weapon states or in France and Britain and is part of research reactor programmes. The total inventory is considerably higher because the HEU in research reactor programmes in nuclear weapon states is rarely safeguarded.

One method of calculating a total is to derive an estimate of the amount in each nuclear weapon state that is not under IAEA safeguards and then to add it to the amount under IAEA safeguards. Because the USA and Russia have historically reprocessed most domestic research reactor fuel and reassigned the recovered uranium to military stocks, the combined civil inventory in those countries is likely to be on the order of 5–10 tonnes of HEU, even though considerably larger quantities of HEU have been used in these reactors (about 40–55 t). The two largest ex-Soviet breeder reactors are ignored. The non-IAEA safeguarded HEU stocks in Britain and France comprise HEU of non-US origin, and these stocks are estimated to amount to about 1–2 tonnes.[35] China's civil stock is believed to be small, or roughly 1 tonne. In total, the weapon states have about 7–13 tonnes, bringing the total to 16–22 tonnes of HEU in civil research programmes worldwide. We assign 20 tonnes to civil research stocks.

These aggregate quantities remain highly uncertain. What is needed is a country-by-country breakdown of civil HEU inventories. Until this becomes available, the picture will remain incomplete.

---

[35] According to Euratom officials, about 87% of the civil HEU under Euratom safeguards in 1993 was of US origin and thus subject to IAEA safeguards, or up to 1.7 t was not under IAEA safeguards but still classified as civil HEU and under Euratom safeguards. Therefore, civil stocks in Euratom not subject to IAEA safeguards are less than about 2 t, and are unlikely to be below 1 t.

A new civil HEU category is material withdrawn from nuclear weapon programmes and placed under IAEA safeguards. This category is treated in more detail in chapters 4 and 15. The USA has taken 10 tonnes of such material and put it under safeguards at Oak Ridge (see chapter 4). South Africa has similarly placed an inventory of more than 400 kg of HEU under IAEA safeguards (see chapter 13). HEU transfers from military programmes to civil ones are expected to increase sharply in the future in the nuclear weapon states.

Part IV
Material inventories and production capabilities in the threshold states

9. De facto nuclear weapon states: Israel, India and Pakistan

Israel is widely believed to have deployed nuclear weapons. Its arsenal continues to provoke demands from Arab nations that it sign the NPT and it also serves to justify and inspire some Middle-Eastern attempts to acquire nuclear weapons.

India and Pakistan have the capability to deploy nuclear weapons quickly, and continue to improve their nuclear weapon capabilities. Their nuclear arms race currently poses the most likely scenario in which nuclear weapons could be used.

Because of the secrecy surrounding nuclear programmes in these countries, little definitive, public information exists about the number of weapons they might possess, or the size of their current or planned stockpiles of separated plutonium and highly enriched uranium.

Nevertheless, during recent years information has emerged about these nuclear programmes. This is expanded upon here to provide an estimate of the stockpiles of fissile materials in these countries.

I. Israel

At a site near Dimona Israel operates several highly secret nuclear facilities for the production of plutonium, and perhaps tritium and enriched uranium, for its nuclear weapon programme. This site, called the Negev Nuclear Research Centre, is known to contain a small production reactor and a plutonium separation facility. Plutonium is Israel's main route to nuclear weapons.

The Dimona facility might also house a small uranium enrichment facility. Israel has pursued laser enrichment and maybe gas-centrifuge technology, at least as research activities, and it may have expanded these activities.

Israel has also been suspected of having diverted about 100 kg of weapon-grade uranium from a US nuclear facility during the 1960s, although Seymour Hersh has presented evidence that the diversion did not occur.[1]

Plutonium production at Dimona

The reactor

Israel's main source of plutonium is a French-supplied reactor. Early reports said that it generated 24 MWth when it began operating in December 1963. The

---

[1] Hersh, S. M., *The Samson Option* (Random House: New York, 1991).

reactor's power, however, appears to be substantially higher, giving Israel a significantly larger plutonium production capability.

The lack of conclusive information about the power levels reached by the Dimona reactor creates the largest uncertainty in estimating Israel's plutonium inventory. The only detailed account of the construction and initial size of the Dimona reactor was published in 1982 by Pierre Péan.[2]

Reactor power. According to Péan, the initial Franco-Israeli agreement in 1956 called for France to supply a nuclear reactor of the EL-3 type, an 18-MWth research reactor that was moderated by heavy water with cooling provided both by the heavy water and by air.[3] However, Péan writes that when the French team charged with building the plutonium extraction facility read the reactor designs, they were surprised that the reactor power was two to three times larger than indicated in the original Franco-Israeli agreement.[4] They also found that the cooling ducts were three times larger than those needed for a 24-MW reactor. With these upgrades in the original design, the head of the construction team and his deputy concluded that the Dimona reactor would be capable of producing as much plutonium as the G1 reactor at Marcoule, France (see chapter 3). This was a 38-MWth gas-cooled graphite-moderated reactor that was capable of producing about 11 kg of weapon-grade plutonium (99% ^{239}Pu) per year if operated at full power 80 per cent of the year (see chapter 3).[5]

Controversy surrounds whether Israel increased the power significantly above 40 MWth. The highest estimated power rating is derived from details about the plutonium separation plant at the Dimona complex made public in 1986 by Mordechai Vanunu, a former nuclear technician at the Dimona separation facility.[6] Based on his statements about plutonium throughputs in specific areas of the plant, independent scientists have calculated that the power reached 150 MWth (see below).[7]

According to a US official, however, who spoke on condition his name would not be used, the consensus of people in the government who have studied the information supplied by Vanunu is that the reactor's power has not reached 150 MWth. He said that the power of the Dimona reactor has probably never exceeded 70 MWth. This latter estimate of the power is consistent with Péan's information about the size of the cooling ducts.

Achieving an increase in power from 40 to 150 MWth would be difficult to accomplish without major modifications to the reactor. Increasing the power

---

[2] Péan, P., *Les deux bombes* [The two bombs] (Fayard: Paris, 1982). Translated into English by the US Congressional Research Service (CRS).

[3] Péan (note 2). In CRS translation: 'Suez or the "Musketeers" of the bomb', p. CRS-4. Details of the EL 3 reactor can be found in International Atomic Energy Agency, *Directory of Nuclear Reactors, Vol. 2: Research, Test, and Experimental Reactors* (IAEA: Vienna, 1959), pp. 295–300.

[4] Péan (note 2). In CRS translation: chapter VI, 'EL 102', p. CRS-11.

[5] Péan (note 2). In CRS translation, p. CRS-12; Turner, S. E., *et al.*, Criticality studies of graphite-moderated production reactors, SSA-125, Southern Sciences Applications, Jan. 1980, table 2.1 (report prepared for the Arms Control and Disarmament Agency).

[6] 'Revealed: the secrets of Israel's nuclear arsenal', *Sunday Times*, 5 Oct. 1986; and Barnaby, F., *The Invisible Bomb* (I.B. Taurus: London, 1989).

[7] *Sunday Times;* and Barnaby (note 6).

can be done by increasing the cooling, which means running water through the reactor faster, but this approach has limits. Enriched uranium can be used and the core can be enlarged, although available evidence tends not to support Israel's implementation of either of these options.

Spector and Smith reported that US Government specialists believe that the reactor operated initially at 40 MWth and might have been enlarged to 70 MWth prior to 1977, when Vanunu began working at Dimona.[8] However, because significant power increases above 70 MWth cannot be excluded, this section includes one estimate in which the power is increased to 150 MWth. This estimate serves as an upper bound (see table 9.1).

Fuel burnup. Vanunu said that about 140 fuel rods are irradiated in the reactor for about three months before being sent to the separation plant. A short irradiation period is consistent with weapon-grade plutonium production.

Vanunu also said that before separating the plutonium from the uranium, the dissolved liquid at one point in the separation process contained about 450 grams of uranium per litre of solution and 170–180 milligrams of plutonium per litre. This statement implies a plutonium concentration of about 0.38–0.4 mg plutonium per gram of uranium. This concentration corresponds to a burnup of about 400 MWd/t.[9] The plutonium would contain about 97–98 per cent plutonium-239.

Uranium and plutonium discharged. Burnup information can be combined with power estimates to derive the amount of uranium in each core. At a power of 40 MWth, a three-month fuel-irradiation period and the above burnup, the reactor would discharge about 9 tonnes of uranium, containing about 3.5 kg of plutonium. For this type of reactor, this uranium probably represents the entire core. Assuming an average of three core reloads a year, the reactor would discharge about 27 tonnes of fuel a year, containing about 10.5 kg of plutonium. Three core loads a year would correspond to a capacity factor of 75 per cent.

At 70 MWth, under the same assumptions, the reactor would discharge about 47 tonnes of fuel a year, containing about 18 kg of plutonium.

Uranium requirement. At power levels of 40–70 MWth, the reactor would have required 810–1410 tonnes of uranium fuel over 30 years of operation.

Little is known about Israel's sources of uranium. Israel acquired 600 tonnes of yellowcake from South Africa in the late 1970s and another 200 tonnes from Europe in the 1960s. Israel is reported to produce indigenously about 10 tonnes of yellowcake a year. It could have other foreign uranium sources as well.

Vanunu implied that at least some of the depleted uranium recovered during reprocessing is recycled back into the reactor, which could have lowered the need for imports and domestic production of natural uranium. The depleted uranium would contain about 0.67–0.68 per cent uranium-235, and may need to

---

[8] Spector, L. S. and Smith, J. R., *Nuclear Ambitions* (Westview Press: Boulder, Colo., 1990), p. 160.
[9] NUS, *Heavy-Element Concentration in Power Reactors*, Report no. SND-120-2 (NUS Corporation: Clearwater, Fla., May, 1977).

be brought to the level of natural uranium.[10] This upgrading could have been accomplished by enriching the material domestically or blending it with enriched uranium obtained internationally.

The separation plant

Vanunu's revelations have provided the public with its only detailed glimpse into Israel's plutonium separation programme.[11] He described in detail an underground facility called Mochon 2, supplied by France, that included six underground levels dedicated to separating plutonium from irradiated fuel by the Purex process, converting the plutonium into metal and shaping the metal into weapon components.

France supplied the separation plant in the late 1950s and early 1960s. However, specifics about the type of equipment and processes are sketchy. According to Mary Davis, a US researcher studying the French nuclear programme, France supplied this reprocessing plant at a time when it was just starting to deploy reprocessing technology itself on an industrial scale. As a result, France kept modifying its own reprocessing facility at Marcoule throughout the early 1960s. Problems were encountered in dissolving the uranium fuel and purifying plutonium in an ion-exchange process,[12] and they shifted from batch to continuous processing after 1963 or 1964, perhaps after supplying the plant to Israel.

In any case, Israel is unlikely to have escaped the problems associated with plutonium separation. Following international improvements in plutonium reprocessing and metallurgy Israel undoubtedly upgraded its own processes. This upgrading probably caused initial delays in spent fuel processing, but resulted in greater capacity afterwards. There is also a report of what is believed to be a criticality accident at the plant in 1966, which killed one person and closed part of the facility for several months.

Throughput information. Vanunu supplied considerable information about the flow of plutonium in various sections of the plant. Two items in particular provide insight into the operations at the plant while he worked there. He said that:

1. The standard flow rate of dissolved uranium and plutonium through one section of the plant is 20.9 litres per hour. The concentration of plutonium is 170–180 milligrams per litre. This corresponds to an annual production of 20.3–21.5 kg of plutonium, assuming that the plant operated continuously for eight months (34 weeks) a year. Vanunu, however, stated that the flow rate normally exceeded the standard rate by 150–175 per cent, implying an annual production of 30–38 kg of plutonium in a eight-month operating cycle.

---

[10] NUS (note 9).
[11] *Sunday Times* and Barnaby (note 6).
[12] Barrillot, B. and Davis, M., *Les dechets nucléaires militaires Français* (Centre de Documentation et de Recherche sur le Paix et les Conflits: Lyon, 1994), pp. 132–33.

2. The metal reduction portion of the plutonium plant (Unit 37) produces an average of nine metallic buttons a week, each of which contains about 130 grams of plutonium. Over an eight-month (34-week) production cycle, the plant would process a total of about 40 kg of plutonium.

Statements that the power of the reactor reached 150 MWth result from calculations based on a steady-state output of 40 kg of plutonium per year.[13] That the separation plant could achieve outputs in the above range is supported by a knowledgeable French nuclear source, who did not dispute most of Vanunu's claims in the *Sunday Times* but said that a yearly capacity of 40 kg a year might be 'a little on the high side'.[14]

Vanunu did not say that operation at the level of 150 per cent to 175 per cent over standard operation was achieved throughout the eight-month campaign, or over the life of the plant. Operation at this level may represent an attempt in certain years to process fuel faster, shortening the campaign.

An alternative explanation is that the years when Vanunu worked at the plant were a time when a backlog of irradiated fuel was processed, resulting in a temporary increase in the plant's annual throughput. Evidence for this hypothesis is that Vanunu was hired as part of a large group of technicians in 1977, and laid off in 1985 with about 180 other workers. Such a backlog could have been caused by a long closure of the separation plant, caused either by an accident or an upgrade programme.

In addition, extrapolations of plant throughput from Vanunu's information about button production ignores any recycling of plutonium from the weapon manufacturing areas of the plant. Depending on the specific design of a nuclear weapon, significant amounts of plutonium scrap are generated during component manufacture. More advanced shapes tend to produce more scrap. Scrap would probably be converted into plutonium oxide, dissolved in nitric acid and, after several more process-steps, turned back into a metal button.[15]

These explanations, if true, would imply that the output of this reprocessing plant is not a reliable indicator of the total plutonium production of the Dimona reactor.

Plutonium inventory

The above discussion illustrates the difficulty of determining Israel's plutonium inventory. As a result, we have developed various estimates based on hypothetical power histories of the reactor, labelled A–E in table 9.1.

In all of these estimates, we assume that the reactor has operated reliably at an average 75 per cent capacity factor since 1965. In reality, the reactor might

---

[13] This estimate can be duplicated if it is assumed that the reactor operates about 75% of the time and produces 0.95 g of plutonium per MWth-day (MWth-d) of heat output.

[14] *Nucleonics Week*, 30 Oct. 1986.

[15] If 75% of the button plutonium ends up in final weapon components, which is a reasonable amount, then 25% of the plutonium would be recycled back into the separation units for processing into buttons. In this case the amount of plutonium from the spent reactor fuel would total no more than 30 kg/y.

Table 9.1. Estimated plutonium production in the Israeli Dimona reactor, 31 December 1994[a]

Years	Power (MWth)	Total Pu[b] (kg)	Number of warheads[c]
A: 1965–94	24	190	38
B: 1965–94	40	320	64
C: 1965–75	40	..	..
1976–94	70	470	94
D: 1965–94	70	560	112
E: 1965–70	40	..	..
1970–77	70	..	..
1978–94	150	880	176

[a] These historical estimates of the amount of plutonium produced in the Dimona reactor are based on information summarized in the text. They provide an upper and lower bound on plutonium production, although our best estimate encompasses the middle three scenarios. Scenario A is a lower bound on plutonium production and uses the lowest known power rating of the reactor, 24 MWth, throughout the period under consideration. Scenario B assumes a higher power rating of 40 MWth that is increased to 70 MWth during the mid-1970s, a time when the number of people working at Dimona is reported to have increased dramatically. Scenario D assumes a continuous power rating of 70 MWth. Scenario E represents an upper bound on plutonium production and arbitrarily assumes a step-wise increase in power from 40 MWth to 150 MWth. The latter increase is considered to be the limit of the power for this reactor.

[b] It is assumed that the reactor operates at full power an average of 75% of the time, and that it produces about 0.97 g of plutonium per MWth-d. Israel may have separated tritium from lithium-6 targets irradiated in the Dimona reactor. Tritium production would have reduced estimated plutonium production somewhat (see text).

[c] It is assumed that each warhead requires 5 kg of weapon-grade plutonium (see text).

have experienced operating difficulties that would have lowered the capacity factor, or it might have been shut down for extended periods for maintenance, safety improvements or power upgrades.

Our best estimate is that up to the end of 1994, Israel had produced about 320–560 kg of weapon-grade plutonium. Our minimum estimate is 190 kg and our maximum estimate is 880 kg.

We assume under our best estimate that the power of the reactor has remained at between 40 and 70 MW since 1994. At the end of 1995, our best estimate is that it had produced 330–580 kg of weapon-grade plutonium. By 2000, Israel could have produced a cumulative total of 370–650 kg of weapon-grade plutonium (see table 9.2).

Estimates of future production are uncertain because the reactor is over 30 years old and might be closed down soon for safety reasons or lack of need. In addition, the USA is reported to have pressed Israel to cease plutonium production.[16]

---

[16] See, e. g., Hibbs, M., 'US wants Israel to cease Dimona plutonium production', *Nucleonics Week*, 26 Sep. 1991.

Table 9.2. Estimated inventories of Israeli weapon-grade plutonium, at the end of 1994, 1995 and 1999

	31 Dec. 1994	31 Dec. 1995	31 Dec. 1999[b]
Weapon-grade plutonium (kg)[a]	320–560	330–580	370–650
Number of warheads	64–112	66–116	74–130

[a] These estimates have not been reduced to account for tritium production, which in any case is assumed to be small.

[b] For future plutonium production projections, it is assumed that the Dimona reactor will maintain a power of 40–70 MWth, and produce about 10.6–18.6 kg of plutonium per year. It is also assumed that each warhead requires an average of 5 kg of weapon-grade plutonium.

Israel's nuclear arsenal

Public estimates of the amount of plutonium used in Israeli nuclear weapons are scarce. Israel is believed to be knowledgeable about sophisticated nuclear weapon design, including thermonuclear weapons and boosted fission weapons, but the details available about its deployed designs are few and contradictory. In any case, Israel is likely to have developed conservative designs that require few, if any, full-scale nuclear tests to demonstrate their reliability or yield. Such a decision could imply that each weapon, whether a pure fission, boosted or thermonuclear design, has a greater amount of plutonium than commonly assumed.

For lack of better information, we assume 5 kg of plutonium per warhead. For comparison, the Trinity explosion in 1945 used 6 kg of weapon-grade plutonium, and modern US weapons average about 3–4 kg of plutonium per warhead. Based on the plutonium production estimates in table 9.1, Israel could have constructed between 64 and 112 warheads up to the end of 1994, and could produce roughly 2–4 weapons per year.

Tritium production

Israel has reportedly produced tritium at the Dimona site. At first, it is reported to have extracted tritium from heavy water irradiated in the reactor. Later, Vanunu said that lithium targets were irradiated in the reactor, producing tritium that was recovered starting in the early 1980s. The Mochon 2 facility both enriched the lithium and extracted the tritium.

Available information does not provide a credible reason for Israeli production of tritium, detail the lithium irradiation strategy in the reactor or allow an estimate of tritium production. The most common need for tritium, beyond research purposes, is for boosted fission weapons. These weapons use a mixture of tritium and deuterium that undergoes fusion and produces a spurt of neutrons that fission the fissile material, dramatically increasing the total fission yield. The fusion yield is typically small in this type of weapon. Tritium could also be used in high-yield boosted fission weapons, although this use is less common.

A typical boosted fission weapon would require on the order of 5 grams of tritium. If Israel has used tritium in all its weapons, an unlikely prospect, it would have needed to produce about 320–560 grams of tritium. In addition, since tritium radioactively decays at a rate of about 5 per cent a year, about 16–28 grams of tritium would be needed each year to replenish the inventory.

The production of 1 gram of tritium in a reactor is equivalent to about 70–80 grams of weapon-grade plutonium. The production of 320–560 grams of tritium is therefore equivalent to about 25–42 kg of weapon-grade plutonium, or about 2.3–4 years of reactor output at 40 MWth and about 1.3–2.3 years at 70 MWth.

The annual replenishment requirement is equivalent to about 1–2 kg of plutonium. At a power of 40 MWth, this requirement represents about 9–18 per cent of the reactor's annual output. At 70 MWth, it is about 5–10 per cent of annual output.

In terms of total plutonium estimates, the impact of tritium production is small, at most 10 per cent, and this assumes that Israel has a large number of boosted fission weapons. This latter assumption is questionable. On a yearly basis, however, the impact of tritium production could be significant. Without knowing the scale of tritium production or the method of irradiating lithium in the reactor, we cannot evaluate the potential significance of tritium production for annual plutonium production.

Uranium enrichment

Little is known about Israel's uranium enrichment programme. Israel is known to have pursued laser enrichment, but the status or exact purpose of this effort is unclear.[17] Some public accounts report Vanunu as saying that Israel was enriching uranium using gas centrifuges, although again there are no details and little confirmation of such a programme.[18]

Vanunu said that one building at Dimona was involved in laser enrichment of uranium, and another one had a secret unit that had been making enriched uranium on a production-scale since 1979–80. This latter building might therefore house a small gas-centrifuge plant. We have no information on its suspected capacity, and are unable to estimate it.

II. India

India has the largest nuclear programme among the developing nations. Within that programme is a sizeable nuclear weapon effort based on producing separated plutonium.

---

[17] Department of Political and Security Council Affairs, UN Centre for Disarmament, *Study on Israeli Nuclear Armament* (United Nations: New York, 1982); Report of the UN Secretary-General, Israeli nuclear armament, UN document A/42/581, 16 Oct. 1987; and *Sunday Times* (note 6).

[18] Spector and Smith (note 8), p. 161.

Although India denies that it has deployed nuclear weapons, there is strong evidence that it could quickly deploy an arsenal of fission bombs. In 1974, India detonated a 'peaceful' nuclear explosive with an explosive yield of about 5–12 kt. Since then, it has continued its nuclear weapon research and development programme, apparently intensifying it in the mid-1980s in response to Pakistan's progress towards nuclear weapons.

Indian officials have often hinted at how quickly they could have nuclear weapons. In 1990, P. K. Iyengar, then head of the Indian Atomic Energy Agency, said, 'In how much time we make it, will depend on how much time we get.'[19] In an interview with one of the authors in 1990, Dr Iyengar said that India could make nuclear explosives in a matter of weeks, including the manufacturing of the plutonium components or 'pits'.

The US Central Intelligence Agency testified before the US Senate in early 1993 that it does not believe that India maintains assembled or deployed nuclear weapons.[20] Nevertheless, the CIA believes India is producing weapon components and could quickly assemble a small nuclear arsenal. It also expects India to gradually increase the size of its *de facto* arsenal.

In addition, India has continued to make advances in weapon design. It is believed to be developing or to have developed a warhead for a ballistic missile, and is interested in thermonuclear weapons.

At the core of India's nuclear weapon programme are two small heavy water-moderated reactors and a plutonium-separation plant at the Bhabha Atomic Research Centre (BARC), near Bombay. Although these reactors are used for civil research, they also provide a steady source of weapon-grade plutonium. Indian Government policy has also been to minimize international safeguards on its nuclear facilities, in effect freeing them for military use if desired. As a result, most of its CANDU power reactors and its oxide spent-fuel reprocessing plants are unsafeguarded and could be used to make plutonium for weapons.

India has not constructed any large-scale uranium enrichment facilities, although it does maintain an enrichment research and development programme at BARC. India has also built a larger enrichment facility in the south of the country near Mysore.

Plutonium production

Cirus and Dhruva

India's two largest research reactors are unsafeguarded sources of weapon-grade plutonium. They are located at BARC and use natural uranium fuel and a heavy water moderator.

---

[19] *Collected scientific papers of Dr P. K. Iyengar: Volume 5, Selected Papers and Speeches on Nuclear Power and Science in India* (Bhabha Atomic Research Centre: Bombay, June 1991), p. 247.

[20] *Proliferation Threats of the 1990s*, Hearing before the Committee on Governmental Affairs, US Senate, 24 Feb. 1993 (US Government Printing Office: Washington, DC, 1993), p. 154.

The 40-MWth Cirus reactor began operation in 1960. India has not provided information about plutonium production in Cirus, but this reactor is known to discharge weapon-grade plutonium. Using standard assumptions, Cirus can produce about 6.6–10.5 kg of weapon-grade plutonium per year, at a capacity factor of 50–80 per cent.[21] Although the Cirus reactor was supplied by Canada on condition that it be used for peaceful purposes only, India maintained after the 1974 test that this did not preclude the use of Cirus plutonium for peaceful nuclear explosives.

The 100-MWth Dhruva reactor, which was developed indigenously, went critical on 8 August 1985 at BARC. Press announcements about the opening appeared on 9 August, the anniversary of the dropping of the atomic bomb on Nagasaki. This 'coincidence' was more significant since Dhruva produces plutonium completely free of international constraints on weapons use.

Because of severe vibrations in the reactor core, Dhruva was shut down soon after starting operation.[22] In December 1986, it began operating at 25 MWth and was reported to be still operating at this level in the spring of 1987.[23] The vibrational problems were subsequently solved, and its power was increased to 80 MWth, and reached 100 MWth in mid-January 1988.[24]

Since the Dhruva reactor is similar to Cirus, we estimate that at full power it discharges about 16–26 kg of weapon-grade plutonium annually, assuming a capacity factor of 50–80 per cent. An official at the reactor told one of the authors in 1992 that its capacity factor is closer to 60 per cent. Iyengar, however, claimed that the annual output of Dhruva is 30 kg a year, although we could not confirm this statement.[25]

CANDU reactors

India also has unsafeguarded CANDU power reactors, including the Madras Atomic Power Station (MAPS), the Narora Atomic Power Station (NAPS) and Kakrapar Atomic Power Station (KAPS), that might have been used to produce plutonium for weapons. But no direct evidence exists that India has done so, or plans to do so.

During normal operation, CANDU reactors produce non-weapon-grade plutonium. Up to January 1993, MAPS-1, MAPS-2, NAPS-1 and NAPS-2 were estimated to have produced almost 1000 kg of reactor-grade plutonium, none of which is subject to international safeguards or restricted to peaceful purposes

---

[21] The plutonium production rate assumes the production of 0.9 g of weapon-grade plutonium per MWth-d of heat output.
[22] 'India's supply of unsafeguarded Pu grows as reprocessing of MAPS fuel begins', *Nuclear Fuel*, 11 Aug. 1986.
[23] Chellaney, B., 'Indian scientists exploring U enrichment, advanced technologies', *Nucleonics Week*, 5 Mar. 1987.
[24] Tefft, S., 'Nuclear emergency planning delays India's first 500 MWe reactor', *Nucleonics Week*, 17 Dec. 1987; and information supplied by the Indian Embassy, Washington, DC, Apr. 1988.
[25] Hibbs, M., 'Indian reprocessing program grows, increasing stock of unsafeguarded Pu', *Nuclear Fuel*, 15 Oct. 1990.

(see chapters 5 and 6).[26] India would undoubtedly know how to use low-quality plutonium in nuclear weapons, but an Indian official at BARC informed a visiting US expert that India is unlikely to use reactor-grade plutonium in its weapons. India's confidence in its weapon design is based on its 1974 test, which used weapon-grade plutonium. It is unlikely to want to build significantly different new designs if it does not have to do so.

During start-up, however, the power reactors typically produce small amounts of high-quality plutonium. During the first several months of operation, each of the power reactors could have discharged low burnup fuel containing roughly 5 kg of high quality plutonium. This plutonium might have been assigned to the weapon programme.

India could, in principle, dedicate one or more of its power reactors to weapon-grade plutonium production, although with a sizeable penalty in fuel costs. Each 235-MWe reactor could in these circumstances produce up to 160 kg of weapon-grade plutonium a year, operating at 60 per cent capacity.[27]

Plutonium separation capability

India has had the capability to separate plutonium since 1964, when it commissioned the unsafeguarded Trombay reprocessing facility at BARC. The Trombay facility was shut down in 1974 for decontamination and reconstruction, and restarted in 1983 or 1984. Trombay is currently sized to handle fuel from both the Cirus and Dhruva reactors, and has a nominal capacity of 50 tonnes of spent fuel per year. According to the Indian Department of Atomic Energy's annual report for 1987–88, Trombay has processed Dhruva's spent fuel.[28]

India can also separate plutonium in the Power Reactor Fuel Reprocessing (PREFRE) facility, located near Bombay, which began operation in 1979. Although designed to separate plutonium from CANDU power reactor fuel, it first processed Cirus's spent fuel.[29] The nominal annual capacity of this facility is usually listed as 100–150 tonnes of CANDU spent fuel a year. According to the BARC *Annual Report 1985–86*, PREFRE has reprocessed RAPS and MAPS fuel. This plant, however, has encountered many operational difficulties, and has not reached its nominal capacity. Indian officials view it as an experimental facility.

---

[26] Sood, D. D., 'The role of plutonium in nuclear power programme of India', Bhabha Atomic Research Centre (BARC), 26 Apr. 1993.
[27] This estimate assumes that about 0.9 g of weapon-grade plutonium are produced per MWth-day. Since the thermal efficiency of a CANDU reactor is about 0.28, a 1000-MWe CANDU reactor would theoretically discharge about 700 kg of weapon-grade plutonium each year at a capacity factor of 60%. See NUS (note 9).
[28] Government of India, Department of Atomic Energy, *Annual Report 1987–88*.
[29] Reference to these campaigns and the starting date of the facility are in Government of India, Department of Atomic Energy, *Annual Report 1980–81*, pp. 4, 31; *Annual Report 1981–82*, p. 26; *Annual Report 1983–84*, pp. 6, 31; and BARC, *Annual Report 1985–86* (BARC: Bombay, 1986).

Separated weapon-grade plutonium inventory

The Indian Government treats information about separated plutonium as classified, whether it is assigned to weapon or civil use. Our estimate of India's stockpile of separated weapon-grade plutonium is therefore difficult to confirm. India also has a stockpile of fuel- or reactor-grade plutonium, which we roughly estimate at about 300–400 kg at the end of 1993 (see table 6.8).

The bulk of India's weapon-grade plutonium is assumed to have been produced in the Cirus and Dhruva reactors. We can estimate this amount from the operating history of these reactors. Assuming that the Cirus reactor produced little plutonium during its first three years of operation, but operated continuously from 1964 onwards, it would have produced 200–280 kg of weapon-grade plutonium up to the end of 1994, assuming an average capacity factor of 50–70 per cent. The mid-point is 240 kg.

After initial start-up problems, the Dhruva reactor began operating at full power in 1988. We estimate that until 1988 the reactor produced about 10 kg of plutonium, and after that about 16.5–23.0 kg a year, at an average capacity factor of 50–70 per cent. By the end of 1994, it would have produced in total about 125–170 kg of plutonium. The mid-point is about 150 kg.

In late 1985 or early 1986, the PREFRE facility began reprocessing spent fuel from the MAPS reactors.[30] Public information about the grade of the plutonium separated at PREFRE is unavailable, although we assume that the vast bulk of the separated plutonium is fuel- or reactor-grade. The only exception is the high-quality plutonium in the fuel initially discharged from the CANDU reactors—perhaps up to 25 kg by the end of 1994 from MAPS-1, MAPS-2, NAPS-1, NAPS-2 and KAPS-1. Adding this amount to the inventory, India is estimated to have produced about 390–415 kg of weapon-grade plutonium by the end of 1994, virtually all of which is already separated.

From this estimate of weapon-grade plutonium production, we must subtract about 10 kg that were consumed in the preparation and conduct of India's 1974 nuclear test, about 10 kg lost during separation and processing, and another 50 kg that has gone into the initial core of the unsafeguarded fast breeder test reactor. (The initial core of this reactor was fabricated before PREFRE began separating lower-quality, unsafeguarded plutonium, and the first core had less than half the design number of fuel elements.[31]) About 20–50 kg of plutonium were also used in the Purnima research reactor; we use a mid-point of 35 kg.

At the end of 1994, the weapon-grade plutonium inventory was therefore about 285–310 kg. Assuming that India would use about 5 kg of weapon-grade plutonium in each weapon, it has enough for 60 nuclear weapons. The uncertainty in this estimate is about 30 per cent. Estimates for the end of 1994 and 1995 are summarized in table 9.3.

---

[30] *Nuclear Fuel*, 1986 (note 22).

[31] On the latter point, see Patri, N., 'Delays, cost of breeder program faulted by India's comptroller', *Nucleonics Week*, 13 May 1993.

Table 9.3. Estimated inventories of Indian weapon-grade plutonium, at the end of 1994 and 1995

Figures are in kilograms.

	31 Dec. 1994	31 Dec. 1995
Production		
Cirus reactor	240	250
Dhruva reactor	150	170
CANDU (first discharges)	0–25	0–30
Total production	**390–415**	**420–450**
Consumption		
1974 test	– 10	– 10
Processing losses (3%)	– 10	– 10
Fast reactor	– 50	– 50
Purnima	– 35	– 35
Total consumption	**– 105**	**– 105**
Total inventory	**285–310**	**315–345**
Mid-point	**300 ± 30%**	**330 ± 30%**

India's inventory of weapon-grade plutonium will grow mainly from continued operation of the Cirus and Dhruva reactors, at a rate of about 28 kg per year at a capacity of 65 per cent. By the year 2000, India might have in total roughly 450 kg of weapon-grade plutonium.

Although unsafeguarded, Cirus plutonium is restricted to peaceful uses through agreement with Canada, which would forbid its use in nuclear weapons. This would lower the amount available for weapons up to the end of 1994 to about 150 kg, or enough for about 30 weapons. At the end of 1995, the amount is estimated to be 170 kg of weapon-grade plutonium, or enough for 34 weapons. By the year 2000, this inventory would be 250 kg, or enough for 50 nuclear weapons.

However, India could have substituted an equivalent quantity of fuel- or reactor-grade plutonium for its Cirus plutonium, thus freeing it for use in weapons. Cirus plutonium is not safeguarded, and thus its location or use is not subject to verification. Nevertheless, estimates of total nuclear weapons based on the use of only Dhruva plutonium correspond more closely to public estimates of Indian nuclear weapon arsenals, which included about 20–25 weapons as of 1994.[32]

Uranium enrichment programme

In November 1986, Indian officials said that India could enrich uranium to whatever level it required.[33] India is believed to have conducted gas-centrifuge

---

[32] Reiss, M., *Bridled Ambition* (Woodrow Wilson Center Press: Washington, DC, 1995), p. 185.
[33] Chellaney (note 23).

research since at least the early 1970s.[34] The experimental gas-centrifuge programme was started at BARC, where there is a pilot plant.[35] In early 1992, the Pakistani *News of Rawalpindi* said that this plant had 100 centrifuges.[36]

No official information is available about the facility at BARC, or about the level and amount of enriched uranium it has produced. In 1986, the experimental facility was reported to have achieved enrichments of less than 2 per cent,[37] but Indian officials expressed confidence at the time that they could produce higher enrichments in whatever quantity required, although they have refused to be more specific.

India is capable of producing maraging steel, a super-strong steel used in centrifuges and which allows for more rapidly spinning centrifuges than aluminum alloys.[38]

India was reported in 1986 to have decided to build a larger facility, called the Rare Materials Plant (RMP) at Ratanhalli near Mysore.[39] In an interview in March 1992 with one of the authors, Dr Iyengar confirmed the existence of this plant.[40] According to Iyengar, the purpose of the centrifuge plant is to develop centrifuge technology further. He said that India did not at that time plan to start serial production of centrifuges. According to other sources, in 1992 or 1993 this plant could enrich up to 30 per cent. However, the plant had frequent breakdowns as a result of corrosion and failure of parts, and one leading Indian scientist said that India has learned that centrifuges are hard to maintain.[41]

The plant appears to be largely indigenous. As a result, it might depend on obsolescent materials. Nevertheless, according to the CIA, the Indian nuclear programme 'must seek machine tools, computers, specialty metals, and other goods associated with the precision manufacturing of reactors and components for uranium enrichment from foreign suppliers'.[42]

The plant is believed by US and other Western officials to have several hundred machines and to have begun producing enriched uranium no earlier than mid-1990. This facility, however, could produce only kilograms of weapon-grade uranium each year, significantly less than the amount needed for nuclear weapons.

The ultimate goal of the centrifuge programme is unclear. One obvious benefit of the programme is that it allows India to claim parity with Pakistan's nuclear capabilities. According to the Department of Atomic Energy's *Annual*

---

[34] 'German nozzle-enrichment knowhow could be passed to India', *Nucleonics Week*, 14 Oct. 1971, p. 8.

[35] Fera, F. and Srinivasan, K., 'Keeping the nuclear option open: what it really means', *Economic and Policy Weekly*, vol. 21, no. 49 (6 Dec. 1986), pp. 2119–20.

[36] This information about the pilot plant was included on a list of facilities that was exchanged by Pakistan and India as part of a bilateral agreement not to attack each other's nuclear facilities. Intended as secret, the document was obtained by the *News of Rawalpindi*.

[37] Fera and Srinivasan (note 35).

[38] 'Public firm produces maraging steel', Delhi Domestic Service in English, 27 Sep. 1987.

[39] Fera and Srinivasan (note 35); and 'The mini-superpower', *Foreign Report*, 14 Jan. 1988.

[40] Interviews in Bombay with Dr P. K. Iyengar and other Atomic Energy Department officials.

[41] 'Leading scientist views Pakistan's nuclear capability', *India Today*, 15 Sep. 1994 (in English). See *JPRS Report: Proliferation Issues*, JPRS-TND-94-019, 17 Oct. 1994, p. 10.

[42] See *Proliferation Threats of the 1990s* (note 20), p. 156.

Report 1990–91, India is also designing a small research reactor that would use enriched uranium fuel.[43] A relatively small number of centrifuges could allow India to indigenously supply fuel for this reactor.

In addition, M. R. Srinivasan, who was chairman of the Atomic Energy Commission until 1992, said in 1994 that India is developing a nuclear-powered submarine.[44] He said India wants to produce enriched uranium to fuel the naval reactor.

Alternatively, the enrichment programme might be part of a long-term option to build thermonuclear weapons. These weapons rely on the nuclear fusion of deuterium and tritium for most of their explosive yield, but the fusion process is started by atomic blasts. Although plutonium is used in the 'primary,' which provides the initial atomic blast, weapon-grade uranium is collocated with the thermonuclear fuel to provide an extra 'kick' to get the fusion reaction going. In addition, HEU permits high-yield, boosted fission weapons.

III. Pakistan

Pakistan has both plutonium separation and uranium enrichment programmes, both of which appear strongly oriented towards obtaining material for nuclear weapons. Its reprocessing programme lacks unsafeguarded spent fuel, although US officials believe that Pakistan is building an unsafeguarded plutonium production reactor and probably a plutonium separation plant. However, its unsafeguarded enrichment plant at Kahuta, near Islamabad, has produced weapon-grade uranium and is the core of its current nuclear weapon programme.

US Senator Larry Pressler, while in Islamabad in early 1992, said that some covert steps by Pakistan in 1990 led the CIA, the State Department and the White House to believe that Pakistan had built an atomic bomb.[45] Bush Administration officials were quoted in the *Washington Post* soon afterwards as saying that Pakistan had all the essential components for at least two nuclear weapons.

In early February, soon after Pressler's visit, Pakistani Foreign Secretary Shahryar Khan told the *Washington Post* that his nation now had the components to assemble at least one nuclear explosive device.[46] Khan said he was speaking candidly to avoid credibility gaps that he suggested had been created by previous Pakistani governments. He also said that his government had stopped producing highly enriched uranium in 1991.

Retired General Mirza Aslam Beg, Pakistan's Army Chief of Staff from 1988 to 1991, said in a 1993 interview in the Urdu-language *Awaz International* newspaper published in London that Pakistan had actually 'crossed the line in

---

[43] Government of India, Department of Atomic Energy, *Annual Report 1990–91*.
[44] Roy, R., 'India-nuclear submarine', Associated Press, 8 Dec. 1994.
[45] MacLachlan, A., Shahid-ur-Rehman Khan and Siddiqi, A. R., 'Pakistani says France will pay $118 million for supply breach', *Nucleonics Week*, 16 Jan. 1992.
[46] Smith, J., 'Pakistan can build one nuclear device', *Washington Post*, 2 Feb. 1992.

1987' when it successfully conducted a 'cold test' of a nuclear explosive.[47] A cold test is usually defined as a test of the complete implosion-system with a dummy nuclear core. It can provide confidence to the military and political leadership that a design will work. By 1987, however, Pakistani scientists had probably had sufficient nuclear material and know-how for a nuclear device for a few years.

In August 1994, former Pakistani Prime Minister Nawaz Sharif said, 'I confirm that Pakistan possesses the atomic bomb.'[48] In response, the Pakistani Government denied that it possessed nuclear weapons, or that it intended to make them, although it admitted it has the capability to make them.

In the spring of 1995, Senator Pressler said he attended a CIA briefing that convinced him Pakistan's programme goes beyond simply having the nuclear capability it already acknowledges. He said, 'Pakistan has about five weapons ready to go.'[49]

The CIA testified before the US Senate in early 1993 that it does not believe that Pakistan maintains assembled or deployed nuclear weapons.[50] Nevertheless, the CIA believes that Pakistan, like India, is producing weapon components and could quickly assemble a small nuclear arsenal. It also expects Pakistan to gradually increase the size of its de facto arsenal. A senior US official said in a July 1994 discussion in Washington that Pakistan was working on miniaturizing its nuclear weapons for deployment on missiles. Such work can be done without the need for full-scale nuclear tests.

Enrichment programme

During the mid-1970s, Pakistan created a well-funded, secret procurement network in the West aimed at obtaining uranium-enrichment gas-centrifuge technology and components, and the equipment to make the centrifuges themselves. This programme has been largely successful, and Pakistan now has an indigenous enrichment effort based at Kahuta. Although Pakistan is still believed to depend on foreign supply for many key centrifuge-related items and materials, it is also thought to have established a reserve of many of the materials necessary to maintain its centrifuge programme.

In the autumn of 1993 a shipment of about 1000 metal 'preforms' for centrifuge scoops bound for Pakistan were confiscated by German customs officials (see centrifuge drawings in appendix A).[51] In Pakistan, the preforms would have been machined into final form. In early 1996 the media reported, and the Clinton Administration confirmed, that China provided the Kahuta plant with about 5000 cobalt-samarium ring magnets for the top suspension bearing of the

---

[47] 'Pakistani quoted as citing nuclear test in '87', Reuter, 24 July 1993.
[48] Barber, B., 'Ex-Premier declares Pakistan has A-bomb', *Washington Times*, 24 Aug. 1994, p. A1.
[49] Constantine, G., 'Pressler maintains Pakistan has a stable of nukes', *Washington Times*, 13 Apr. 1995, p. A13.
[50] See *Proliferation Threats of the 1990s* (note 20).
[51] Hibbs, M., 'Preform export controls key issue in German–Pakistan centrifuge case', *Nuclear Fuel*, 20 June 1994.

centrifuge.[52] The media reports said that the shipments occurred in three shipments, between December 1994 and mid-1995, and were worth about $70 000.[53]

Pakistan is believed to have produced weapon-grade uranium in sufficient quantity for nuclear weapons beginning in the mid-1980s. Although it is believed to have halted HEU production by mid-1991, it was still producing low-enriched uranium. LEU production has continued to this day, according to US officials.[54]

Pakistan's efforts to produce weapon-grade uranium

After almost a decade of concerted effort, Pakistan announced in 1984 that it was capable of producing low-enriched uranium at its pilot centrifuge plant at Kahuta.[55] By mid-1986, US intelligence had concluded that Pakistan had produced weapon-grade uranium at this facility.[56]

According to a declassified 1986 memorandum for Henry Kissinger, who received the document in his capacity as a member of President Reagan's Foreign Intelligence Advisory Board, Kahuta had a nominal capability sufficient to produce 'enough weapons-grade material to build several nuclear devices per year'.[57] The memorandum, however, said that Pakistan was not believed to have assembled any nuclear explosive devices.

The *New York Times Magazine* reported in March 1988 that US officials had by then concluded that Pakistan had enough weapon-grade uranium for four to six nuclear weapons.[58] This amount is roughly equivalent to 100 kg of weapon-grade uranium.

Pakistan was reported to be building another enrichment facility at the town of Golra, some 10 km west of Islamabad.[59] A 1987 report said that a centrifuge hall had reportedly been completed, although Western diplomats say the several thousand centrifuges needed have not yet 'been installed'.[60] Pakistan has reportedly proceeded with the construction of this plant, although it is not known how far it has progressed.

---

[52] Gertz, B., 'China's nuclear transfer exposed', *Washington Times*, 5 Feb. 1996, p. A1; and Smith, J. R., 'US aides see troubling trend in China–Pakistan nuclear ties', *Washington Post*, 1 Apr. 1996, p. A14.
[53] Smith (note 52); and Reuter, 'US weighs new facts from China on nuclear sales', 13 Apr. 1996.
[54] Smith, J. R., 'Production concerns may delay US arms shipment to Pakistan', *Washington Post*, 15 Feb. 1996.
[55] 'Scientist affirms Pakistan capable of uranium enrichment, weapons production', *Nawa-I-Waqt*, 10 Feb. 1984, magazine supplement, pp. F1-F8. Translated in FBIS/NDP, 5 Mar. 1984, pp. 32–45.
[56] Woodward, B., 'Pakistan reported near atom arms production', *Washington Post*, 4 Nov. 1986, p. A1; and Smith, H., 'A bomb ticks in Pakistan', *New York Times Magazine*, 6 Mar. 1988. In early 1986, Pakistan was reported to have enriched uranium to over 30 per cent. This was the conclusion of Western intelligence agencies which analysed the uranium in dust from the immediate vicinity of the Kahuta facility. See Koch, E. and Henderson, S., 'Pakistan getting atomic bomb through back door', *Der Stern*, no. 19 (30 Apr. 1986); and 'Inside Kahuta', *Foreign Report*, 1 May 1986.
[57] National Security Archives, *US Nuclear Non-proliferation Policy: 1945–91*, Document no. 02328, 1992.
[58] See Smith (note 56).
[59] Henderson, S., 'Pakistan builds second plant to enrich uranium', *Financial Times*, 11 Dec. 1987.
[60] See Henderson (note 59).

A US official said in an interview with one of the authors in March 1994 that Pakistan is believed to have small, alternative centrifuge locations where it is trying to replicate its capability. But he added that this activity is difficult to confirm.

Prime Minister Benazir Bhutto was reported to have ordered a halt to any weapon-grade uranium production before her visit to Washington in June 1989, a step that the USA was able to verify prior to her visit.[61] In reaction to increased tensions with India over Kashmir, however, Pakistan decided by the spring of 1990 to resume producing weapon-grade uranium. Production evidently continued until sometime in 1991, according to Foreign Secretary Shahryar Khan's February 1992 interview with the *Washington Post*.

The Kahuta centrifuges

A. Q. Khan, the father of the Pakistani enrichment programme, is widely believed to have stolen the designs for several centrifuges from Urenco, the Anglo-Dutch-German enrichment consortium, along with the technical specifications for manufacturing several of them. Khan also gained access to details about Urenco suppliers that Pakistan used in creating its procurement network. Khan had access to all this information when he worked from 1972 to 1975 at a Dutch engineering firm whose parent company played an important role at the Urenco pilot enrichment facilities at Almelo, in the Netherlands.

Pakistan is believed to have obtained sensitive design information for early-generation German centrifuges, known as G-1 and G-2, and Dutch prototype machines, referred to as SNOR and CNOR. According to the Dutch Government investigative report on the Khan case, Pakistan is also believed to have acquired several SNOR and CNOR machines. These machines have a separative capacity of less than 5 SWU per year.

Despite this head start, Pakistan encountered many difficulties in building reliable gas centrifuges. Its efforts to duplicate and operate the Netherlands machines evidently ended in failure after it built several thousand of them during the early 1980s. It was more successful with its machine based on the G-2 design, however, and by the mid-1980s Pakistan was able to mass-produce them. Pakistan's enrichment programme is believed to rely on this type of machine, although it still might be operating a certain number of older models, such as G-1 models.

The G-2 has a capacity of about 5 SWU a year. It contains two maraging-steel rotor tubes connected by a maraging-steel bellows, which acts to reduce damaging vibrations in supercritical centrifuges. Mastering the construction of bellows is likely to have presented special difficulties for Pakistan, and might have been a major reason for its long delay in developing a successful machine. Bearings might have also caused problems.

In an interview in the spring of 1991, a US official with access to intelligence information on Pakistan's programme said that Kahuta had close to 3000 opera-

[61] Hussain, M., 'Nuclear issue: ball now in Pakistan's court', *Nation*, Lahore, 29 Nov. 1990.

ting machines at any given time. At an average of 3–5 SWU per machine, Kahuta has a capability of about 9000–15 000 SWU per year. This is enough to produce about 55–95 kg of weapon-grade uranium a year, assuming that about 0.5 per cent of the uranium-235 remains in the tails.[62] We assume a high tails assay, since we believe Pakistan has more than sufficient natural uranium to sustain this type of production.[63]

Pakistan has the manufacturing capability and know-how to increase the number and separative output of its machines. What it exactly is doing remains unknown. A US official said in the early 1990s that Pakistan was concentrating on developing more advanced machines and replacing older centrifuges rather than increasing the number in operation. The 1993 scoop preform export case implies that subsequent Pakistani centrifuges are derived from early Urenco designs, since the design of the preforms resembles those used in early Urenco machines.[64] However, this information does not preclude that the newer machines have a significantly greater output.

The number of operating machines at any one time is highly uncertain. Part of the reason for confusion is that Pakistan has installed considerably more machines than it has successfully operated. In 1986 it was reported that Kahuta had 14 000 centrifuges.[65] US Government officials confirmed that Pakistan might have installed this number of centrifuges, but they said that a more accurate count of the number operating in 1986 or 1987 was closer to 1000. One added that Pakistan's centrifuge 'junk pile was sizeable'.

The lower estimate of the number operating in 1986 is also consistent with reports published in Pakistan by Islamabad's English-language daily newspaper, the *Muslim*. It reported that Kahuta was 'rumored to have 1000 centrifuges, against a planned capacity of 2000 to 3000 centrifuges'.[66]

Estimates, however, of Kahuta's planned capacity vary. The largest appears to be from the Russian Foreign Intelligence Service which estimated in 1993 that when Kahuta reaches its planned capacity, it could produce enough HEU for 12 nuclear weapons a year.[67] This capacity would correspond to a plant with about 6000–8000 G2 machines of the type discussed above, assuming 15–20 kg of WGU per weapon.

---

[62] Pakistan has a uranium mining capability sufficient to supply the Kahuta enrichment plant, and it can produce uranium hexafluoride.

[63] Going to a higher tails assay requires significantly more natural uranium feed. For example, production of 25 kg of weapon-grade uranium requires 4600 kg of natural uranium at 0.2 per cent tails assay, and 11 000 kg at 0.5 per cent tails assay.

[64] See Hibbs (note 51).

[65] See Koch and Henderson, 'Inside Kahuta' (note 56). The centrifuges were reportedly housed in two halls at Kahuta. One hall held centrifuges based on a Netherlands design, and made from special steel and aluminium. Each centrifuge stands about 2.5 m tall and is about 11 cm in diameter. The other hall was reported to hold centrifuges based on the German design, which are made of maraging steel, are only half as tall as the Dutch model and are 15 cm wide.

[66] Ameen, A. F., 'The mythical bomb', *Muslim*, 5 Aug. 1986. Quoted in Khan, S., 'Fear of US aid cut-off said to have deterred Pakistan's bomb program', *Nuclear Fuel*, 11 Aug. 1986.

[67] Russian Federation, Foreign Intelligence Service Report, *A New Challenge After the Cold War: Proliferation of Weapons of Mass Destruction*, Moscow, 1993. English translation in Foreign Broadcast Information Service, *JPRS Report*, JPRS-TND-93-007, 5 Mar. 1993, p. 32.

Inventory of weapon-grade uranium

Based on the above information, table 9.4 gives an estimate of the annual production of weapon-grade uranium at Kahuta up to the time when Pakistan stopped making weapon-grade materials in mid-1991. Up to mid-1991, Pakistan is estimated to have produced about 160–260 kg of weapon-grade uranium, using natural uranium feed and a tails assay of 0.5 per cent. The midpoint is 210 kg of weapon-grade uranium.

In this estimate, we assume that weapon-grade uranium production began in earnest in 1986 and continued uninterrupted until 1989. After about a one-year pause, production resumed, stopping again in 1991. We also assume that Pakistan linearly increased Kahuta's enrichment capability during this period from an initial capacity of 1000 machines in 1986 to 3000 machines in 1990, each with an average capacity of 3–5 SWU per year. This estimate of weapon-grade uranium production has an uncertainty of plus or minus 25 per cent.

There have been reports that Pakistan obtained enough weapon-grade uranium from China to make two nuclear bombs, although the authenticity of this information cannot be determined.[68] These reports cannot be dismissed, however, because a companion report that China supplied Pakistan with a nuclear weapon design has been confirmed to one of the authors by US officials on several occasions.[69] But in the absence of confirmation of the weapon-grade uranium, we have not added this amount to our estimate of Pakistan's inventory.

US officials interviewed in late 1994 and early 1996 said that Pakistan had not produced any material enriched to over 20 per cent since 1991. This cut-off, however, is not verified, and questions about its exact nature persist. To make matters more confusing, Pakistani officials have sometimes stated that Kahuta has stopped producing uranium enriched to over 90 per cent, but left open the question of whether non-weapon-grade HEU was being produced.

Little specific information is available about the amount of weapon-grade uranium in Pakistan's nuclear weapons. A US official said that the design provided by China has a solid core of weapon-grade uranium. This type of design need not require more than about 15 kg of weapon-grade uranium, although a larger quantity could be used to reduce the weapon's size or to increase its reliability and yield. In addition, some small fraction of WGU is lost to waste during manufacturing. We therefore assume 20 kg of weapon-grade uranium per weapon. Under this assumption, at the end of 1991 Pakistan had enough material for roughly 8–13 nuclear weapons.

---

[68] 'Pakistan's atomic bomb', *Foreign Report*, 12 Jan. 1989. Pakistan is known to have acquired about 1 kg of weapon-grade uranium as part of a deal with China for the supply of a 27 kW reactor. This reactor, at the Pinstech nuclear centre, went critical in 1989. It is under IAEA safeguards, precluding its military use. This earlier shipment, if true, would correspond to about 30–40 kg of weapon-grade uranium.

[69] Smith (note 52).

Table 9.4. Estimated production of weapon-grade uranium at the Pakistani Kahuta centrifuge enrichment plant, 1986–91

		Weapon-grade uranium	
Year	Capacity (SWU/y)	Annual production[a] (kg)	Cumulative production (kg)
1986	3 000–5 000	19–31	19–31
1987	4 500–7 500	28–47	47–78
1988	6 000–10 000	38–63	85–141
1989	7 500–12 500	16–28[b]	101–169
1990	9 000–15 000	28–47[b]	129–216
1991	9 000–15 000	28–47[c]	157–263

[a] Tails assay of 0.5% and natural uranium feed.

[b] It is assumed that no weapon-grade uranium was produced from the beginning of May 1989 until after June 1990 as a result of a moratorium instituted by then Prime Minister Benazir Bhutto (see text).

[c] Assumes weapon-grade uranium production in the first half of 1991 only.

This estimate is consistent with a late-1990 report in the *Press Trust of India* which quotes a senior Indian official as saying that Pakistan had nine nuclear bombs.[70] In September 1991, US officials asserted that Pakistan had the parts and capability to assemble up to six nuclear bombs.[71] The Russian Foreign Intelligence Service estimated that in 1993 Pakistan had 4–7 devices.[72] Our estimate implies that Pakistan may be able to increase the size of its de facto arsenal with its existing stockpile of fissile material.

What if the cut-off ended?

If Kahuta resumed its production of weapon-grade uranium at 1991 levels, it would be capable of producing an estimated 55–95 kg of weapon-grade uranium a year, or enough for roughly 3–5 weapons. This estimate assumes that natural uranium is 'fed' into the plant, and the waste contains depleted uranium with 0.5 per cent ^{235}U.

According to US officials, however, Pakistan has continued making low-enriched uranium at Kahuta since 1991. They were unable to provide information about the enrichment level of this enriched material, or the amount produced so far.

If low enriched uranium were used as feedstock, the weapon-grade uranium output would increase several-fold. Since such a strategy would allow Pakistan to produce weapon-grade uranium much faster than normal, Pakistan could, after a relatively short period, produce almost as much weapon-grade uranium as if no 'freeze' in weapon-grade uranium had occurred.

[70] 'Indian politician claims Pakistan has 9 nukes', *Washington Times*, 27 Nov. 1990.
[71] Levine, S., 'Bhutto says Pakistan can build nuclear weapons', *The Guardian*, 2 Sep. 1991.
[72] See *JPRS Report* (note 67).

For example, if Pakistan produced only 5 per cent enriched uranium from mid-1991 until the end of 1995, it could have accumulated about 5620–9360 kg of low-enriched material during this period, where the tails assay is taken as 0.3 per cent and the plant is assumed to have a capacity of 9000–15 000 SWU per year. At a tails assay of 0.5 per cent, the estimates would be about 35 per cent greater.

If the freeze ended now and Pakistan decided to enrich this 5 per cent material to 90 per cent, it could produce about 250–420 kg of weapon-grade uranium in about 13 months, assuming a tails assay of 1 per cent. If 20 per cent enriched uranium was produced instead (0.3 per cent tails), and was then further enriched to 90 per cent (1 per cent tails), total production would be about 225–375 kg WGU. This WGU would only take a little over four months to produce, however. In comparison, if the freeze had never occurred, Pakistan could have produced about 250–425 kg of weapon-grade uranium by the end of 1995, assuming a tails assay of 0.5 per cent. Although the freeze is important, Pakistan can overcome its effect relatively rapidly if it chooses to do so.

Plant renovation

The centrifuges at Kahuta are probably nearing the end of their lifetime and are due for replacement. Early-generation machines are unlikely to last more than 5–10 years, and Pakistan is believed to have installed most of Kahuta's machines between 1985 and 1990.

Because centrifuge R&D activities have undoubtedly continued at Kahuta, replacement machines could have a larger separative output. Little is publicly known about the types of more advanced design that Pakistan has developed. Pakistani scientists would be expected to concentrate on increasing the length of the rotor and its peripheral speed, while striving to minimize machine failure rates. To increase rotor speed, Pakistan may have shifted away from all-metal rotors to either a metal rotor reinforced by an overwrap of a composite material or to an all-composite rotor design.

In any case, Pakistan should have been able to improve the separative output of replacement centrifuges by 50 per cent over their predecessors.[73] Because early machines are assumed to have an output of 3–5 SWU/y, the replacement machines under this assumption would have a capacity of 4.5–7.5 SWU/y. If the number of machines was kept constant, the plant's total output would increase after renovation to about 13 000–22 000 SWU/y.

Pakistan could have pursued other replacement strategies, perhaps not improving their separative output but building sturdier machines. It could also

---

[73] This improvement could be accomplished by increasing the length of the rotor by 50%, i.e., by adding a third rotor segment to an existing 2-segment machine, or alternatively, by increasing the peripheral speed of an existing rotor, with a diameter of 14.5 cm, from 450 m/s to 515 m/s. This latter calculation assumes that the separative output increases in proportion to the cube of the velocity. This higher speed could be accomplished with a metal rotor with a composite overwrap, although all carbon-fibre rotor tubes could be used as well.

have increased their output more than assumed above. Lack of public information prohibits a more precise estimate.

Plutonium production

Reactor

Because Pakistan lacks a significant supply of unsafeguarded irradiated fuel, intelligence reports have stated, and experts have speculated for many years, that Pakistan has been trying to build a nuclear reactor which would generate significant amounts of plutonium for use in nuclear weapons. Pakistan may want plutonium because it would enable Pakistani nuclear weapon experts to build smaller warheads for missiles. The 1993 Russian Foreign Intelligence Service Report said that Pakistan was building a 70 MWth reactor and had finished about half the construction and assembly.[74]

In 1994, *Nucleonics Week* provided additional details about the reactor.[75] The reactor is under construction at Khushab in Pakistan's Punjab province. It is estimated to have a capacity of 50–70 MWth, to be fuelled with natural uranium and moderated with heavy water or graphite.

The Washington Post reported that Prime Minister Benazir Bhutto had confirmed in an interview the existence of this reactor, although she said it was for an 'experimental purpose'.[76] The article quoted US intelligence officials saying that they estimate the reactor's power as 40 MWth and believe it is moderated by heavy water.

According to a US official interviewed in December 1994, the reactor could be finished in 1996. He added, however, that a considerable amount of work remained to be done. A US official confirmed in November 1995 that the reactor was still being constructed. A US Defense Department publication estimated that the reactor is expected to become operational in the late 1990s.[77] If completed, the roughly 50-MWth reactor could produce about 10–14 kg of weapon-grade plutonium per year, assuming a capacity factor between 60 and 80 per cent.

The future safeguards status of this reactor is unclear, although the expectation is that Pakistan does not intend to place the reactor under IAEA safeguards. In apparent anticipation of this outcome, US officials stated in the spring of 1995 that the US Government is seeking assurances from Pakistan that it will not use any plutonium from this reactor in nuclear weapons. The status of this effort is unknown.

---

[74] See *JPRS Report* (note 67).
[75] Hibbs, M., 'Bhutto may finish plutonium reactor without agreement on fissile stocks', *Nucleonics Week*, 6 Oct. 1994.
[76] Smith, J. and Lippman, T., 'Pakistan building reactor that may yield large quantities of plutonium', *Washington Post*, 8 Apr. 1995, p. A20.
[77] Office of the Secretary of Defense, *Proliferation: Threat and Response* (US Government Printing Office: Washington, DC, Apr. 1996).

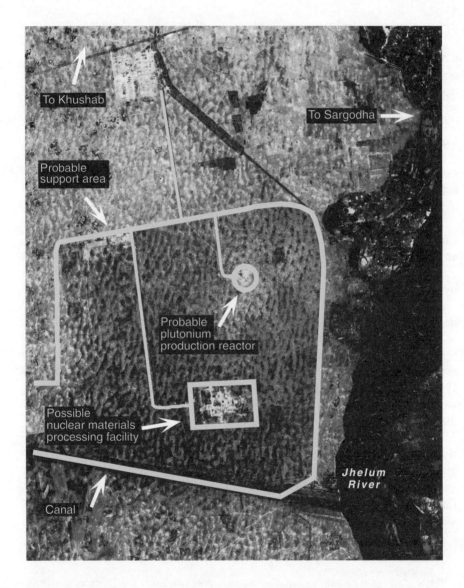

Figure 9.1. SPOT image showing a probable plutonium reactor complex near Khushab, Pakistan

Source: Photograph courtesy of SPOT Image Corporation, Reston, Va.

Possible Chinese assistance has complicated the discussion about whether safeguards must be applied to the Khushab reactor. Many reports state that China has supplied Pakistan equipment or technical assistance for this reactor, although the precise types of aid are not specified. Some US officials interviewed by one of the authors discount the level of assistance and believe that the reactor is essentially indigenously produced. Others emphasize the importance of the Chinese assistance.[78] Because China is a signatory to the NPT it may have to require Pakistan to apply IAEA safeguards to the reactor, if it supplied certain types of key items or 'know-how' after acceding to the NPT in 1992.

Plutonium separation

Pakistan has established a small capability to separate plutonium at a facility, called 'New Labs,' at the Pinstech complex, near Rawalpindi. A US official believes that Pakistan probably did some experimental separation of plutonium, although on a very small scale.

This facility is based on blueprints delivered by France, with key equipment bought from a variety of suppliers. A French manufacturer of remote manipulators, who had supplied New Labs in 1977, planned a visit in 1983 with the intent to sell more.[79] After a request from the US Government, the French Government stopped the sale.

Pakistan also has a partially completed plutonium separation plant near Chasma, in central Pakistan. Pakistan originally tried to acquire this plant on a turn-key basis from France in the 1970s. France, however, decided not to fulfil the contract in 1977, under pressure from the United States. Pakistan may have continued to build this facility on its own, at least finishing the civil works. A recent media account says that the Chinese are helping Pakistan finish the plant.[80]

Bhutto told the *Washington Post* in 1995 that Pakistan lacks the capability to separate the plutonium from the Khushab reactor currently under construction.[81] However, the secrecy surrounding this reactor suggests that Pakistan plans to separate the plutonium for nuclear weapons. Where plutonium separation would occur could not be determined. Whether it would be done at New Labs or at Chasma remains unknown.

A 21 May 1992 SPOT image of the reactor site shows another high-security facility a few kilometres south of the reactor but within the common security perimeter, as illustrated in figure 9.1. There is insufficient information to indicate the purpose of the facility, although a possible candidate is a nuclear materials processing plant, possibly even a reprocessing plant.

---

[78] Smith (note 52).
[79] CIA, Directorate of Intelligence, *French Nuclear Reactor Fuel Reprocessing Program: An Intelligence Assessment*, Sep. 1984, unclassified version released under the Freedom of Information Act.
[80] Gertz, B., 'China aids Pakistan's plutonium plants', *Washington Times*, 3 Apr. 1996.
[81] Smith and Lippman (note 76).

10. North Korea*

I. Introduction

How much plutonium does North Korea have? What is the true purpose of this plutonium? Is it part of a nuclear weapon programme? Or does North Korea have little separated plutonium, engaging in an elaborate attempt to trade off its programme for economic and diplomatic concessions from the USA, Japan and South Korea? Few know. There is so little factual information that answers to questions about motivation and past activities are strongly influenced by where an analyst sits.

Analysts responsible to defence departments or intelligence agencies in the USA and South Korea do not want to underestimate the amount of plutonium North Korea possesses, or wrongly attribute the plutonium to a civilian application. But when these 'assessments' are 'leaked' to the news media, they are often portrayed as facts. In late 1993, for example, when the CIA leaked its assessment that North Korea had probably acquired enough plutonium for one or two nuclear weapons, this leak became the underpinning of most subsequent claims that North Korea has enough plutonium for such an arsenal.

Few doubt that North Korea's original intention was to produce large quantities of plutonium for nuclear weapons. But great uncertainty surrounds assessments that North Korea separated sufficient plutonium to build nuclear weapons, or actually built them. Determining the true state of affairs will not be easy and will require North Korea to provide substantially more information than it has done so far.

The Agreed Framework between the United States of America and the Democratic People's Republic of Korea, signed on 21 October 1994 in Geneva, has provided a means to ease tensions on the Korean Peninsula and answer these questions, but not quickly. North Korea has 'frozen' its activities at its most controversial nuclear facilities, and it has agreed to dismantle them after about 10 years. In return, the USA will ensure that North Korea receives two 1000-MWe light water reactors. North Korea has agreed that after roughly five years, but before it receives significant nuclear components for the light water reactors, it will allow the IAEA to seek answers to questions about past plutonium diversion.

II. An unsafeguarded reactor emerges

The plutonium at issue comes from a highly secret nuclear programme that started many years ago outside any international inspections. The first concrete

---

* An earlier version of this chapter appeared in *Science and Global Security*, vol. 5 (1994), pp. 63–87. A shortened and less technical version was published in *Bulletin of the Atomic Scientists*, vol. 50, no. 5 (Sep./Oct. 1994).

sign of North Korea's plutonium production programme was the US intelligence discovery in the early or mid-1980s that North Korea was constructing a small reactor at Yongbyon, a nuclear complex about 100 km north of Pyongyang. This reactor is reported to have a nominal capacity of between 20 and 30 MWth, and to rely on gas cooling and graphite for moderation. This type of reactor is ideal for producing weapon-grade plutonium; its size is sufficient to produce enough for one nuclear weapon a year.

In response to Western concerns about this reactor, Russia successfully pressed North Korea to sign the NPT on 12 December 1985. International interest then faded in this reactor, even when it started operating in 1986.

In 1989, however, the press reported the existence of a large, narrow structure at Yongbyon suspected of being a plutonium separation plant.[1] Thus, the CIA assessment assumes that North Korea first produced plutonium in the small reactor, unloaded much of the irradiated fuel in the core and subsequently separated the plutonium from the fuel in the separation plant.

Reports soon followed that North Korea was also building a much larger gas-graphite reactor for plutonium production at Yongbyon.[2] These reports also mentioned the existence of a high-explosive testing site at Yongbyon, which was suspected of being part of a rudimentary nuclear weapon development programme. Extensive high-explosive testing would be necessary to build a nuclear weapon with plutonium, and any evidence of high-explosive tests at a nuclear research centre immediately increases suspicions.

Aggravating the situation was North Korea's continuing refusal to allow IAEA safeguards on all its nuclear activities, as required by the NPT. After a few more years of diplomatic wrangling, on 30 January 1992 the IAEA and North Korea finally signed a safeguards agreement. It entered into force on 10 April 1992 and the political crisis at long last appeared to be subsiding.

III. Initial safeguards declaration

On 4 May 1992, North Korea provided the IAEA its initial report of all nuclear material subject to safeguards. One of the IAEA's first inspection tasks was to verify the information in North Korea's initial declaration and to assess its completeness.

From 11 May to 16 May, an IAEA delegation, headed by IAEA Director General Hans Blix, visited North Korea, where its nuclear officials briefed the delegation on its nuclear programme.

Among the first sites this delegation visited was the unfinished plutonium separation facility at Yongbyon, which North Korea calls the 'Radiochemical Laboratory'. North Korean nuclear officials said that this facility had separated

---

[1] See, e.g., Fialka, J. J., 'North Korea may be developing ability to build nuclear weapons', *Wall Street Journal*, 19 July 1989, p. A16.
[2] Bermudez, J. S., 'N. Korea—set to join the nuclear club?', *Jane's Defence Weekly*, 23 Sep. 1989, pp. 594–97; and Bermudez, J. S., 'North Korea's nuclear programme', *Jane's Intelligence Review*, Sep. 1991, pp. 404–11.

about 100 grams of plutonium in a single campaign in the spring of 1990. North Korea said that the plutonium was from a few damaged fuel rods taken out of the small reactor. The fuel's outer metal casing or 'cladding' was damaged during the operation of the reactor, and therefore the fuel was removed from the core.

Blix said in Beijing following this visit that about 80 per cent of the construction work of the Radiochemical Laboratory was complete, but only about 40 per cent of the equipment was installed.[3] North Korean officials also told the IAEA that the rest of the equipment had been ordered but not yet delivered.

Blix said in his Beijing interview that North Korea called this plant a Radiochemical Laboratory because it was incomplete and used only for testing. He added that, if the plant was completed, 'I have no doubt that it would have been considered a reprocessing plant in our terminology.'[4]

During this initial visit, North Korean officials also told the IAEA that their scientists had first separated grams of plutonium in 1975 at the Isotope Production Laboratory. The plutonium was produced in the Russian-supplied IRT research reactor that started operation in 1975. This reactor was placed under IAEA safeguards in 1977.

The IAEA also toured three gas-graphite reactors, two of which were unfinished. At Yongbyon, the IAEA delegation visited the small 20–30 MWth operating reactor and an estimated 200-MWth gas-graphite reactor under construction. The delegation also flew to the construction site of an estimated 600–800 MWth gas-graphite reactor at Taechon in North Pyongan Province.

North Korea told the IAEA that all three reactors are part of a nuclear electricity production programme. North Korea refers to these reactors by their electrical power—5 megawatt-electric (MWe), 50 MWe and 200 MWe, respectively.

In a press announcement released at the end of the IAEA's visit, North Korea offered a standing invitation to IAEA officials that they could visit any site in North Korea, even if that site was not included in the initial report.

Inconsistencies appear

Starting in the summer of 1992, the IAEA's analyses began to identify a series of inconsistencies in North Korea's initial declaration. In particular, the IAEA became suspicious that the amount of plutonium declared by North Korea to the IAEA was smaller than the actual amount separated in the Radiochemical Laboratory.

During its initial inspections of the Radiochemical Laboratory, the IAEA had collected 'samples' of separated plutonium, of material caught up in separation processing steps and of different types of nuclear waste generated during the various separation operations.

[3] IAEA, 'Transcript from the press briefing by Dr Hans Blix, Director General of the IAEA', Beijing Hotel, Beijing, 16 May 1992.
[4] See IAEA (note 3).

According to an IAEA official, the facility operators were willing but not prepared technically to take samples from waste temporarily stored at the site. When inspectors asked for samples of the highly radioactive 'fission product' waste, North Korean technicians had to improvise a way to get into that waste. North Korean officials told the IAEA that this procedure led to some facility operators being exposed to radiation doses above the permitted levels.

The samples were analysed by the IAEA's laboratory at Seibersdorf, Austria, and by the IAEA's affiliated laboratories in Europe and the USA. These analyses uncovered discrepancies from North Korea's initial declaration that led to the IAEA's suspicions that North Korea had separated more plutonium.

One set of analyses focused on checking North Korea's declaration that it separated plutonium during only one campaign in 1990. This method depends on measuring the amount of americium-241 in 'smear' or 'swipe' samples from the 'hot' insides of glove boxes at the end of the separation process, where freshly purified plutonium oxide is handled. Since americium-241 is a decay product of plutonium-241, the amount of americium-241 in the samples can indicate the period of time that has passed since the plutonium was originally separated. IAEA analyses suggest that there were distinct separation efforts in 1989, 1990 and 1991.

A second inconsistency emerged when the IAEA tried to verify that the declared plutonium and the waste had originated from the same irradiated fuel rods. The IAEA compared the isotope ratios of the plutonium remaining in several waste samples and glove boxes to the ratios in the separated plutonium. Ideally, both the plutonium and the trace quantities of plutonium in the waste samples and glove boxes should have the same ratio of principal plutonium isotopes—^{239}Pu, ^{240}Pu and ^{241}Pu. They found that the fraction of ^{240}Pu in the various waste samples and the glove boxes was inconsistent with the fraction in the separated material. This inconsistency implies that additional fuel rods were processed.

A third inconsistency involved the irradiation history of the fuel declared by North Korea to have been reprocessed in 1990. The isotopic ratios of the sampled separated plutonium and the sampled plutonium from the waste were inconsistent with the declared irradiation history of the reprocessed fuel. The IAEA calculated that the irradiation level of the fuel was higher than that claimed by North Korea. This inconsistency implies that the total amount of separated plutonium is higher than declared by North Korea.

An IAEA official familiar with the inspections said that the inspectors believed North Korea had separated more than it declared. However, they could not determine if North Korea had separated grams or kilograms of plutonium. The inspectors developed two possible scenarios to explain the inconsistencies. The first assumed that more fuel or natural uranium targets from the IRT research reactor were reprocessed than declared, resulting in, at most, up to a few kilograms of separated plutonium. The other scenario assumed that additional fuel rods were taken out of the 5-MWe reactor and reprocessed in the

Radiochemical Laboratory, possibly resulting in many kilograms of separated plutonium.

North Korea denied all these accusations. It stuck to its initial story and accused the IAEA of misunderstanding the actual situation. North Korea insisted that virtually the entire first core remained in the reactor. For North Korea to have separated enough plutonium in the Radiochemical Laboratory to make a nuclear bomb, it would have had to remove much of the fuel in this first core.

North Korean officials attempted to resolve these inconsistencies, although they rarely provided the IAEA with sufficient documentation to support their statements. For example, North Korea said that the second inconsistency mentioned above, namely differences in the plutonium isotopic ratios between the waste and product samples, could be explained by the 1975 separation of plutonium from IRT reactor fuel. North Korea said that the waste from this separation had become mixed with the newer waste. With regard to the third inconsistency, it said that the IAEA did not properly account for variations in the irradiation of the spent fuel, leading it to inaccurately estimate plutonium production.

Some US and IAEA officials believe that a few of the inconsistencies could be explained if North Korea's explanation of the reactor's operation is accepted. But others cannot be explained away, particularly those involving measurements in the glove boxes. In addition, intelligence agencies provided the IAEA with other information that further increased its suspicions that North Korea separated significantly more plutonium than it declared.

Suspect waste sites

In autumn 1992, the IAEA began to receive information from member states on undeclared sites at Yongbyon, namely two camouflaged nuclear waste sites. Moreover, according to a US official, this information showed that these sites were camouflaged not long before the beginning of the IAEA inspections.

The IAEA wants to inspect these two undeclared sites to verify its suspicions that they may contain radioactive waste generated during the process of separating plutonium; such inspections might provide proof that North Korea hid plutonium from the inspectors.

US satellite photographs, taken over many years, show what appear to be two camouflaged nuclear waste sites near the Radiochemical Laboratory. They are big enough to handle large quantities of liquid and solid nuclear waste. One set of photographs shows a suspected outdoor waste facility believed to be associated with the IRT research reactor. In early photographs, the facility's layout resembles that of waste sites associated with Soviet-supplied research reactors. These sites have a distinctive pattern of round and square holes in an above-ground concrete structure for liquid and solid nuclear wastes. One Western official said that it closely resembled a site in Iraq next to its Soviet-supplied research reactor. Later photographs showed the same site, now covered by earth and landscaped and effectively hidden from inspectors or satellite surveillance. The declared nuclear waste site located nearby is new and barely used.

The second suspected site is a building reported to be about 50–70 m long and 20 m wide. It is about 150 m east of the Radiochemical Laboratory and separated from it by a small ridge. Early photographs show a two-story building, but in later shots the building has only one story because soil has been pushed up around the lower story to turn it into a basement. The pictures also reveal two trenches, which had been built by the early 1990s and which had connected the Radiochemical Laboratory with the building, suggesting the laying of pipes between the buildings. The IAEA wants to determine if the basement contains waste tanks holding reprocessing waste from the Radiochemical Laboratory.

In September, 1992 before the IAEA received this intelligence information, IAEA inspectors had visited the building, taking advantage of North Korea's invitation to visit undeclared sites. During this visit, the inspectors saw what appeared to be a one-story building under military control, but they did not see any evidence of a basement. However, they did not have any inspection equipment that would have enabled them to find a hidden basement.

In late 1993 and early 1994, the IAEA asked North Korea several times for access to the two potential waste sites to take samples. In the case of the one-story building, the IAEA asked for access to the spaces under the floor of the building and for permission to take samples from the building's below-ground level. Retreating from its earlier offer, North Korea refused to allow inspectors to visit either site, claiming that they were non-nuclear military sites. North Korea said that the IAEA-proposed inspections sought to confirm espionage information and would create a precedent for inspection of any of its military sites, jeopardizing the country's supreme interests.

By February 1993, the IAEA and North Korea had reached an impasse. The IAEA concluded that it could not fulfil its responsibilities under the safeguards agreement to confirm the correctness and completeness of North Korea's initial report on the inventory of plutonium.

Special inspections

Faced with North Korea's refusal to satisfactorily resolve the inconsistencies, on 25 February 1993 the IAEA Board of Governors' demanded 'special inspections' of the two suspected nuclear waste sites at Yongbyon. The Board then set a 25 March deadline for North Korea to agree to the special inspections.

Before the deadline arrived, North Korea announced on 12 March 1993 that it was withdrawing from the NPT under Article X of the Treaty, which gives signatories the right to withdraw with three-months notice if supreme national interests are threatened. Then in a bizarre twist, North Korea suspended its withdrawal from the NPT in early June, just days before it was intended to go into effect.

For the next year, North Korea engaged in a series of negotiations with the IAEA and the United States. These negotiations, however, were unsuccessful in ending the crisis, re-establishing safeguard inspections or verifying North Korea's initial declaration.

IV. Reactor defuelling

In April 1994, North Korea announced it had shut down its small reactor to refuel the core, which it said would begin as early as 4 May. US Secretary of Defense William Perry said publicly that this core contained enough plutonium for four or five nuclear bombs.

On 12 May, North Korea informed the IAEA that it had already started unloading the reactor without the safeguards measures the IAEA had requested, precipitating yet another crisis. The IAEA wanted to view the core unloading to ensure that North Korea did not divert fuel. It also wanted to select for later measurements a few hundred irradiated fuel rods in specific verified locations in the core to determine whether the fuel had been in the core since the reactor started in 1986, as North Korea claims.

Although North Korea decided to allow the IAEA to observe the fuel unloading, it refused to allow the IAEA to select and secure fuel rods for later measurements. During consultations in Pyongyang from 25–27 May, North Korea proposed to the IAEA that it could sample fuel rods after they were placed in the spent-fuel pond. IAEA inspectors refused this offer. They said North Korea's proposal would not allow the inspectors to know independently the location of the selected rods within the core. 'Without such identification', according to the Agency, 'future measurements would be meaningless and the Agency's ability to verify non-diversion would be lost'.

In a letter to the members of the United Nations Security Council of 27 May, Blix reported that 'the fuel discharge operation at the reactor was proceeding at a very fast rate which was not in line with information previously conveyed to the Agency'. Blix warned that if the fuel discharge continued at the same rate, the Agency would lose within days the ability to select fuel rods for later measurements in accordance with Agency standards.

Earlier, North Korea had told the IAEA that it planned to take two months to unload the reactor. Several US and IAEA officials believed that North Korea would take considerably longer to refuel the core. However, Blix said in his 27 May letter that North Korea had already unloaded almost half of the fuel in the core. *The Washington Post* reported on 1 June that US officials said North Korea was using a new, faster unloading machine that was previously unknown to Western intelligence. An IAEA official said in an interview that the machine was delivered to the reactor a few weeks before unloading began. Another

IAEA official said that North Korea had also accelerated its unloading operation to 24 hours a day.

On 2 June, the IAEA stated that the rapid unloading had made it impossible to select the desired fuel rods for later measurements. In a letter to the Security Council on the same day, Blix wrote that the IAEA's ability to determine past diversion of plutonium had been 'seriously eroded'. He added that because of North Korea's refusal to allow special inspections of the suspect waste sites and its unloading of the reactor core without the IAEA's required verification measures, the IAEA 'cannot achieve the overall objective of comprehensive safeguards in [North Korea], namely, to provide assurance about the non-diversion of nuclear material'.

Because of North Korea's actions, the US Government announced on 3 June that it was breaking off further bilateral negotiations with North Korea and moving to impose UN Security Council sanctions against the regime. One week later, however, the United States was struggling to develop a sanctions resolution that could overcome Chinese opposition to sanctions and Japanese reservations about tough sanctions. As a result, the United States shifted to an alternate strategy that sought support for a gradual imposition of sanctions. A vote in the Security Council, however, was not expected for several weeks.

On 10 June, the IAEA Board of Governors moved to impose its own sanctions by voting to suspend about $250 000 a year in technical aid to North Korea. In response, North Korea formally withdrew from the IAEA. However, this withdrawal did not undo its commitment to allow IAEA inspections required under the NPT.

Despite the escalation in tensions, North Korea permitted IAEA inspectors to maintain their watch of the spent fuel. By this time, almost all the fuel had been discharged from the reactor and transferred to an adjacent spent-fuel storage facility via a tunnel.

Fearing that economic sanctions would only increase the likelihood of war on the Korean Peninsula, former President Jimmy Carter went to North Korea to mediate an end to the crisis with its leader Kim Il Sung. An immediate result of Carter's personal diplomacy with Kim Il Sung was that the crisis moved back to negotiations.

A freeze on nuclear activities

Following up on Carter's visit, President Clinton announced on 22 June that North Korea had agreed to 'freeze' its nuclear programme, effectively ending the latest stand-off over North Korea's nuclear programme. In return, Clinton said that the United States would suspend its drive to impose sanctions on North Korea and would resume its bilateral negotiations with North Korea to resolve the nuclear issues.

The freeze announced by North Korea included a commitment not to reload the small reactor with fresh fuel or to reprocess the discharged fuel while bilateral negotiations proceed. North Korea also agreed that IAEA inspectors could remain at Yongbyon to verify that reloading or reprocessing did not occur. North Korea also agreed to continue maintaining the 'continuity of safeguards', a commitment it had made many times in the past. Not included in this 'freeze' was North Korea's agreement to allow the IAEA to determine if it diverted any plutonium in the past.

The sudden death of Kim Il Sung on 8 July 1994 complicated the resolution of this crisis. Nevertheless, the United States and North Korea reached an agreement on 12 August that (*a*) reaffirmed North Korea's commitment to maintain the freeze on its programme while bilateral negotiations proceed; and (*b*) established a list of elements that should be included in any final resolution of the nuclear issue. These elements include replacing its two larger gas-graphite reactors with modern, more proliferation-resistant light water reactors financed with Western assistance, foregoing reprocessing, verifiably sealing the Radiochemical Laboratory and remaining a member of the NPT.[5]

The Agreed Framework

After a few more months of negotiations, North Korea and the United States concluded an 'Agreed Framework' aimed at structuring an overall resolution of the North Korean nuclear issue during the next decade.[6] The centrepiece of this agreement is an unprecedented trade of reactor technologies. The USA has agreed to lead an international consortium to provide two 1000-MWe LWRs to North Korea. In exchange, North Korea has agreed to freeze and eventually dismantle its gas-graphite reactors and associated spent-fuel reprocessing plant, not to reprocess a stock of plutonium-laden spent fuel and to comply fully with its safeguards obligations under the NPT.

Negotiators concluded their work during the week of 17 October, and the four-page agreement was signed on 21 October in Geneva. The agreement includes a number of key elements:

1. The United States has agreed to make arrangements for the provision of two LWRs, generating a total of about 2000 MWe by the year 2003. Although the agreement does not obligate the United States to provide the reactors or finance the project itself, the USA is obligated to 'organize under its leadership an international consortium to finance and supply the LWR project'. The USA promised to make 'best efforts' to conclude these supply contracts within six months. South Korea and Japan agreed to pick up the bulk of the costs. If the consortium fails for reasons beyond the control of North Korea, President Clinton promised in a 20 October 1994 letter to Kim Jong Il, the supreme leader of North Korea, to use the 'full powers of my office to provide, to the

---

[5] Agreed Statement Between the United States of America and the Democratic People's Republic of Korea, Geneva, 12 Aug. 1994.

[6] Agreed Framework of 21 October 1994 between the United States of America and the Democratic People's Republic of Korea, IAEA document INFCIRC/457, 2 Nov. 1994.

extent necessary, such a project from the United States, subject to approval of the U.S. Congress'.

2. North Korea has agreed not to restart its small gas-graphite reactor, plutonium separation facility and other related facilities, and to stop construction of its two larger gas-graphite reactors. North Korea further agreed to dismantle these facilities when the LWR project is completed, estimated initially as 2003. North Korea also agreed to allow the IAEA to monitor the status of the freeze, and it agreed to provide full cooperation to the IAEA for this purpose.

3. North Korea agreed to cooperate with the United States to 'store safely' its spent fuel from its small gas-graphite reactor during the construction of the LWRs and to dispose of the fuel in a safe manner that does not involve reprocessing in North Korea, interpreted to mean North Korea will ultimately send the fuel to another country.

4. North Korea has also agreed to remain a state party to the NPT and to come into full compliance with its safeguards agreement. Under the framework, the IAEA will inspect those facilities in North Korea not subject to the freeze to the extent required 'for the continuity of safeguards' until conclusion of the LWR supply contract. Once the supply contract is concluded, the IAEA will be able to resume *ad hoc* and routine safeguard inspections at the facilities not subject to the freeze.

5. Once a significant portion of the LWR is completed, but before delivery of key nuclear components—believed in 1994 to be in five years' time—North Korea will come into full compliance with its safeguards agreement, presumably encompassing 'special inspections' of two undeclared facilities and permitting the IAEA to verify the accuracy and completeness of North Korea's initial declaration of nuclear material. This implies that North Korea must satisfy the IAEA that it does not have an undeclared stockpile of separated plutonium.

6. Until the first LWR is completed, heavy oil for heating and electricity production to offset the energy lost because of the freeze of the graphite reactors will be provided, starting in early 1995 and reaching a level of 500 000 tonnes a year.

Although a settlement appears at hand, a final resolution to the crisis is far from realization. North Korea has made agreements with the IAEA and South Korea before, only to renege. The Agreed Framework is just that, and many complicated issues need to be further negotiated.

Implementation of the Agreed Framework

As of the summer of 1996, the implementation of the Agreed Framework was occurring slowly. Chief among the reasons was that suspicions between the United States and North Korea remained high. In addition, North Korea has consistently refused to live up to its commitment in the agreement to open a new dialogue with the South, fuelling suspicions about North Korea's ultimate objectives.

Although many complicated issues remain to be settled, progress has been made in a number of key areas.

Nuclear freeze. The North has not refuelled the small reactor and has ceased construction at the reprocessing plant and the two other gas-graphite reactors. It has also permitted the IAEA to monitor the freeze which involves 30 buildings at five facilities. However, the IAEA continues to have conflicts with North Korea about the specific monitoring techniques that it can apply inside the 'frozen' facilities. For example, the IAEA has said that it could not adequately monitor the waste tanks inside the reprocessing facility to ensure that waste is not moved.

In addition, the IAEA has been unable to reach agreement with North Korea on the types of information which it must preserve to allow the IAEA to verify the correctness and completeness of its initial safeguards declaration. A great deal can be learned from a sophisticated analysis of the spent fuel rods, although this information degrades with time, making records more important.

LWR contract talks. Under the agreement, supply contracts for the LWRs were supposed to be concluded by April 1995, but delays pushed that date forward to December 1995. The Korean Peninsula Energy Development Organization (KEDO), a multilateral consortium led by the United States, Japan and South Korea that will finance and build the LWRs, was not officially formed until March 1995. After many months of negotiations, during which North Korea refused to accept South Korean reactors, the United States and North Korea agreed in June 1995 that a US company will serve as 'program coordinator' but that KEDO will choose the main contractor to build the reactors. Soon afterwards, KEDO selected a South Korean company to build 'an advanced version of US-origin design and technology'. Negotiations of the actual supply agreement started in October, and the agreement was signed by KEDO and North Korea on 15 December 1995. Delays in reaching agreement were caused by North Korea's demand for additional infrastructure items in the contract, several of which KEDO refused to provide. For example, North Korea's existing power grid cannot handle the electricity produced by two 1000-MWe reactors, and thus North Korea wants the contract to include upgrades to its electrical distribution system. Additional stumbling blocks to concluding the supply agreement included the precise role of South Korea in the actual construction work and the firmness of delivery schedules for reactor components. The reactor site will be at the Kumho area near Sinpo City in north-eastern North Korea.

Additional details of the supply agreement. The supply contract clarified several outstanding issues in the Agreed Framework.

1. In addition to building the two LWRs, KEDO agreed to supply the enriched fuel for the initial loading of the two reactors, provide training equipment, train North Korean personnel in operating the reactors, provide spare parts for the first two years of plant operation, build low- and medium-waste storage buildings, and build roads and barge-docking facilities at a nearby harbour facility necessary for the construction of the LWRs.

2. KEDO agreed to assist North Korea in obtaining commercial contracts for the enriched uranium fuel and spare parts for the life of the LWRs. In a letter to North Korea that is attached to the agreement, KEDO also agreed to 'use its good offices' to assist the country to obtain commercial loans to upgrade its electricity power grid.

3. North Korea agreed to repay KEDO for the two LWRs on a long-term, interest-free basis in cash, cash equivalents, or through the transfer of goods. The amount North Korea would repay to KEDO was to be negotiated later. However, the total cost of the two LWRs is about $5 billion, as of July 1996.

4. In addition to better defining what KEDO actually supplies, the supply agreement also contains several conditions that strengthen the non-proliferation constraints on North Korea, including restating the ones contained in the Framework Agreement. The North agrees that the reactors are exclusively for peaceful, non-explosive purposes, and that it will not reprocess or increase the enrichment level of any LWR fuel. If KEDO requests, North Korea will relinquish ownership of the LWR spent fuel and transfer it out of the country as soon as technically possible. North Korea further agrees not to transfer any nuclear equipment, technology or material obtained through this agreement to any other country without KEDO's agreement.

5. The agreement made significant progress in defining two controversial phases in the Agreed Framework. 'Key nuclear components' are defined as the components controlled under the Export Trigger List of the Nuclear Suppliers Group. The definition of when 'a significant portion of the LWR project is completed' was better defined, although not all details were resolved. Included in this definition is the 'completion of buildings, fabrication of major reactor components for the first LWR unit to the point suitable for the introduction of components of the nuclear steam supply system', and 'civil construction and fabrication and delivery of components for the second LWR unit according to project plans and schedules'. Whether all this construction can be completed in five years, as originally expected, is doubtful.

Protocols to supply contract. Implementation of the supply agreement requires KEDO and North Korea to negotiate 10–12 protocols, and by the summer of 1996 three protocols had been signed. These three grant KEDO an independent communication system, establish procedures and routes to move people and equipment to the North Korean site, and create privileges, immunities and consular protection governing employees of KEDO and its contractors while working in North Korea. Two additional protocols concerning site takeover and the provision of North Korean labour, goods and services were finalized in early autumn 1996. The plan in September 1996 was that groundbreaking would start before the end of 1996. With all the delays, however, and continued tension between North Korea and South Korea, ground breaking is not expected to start until 1997. In addition the project is no longer expected to be completed in 2003.

Alternative energy. Following US accusations that North Korea diverted some of the first shipment of 50 000 tonnes of heavy fuel oil to unauthorized uses, North Korea permitted the USA to install a monitoring system at the facility where it receives the oil. Since then, oil shipments amounting to about 20 per cent of its annual fuel needs have resumed as originally scheduled, although KEDO struggles to obtain adequate funds to pay for the heating oil.

Spent-fuel storage. The USA has been working with North Korea to place about 8000 spent-fuel rods from the 5-MWe reactor into airtight stainless steel canisters or 'cans'. The spent fuel rods are loaded into cans inside the spent fuel pond. Afterwards the water is purged from the cans and replaced by a combination of argon gas and a small amount of oxygen, a scheme that is expected to permit safe storage for 4–7 years. About 650 spent fuel rods are in dry storage adjacent to the spent fuel pond, and these rods will need to be handled carefully when they are canned. The cans are being sealed and tagged and placed in the spent fuel pond until they can be shipped out of the country. According to the LWR supply agreement, the spent fuel rods will start to leave North Korea when key nuclear components for the first LWR start to arrive in the country. All rods will have left North Korea when the first LWR is completed. The rod's final destination has not been determined. Safely storing the spent fuel has occurred more slowly than expected because of delays caused by the US Congress in approving funds to hire contractors, by unexpected problems in getting all the equipment to the site, and by difficulties encountered in clarifying the pond water and in removing sludge from the bottom of the pool. The pond must be clear enough to permit the operators to see the rods. US experts also had to remove many centimetres of sludge from the bottom of the pool that had caused filters to plug and pumps to break. The canning operation started in late April 1996 and is expected to take about one year to complete. During the canning operation, the IAEA is taking measurements of the rods to verify that they are all irradiated. However, North Korea has refused to allow the IAEA to perform any measures that would provide information on the total amount of plutonium in the rods. These latter measurements may not be performed for more than five years and may require opening the canisters. If any cans develop leaks the fuel will be unloaded and recanned.

Routine and ad hoc inspections. Following the signing of the LWR supply contract in January 1996, North Korea notified the IAEA that it could resume routine and ad hoc inspections of 'non-frozen' facilities.

Despite all the progress, the issue of how much plutonium is in North Korea's possession remains unsettled. Sections V–VII evaluate that question.

V. Plutonium production reactors

The small operating reactor and the two others that were under construction prior to the freeze use a design that depends on uranium fuel, carbon-dioxide gas cooling and graphite moderation ('Magnox' or gas-graphite reactor). Designs of this type of reactor are largely unclassified and the reactors themselves straightforward to build. North Korea appears capable of building this type of reactor without significant foreign assistance. The actual extent of foreign assistance remains unknown.

The disadvantage of this type of reactor is that it discharges irradiated spent fuel that typically requires plutonium separation. The spent fuel is difficult to store safely for an extended period or to dispose of in a geological repository. The outer casing of North Korea's fuel uses a magnesium alloy. This type of cladding breaks down when stored in water or exposed to moisture, eventually exposing the uranium metal to air. This allows radioactive material to escape. Under certain conditions, it can lead to the uranium metal spontaneously burning when exposed to air. If the fuel burns, a significant fraction of the radioactive materials can be released into the environment.

North Korea's choice of gas-graphite reactors has increased suspicions that its initial goal was a nuclear arsenal. North Korean officials have told the IAEA that their reactors are only for electricity production, and they chose to build gas-graphite reactors because they could develop them without foreign assistance. However, the gas-graphite models, developed in the 1950s, can easily produce weapon-grade plutonium.

The 20–30 MWth reactor

Construction of this reactor started in 1980, and the reactor began operation in 1986.[7] According to US officials, the reactor had start-up problems during its first few years of operation, although by 1990 or 1991 they said it was operating at 20 to 30 MWth. One official added that US intelligence agencies did not know precisely the total energy output of this reactor during its first several years of operation, before safeguards were applied in 1992.

North Korea decided to talk about the energy output of this reactor in terms of its 5 MW of electricity production. This decision is an apparent attempt to direct attention away from its possible military purpose. The total energy output of a reactor, and thus its output of plutonium, depends on its thermal power. Although estimates of its total power vary between 20 and 30 MWth, a midpoint of 25 MWth is used in subsequent discussions.

This reactor produces weapon-grade plutonium. Although the fuel remains in this reactor for several years, the total irradiation of the discharged fuel is small. However, the fuel is irradiated long enough to ensure that sufficient weapon-

---

[7] So Yong-ha, 'Capacity for nuclear weapons development' [in Korean], *Hoguk* (Seoul), July 1989, pp. 119–22. English translation in Foreign Broadcast Information Services, *Daily Report–East Asia (FBIS-EAS)*, FBIS-89-148, 3 Aug. 1989, pp. 23–26; and IAEA (note 3).

grade plutonium is produced to warrant recovery. For this type of reactor, weapon-grade plutonium would correspond to burnups less than about 900–1000 MWd/t.

North Korea's reactor must first be shut down before the fuel can be unloaded, although perhaps a small number of elements could be unloaded 'on-line,' that is, while the reactor operates.

Just prior to its 1994 unloading, the reactor core contained about 48 tonnes of natural uranium fuel. The fuel is in the form of short fuel rods, each of which is roughly 50 cm long and 3 cm in diameter and has a mass of about 6.2 kg. The core contained a total of about 7700 fuel rods, located in 812 fuel channels in the core. Each channel can hold up to 10 fuel rods stacked one on top of the other. According to a US official, the reactor is designed to hold a total of about 8000 fuel rods, but the reactor contained fewer than the maximum because damaged rods had been removed earlier and not all of them had been replaced. The fuel unloading machine refuels the reactor through the top of the core.

In terms of plutonium discharges, there are two periods of interest. The first corresponds roughly to the period before the IAEA started its inspections in 1992. The second is the period afterwards, and in particular the defuelling in the spring of 1994, which was the only major refuelling after inspections began.

Pre-safeguards plutonium discharge

1989 refuelling. The most credible worst-case estimate assumes that North Korea unloaded the first core several years ago, but that the reactor did not operate at full power.

In December 1993, the public learned about a long shut-down of this reactor that could have enabled North Korea to unload all of the fuel in the reactor core. In a television statement, then Secretary of Defense Les Aspin said that 'in 1989 the North Koreans shut down their reactor for 100 days, and that would have given them enough time' to extract some fuel. He continued: 'Depending upon how much plutonium they processed and their capabilities of putting that together into a bomb, they might have gathered enough plutonium for a bomb, maybe a bomb and a half at the outside, perhaps.'[8]

A US official said in June 1994 that Aspin's 100-day statement represented an order of magnitude estimate of the length of the reactor shut-down. He said that the actual length of the shut-down was significantly less, about 70 days. Estimating the amount of fuel (and the contained plutonium) that North Korea could have unloaded from the core during the 1989 shut-down requires several pieces of information. Since much of this information is unknown, any estimate is highly uncertain.

A 70-day shut-down appears long enough to unload the entire core and put in new fuel, as events in the spring of 1994 have shown. Before this recent reloading, North Korea told the IAEA that it would take about two months to

---

[8] Secretary of Defense Les Aspin, statement on the *MacNeil/Lehrer* television news show, 7 Dec. 1993.

change the core. With two machines unloading the fuel, North Korea took less than one month to unload almost all the fuel.[9] Thus, an upper bound is that the entire core was replaced in 1989.

US and IAEA officials have expressed scepticism about so rapid an unloading for the 1989 refuelling. Based on a comparison with British Calder Hall reactors, a 70-day shut-down might have provided only enough time to unload about half the core.[10]

These estimates therefore imply that about one-half to all the core, or roughly 25–50 tonnes of spent uranium fuel, could have been unloaded in 1989.

The plutonium in the samples taken by the IAEA at the Radiochemical Laboratory had isotopic compositions of slightly more than 97.5 per cent ^{239}Pu, or equivalently about 2.25 to 2.5 per cent ^{240}Pu. Based on unclassified US Government studies of gas-graphite reactors, irradiated fuel that contains such a fraction of plutonium isotopes has a burnup of a little less than 300 to almost 330 MWd/t. Fuel with this burnup contains about 0.27–0.30 kg of weapon-grade plutonium per tonne of uranium fuel.[11] In subsequent calculations, the mid-point of this range is used. We consider two distinct cases:

1. The first case assumes that the entire core was unloaded in 1989. If the values calculated above represented the average burnup of *all* the fuel in the core in 1989, the 50-tonne core would contain about 14 kg of weapon-grade plutonium. To produce 14 kg of plutonium in the core, the reactor would have had to operate at the equivalent of 25 MWth about 55 per cent of the time from 1986 through 1989. This level of operation could be consistent with this reactor's actual operation although it implies few of the start-up problems mentioned by US and IAEA officials.

According to US officials, the reactor did not operate this well. They have said that the plutonium sampled by the IAEA was from fuel irradiated in regions of the core that had higher than average burnups, implying that the

---

[9] According to an IAEA official, North Korea started unloading the fuel about 10 May 1994 and finished unloading the bulk of the fuel by 15 June. A few of the fuel rods were stuck in the reactor channels, and these rods took several more days to unload.

[10] The North Korean gas-graphite reactors are similar in design to the British Calder Hall reactors. Each of these reactors had an initial design power of 180 MWth and used off-line refuelling. Each reactor had two discharging and two charging machines, and refuelling took six weeks. IAEA, *Directory of Nuclear Reactors: Vol 1. Power Reactors* (Vienna: IAEA, 1959), pp. 125–30. A Calder Hall reactor core contained 127 t of natural uranium in about 10 200 fuel rods, where each fuel rod was about 100 cm long and 3 cm in diameter. Based on this information, each Calder Hall discharging machine could remove about 1.5 t of fuel each day, or about 120 rods per day. Assuming a similar fuel rod unloading rate for the North Korean discharge machine, the North would have needed about 65 days to unload all 48 t of fuel (a rate of 0.75 t per day). It takes longer to unload the North Korean reactor according to this estimate, because the North Korean reactor is assumed to have had only one fuel unloading machine in 1989 and has about 80% as many fuel rods as a Calder Hall reactor. If North Korea had only one machine in 1989, the core would have been reloaded with this same machine. Assuming that reloading took from half as long to as long as unloading, North Korea could have changed the core in about 100–130 days. If the 1989 shut-down lasted 70 days, roughly one-half to two-thirds of the core could have been changed. Without more specific information about the North Korean discharge machine, however, this estimate is highly uncertain.

[11] Turner, S. E. et al., *Criticality Studies of Graphite-Moderated Production Reactors*, Report prepared for the U.S. Arms Control and Disarmament Agency, SSA-125, Southern Sciences Applications, Washington, DC, Jan. 1980.

average burnup for the entire core was lower. One official said that the estimated average burnup for the core was no more than about 200 MWd/t at the time the reactor was shut down in 1989. In this case, the core contained a total of about 9.5 kg of plutonium (1.5 per cent ^{240}Pu; see appendix A.) An average burnup of 200 MWd/t corresponds to the reactor operating at 25 MWth about 35 per cent of the time since it started in 1986.

Based on a series of interviews in 1996 with US and IAEA officials, the more credible scenario appears to be that the average burnup of the core is about 200 MWd/t. As a result, the estimate for this case is 9.5 kg.

2. If only 25 tonnes of fuel were discharged in 1989, then this fuel contained an estimated 7 kg of plutonium, assuming that it had an average burnup of about 300 MWd/t. In this case, the burnup of the fuel remaining in the core is significantly less than the average burnup of the discharged fuel. Reactor operators typically remove the higher burnup fuel first if they are unloading only a portion of the core, because this fuel contains a higher concentration of plutonium.

If half the core had a burnup of about 300 MWd/t, then the other half would have a burnup only roughly one-third as high, giving an average burnup of all the fuel in the core of about 200 MWd/t. At this average burnup, as mentioned above, the core would have contained in total about 9.5 kg of plutonium.

Under normal conditions, burnups vary greatly depending on the location of the fuel in the core. The burnup of fuel in the centre of the core is several times greater than that of fuel near the periphery. In addition, according to an IAEA official, North Korea said that the fuel irradiation had not been symmetrical in the core during the first years of the reactor's operation. This official said that the control rod pattern, which determines the irradiation level or 'neutron flux' in different regions of the core, was asymmetrical. He said that North Korea has declared that the reactor operated with the control rods partially inserted, depressing the neutron flux in the top of the reactor core and increasing it in the bottom half of the reactor. In addition, chemical impurities in the fuel might have also caused significant asymmetries in the neutron flux in the core. Although North Korea has given the IAEA this information, it has not provided verifiable operating records or information to substantiate these claims.

In summary, we estimate that the 25–50 tonnes of fuel possibly discharged in 1989 contained about 7–9.5 kg of weapon-grade plutonium.

Maximum discharge estimate. The press has reported reactor shut-downs other than in 1989. One 1994 press report said that North Korea had also shut down the reactor for a month in 1990 and for 50 days in 1991.[12]

US officials have repeatedly discounted the possibility that North Korea discharged significant quantities of spent fuel in 1990 or 1991. Nevertheless, if additional refuelling occurred, a crude estimate of average annual plutonium production gives an upper bound on the reactor's plutonium output. When

[12] Reuter, 'N. Korea may have enough plutonium for 3 to 4 bombs', 7 Aug. 1994.

operating at a thermal power of 25 MWth an average of 80 per cent of the time, this reactor could produce on average about 6.6 kg of weapon-grade plutonium a year, assuming a fixed plutonium conversion rate of 0.9 g/MWth-d (see appendix A). If this reactor operated consistently at this capacity factor for five years until 1991, it could have produced 33 kg of weapon-grade plutonium. Few, however, believe that this reactor operated so consistently, or at such a high capacity factor in the period 1986–91.

Minimal plutonium discharge estimate. What about a lower bound for the amount of plutonium in previously discharged fuel? A lower bound corresponds to North Korea's declaration that it only removed a few hundred damaged fuel rods, containing several hundred grams of plutonium.

North Korea has declared that it processed some fuel rods in the Radiochemical Laboratory. It reportedly said that these rods contained a total of about 0.13 kg of plutonium, of which about 0.09 kg were recovered. The rest went into the waste or processing equipment. Using the above assumptions, this amount of plutonium would correspond to a total of about 450 kg of uranium fuel, or about 70 fuel rods. This estimate would correspond to the reactor operating poorly the first several years, followed by a period of operation near full power.

Post safeguards plutonium discharge: spring 1994 defuelling

In April 1994, North Korea shut down the small reactor for refuelling. US and IAEA estimates of the amount of plutonium in the core vary from about 20 kg to over 30 kg of weapon-grade plutonium.

Under the previous assumptions about the fuel irradiation levels, this 48-tonne core would contain about 9 kg of weapon-grade plutonium. However, the irradiation levels of the fuel rods discharged last spring are believed to be significantly higher than they were for fuel discharged in 1989.

If this core was the first one, the fuel would have been in the reactor for up to eight years. Such a long residence time would require a higher burnup. At another extreme, if this core is the second one, the fuel would have been in the core since 1989. This case also implies a higher fuel burnup because the reactor is believed to have operated significantly better after 1989.

Reports about the average burnup of this fuel vary. One report says that the burnup of the rods varies between 200 MWd/t and 1200 MWd/t, with an average of 800 MWd/t. US officials have said that the average burnup is about 600–650 MWd/t. At these burnups, the spent fuel would contain about 26–28 kg of plutonium.

We believe that the best estimate is about 25–30 kg. Determining a more accurate figure will require additional information.

Other gas-graphite reactors

The 200-MWth reactor

Until the freeze took effect, North Korea was building a 200-MWth, or 50-MWe gas-graphite reactor at Yongbyon. The earliest possible start-up date was late 1995, although Secretary of Defense William Perry stated in the spring of 1994 that the reactor would have taken a few more years to complete.[13] Historically, construction tended to wax and wane, depending on a range of factors including the availability of concrete. If North Korea had decided to speed construction prior to the freeze, Western intelligence agencies believe it could have done so.

The reactor building, according to inspectors, is largely finished, although the inside would have required more work. According to a US official, North Korea started installing electricity production equipment in this reactor just prior to the beginning of IAEA inspections. The addition of electrical generating equipment apparently slowed the construction of this reactor.

Many analysts believe that North Korea originally intended this reactor to be its main source of plutonium for its nuclear weapon programme, while the small reactor would have provided plutonium for only the first few weapons. A 200-MWth reactor, operated an average of 60 to 80 per cent of the year at full-power, could produce about 40–53 kg of weapon-grade plutonium a year. This amount is sufficient to make 8–10 implosion-type weapons a year.

This estimate contains uncertainty. For example, the actual thermal output could be higher or lower than estimated. Nevertheless, it illustrates the plutonium production potential of this reactor.

The 600–800 MWth reactor

North Korea could have also completed a 200-MWe reactor of the same design at Taechon. The IAEA was told in 1992 that this reactor could have been finished in 1996.[14] The thermal power of this reactor is estimated to be about 600–800 MWth. A mid-point of 700 MWth is assumed in the subsequent discussion. If the reactor was operated to produce weapon-grade plutonium at a capacity factor of between 60 and 80 per cent, it could produce about 140–180 kg of weapon-grade plutonium a year.

Except in the most dire interpretations, few believe that this reactor was intended to produce weapon-grade plutonium for nuclear weapons. This reactor would have probably been optimized to produce electricity, meaning that the reactor would produce plutonium that is not weapon-grade. However, it could have served as a backup production reactor in case the other reactors did not produce enough weapon-grade plutonium, or failed for some reason. If this

---

[13] IAEA (note 3); and 'Remarks by Secretary of Defense William Perry to the Asia Society: US Security Policy in Korea', National Press Club, Washington, DC, 3 May 1994.
[14] IAEA (note 3).

reactor were optimized for electricity production and had a capacity factor of 60–80 per cent, it could produce about 90–120 kg of reactor-grade plutonium a year.[15] The plutonium production rate is lower in this case, because the fuel is exposed to many more neutrons, resulting in a significant amount of the plutonium fissioning.

How much plutonium could the three gas-graphite reactors have produced?

The most straightforward estimate of plutonium production calculates annual plutonium production in terms of the reactor power, its capacity factor and a fixed plutonium conversion factor of about 0.9 g/MWth-d for weapon-grade plutonium and 0.6 g/MWth-d for reactor-grade plutonium (see appendix A). In this estimate, a capacity factor of 70 per cent is applied to each reactor. Although this capacity factor is at the high end of the range for this type of reactor, it is achievable over the long term.

This projection assumes that the 200-MWth reactor would have started operation in 1997 and the 600–800 MWth reactor would have started in 1998. In both cases, these start-up dates are somewhat later than those North Korea provided to the IAEA. During their first three years of operation, their plutonium production rate is assumed to increase linearly to full output. Two different scenarios are presented:

1. The first scenario assumes that the 25-MWth and 200-MWth reactors produce weapon-grade plutonium and the 600–800 MWth reactor discharges reactor-grade plutonium. At a 70 per cent capacity factor, the two smaller reactors would produce 5.7 kg and 46 kg respectively of weapon-grade plutonium a year at steady state. The largest reactor is assumed to have an average power of 700 MWth and to discharge about 105 kg of reactor-grade plutonium per year at steady state.

Under this scenario, North Korea could have produced about 250 kg of plutonium by the end of the year 2000 and almost 2000 kg of plutonium through 2010. If only the two smaller reactors are considered, they would have produced a total of about 150 kg of weapon-grade plutonium through 2000 and about 730 kg of weapon-grade plutonium through 2010, or enough plutonium for 30 and 145 nuclear weapons in 2000 and 2010 respectively. The prospect of such a large amount of weapon-grade plutonium is part of the reason the Agreed Framework is so important.

2. The second scenario assumes that all three reactors produce weapon-grade plutonium. The other assumptions remain the same. As of 2000, the three reactors would have produced about 300 kg of weapon-grade plutonium, or enough for about 60 nuclear weapons. By 2010, they would have produced about 2650 kg of weapon-grade plutonium, or enough for over 500 nuclear weapons.

---

[15] This estimate assumes a burnup of 4000 MWd/t and a Pu conversion factor of about 0.6 g/MWth-d. Under these conditions, the reactor would discharge on average about 40–50 t of fuel each year.

Plutonium production in LWRs

Although LWRs have not been used to make plutonium for nuclear weapons, they could be used to make either reactor-grade plutonium or weapon-grade plutonium for use in nuclear weapons. Thus, it is important to estimate the amount of plutonium that will accumulate in North Korea once its LWRs are operating.

Reactor-grade plutonium

Under normal operation, economically optimized to make electricity, an LWR will discharge reactor-grade plutonium. Two 1000-MWe pressurized water reactors similar to those to be built in North Korea will discharge about 490 kg of reactor-grade plutonium each year.[16] This estimate assumes that the two reactors achieve average annual capacity factors of 70 per cent and discharge fuel with a burnup of about 30 000 MWd/t. This estimate could be higher or lower, depending on the actual burnup and capacity factors achieved, but it represents a typical case for this type of reactor.

The two LWRs will discharge over twice as much plutonium as the three gas-graphite reactors, even if all three gas-graphite reactors produce only weapon-grade plutonium. Although LWRs produce less plutonium per megawatt-day of operation than gas-graphite reactors, the two LWRs have roughly six times the power of the three gas-graphite reactors. By the end of 2010, the two LWRs could discharge about 2800 kg of plutonium.[17] By the end of 2020, the amount would be 7600 kg of plutonium.

Comparing quantities of plutonium, the three gas-graphite reactors would have discharged significantly less plutonium than the LWRs during their lifetime. Operating in a way to produce the maximum amount of plutonium, namely weapon-grade plutonium, the three gas-graphite reactors would discharge about the same amount as the LWRs by 2010 and almost 40 per cent less by 2020.[18]

Weapon-grade plutonium

An LWR can also discharge significant quantities of weapon-grade plutonium. The operator would need to unload the reactor earlier than scheduled, greatly

---

[16] This estimate of annual discharge of reactor-grade plutonium from two 1000-MWe LWRs assumes a core load of about 84 t of enriched uranium, and a refuelling of one-third core each year. It further assumes a steady-state average burnup of 30 000 MWd/t, which implies an average capacity factor of about 0.7. See appendix B. For 30 000 MWd/t, the spent fuel contains 8.7 kg/t.

[17] Same assumptions as above (see note 15). The projection also assumes a steady state average burnup of 30 000 MWd/t after about three years. The first and second refuelling are assumed to have average burnups of almost 15 000 and 25 000 MWd/t, respectively. See appendix B. For 30 000 MWd/t, 8.7 kg/t; for 25 000 MWd/t, 7.9 kg/t; for 15 000 MWd/t, 6.6 kg/t. For the sake of simplicity, both reactors are assumed to start up in 2004, and discharge their first fuel in 2005.

[18] The estimated cumulative amounts of weapon-grade plutonium for the gas-graphite reactors are about 2600 kg in 2010, and 4800 kg in 2020. If the largest gas-graphite reactor produced reactor-grade plutonium, these cumulative estimates would have been lower. The estimate assumes the same capacity factor as that of the LWRs, or 70%.

under-irradiating the fuel. For example, if the spent fuel has a burnup of about 2500 MWd/t, it would discharge weapon-grade plutonium at a rate of about 1.5 kg per tonne of spent fuel.[19] Thus, if one-third of the core were discharged after such a short irradiation period, the discharged fuel would contain about 40 kg of weapon-grade plutonium, where the core contains about 85 tonnes of fuel as before. The two reactors could in this way discharge about 80 kg of weapon-grade plutonium.

One method by which to obtain more weapon-grade plutonium is to unload the reactor within a few months of the first commencement of full-power operation. In this case, over 100 kg of weapon-grade plutonium could be obtained from each reactor.

Whether North Korea would misuse LWRs is impossible to determine. Certainly, any diversion of a significant amount of spent fuel would be discovered soon afterwards. A diversion would undoubtedly lead to a cut-off of additional fresh fuel and a reactor shut-down, ending any future misuse. It would also probably lead to demands to remove the spent fuel from the North.

VI. Plutonium separation

Producing plutonium is only the first step in making a nuclear weapon. The plutonium must be chemically separated from the irradiated fuel. During processing, some fraction of the plutonium is lost to the waste.

North Korea has worked on the processes associated with separating plutonium for many years. Its knowledge of plutonium chemistry appears extensive.

Early efforts

IAEA Director Blix told the US House of Representatives Committee on Foreign Affairs on 22 July 1993 that North Korea carried out 'experiments quite a number of years ago in which they identified plutonium'. US officials say North Korea's early laboratory-scale plutonium separation was carried out in hot cells. The Soviet Union supplied the hot cells during the 1960s or 1970s as part of a deal to supply an IRT research reactor. According to IAEA officials, North Korea orally declared to the IAEA that it separated grams of plutonium in these hot cells in 1975 from irradiated fuel from its IRT reactor. This campaign occurred before the IAEA applied safeguards to the IRT reactor in 1977.

North Korea told the IAEA during its initial visit in May 1992 that it scaled up from laboratory experiments to an industrial-scale plant without building a pilot plant. North Korea told the IAEA that it has often followed such a course of action in its industrial development. Although Western officials believe that North Korea could have jumped from hot cells to industrial-scale production,

---

[19] *Heavy-Element Concentration in Power Reactors.* Report no. SND-120-2 (NUS Corporation: Clearwater, Fla., May 1977).

questions remain about the developmental history of North Korea's plutonium separation programme. In particular, some believe that North Korea operated a pilot separation plant that the IAEA and intelligence services have not discovered.

The Radiochemical Laboratory

The plutonium separation plant at Yongbyon is sizeable—180 m long and equivalent to six stories high. US officials estimate that this facility could theoretically process up to several hundred tonnes of spent fuel a year if fully operational. Its theoretical capacity is viewed as sufficient to handle all the spent fuel from the three gas-graphite reactors.

North Korea is suspected of having obtained some important items from abroad. Basic knowledge about reprocessing technology and chemistry is reported to have come from Russia and perhaps China many years ago. An inspector stated that the Radiochemical Laboratory looked similar to the Eurochemic reprocessing facility that operated in Belgium from 1966 until the mid-1970s (see chapter 6). Information about this plant has been largely declassified.

Some reprocessing chemicals are reported to be of foreign origin. In addition, *The Washington Post* reported on 2 April 1994 that North Korea obtained stainless steel tanks from Japan. North Korea is also thought to have imported leaded glass for its hot cells.

Despite these imports, North Korea is unlikely to have depended on imports to build its plutonium separation capabilities, as countries such as Iraq have done. North Korea both values self-reliance and suffers a severe shortage of funds.

The first line

In 1992, the plant had one operating 'line' which included equipment to dissolve the fuel, extract the plutonium and convert the plutonium into a pure form. According to one inspector, as of the end of 1992 the facility's waste reduction processing section was not finished. Typically, plutonium separation facilities treat liquid waste to recover acids and other chemicals, thus reducing the total volume of liquid waste that goes to waste tanks.

There is little hard information on the capacity of the first line to process spent fuel. Most believe that the capacity was large enough to have processed all the fuel in the core of the small reactor before inspections started.

A preliminary estimate of this line's nominal annual capacity can be derived by assuming that the first line was sized to process all the fuel from the 25-MWth and 200-MWth reactors. As part of defining a nominal or maximal capacity for the first line, the following are assumed: (*a*) the reactors produce weapon-grade plutonium; (*b*) the reactors achieve a capacity factor of 80 per cent; and (*c*) the reprocessing line is large enough to handle the situation in which the fuel burnup is less than expected, or only 400 MWd/t, corresponding to a maximal spent-fuel output of the reactors. In this scenario, the line would need to be able to process on average about 160 tonnes of fuel per year.[20] If this estimate is accurate, the first line could reprocess 48 tonnes of spent fuel from the 25-MWth reactor in as little as about 3.5 months. This estimate represents the shortest time to reprocess this fuel. The actual throughput of this plant could be considerably less, in which case the plant would require more time to process a given quantity of fuel.

If the nominal fuel burnup is 600 MWd/t, the nominal throughput of the first line would be 110 tonnes a year, where the rest of the assumptions remain the same. In this case, the line could reprocess 48 tonnes of spent fuel in about five months.

No plutonium separation process is 100 per cent efficient. Although North Korea declared that it separated about 90 grams of plutonium, it also declared that it had lost about 40 grams of plutonium to the various waste streams. A loss rate of 30 per cent is high, but possible when first starting a plant. The operators, however, should have been able to reduce the plant's plutonium losses. Only a 10 per cent loss rate is therefore applied to the plutonium estimated as processed in the Radiochemical Laboratory.

The second line

Following an IAEA inspection in March 1994, inspectors reported that they were surprised to see a nearly-completed second separation line at the Radiochemical Laboratory.[21] This line is nearly identical to the first line and, therefore, if finished, would have roughly doubled the capacity of the separation plant, more than sufficient for the third and largest reactor. Alternatively, the second line could be a backup to the first line, in case it fails. Redundancy is common in newer plutonium separation plants.

Although the IAEA first learned about the second line during its initial inspections in 1992, an IAEA official said in a March 1994 interview with one of the authors that what surprised the inspectors in the March 1994 inspection was that 'a great deal of construction activity was going on'. The IAEA did not think significant construction had been occurring at the Radiochemical Laboratory. He added, however, that North Korea had not allowed the inspectors to inspect the Radiochemical Laboratory adequately since the spring of 1993.

This official said that at the time of the March 1994 inspection North Korea had installed the steel components of the second line. But he said that North Korea had not at that time installed the instrumentation necessary to monitor and operate the complex plutonium separation process. This assessment is consistent with statements made by US intelligence officials in the summer of 1994 that plutonium separation did not occur in the second line.

---

[20] This estimate is derived by dividing the average burnup into the total annual energy output of the two reactors, where the capacity factor is 80%. The equation is:
[(200 MWth + 25 MWth) x 365 d x 0.8]/400 MWd/t.

[21] Hibbs, M., 'Second, hidden reprocessing line feared opened at Yongbyon plant', *Nucleonics Week*, 24 Mar. 1994.

LWR fuel reprocessing

Unlike gas-graphite fuel, LWR spent fuel can be stored indefinitely in water. LWR fuel is composed of enriched uranium-oxide, surrounded by zirconium cladding which allows safe storage in water. Thus, North Korea has no need of a reprocessing plant to handle this fuel.

In addition, the LWR supply contract says that North Korea will not reprocess its LWR fuel. In addition, North Korea promised in the framework agreement to take steps to implement the 1991 North–South Joint Declaration on the Denuclearization of the Korean Peninsula, which includes a condition banning reprocessing and enrichment plants on the Korean Peninsula.

Nevertheless, North Korea is capable of building a small reprocessing plant to separate plutonium from light water reactor fuel. Given North Korea's experience with reprocessing and despite the technical challenges it would face, it could secretly build a facility able to separate tens of kilograms of plutonium a year. It also might be able to operate such a facility in secret, at least for a few months. As a result, any spent fuel will need careful and nearly continuous monitoring by the IAEA.

VII. How much plutonium does North Korea have?

Despite all the information available, the answer to the original question remains highly uncertain.

Grams or kilograms of separated plutonium?

A lower bound on plutonium separation corresponds to North Korea's declaration of 100 grams of separated plutonium. The most credible upper-bound or worst case estimate is that North Korea discharged irradiated fuel containing roughly 7 to 9.5 kg of weapon-grade plutonium from the core of the 25-MWth reactor in 1989. In this case, the fuel would have been processed in 1989–91 and most of the plutonium extracted. Assuming that 90 per cent of the plutonium was recovered, a total of 6.3–8.6 kg of separated plutonium was recovered. This is rounded to 6–9 kg.

A first nuclear weapon can require up to 8 kg of separated weapon-grade plutonium. This quantity is about double the amount actually needed in the device, because the manufacturing process leads to losses at each step. But most of this plutonium can be recovered and used in later weapons. North Korea might therefore have enough separated plutonium for one and perhaps two nuclear weapons. Obtaining two weapons would mean that each device could not contain more than 4 kg of plutonium, assuming minimal losses.

How much more?

Regardless of whether North Korea unloaded more fuel than it declared to the IAEA, the spent fuel being unloaded contains an estimated 25–30 kg of weapon-grade plutonium.

Making the same assumptions as above, North Korea has enough plutonium in this spent fuel for roughly five or six nuclear weapons, assuming 5 kg per weapon. This plutonium, however, is in irradiated fuel and must first be separated before it could be used in nuclear weapons.

How much more in the future?

Under the Agreed Framework, North Korea will not produce any more plutonium until its light water reactors operate, no sooner than 7–8 years. After the LWRs start up, North Korea will accumulate plutonium in spent fuel at the rate of about 490 kg per year. Because this quantity is so large, North Korea will need to provide nuclear transparency to ensure that diversion does not occur.

VIII. What about enrichment activities?

Little evidence of any uranium enrichment activities has emerged. CIA analysts believe that North Korea has not pursued any significant enrichment activities, because of shortages of hard currency and technical capabilities. It may have conducted some research into chemical enrichment using know-how obtained in Japan. Determining small-scale enrichment activities is difficult, however, and therefore, questions remain about North Korea's progress in enrichment.

IX. Has North Korea built nuclear weapons?

Many experts believe that North Korea is capable of developing an implosion-type nuclear device, although little direct evidence of such a programme exists. Despite the CIA's worst-case plutonium assessments, it said that North Korea may or may not have a nuclear explosive device. CIA officials have also said that North Korean scientists did not receive any training in nuclear weapon technologies from Russia or China. This statement implies that North Korea would have to develop nuclear weapons largely on its own.

US intelligence officials believe that North Korea has clandestine nuclear weapon manufacturing sites which have so far eluded Western detection. Several indicators support this view. One indicator, for example, is North Korea's high-explosive testing at Yongbyon. In answer to questions about this high-explosive testing, however, North Korean officials told the IAEA that they were using high explosives to shape metals. Several countries are pursuing this technique for metals that cannot be shaped conventionally.

There are a few other indicators of a nuclear weapon programme. North Korea has reportedly shown an interest in acquiring instrumentation for conducting non-nuclear tests associated with a nuclear weapon programme. Usually these tests involve high explosives.

The CIA has estimated that North Korea could have only a first-generation implosion design. It estimates that the mass of a device within North Korea's capabilities would probably be greater than 500 kg but less than 1000 kg.

Little is known about how North Korea might deliver a nuclear device. Delivery by aircraft is possible, but air defences in North Asia are substantial and would pose a serious threat to any North Korean aircraft. Certainly, North Korea could deliver a nuclear device by ship or truck. According to CIA assessments, the North Korean device would not fit on a SCUD missile. It could fit on the Rodong missile, which North Korea is currently developing. This missile, which was first flight-tested in late May 1993, has an estimated range of over 1000 km, but it is reported to be a few years from deployment.

11. A special case: Iraq

I. Introduction

Iraq is a special case. It is the only country forbidden to possess separated plutonium and HEU. Under UN Security Council Resolution 687, adopted by the Security Council on 3 April 1991, Iraq was obliged to pledge not to acquire or develop nuclear weapons, nuclear weapon-usable material or the means to make them. In accordance with this resolution, Iraq's nuclear, chemical and biological weapon capabilities have been systematically destroyed. However, Iraq retains a formidable amount of 'know-how' concerning how to make nuclear weapons with its original cadre of scientists and technicians and, perhaps, small quantities of equipment and materials. A major concern is that Iraq will try to resurrect its programme if given the chance, especially if Saddam Hussein remains in power. Standing against this possibility are economic sanctions imposed by the Security Council following Iraq's invasion of Kuwait in August 1990 and an on-going monitoring and verification system.

Uncovering the major aspects of Iraq's former nuclear weapon programme has taken many years, involving many reassessments as new information has been uncovered by international inspections or declared by Iraq. Major revelations about the Iraqi nuclear weapon effort followed the defection in August 1995 of General Hussein Kamel, Saddam Hussein's son-in-law and the former head of the Ministry of Industry and Military Industrialization (MIMI), which in the 1980s supervised Iraq's weapons of mass destruction programmes. One of the most startling pieces of new information was that Iraq had launched a 'crash programme' after its invasion of Kuwait in August 1990, that was intended to turn its safeguarded HEU into a nuclear weapon.

These revelations about Iraq's programme have resulted from inspection rights granted by the United Nations Security Council to the UN Special Commission (UNSCOM) and the International Atomic Energy Agency Action Team. Under this mandate, inspectors have systematically uncovered large portions of Iraq's nuclear weapon programme, including an enrichment programme that was expected to produce large amounts of HEU by the mid-1990s.[1] This knowledge has been gained despite efforts by Iraqi officials and

---

[1] Albright, D. and Hibbs, M., 'Iraq's nuclear hide and seek', *Bulletin of the Atomic Scientists*, vol. 47, no. 7 (Sep. 1991), pp. 14–23; Albright, D. and Hibbs, M., 'Iraq's bomb: blueprints and artifacts', *Bulletin of the Atomic Scientists*, vol. 48, no. 1 (Jan./Feb. 1992), pp. 30–40; Albright, D. and Hibbs, M., 'It's all over at Al Atheer', *Bulletin of the Atomic Scientists*, vol. 48, no. 5 (June 1992), pp. 8–10; Albright, D. and Hibbs, M., 'Iraq's quest for the holy grail: What can we learn?', *Arms Control Today*, vol. 22, no. 6 (July/Aug. 1992); Albright, D. and Hibbs, M., 'Supplier-spotting', *Bulletin of the Atomic Scientists*, vol. 49, no. 1 (Jan./Feb. 1993), pp. 8–9; Albright, D., 'Engineer for hire', *Bulletin of the Atomic Scientists*, vol. 49, no. 10 (Dec. 1993), pp. 29–36; and Albright, D. and Kelley, R., 'Has Iraq come clean at last?' *Bulletin of the Atomic Scientists*, vol. 51, no. 6 (Nov./Dec. 1995).

scientists to stall, obfuscate, cover up and lie about the extent of their programme. Consequently, few believe that Iraq has disclosed everything, even though, since Kamel's defection, Iraq has provided vast amounts of new information about its nuclear programme and has appeared to cooperate with nuclear inspectors. Nevertheless, a large body of detailed information now exists about the Iraqi nuclear weapon programme.

The IAEA's major findings are presented in a series of reports available from the United Nations.[2] This assessment is based on these reports, as well as on Iraqi documents and declarations, and interviews with Action Team inspectors. It is current as of August 1996. This chapter first briefly discusses Iraq's former nuclear weapon programme and the risk that it will be revived, along with the

[2] Consolidated report on the first two IAEA on-site inspections under Security Council Resolution 687 (1991) of Iraqi nuclear capabilites, UN Document S/22788, 12 July 1991; Report on the third IAEA on-site inspection in Iraq under Security Council Resolution 687 (1991), 7–18 July 1991, UN Document S/22837, 25 July 1991; Report on the fourth IAEA on-site inspection in Iraq under Security Council Resolution 687 (1991), 27 July–10 August 1991, UN Document S/22986 and Corr. 1, 28 Aug. 1991; Report on the fifth IAEA on-site inspection in Iraq under Security Council Resolution 687 (1991), 14–20 September 1991, UN Document S/23112, 4 Oct. 1991; Report on the sixth IAEA on-site inspection in Iraq under Security Council Resolution 687 (1991), 22–30 September 1991, UN Document S/23122, 8 Oct. 1991; Report on the seventh IAEA on-site inspection in Iraq under Security Council Resolution 687 (1991), 11–22 October 1991, UN Document S/23215, 14 Nov. 1991; Report on the eighth IAEA on-site inspection in Iraq under Security Council Resolution 687 (1991), 11–18 November 1991, UN Document S/23283, 12 Dec. 1991; Report on the ninth IAEA on-site inspection in Iraq under Security Council Resolution 687 (1991), 11–14 January 1992, UN Document S/23505, 30 Jan. 1992; Report on the tenth IAEA on-site inspection in Iraq under Security Council Resolution 687 (1991), 5–13 February 1992, UN Document S/23644, 26 Feb. 1992; Report on the eleventh IAEA on-site inspection in Iraq under Security Council Resolution 687 (1991), 7–15 April 1992, UN Document S/23947, 22 May 1992; Report on the twelfth IAEA on-site inspection in Iraq under Security Council Resolution 687 (1991), 26 May–4 June 1992, UN Document S/24223, 2 July 1992; Report on the thirteenth IAEA on-site inspection in Iraq under Security Council Resolution 687 (1991), 13–21 July 1992, UN Document S/24450, 16 Aug. 1992; Report on the fourteenth IAEA on-site inspection in Iraq under Security Council Resolution 687 (1991), 31 August–7 September 1992, UN Document S/24593, 28 Sep. 1992; Report on the fifteenth IAEA on-site inspection in Iraq under Security Council Resolution 687 (1991), 8–18 November 1992, UN Document S/24981, 17 Dec. 1992; Report on the sixteenth IAEA on-site inspection in Iraq under Security Council Resolution 687 (1991), 5–8 December 1992, UN Document S/25013, 24 Dec. 1992; Report on the seventeenth IAEA on-site inspection in Iraq under Security Council Resolution 687 (1991), 25–31 January 1993, UN Document S/25411, 13 Mar. 1993; Report on the eighteenth IAEA on-site inspection in Iraq under Security Council Resolution 687 (1991), 3–11 March 1993, UN Document S/25666, 26 Apr. 1993; Report on the nineteenth IAEA on-site inspection in Iraq under Security Council Resolution 687 (1991), 30 April–7 May 1993, UN Document S/25982, 21 June 1993; Consolidated report on the twentieth (25–30 June 1993) and twenty-first (24–27 July 1993) IAEA on-site inspections in Iraq under Security Council Resolution 687 (1991), UN Document S/26333, 20 Aug. 1993; Report on the twenty-second IAEA on-site inspection in Iraq under Security Council Resolution 687 (1991), 1–15 November 1993, UN Document S/1994/31, 14 Jan. 1994; Report on the twenty-third IAEA on-site inspection in Iraq under Security Council Resolution 687 (1991), 4–11 February 1994, UN Document S/1994/355, 25 Mar. 1994; Report on the twenty-fourth IAEA on-site inspection in Iraq under Security Council Resolution 687 (1991), 11–21 April 1994, UN Document S/1994/650, 1 June 1994; Report on the twenty-fifth IAEA on-site inspection in Iraq under Security Council Resolution 687 (1991), 21 June–1 July 1994, UN Document S/1994/1001, 26 Aug. 1994; Report on the twenty-sixth IAEA on-site inspection in Iraq under Security Council Resolution 687 (1991), 22 August–7 September 1994, UN Document S/1994/1206, 28 Oct. 1994; Report on the twenty-seventh IAEA on-site inspection in Iraq under Security Council Resolution 687 (1991), 14–21 October 1994, UN Document S/1994/1443, 29 Dec. 1994; Report on the twenty-eighth IAEA on-site inspection in Iraq under Security Council Resolution 687 (1991), 9–20 September 1995, UN Document S/1995/1003, 1 Dec. 1995; Report on the twenty-ninth IAEA on-site inspection in Iraq under Security Council Resolution 687 (1991), 17–24 October 1995, UN Document S/1996/14, 10 Jan. 1996; and Report on the thirtieth IAEA on-site inspection in Iraq under Security Council Resolution 687 (1991), report pending.

inspection effort aimed at deterring such a revival. Iraq's fissile material production programme up to the 1991 Persian Gulf War is then summarized. The remainder of the chapter focuses on the specific uranium enrichment methods and plutonium production strategies that Iraq pursued, including the crash programme.

II. The Iraqi nuclear weapon programme

Iraq has stated that it formally decided to build nuclear weapons in 1988, although many related activities had taken place earlier. Under the 1988 plan, Iraq intended to have its first weapon by the summer of 1991. Iraq had worked on developing the capability to make fissile material for many years prior to this date, and has explained that the decision at that particular time reflected the expectation that indigenously produced HEU would become available within a few years.

Iraq intended its nuclear weapons to be put on ballistic missiles, but the conceptual nuclear weapon design of mid-1988 was too heavy for delivery by Iraqi missiles. Subsequently, the group in charge of the nuclear weapon programme was ordered to reduce the weight of the design to about 1 tonne or less.

Questions remain about the status of Iraq's nuclear weapon programme at the time of the Allied bombing campaign in 1991, when most activities were halted. Nevertheless, the Action Team inspectors have concluded that, with the accelerated effort under the crash programme (described in section VII), Iraq could have finished a nuclear explosive design by the end of 1991 if certain technical problems were overcome. However, it would have needed significantly longer to test a design for the Al Hussein missile which, for example, would have required a warhead with a diameter of 70–80 cm, much smaller than the approximately 100–120-cm diameter of the design nearing completion.

Iraq was also developing a larger missile, based on the Al Abid satellite launcher, able to carry a 1 tonne warhead with a diameter of 125 cm a distance of 1200 km. According to the IAEA Action Team, however, Iraq said that this missile would not have been completed until 1993.

Iraq was also planning to build a nuclear testing site. At the time of the Allied bombing campaign, Iraq had picked candidate sites in the south west of the country, but it had not performed a site investigation. In addition, a senior Iraqi official told inspectors that Iraq did not plan to conduct a test before it had accumulated a few nuclear weapons. Iraq has stated that it planned to develop confidence in its weapon designs through an extensive experimental testing programme that stopped short of a full-scale nuclear test.

The future risk

Despite international sanctions, the Persian Gulf War and the ensuing Action Team inspections, Iraq retains many capabilities and possesses considerable nuclear know-how. This knowledge would allow it to restart its nuclear weapon programme at fairly short notice. The US CIA testified to Congress in 1993 that it believes that Iraq 'probably still has more than 7000 nuclear scientists and technicians and may harbour weapons-related equipment and technology'.[3] Although the number of key individuals would probably number only a few hundred, this cadre retains a formidable amount of information about creating a nuclear weapon programme. If both sanctions and international inspections end, the CIA estimates that Iraq could produce enough fissile material for an atomic bomb in five to seven years.[4] Access to fissile material obtained illicitly overseas could shorten this time-frame considerably.

Immediately following the defection of General Kamel, Iraq handed over more than 500 000 pages of previously hidden documents, 17 tonnes of high-strength 'maraging steel' and a stock of carbon fibre sufficient for more than 1000 gas centrifuges to the IAEA. Iraq subsequently stated that many of its nuclear weapon-related teams of experts had remained together, working in what one inspector referred to as 'unreal career fields.' Although Iraq says that these teams are working on non-proscribed activities, suspicion remains that the teams are being kept together to facilitate restarting the weapon programme. As of August 1996, the investigation of Iraq's past activities continues, with Iraq continuing to release new information.

The extent of Iraq's work on nuclear weapons after the Persian Gulf War remains unknown. Some information implies that Iraq was reconstituting its weapon programme in the period after the war but before UN Security Council inspections were fully functioning, in about June 1991. In addition, Iraq's well-documented efforts to hide its past activities from inspectors, its apparent decision to keep its technical teams together, and its long refusal to provide complete declarations of past activities have created intense, on-going suspicion of its weapon-related activities since the summer of 1991, and of its intentions once sanctions are removed.

Because of continuing concerns about Iraq's ambitions, the UN Security Council continues to maintain economic sanctions on Iraq. Faced with a worsening economic situation for the Iraqi people, the UN Security Council decided in May 1996 that a part of the sanctions would be relaxed under strict Security Council oversight. Iraq can sell up to $2 billion of oil every six months and use a significant portion of the proceeds to import food, medicine, and other humanitarian supplies under strict UN supervision. The rest of the funds will pay for UN inspections and war reparations. The Security Council will carefully control all funds gained from the sale of Iraqi oil to ensure that Iraq does not misuse any part of it to import any items in violation of the on-going economic embargo.

To ensure greater transparency of Iraqi industrial activities, the UN Security Council decided several years ago that the IAEA Action Team will conduct long-term monitoring of Iraq to assure that it complies with its obligations not

---

[3] *Proliferation Threats of the 1990s*, Hearing before the Committee on Governmental Affairs, US Senate, 24 Feb. 1993 (US Government Printing Office: Washington, DC, 1993).

[4] *Proliferation Threats of the 1990s* (note 3).

to acquire or develop nuclear weapons or nuclear weapon-usable materials. The IAEA's ongoing monitoring and verification (OMV) programme is based on extensive and detailed Iraqi declarations of its nuclear and nuclear-related activities, the monitoring of declared activities and the ability to search for covert activities.[5] Once sanctions are removed, the IAEA will also monitor Iraq's imports of sensitive items and match these declarations with those provided by the nations exporting sensitive items to Iraq.

The goal of the OMV programme is to detect an Iraqi attempt to acquire nuclear materials or other essential nuclear weapon components before a nuclear weapon can be developed. While not designed to discover prohibited theoretical studies, small-scale laboratory research or prototype machines, such as single centrifuges, it represents the most stringent international nuclear inspection programme ever instituted and a powerful deterrent to Iraq.

It is impossible to know how Iraq would acquire fissile materials for nuclear weapons if it decided to do so. Scenarios vary from reconstituting one of its enrichment efforts to acquiring fissile material in the former USSR. Because Iraq's nuclear weapon programme before the Persian Gulf War was so vast, it could probably produce fissile material and build a nuclear weapon considerably faster than many other developing states. As a result, the international community will need to scrutinize carefully Iraq's activities, perhaps for decades. The key to this scrutiny is the OMV programme, which will need to maintain a capability to detect a wide-range of fissile material production or nuclear weapon-related activities and to monitor the activities of former members of the Iraqi nuclear weapon programme.

III. Summary of Iraqi fissile material production plans before 1991

Initial efforts to obtain plutonium

Iraq's efforts to obtain plutonium are believed to date from the 1970s. At this time Iraq concentrated on acquiring nuclear facilities overseas that would have been safeguarded, since Iraq had signed the NPT in 1968. Nevertheless, the current understanding is that these facilities could have been used to produce unsafeguarded plutonium.

In 1976, Iraq succeeded in buying from France a 40-MWth materials test reactor called the Tammuz-1, or Osirak, reactor. However, in June 1981, just prior to the reactor's initial operation, Israel bombed the reactor because it was convinced that it would be used to produce plutonium for nuclear weapons. Although Iraq continues to deny it intended to make unsafeguarded plutonium in this reactor, it has declared that in the second half of 1979 Iraqi scientists estimated maximum production in this reactor at 2 kg of plutonium per year.

---

[5] Stokes, P., 'IAEA on-going monitoring and verification in Iraq', Sep. 1994, IAEA Action Team, Vienna; and United Nations, *The United Nations and the Iraq–Kuwait Conflict, 1990–1996* (United Nations: New York, 1996).

Unofficial French estimates are reported to be about double this quantity, and Israeli estimates are about four times as high.

Following the bombing of the Tammuz-1 reactor, Iraq decided to pursue the following two paths: (*a*) to replace the Tammuz-1 reactor or to develop a heavy water or enriched uranium reactor and associated plutonium separation capability; and (*b*) to develop uranium enrichment production capacity. It also decided that any unsafeguarded efforts would be pursued in utmost secrecy.

The post-1981 plutonium path

At first, Iraq concentrated on trying to replace the Tammuz-1 reactor, but by 1985, it realized that it could not buy a replacement. A secret project to build a 40-MWth heavy water, natural uranium reactor was therefore launched. Iraq pursued this project until the late 1980s when it de-emphasized plutonium in favour of uranium enrichment, which at that time appeared more promising. However, Iraq continued its efforts to learn how to separate plutonium from irradiated fuel and to make heavy water, two indicators that it held fast to at least its ambitions for a nuclear reactor.

The HEU path

Even before the Israeli bombing of the Osirak reactor, Iraqi scientists had been evaluating the development of uranium enrichment technologies. However, Iraq has declared that a decision by the Iraqi leadership to pursue these options came after the June 1981 bombing. Dr Jaffar D. Jaffar, the father of the enrichment programme, told inspectors that the attack on the reactor was a good catalyst for the enrichment programme. In fact, he stated that without the attack the programme could not have been implemented.

An Iraqi evaluation finished in the second half of 1981 concluded that electromagnetic isotope separation (EMIS) was the most appropriate technology for Iraq and that gaseous diffusion was the next most appropriate option. Gas-centrifuge technology was viewed as too difficult to accomplish at that time. (See chapter 2 for a brief description of these enrichment technologies.)

Iraq has stated that in late 1981 it formally decided to establish an enrichment technology to produce HEU. EMIS was adopted as the primary option, with the goal of building production units each able to achieve 15 kg per year of 93 per cent enriched uranium. Gaseous diffusion was selected as the second option, with a goal of producing LEU that could be used as a feedstock for EMIS, dramatically increasing overall HEU production.

EMIS

By the end of 1987, Iraq had contracted with a Yugoslavian firm to build its first EMIS production facility at Al Tarmiya, north of Baghdad. This contract occurred just before Iraq formally decided to start work on building a nuclear weapon. Goal quantities of weapon-grade uranium for this site were set at 15 kg per year using natural uranium feed.

Iraq recognized the importance of LEU feed in raising the output of the plant. Iraqi estimates of the output using LEU feed (2.5 per cent) vary between roughly 40 and 50 kg of weapon-grade uranium per year. The variation reflects different plant designs and performance uncertainties. Towards this end, Iraq was seeking unsafeguarded LEU on the international market during the late 1980s. However, Iraqi officials told inspectors that the search was unsuccessful.

Also in late 1987, Iraq decided to build a replica of Al Tarmiya at Ash Sharqat, about 200 km north-west of Baghdad. This facility, which was built by Iraqis only, was originally viewed as a second production site that would come into operation roughly at the same time as Al Tarmiya. In the late 1980s, this plan was modified to one in which Ash Sharqat would come into operation after Al Tarmiya was fully operational.

Al Tarmiya faced repeated delays and technical problems, and by the time of Iraq's invasion of Kuwait, it was at least a year behind schedule. At that time, Al Tarmiya was not expected to produce its first goal quantity of weapon-grade uranium, or 15 kg, until at least 1992, assuming that the plant would function well and that a stock of low-enriched uranium would be used (see below). If natural uranium was used, the date for the production of the first goal quantity would have been 1993 or later. Civil construction at Ash Sharqat was to be completed in late 1990, after which the plant would have been kept on standby awaiting the installation of separators in the mid-1990s.

In total, both plants could have produced 25–100 kg of weapon-grade uranium if they had operated successfully. However, problems in the separators made such a possibility unlikely for many years.

Gaseous diffusion and chemical enrichment

In 1988, it became apparent to the Iraqi leadership that the gaseous-diffusion programme was not progressing well, and Iraq decided to de-emphasize this effort and instead concentrate on chemical enrichment as a source of LEU feedstock for the EMIS programme. By 1990, Iraq hoped to build a chemical enrichment plant to produce about 5 tonnes per year of 4 to 8 per cent enriched uranium. A time schedule for the major milestones in this approach is unavailable.

Gas-centrifuge enrichment

In 1987, a split developed in the enrichment programme, and the research group in charge of developing gaseous-diffusion technologies was transferred from the Al Tuwaitha Nuclear Research Centre to a new site on the northern edge of Baghdad near Rashdiya, later named the Engineering Design Centre (EDC). This change reflected both personality conflicts between Jaffar and the

Figure 11.1. Map of Iraq showing the approximate locations of the main inspection sites.

Source: Based on Kokoski, R., SIPRI, *Technology and the Proliferation of Nuclear Weapons* (Oxford University Press: Oxford, 1995), figure 20, p. 106.

leader of the centrifuge effort and the centrifuge group's belief that the EMIS programme was progressing too slowly and would fail to overcome certain technical problems.

When the diffusion group was transferred, it was also assigned the responsibility for developing gas centrifuges. It later concentrated on centrifuges when diffusion was de-emphasized.

This group managed to acquire extensive overseas cooperation in designing and building gas centrifuges, so much so that inspectors have characterized the assistance as the key to progress in the centrifuge programme. Despite such help, at the time of the Persian Gulf War, Iraq was still a few years from an operating plant able to produce goal quantities of weapon-grade uranium, declared by the centrifuge programme as 1000 centrifuges producing 10 kg of weapon-grade uranium per year.

The crash programme

By the time Iraq invaded Kuwait, Iraq still lacked an indigenous source of fissile material; its enrichment plants were still far away from producing HEU. In mid-August 1990, the Iraqi leadership ordered the diversion of its stock of safeguarded HEU fuel. Iraq's initial plan was to extract the HEU from the fuel, further enrich a portion of it and build a nuclear weapon. The goal was to execute this plan within six months, although by the time of the Allied bombing campaign in mid-January 1991, which stopped the effort, Iraq had fallen several months behind and was unlikely to finish a nuclear explosive device until at least the end of the year. A nuclear warhead for a ballistic missile would have taken significantly longer.

Whether Iraq planned to obtain HEU for additional devices remains, as of August 1996, a matter for speculation. However, Iraq is suspected to have developed such plans.

Iraq is believed to have continued certain aspects of the crash programme after the Gulf War. Information as of the summer of 1996 is still sketchy. Iraq denies such efforts, but is now believed to have tried to reconstitute a portion of its nuclear weapon programme until it was stopped by early Action Team inspections.

IV. The Iraqi enrichment programme

Iraq's enrichment programme was both vast and clandestine. It included at least five enrichment approaches which were pursued at a cost of more than $1 billion.

As mentioned above, the first efforts concentrated on gaseous diffusion and EMIS. In the early 1980s, Iraq also started a rudimentary laser enrichment programme. Later, both gas centrifuges and chemical enrichment were added to the development effort. At the time of the Persian Gulf War, Iraq's two most successful programmes involved EMIS and gas centrifuges. Laser enrichment and gaseous diffusion had been de-emphasized, but chemical enrichment was approaching the pilot stage.

The EMIS programme

Iraq's most developed enrichment technology was electromagnetic isotope separation. The EMIS machines developed by the United States during World War II were called calutrons and produced the HEU that destroyed Hiroshima (see chapter 4). After the war, the United States declassified much of the EMIS technology, much of which was used by Iraq. Some aspects of Iraq's separators, particularly the magnet design, were more advanced than the Manhattan Project design. Therefore, to preserve this distinction the Iraqi machines are not called calutrons. Iraq refers to them as 'Baghdadtrons.'

Iraq's 1981 evaluation of uranium enrichment technologies saw many advantages in EMIS and few disadvantages. The advantages included that (a) EMIS was well documented in the open literature; (b) the basic scientific and technical problems associated with the operation of EMIS separators were straightforward to master; (c) the computational software and main equipment were not on international export control lists, making procurement easy; (d) the design and manufacture of the main equipment for prototypes could be accomplished indigenously; (e) the feed material would be relatively easy to produce and handle; (f) final enrichment would be accomplished in two stages in machines that act independently of each other, so one or more separator outages do not affect the operation of other separators; and (g) LEU feed could be used for a substantial increase in productivity.

A special advantage was that EMIS involves large and static equipment rather than high-speed moving parts. This type of technology was better suited to the relatively inexperienced group of Iraqi engineers and scientists who had to design and manufacture the components in the early 1980s.

Iraq also determined several disadvantages in choosing EMIS. An extensive research and development phase was necessary before significant production could start. The process also has high costs, is labour intensive and uses uranium inefficiently.

In an attempt to lessen the process's dependence on armies of people to produce enriched uranium, Iraq sought to increase automation in its production-scale facility using on-line computers. Nevertheless, Iraq would have needed to overcome considerable challenges before it could have produced significant quantities of weapon-grade uranium using this method.

Development of the EMIS programme

Iraq carried out a large research and development effort on all aspects of EMIS at both Al Tuwaitha and Al Tarmiya and was implementing its EMIS programme in three overlapping phases.

Phase one. The first phase, which lasted from 1982 to 1987, involved basic R&D of all aspects of EMIS. During this phase, Iraq built its first separator, the R40 separator, in which ion beams travelled a circular path with a radius of 400 mm and had a current of roughly 1 milliampere. The first separation of uranium occurred in January 1986. Table 11.1 describes the major separators.

Phase two. Phase two, which started in 1983, reached an experimental stage in 1987 and continued until the Allied bombing campaign in January 1991. The original design idea, later abandoned for two stages of separators (see below), was to construct a 'unit cell' of two single-ion source separators, the first with a 1000-mm radius (R100) and the second with a radius of 500 mm (R50), that could produce HEU from natural uranium feed. Another R100 separator, with two ion sources, was also constructed to test operations with multiple sources. During this phase, other types of experimental separator were developed in order to test new magnet designs and multiple ion-sources for the production-scale separators slated for phase three.

A pilot plant to treat uranium solutions from liners and collectors was completed at Al Tuwaitha in the late 1980s. Al Tuwaitha also initially produced the uranium tetrachloride feedstock. Later, Iraq established an industrial-scale plant at Al Jesira, near Mosul, in 1990 to make and recycle feed material from Al Tarmiya separators.

The R100-1, R100-2, R100-3 and R50 separators were put into operation in 1987 at Al Tuwaitha in building 80 and began enriching uranium in the spring of 1988. Separators R100-1 and R100-2 had one source each and R100-3 had two sources. These sources achieved average collected beam currents of about 85–120 milliamps. Although designed to reach enrichments of about 12 per cent, they typically produced less than 8 per cent enriched material. Separator R50 had one source and was designed to reach enrichments of 18 per cent using natural uranium feed, or about 80 per cent enriched uranium using feed of 12 per cent enriched uranium from the R100 separators.

Based on Action Team analysis of monthly Iraqi progress reports for these separators, which were seized by inspectors in September 1991 in Baghdad, these machines operated through 1990, but never obtained their design capacities. Iraqi declarations in 1991 that the R100 separators achieved an availability of 35 per cent were misinterpreted initially by the inspectors as the fraction of time these machines operated at design values. However, 'availability', as used by Iraq, referred to the amount of time a separator spent collecting enriched and depleted uranium during a 'run', not the fraction of a year nominal output was achieved. Each run was divided into four phases under normal conditions: preparation of operations, elementary operations, continuous operations and halting of operations. Typically, the third phase started when the collector pockets were opened and enriched material collected. The availability refers to the fraction of time in a run that a separator spends in phase three, regardless of the amount of enriched uranium produced.

Table 11.1. Selected EMIS separator design specifications

System	Number of ion sources	Design current at collector per source (mA)	Design enrichment (%)	Separator availability[a] (%)
Al Tuwaitha				
R40	1	~1	Unknown	Unknown
R100	1–2	80–120	*12*	*30–40* (actual)
R50	1	25–40	*80* (12% feed)	*< 30* (actual)
Al Tarmiya				
R120	4	150 (not achieved)	*18*	*55* (goal)
R60	2	50–100 (not achieved)	*93* (18% feed)	*45* (goal)

[a] In this case, availability refers only to the fraction of a run during which a separator was enriching uranium. Most of the time was spent in start-up procedures. The time between runs, which could be substantial, is ignored.

Source: Text and IAEA Action Team.

For example, according to the progress report for November 1990, which was one of the best months for the Al Tuwaitha machines, the three R100 separators operated throughout the month, accomplishing a total of 26 runs, of which 14 operated in phase three. The other runs encountered problems, such as short circuits in the ion source, that curtailed operation before phase three was reached. The other runs also encountered numerous breakdowns and shutdowns, but operation could be resumed without starting over. The average availability, or the fraction of time each spent in phase three, was about 35 per cent. But when the separators reached phase three, they were not operating at design values. In November, the three separators produced only about 44 grams of material with an average enrichment of 6.5 per cent. This amount corresponds to maintaining design current levels an average of only 19 per cent of the month, ignoring the additional fact that design enrichments were not achieved.

A similar phenomenon occurred throughout 1990, during the third year of operations. Action Team analysis of the monthly progress reports for these separators indicates that during 12 months of operation the three R100 separators produced about 360 grams of enriched uranium with an average enrichment of about 7 per cent. If each separator operates at full output an average of 35 per cent of the time, these three separators should produce about 75 grams of 7 per cent enriched uranium per month, or about 900 grams per year. Since the separators produced only about 40 per cent of this amount, they were effectively operating at design currents an average of only 15 per cent of the time during 1990.

As noted above, these separators were beginning to work somewhat better by late 1990, operating at design currents an average of 20 per cent of the month. The reason was that improvements continued to be made in basic aspects of machine operation. For example, during November 1990, following an experiment to measure the characteristics of the ion beam, the operators discovered a better position for the collectors in the R100-3 separator that dramatically improved its separation of the uranium. However, this progress report does not suggest that the separators were on the verge of achieving design currents significantly longer than the 1990 average of 15 per cent. All the separators continued to require extensive maintenance and to encounter technical malfunctions in the ion sources and other components.

In total, Iraq said that the Al Tuwaitha machines produced about 640 grams of enriched uranium with an average enrichment of 7.2 per cent from the spring of 1988 through 1990. Of this quantity, 7.3 grams comprised HEU enriched to over 20 per cent, with an average enrichment of 24 per cent. The highest enrichment level reached was slightly over 40 per cent, although only about 0.06 grams of this material was declared to have been produced in this way. All uranium enriched above 20 per cent was produced in 1989 in the R50 separator from natural uranium feed by drastically increasing the machine's separation factor. The penalty was that only minute quantities of material could be produced.

Phase three: Al Tarmiya and Ash Sharqat. The last phase of the EMIS programme aimed at reaching industrial-scale production. According to an Iraqi document seized by inspectors in 1991, Iraq signed a contract with an unidentified foreign (other sources say Yugoslavian) firm in April 1987 to build a few tens of buildings at Al Tarmiya for a cost of about $110 million. Another $55 million were set aside to build the chemical recovery buildings and the electrical power supply station. At the time the contract was sign, the construction project without the chemical recovery buildings was expected to take 37 months, or until May 1990. Finishing the recovery buildings would add about four months to the project. This part of the project was delayed, and the recovery buildings were unfinished as of January 1991.

In total, the site covered about 800 000 square metres. The EMIS separators received electrical power from a sub-station about 0.5 km from Al Tarmiya. The power lines were buried to conceal the amount of electricity consumed by Al Tarmiya.

Al Tarmiya was to house 90 separators, 70 with an ion radius of 1200 mm (R120), and 20 with a radius of 600 mm (R60). The R120 separators had four ion sources with total design currents of 600 milliamps and pole pieces with diameters of 4.5 metres. The R60 machines had two ion sources with total design currents of 100–200 milliamps.

Installation of the separators at Al Tarmiya was to occur in phases. Table 11.2 shows an Iraqi plan dated late 1988 or early 1989 for the installation of the R120 separators. Iraq has stated that it did not prepare a revision of this schedule, despite falling behind.

The first eight R120 separators were installed between February 1990 and September 1990. In July 1990, work started on installing the second line of 17 separators, but by January 1991, none had been installed (see table 11.3).

Table 11.2. An R120 separator deployment schedule for 70 machines, declared by Iraq (but not achieved)

Phase	No. of separators to be installed	Starting date	Completion date
First	7	1 Nov. 1989	15 Feb. 1990
Second	17	15 Mar. 1990	1 Dec. 1990
Third	10	1 Jan. 1991	1 June 1991
Fourth	18	1 July 1991	1 Apr. 1992
Fifth	18	1 May 1992	1 Feb. 1993

Source: IAEA Action Team.

The R60 separators were to be installed in parallel in four phases of five separators each. The installation of the first five started in November 1990, but none were installed by January 1991. Iraq has stated that not all of the major components had been manufactured by this date.

The separators at Al Tarmiya did not perform well. Monthly progress reports for Al Tarmiya show that the R120 machines were experiencing serious operating problems.

Although the Iraqis told the Action Team that these machines should have an availability of 55 per cent, they were not even close to this level at the time of the Persian Gulf War. The major priority of the EMIS programme was in fact improving the separators' availability.

According to Iraqi declarations in 1991, peak production in the eight R120 separators occurred in November 1990 when they produced a total of about 150 grams of enriched uranium with an average enrichment of about 4 per cent, far below design levels for this separator. Operating at design values for an average of 55 per cent of time, these eight separators (each with four sources) should produce about 115 grams of uranium-235 in about 640 grams of 18 per cent enriched uranium (or equivalently, 2900 grams of 4 per cent enriched uranium). Thus, in November, the separators produced at about 5 per cent of design value at an enrichment level of 4 per cent.[6] During other months, the performance was worse.

In total, Iraq declared in 1991 that Al Tarmiya produced about 685 grams of enriched uranium with an average enrichment level of 4 per cent. All of this material was declared as less than 10 per cent enriched. The chemical processing was done both at Al Tarmiya and Al Tuwaitha, because the new chemical facilities at Al Tarmiya were not finished at the time.

---

[6] In 1996, the Action Team said that the Iraqis provided a summary of enriched uranium output at Al Tarmiya that shows that less enriched material was produced than declared by Iraq in 1991. The new information shows that October was the most productive month, when about 120 g were produced. The Action Team has asked Iraq to resolve the two inventory declarations, although the conclusion in the text above remains unchanged and in fact would become lower if the new data were substituted.

Table 11.3. Actual R120 separator deployment schedule at Al Tarmiya

Phase	No. of separators to be installed	Starting date	Completion date
First	8	Feb. 1990	Sep. 1990
Second	17	July 1990	None installed by Jan. 1991
Third	Not started		
Fourth	Not started		
Fifth	Not started		

Source: IAEA Action Team.

Jaffar expressed pride in his programme to the inspectors, telling them in 1991 that Iraq's system was, or would soon be, superior to the US calutron.[7] Whether Iraq's EMIS programme would have succeeded remains unclear to this day. In any case, the Al Tarmiya facility was functioning as an advanced research and development facility for separators when it was bombed by the Coalition forces in January 1991.

Phase three: Ash Sharqat. Iraq was building a twin facility at Ash Sharqat, 200 km north-west of Baghdad, and construction was about 80–90 per cent complete when the site was bombed. Iraq had expected to finish civil construction and 'check-out' the buildings by the end of 1990. It declared after the war that no separators had been installed and no uranium had been used in the chemical buildings.

Iraq's original plan (dated to the end of 1988 or early 1989) called for installation of the first half of the separators at Al Tarmiya, followed by the installation of the first half of those at Ash Sharqat. Then the second half would be installed at Al Tarmiya before completing installation at Ash Sharqat. In the beginning of 1990, the plan was changed to full installation of separators at Tarmiya and then full installation at Ash Sharqat. In the meantime, Ash Sharqat was to be kept running at a 'low pace.'

In explaining the change in their plans, the Iraqis stated that the logistics involved in the deployment at two separate sites would have been too difficult. Housing was in short supply at Ash Sharqat, which is an isolated site far from population centres. Training personnel for two sites simultaneously was also beyond Iraq's means at that time; in fact, Al Tarmiya was already encountering a shortage of trained personnel.

The motivation for bringing two sites into near simultaneous operation is simple to understand. If one were to be identified and destroyed, the other might survive undetected. Adding to the chance that Ash Sharqat would remain undetected was the fact that it was built exclusively by Iraqis, unlike

[7] IAEA Action Team, The Iraqi Electro-Magnetic Isotope Separation Program: Results of the Third Inspection Activities (7–18 July 1991), IAEA, Vienna, undated, p. 1.

Al Tarmiya. To make indigenous construction at Ash Sharqat easier, the Iraqi contractor used Al Tarmiya's plan with some simple adjustments in the buildings and layout.

Phase three: 1987 document. Iraq's strategy for the production, installation, operation and material flow for Al Tarmiya was laid out in a September 1987 document, entitled 'New procedures for setting up and operating the third phase of a separation system.'[8] This document evaluated various options for producing the first goal quantity of enriched uranium, set at 15 kg of weapon-grade uranium. The report contains a plan for deploying and bringing into operation all 90 separators at Al Tarmiya over a roughly 40-month period starting in August 1989 and ending about January 1993. (For the R120 separators, the schedule is similar to that in table 11.2, although the number in each phase is somewhat different.)[9] By the end of this period, the separators would have produced the first goal quantity of weapon-grade uranium, using natural uranium feed. After this date, the study estimated production at about 12.7 kg of weapon-grade uranium per year.

The study estimated that if a stock of 1.7 tonnes of 2.5 per cent enriched uranium feed was used in the R120 separators and then further enriched in the R60 machines, Al Tarmiya could produce the first goal quantity after about 24 months, or by about August 1991. This date corresponds roughly to when Iraq hoped to finish its first nuclear weapon under its 1988 plans. The study did not provide the potential source of this LEU. Its author has denied knowing where the LEU would have originated; he said that the size and enrichment level of the LEU stock was given to him to use in this study. He said in particular that he did not know that Iraq was in possession of a stock of safeguarded LEU obtained from Italy (1.77 t of 2.6 per cent enriched material), which is very close in both quantity and enrichment to the stock mentioned in his study. In any case, it appears that in 1987 Iraq intended to acquire a stock of LEU, probably unsafeguarded, to speed up the production of HEU at Al Tarmiya.

In this case, the study planned to first feed the R120 separators with this LEU material. Once this material was exhausted, natural uranium would be used as feedstock.

The author of the 1987 report assumed that annual production of HEU using LEU feed would reach about 27 kg of 93 per cent enriched uranium per year once all 90 separators were installed. This annual value could have been raised to about 40–50 kg per year if the R60 currents were successfully raised to 100 amps.

The schedules in this report were not met in practice (see below), and Iraq did not divert its stock of safeguarded LEU as this report strongly implies it

---

[8] This document had been one of those seized by inspectors in Sep. 1991 in Baghdad, but for several years Iraq dismissed the document as the speculations of an engineer. Only in Feb. 1996 did Iraq state that this document laid out the strategy for implementing phase three of the EMIS programme. IAEA Action Team.

[9] The deployment schedule was also in five phases of 6, 16, 16, 12 and 20 separators.

planned to do. The delays at Al Tarmiya and the question of whether it could have achieved nominal outputs are analysed below.

Annual output at Al Tarmiya

As mentioned above, the design capacity of Al Tarmiya was to be about 13 kg of weapon-grade uranium a year, using natural uranium feed. Western experts differ on the level of output that would actually have been achieved. The transition from prototype machine operation to industrial production can be difficult and extrapolations are at best fraught with uncertainty. Al Tarmiya's actual production rate is even more difficult to estimate because Iraq had been unable to produce HEU, operate any of its machines regularly because of frequent breakdowns or achieve design currents when machines did operate.

Soon after the inspections began in 1991, IAEA inspectors estimated Al Tarmiya's output at about 12–15 kg of weapon-grade uranium per year.[10] However, a few inspectors believed that Al Tarmiya could have produced considerably more, perhaps over 20 kg of weapon-grade uranium a year.

All of these estimates were based on initial inspections of equipment and information supplied during initial declarations. These estimates assume consistent separator operation at design currents, successful computer automation of machine operation and highly efficient changing of liners and collectors. Jaffar also led the IAEA inspectors to believe that Iraq had solved many of the problems in starting and operating separators.

The rationale for these estimates is clear. At the design levels for the ion currents and the machine availability stated in table 11.1, the four collectors of a R120 separator would receive about 2000 grams per month of enriched and depleted uranium. This uranium would contain in total about 14 grams of ^{235}U, if natural uranium were used as feed. For seventy R120 machines, the yearly collection would total almost 12 kg of ^{235}U. This ^{235}U would be contained in about 65 kg of 18 per cent enriched material, assuming complete separation into light and heavy fractions of uranium. If we ignore losses in the tails and assume complete recovery of this material and further enrichment of all of this material to weapon-grade in R60 machines, one arrives at about 13 kg of weapon-grade uranium per year.

Existing estimates above 20 kg per year assume that the R120 separators would have achieved availability factors approaching 100 per cent. Inspectors said that Iraqis also told them in the summer of 1991 that they planned to change the sources, collectors and vacuum liners considerably faster than originally planned. Recent information shows that the major effort was to achieve the design values listed in table 11.1, including an availability of only 55 per cent. However, Jaffar stated recently that if these design levels were achieved, he planned to seek improvements in ion sources, higher currents and collection efficiency rather than availability.

---

[10] IAEA (note 7); and interviews by one of authors with inspectors, summer 1991.

An important part of the availability factor is the reliability of the machines themselves, an aspect of separator operation which was causing serious problems. Based on the frequent breakdowns of the separators at Al Tarmiya throughout 1990, Iraq could expect difficulty in achieving design values at Al Tarmiya, let alone exceeding them.

Unless Iraq improved the performance of its separators, Al Tarmiya might have produced as little as a few kilograms of weapon-grade uranium per year. Achieving design levels would have required Iraq to overcome the problems plaguing its ion sources, collectors and other components and to assemble sufficient trained personnel. Iraq is believed to have been capable of eventually overcoming these problems, but doing so may have caused serious delays in achieving full operation at Al Tarmiya.

Minimum time to first goal quantity

The problems in the EMIS programme would have most directly affected the time that Iraq needed to acquire its first goal quantity of HEU. One way to determine the minimum time to produce a goal quantity is to assume that design values could have been achieved and to base the estimate on the expected deployment dates of the separators in table 11.2, after accounting for known delays.

By the time of the Allied bombing campaign, the deployment plans were at least 6–12 months behind schedule. Thus, under this estimate, all 90 separators would have been deployed no sooner than between June/July 1993 and January 1994. Using natural uranium feedstock, this date corresponds to the production of the first goal quantity of HEU. If LEU feed was used, this date would be in the first half of 1992.

The deployment schedule would probably have been further delayed because the separators encountered so many unexpected start-up problems. Iraqis have stated that phase one of the separator installation was actually serving as an R&D stage, and not a production phase as originally planned.

Action Team analyses of Iraqi documents and interviews of Iraqis raise serious questions about whether Iraq could have met the schedule assumed above. In particular, the monthly progress reports from 1989 and 1990 on machine operation at Al Tarmiya and Al Tuwaitha list persistent problems in operating the separators. Throughout 1990 at Al Tarmiya, currents at the collectors remained significantly below design levels, often not exceeding 100 milliamps at a collector. Total current for all four sources in a separator rarely exceeded 400 milliamps, and was often considerably less. The most serious problems included unreliable ion sources and inadequately manufactured components for the vacuum liners and chambers. Until these problems were overcome, enriched uranium production would remain low, no matter how many separators were installed.

Despite years of development, the programme was facing serious problems. In total, the EMIS programme had produced only about 1300 grams of low-enriched uranium and virtually no HEU.

However, EMIS technology is relatively straightforward, and Iraq is judged to have been able to overcome these problems with sufficient time and expense. Thus, there is no reason why Al Tarmiya and later Ash Sharqat could not have achieved near full operation, if Iraq had not invaded Kuwait. But it is highly unlikely that its first goal quantity could have been produced before 1992 in the case of LEU feed or before mid-1993 using natural uranium feed. Because of the problems in the programme, these dates would have probably been significantly later, by at least a year or two.

The gaseous-diffusion programme

Iraq decided in late 1981 to launch an effort to build a gaseous-diffusion plant. It saw several advantages; for example, the technology was commercially proven. However, Iraqi scientists lacked the basic 'know-how.' Furthermore, gaseous diffusion involves uranium hexafluoride, which is difficult to produce and handle; and it requires a large number of items which were internationally controlled and could not be built indigenously. Despite these drawbacks, Iraq decided that gaseous diffusion was to be its second option after EMIS. Its goal was to construct a cascade of several hundred stages to produce 4–5 tonnes per year of 3–4 per cent enriched uranium as feedstock for the EMIS programme.

Gaseous-diffusion work started at Al Tuwaitha in 1982, focusing initially on basic scientific and technical questions. A key part of this initial effort was the investigation of a suitable porous barrier tube, which is the basic enrichment element and considered highly classified by countries that have mastered gaseous diffusion. In 1985, as work on the barrier progressed, Iraq increased its work on other components of the separation stage, such as compressors, heat exchangers and control systems.

The first milestone of the gaseous-diffusion programme was to operate a single working stage model with uranium hexafluoride gas and all auxiliary systems and components. Iraq commissioned one separation stage that used a surrogate material for uranium hexafluoride.

In mid-1987, the team, named Group 1, moved to an industrial site near Rashdiya, north of Baghdad, where it created a large enrichment research and development centre. The creation of this centre was motivated by a perceived need to bypass Jaffar's authority and influence and to report directly to General Kamel. Group 1 leaders have stated to inspectors that by that time the EMIS programme was encountering serious technical problems, but Jaffar was primarily committed to that programme. When this transfer was approved by Kamel, Group 1 also received a mandate to develop gas centrifuges (see below).

The gaseous-diffusion programme was also facing difficulties. Iraq has declared that it cut back its work on gaseous diffusion in 1987. The plan to introduce uranium hexafluoride into the experimental stage was never implemented, and the project involving the single stage was cancelled in 1987. Iraq subsequently focused on the development of small-size cascades utilizing small compressors that could be manufactured domestically or bought overseas in sufficient quantities.

Although Iraq attempted to make many types of barrier tube, it succeeded only with one type made of an anodized aluminum. In 1988 Iraq successfully separated uranium hexafluoride in a laboratory set-up with this type of barrier. However, an anodized aluminum barrier is usually considered too fragile for use in an industrial enrichment plant. Nevertheless, Iraq made several hundred such tubes and conducted several successful experiments that demonstrated the corrosion resistance of the tube to uranium hexafluoride. Although Iraq achieved measurable uranium isotopic separation, the amounts of enriched uranium involved in such experiments would have been very small.

Iraq's work on compressors focused mainly on the foreign procurement of suitable machines for experimental facilities. Iraq has stated that plans to develop the capability to produce compressors indigenously or to secure large numbers of them from overseas suppliers for a cascade were not carried out.

Little progress was made in the gaseous-diffusion effort. Iraq has declared that the scale of the technological and manufacturing requirements for this effort led to its cancellation in 1989. It has reported that it dismantled the last of the equipment in 1991.

The gas-centrifuge programme

Iraq has said that it launched its gas-centrifuge programme in August 1987 with the creation of the Rashdiya centre. The prior experience with gaseous-diffusion technology was useful, particularly work relevant to the corrosion properties of uranium hexafluoride gas, which is also the material used in gas centrifuges. A gas-centrifuge effort had not been envisioned under the 1981 plan. The Iraqi planners viewed gas centrifuges as even more difficult than gaseous diffusion because of the special materials involved and the high rotational speed of the key parts.

Iraq was able to make significant progress on this technology only after it obtained extensive foreign assistance from European gas-centrifuge experts. Iraqis have stated that in the spring of 1988, Iraqi centrifuge experts approached a German firm, H&H Metalform GmbH, for help in making gas centrifuges. This firm was already assisting Iraq's ballistic-missile manufacturing programme and was developing the technology to produce maraging-steel tubes for gas centrifuges by flow-forming. More importantly, H&H was well connected to a group of German centrifuge experts.

H&H quickly became Iraq's most important contact for assistance in obtaining centrifuge expertise and manufacturing equipment from abroad. In essence, it acted as a 'funnel' for a wide variety of important components, materials, 'know-how', design information and manufacturing equipment.

Iraq's declared goal for the gas-centrifuge project was a 1000-machine centrifuge plant able to produce 10 kg of weapon-grade uranium per year. The initial plan was to build first a workable centrifuge, then assemble experimental cascades and finally make a 1000-machine cascade. The timetable for this effort at the time of the Persian Gulf War is not certain. However, Iraq has stated that it planned to commission the 1000-machine cascade in 1994.

Centrifuge development

The Iraqi effort focused on overcoming a weak technological base and creating the technical infrastructure to research, develop and manufacture gas centrifuges. Iraq worked on two types of centrifuge. The first type, the 'Beams', or 'oil' centrifuge, was dropped in 1989 as a result of its inefficiency and of steady progress on the 'Zippe-type' or 'magnetic' centrifuge.

However, Iraq appears to have planned to build both types of centrifuge at the start of its effort and also appears to have realized that the Zippe-type was the more effective and practical one to develop (see chapter 4). Nevertheless, the Beams-type centrifuge was more accessible to Iraq, and it provided valuable experience in building and operating centrifuges. One Iraqi stated that without the experience with the oil centrifuge, the team may not have been able to properly understand the Zippe-type centrifuges, or the highly specialized design assistance provided by German centrifuge experts in 1988, courtesy of H&H Metalform.

Beams or 'oil' centrifuge. The centrifuge programme started working on an unclassified oil centrifuge that was developed by Jesse Beams in the United States in 1930s and 1940s. According to Iraqis, valuable mechanical design information for this centrifuge was available in the open literature. By the end of 1987, the team had built its first machine, called the GS-1 (gaseous separator-1). During the next two years, Iraq made many modifications to this original subcritical design as it struggled to overcome problems of excess vibration, inadequate seals and excessive power consumption.

Most of this work was carried out in building 22 at Rashdiya, which was finished in the first quarter of 1988. Inside this building, the team built two 'pits.' Each pit was about 6.5 metres deep, extending 2.5 metres below ground level and about 4 metres above the floor. Forty-centimetre thick concrete walls, to which the test centrifuges were attached, surrounded the above-ground portions of the pit. The Iraqis had calculated that this wall thickness was required in case a heavy centrifuge jacket (60–80 cm long) was to break away from its fixtures while spinning at high speed and crash into a wall. One pit was designed to conduct mechanical tests, and the other one was designed to use process gas in the centrifuges.

Because of known weaknesses in Iraqi manufacturing capabilities, the centrifuge group decided to supervise the production of high-precision components at indigenous manufacturing installations and to check the quality of the components jointly with the manufacturer. Nevertheless, the reject rate for components remained high because of low machine quality, inexperienced staff and the shortage of suitable cutting tools and fixtures.

Raw materials, such as duraluminum for the centrifuge cylinders, vacuum oil and stainless steel, were imported in sufficient quantities for research and development purposes. Motors and frequency converters were also imported. European firms were Iraq's main suppliers.

The centrifuge programme was initially plagued by poor diagnostic instrumentation and a lack of theoretical understanding of rapidly rotating machinery. Towards the end of this effort in 1989, Iraq had overcome many of these problems, mainly through importing better bearings and more sophisticated computer codes and balancing machines. For example, until the programme obtained a sophisticated balancing machine from the German company Rutlinger, it did not realize that the centrifuge's inability to spin at design values was the result of improperly machined rotating components which were unbalanced. After correcting these problems, the centrifuge achieved machine speeds in excess of 50 000 revolutions per minute (rpm).

Iraq has declared that it designed, but did not build, 10- and 50-machine cascades. It also said that in 1988 it envisioned building a 4000-machine cascade of oil centrifuges at a site south of Al Taji. This site was also originally envisioned to be the location of an oil-centrifuge manufacturing facility. With a goal quantity of 10 kg per year of weapon-grade uranium, each machine would have needed to achieve an output of roughly 0.5 SWU per year

The process and mechanical difficulties inherent to this type of machine were never completely overcome. Nevertheless, by mid-1989 Iraq was conducting separation tests using a gas composed of freon and carbon dioxide. This mixture, unlike uranium hexafluoride, does not react chemically with the oil in this type of centrifuge. Although the mixing of the process gas and the oil remained a problem, Iraq has stated that this problem was manageable. However, it said that uranium hexafluoride gas was not used mainly because the operating speed of these centrifuges while filled with process gas was too slow (about 21 000 to 25 000 rpm) for a noticeable separation of uranium isotopes.

Parallel to the Beams centrifuge effort, the research group was making steady progress on the magnetic centrifuge, which was more efficient and practical than the oil centrifuge. By mid-1989, Iraq decided to cancel the oil centrifuge project, but allowed it to continue until late 1989 in order to provide training on centrifuge operation.

Subcritical Zippe-type or 'magnetic' centrifuge. Work on this machine started formally in the second quarter of 1988. A major break occurred in August 1988, when the centrifuge expert Bruno Stemmler, accompanied by Walter Busse and H&H personnel, travelled to Baghdad and gave the Iraqis two secret assembly drawings of a subcritical centrifuge and a supercritical centrifuge with two coupled sections. These machines had been developed at MAN Technologie AG in Munich, Germany in the late 1960s and early 1970s.

Both Stemmler and Busse had started working at MAN during this period. Stemmler remained an employee until about 1990; Busse retired a few years earlier.

During this meeting and subsequent meetings in Baghdad and Germany, Stemmler provided many detailed drawings of centrifuge components from several different models. He also gave the Iraqis about 70 classified technical reports, actual components, production data and procurement information for early-generation Zippe-type centrifuges. These technical reports contain many details for the production and operation of early-generation Urenco centrifuges.

Until Iraq invaded Kuwait, Iraq continued to receive a steady stream of outside assistance, mostly facilitated by H&H. Starting in the spring of 1989, a third German expert, Karl Heinz Schaab, started providing important centrifuge assistance. According to the Iraqis, he provided components, such as carbon-fibre tubes and samples of bellows and baffles, carbon-fibre rotor manufacturing assistance, about 90 secret technical reports, and components and equipment important to the mechanical testing of subcritical machines. The Iraqis valued Schaab, because, unlike the other foreign experts, he was 'very good with his hands.' He assembled some of the components of Iraq's first mechanical test stand and the first subcritical rotor prototype submitted to mechanical tests, and witnessed the stand's initial operation.

In addition to this type of assistance, Iraq also participated in specialized training courses in Europe. The Iraqis have stated that the goal of these courses was to acquire specific information relevant to centrifuges, although the training companies did not know the true Iraqi purpose for attending.

The most important course was held at INTERATOM in Germany in 1989. INTERATOM makes centrifuge cascades for Urenco. The training course involved about 20 Iraqis, each of whom was trained for periods between 4 days and 10 weeks in the second half of 1989. During these courses, the Iraqis learned about 'dual-use' vacuum technology, advanced welding, piping design and the materials science of maraging steel. In secret, without the knowledge of INTERATOM, Iraqis have stated that they entered classified areas of the company where gas-centrifuge cascade piping was located and copied exact dimensions and arrangements of pipes, valves and tri-flanges. INTERATOM cancelled a second series of training courses scheduled for 1990 because of growing suspicions in the West German Government about Iraqi intentions.

After unsuccessful attempts to manufacture oil-centrifuge components, Iraq realized it would have difficulty manufacturing components for magnetic centrifuges. It was therefore decided in late 1988 to place orders for the manufacture of many key centrifuge components with European firms in sufficient quantity for 50 machines, which were needed for the Iraqi prototype development programme.

Some of the components were procured directly from European suppliers; others were provided as demonstration tests during negotiations for the purchase of machine tools in Europe. Although tighter export controls prevented

Iraq from obtaining components for all 50 machines, it acquired enough of them to run its research programme.

With all this inadvertent and deliberate assistance, Iraq created several reliable designs of subcritical centrifuges during 1989 and 1990 that were based on both maraging-steel and carbon-fibre rotors. In this effort, the available designs were critical, but not sufficient. The designs did not specify all the parts, requiring Iraq to work out the detailed specification of components without foreign assistance. In addition, Iraq had to create a testing programme and a small-scale manufacturing programme for many centrifuge components.

In 1990, the centrifuge team may have been starting to opt for carbon-fibre rotor designs rather than maraging-steel ones. Schaab had already supplied about 50 finished carbon-fibre rotors and was helping Iraq to develop an indigenous capability to make them. Meanwhile, Iraq had not yet mastered the manufacture of maraging-steel rotors on H&H-supplied 'flow-forming' machines, nor had they received an adequate number of rotors from H&H.

By March 1990, Iraq had received from overseas and domestic sources most of the components for its first prototype centrifuge. Within about a month, the first mechanical test stand was operating in building 10 at Rashdiya using a carbon-fibre rotor provided by Schaab. The Iraqis have said that Schaab personally assisted them in assembling this first prototype.

After a few more months the test stand's rotor was spinning at up to 60 000 rpm, or at a wall speed of about 450 metres per second (m/s). In July, Iraq also started a test stand utilizing uranium hexafluoride gas.

These two test stands continued to operate until the end of 1990, when they were disassembled and hidden. It is known that the production of enriched uranium was not a goal at this point, because the slightly enriched product from the test stand was mixed with the tails and reused, rather than kept separate.

Iraq has said that it did not build additional test stands using uranium hexafluoride. It has stated that it planned to operate two optimized centrifuges in parallel or in series, but it had not optimized the centrifuges as of the end of 1990 and thus had not yet conducted this particular experiment.

The capacity of Iraq's test machine during one test run at the end of 1990 reached 1.9 SWU per year. Results in previous months were less successful. With additional machine optimization, inspectors believe that this design could have achieved a separative output as high as 2.7 SWU per year. Because the rotor was made of carbon-fibre, it could have spun faster than 450 m/s, which is near the limit for maraging-steel rotors. Thus in theory, it could have reached a higher separative work output. Iraq, however, has stated that it expected an output of only 2 SWU per year, even when using carbon-fibre rotors. One centrifuge expert said that he expected the rotor to spin at only 400 m/s, which is consistent with the declared separative output.

Centrifuge manufacturing

In late 1988, Iraq decided to build a dedicated centrifuge manufacturing site and obtain the necessary manufacturing equipment and raw materials abroad. This decision was controversial because at the time Iraq lacked a proven centrifuge prototype. As a result, Iraq had to order raw materials and equipment with the knowledge that the design would change and thus render some of the orders obsolete. But this approach permitted the programme to obtain many key items before the international community understood Iraq's true intentions and tightened export control policies.

Although the first site considered for the plant was near Al Taji, the team selected Al Furat, about 30 km south of Baghdad. This site had existing buildings, workshops and utilities and thus could start operations more quickly than the one at Al Taji. The plan was to build 'clean environment' workshops where the temperature and cleanliness would be carefully controlled, a requirement for manufacturing and assembling magnetic centrifuges.

The civil works at the site were nearing completion at the time of the war. More importantly, as of August 1990, Iraq was on the verge of acquiring enough manufacturing machine tools and associated fixtures to complete its capability to make centrifuges. Only the UN Security Council embargo imposed after Iraq invaded Kuwait kept it from receiving the rest of manufacturing equipment.

Iraq declared in 1991 that Al Furat would be capable of making 200 centrifuges a year, although IAEA inspectors at the time believed that annual production would reach 600 machines a year and under certain circumstances production could reach 2000 centrifuges a year.

In 1996, Iraq changed its story and said that it had envisioned Al Furat's production starting at 1000 centrifuges a year in 1992, reaching 4000 machines a year after a future expansion in 1994 and 1995. The Iraqis have said that they expected a very high failure rate, so the actual number of centrifuges that would have been produced remains an open question.

Because the Iraqi construction industry was unfamiliar with making cleanroom facilities, Iraq contracted with German and British firms to make two clean-room buildings. The contract to construct Building B01, which was to house dimensional quality control, mechanical testing of single centrifuges, final assembling of centrifuges and a hall for a 100-machine cascade, went to INTERATOM and was signed in October 1989. The original time schedule called for the building to be completed in October 1990. Iraq decided to build the civil works of building B02 itself, which was to produce maraging-steel rotors and lower bearings, but it awarded a contract to construct its mechanical and electrical utilities to ITSC, a British company. This contract was signed in late September 1989, and the building was expected to be finished in December 1990.

Meanwhile, by late 1989 the centrifuge programme was advancing. Rashdiya was making components that had not been ordered abroad, such as the lower bearings and the motor. The programme was starting to receive key components, testing equipment and manufacturing items for the prototype centrifuge from abroad. Because buildings B01 and B02 were many months from completion, the programme leaders decided to modify building 10 at Rashdiya for centrifuge assembling and testing. This building held the centrifuge test stands mentioned above.

In early 1990, building B03 at Al Furat, which was closer to completion than either B01 or B02, was outfitted with machines to make molecular pumps and the outer casings. In addition, B03 housed research and development efforts for maraging-steel rotor production, even though Iraq may have been moving away from maraging-steel rotors.

Iraq apparently planned to finish Al Furat, although its expected completion date remains unclear. The centrifuge programme may have also turned over the operation of Al Furat to other Iraqi industrial organizations while maintaining staff participation for supervision. In addition, Rashdiya would have probably continued to manufacture certain components.

Iraq did not reach the point at which it could produce a sizeable number of centrifuges. Moreover, the embargo significantly impaired its ability to make centrifuges. Nevertheless, Iraq had made great progress in creating an indigenous centrifuge manufacturing capability in just a few short years.

When completed, Iraq's manufacturing facilities may have generated far more reject parts than adequate ones. Iraqi centrifuge experts in fact expected a significantly high reject rate. They planned to compensate for this weakness by trying to make thousands of centrifuges a year. If Iraq could have continued to procure sufficient raw materials and other necessary items, it may have succeeded in producing adequate centrifuges at a rate of many hundreds per year by the early 1990s.

Carbon-fibre manufacturing

Because Iraq was encountering problems in making maraging-steel rotors, it decided in 1989 to develop the capability to make carbon-fibre rotors. Schaab had supplied about 50 carbon-fibre tubes. He made these tubes on a manual winding machine at his company ROSCH in Kaufbeuren, Germany, south-west of Munich.

In late 1989, Iraq said it had signed a contract with Schaab and the Swiss firm ALWO for an automated carbon-fibre winding machine. The final construction of this machine was delayed by months, because of problems in obtaining a German export licence for a computer-controlled subcomponent of the winding machine from Siemens. According to Iraqis, the embargo following Iraq's invasion of Kuwait prevented them from receiving this machine, and they cancelled payment of the machine. They have said that they do not know the machine's whereabouts.

The 1989 contract with Schaab and ALWO included a provision for a two-week training course and the production of 50 finished rotor tubes. The training would have been held at ROSCH in Germany or at an unspecified place in Austria. The expected date of the training was the autumn of 1990.

In September 1990, after the start of the crash programme, Iraq declared that it had ordered an identical winding machine from Schaab and ALWO through a intermediary company in Singapore (see section VII on the crash programme). The training session again included the production of 50 rotors and was planned for either Austria or Latin America. The second winder was shipped to Jordan via Singapore, where it was embargoed. It has been inspected by the Action Team. The Iraqis have stated that the training session did not occur.

Based on the available information, Iraq knows how to make carbon-fibre rotors, at least in theory, but has not had the opportunity or capability to make its own rotors either before the war or since its end. Iraq possessed a plentiful supply of the basic building blocks for rotors, namely carbon fibre and resin, until after General Kamel's defection when at least the carbon fibre was handed over to the inspectors. If it also had received a winding machine, it would have had five years to master the production of the rotor.

Cascades for subcritical Zippe-type centrifuges

In 1989, Iraq originally intended to house a 1000-machine cascade (or two 500-machine cascades) along with a uranium hexafluoride production plant at a site south of Al Taji. After conducting initial studies, Iraq maintains that it did not do any construction work at this site, although it apparently intended to do so.

Building B01 at Al Furat was selected to house an experimental cascade of 36 machines able to produce 3 per cent enriched uranium. This plan was modified to include the possibility of installing up to 120 centrifuges in a cascade, and the cascade hall was enlarged accordingly. This larger cascade would have produced about 1 kg per year of weapon-grade uranium.

However, because construction at Al Furat was delayed, the centrifuge team decided to build a new building at Rashdiya to house the experimental 120-machine cascade. This facility, called building 21, was unfinished at the time of the Allied bombing campaign. An important benefit of this change would have been the elimination of the need to move key personnel between Rashdiya and Al Furat, allowing personnel to be concentrated at Rashdiya. For a similar reason, a decision was made to locate uranium hexafluoride production at Rashdiya and not at the Al Taji site.

The Rashdiya cascade hall was designed to be larger than the similar one at Al Furat, leading to suspicions that Rashdiya would contain a significantly larger cascade. Iraq denies this allegation.

Supercritical design

Iraq has declared that it did not build any supercritical designs, which are characterized by having longer rotors than subcritical machines. European supercritical centrifuges have rotors made of two or more roughly 50-cm long rotor tubes connected together by a flexible maraging-steel joint (called a 'bellows') that acts like a spring. The bellows can be very difficult to make. The supercritical centrifuge itself is complicated to assemble and balance.

What Iraq obtained. As mentioned above, Stemmler had provided Iraq with a general assembly drawing of a two-rotor, maraging-steel machine. Iraq, however, says it did not intend to build one. A major surprise following Kamel's defection was Iraq's declaration that in August 1989 Iraq obtained from Schaab a MAN design for a 3-metre long supercritical centrifuge and some detailed drawings of some of the key components. Such a machine would have had a separative capacity of about 20–30 SWU per year, depending on the rotor's speed of rotation, and represents a sophisticated, mid-1980s Urenco design. Iraq did not obtain a complete set of the 3-metre machine designs; some component designs were missing.

The Iraqis said that Schaab also supplied three samples of bellows, a number of carbon-fibre baffles and many classified technical reports about mid-1980s centrifuges. Combined with its acquired drawings and knowledge of materials, these technical reports would greatly expedite the production and operation of modern centrifuges.

Iraqi declarations and resulting questions. Iraq has consistently downplayed the level of its work on supercritical centrifuges. It has declared that it concentrated on subcritical designs and its work on the supercritical 3-metre centrifuge design was progressing more slowly. It says it barely worked on the two-rotor design at all.

Questions remain, however, about the scope of Iraq's work on supercritical centrifuges. For example, why would Iraq ignore a G2 design which is considerably easier to construct than a 3-metre design? Other questions result from apparently contradictory information supplied by Iraq.

Iraq has admitted compiling a 3-metre design and doing some simple theoretical analyses of it. Iraq also enlarged the doors and raised the roof height of the cascade hall in building 21 at Rashdiya to permit the installation of the 3-metre machines. It also said that it enlarged the size of the doors to a building at Al Furat where machine testing would have occurred.

Questions also remain about Iraq's efforts to procure bellows manufacturing equipment. Iraq has stated that in about October 1989 Schaab offered to arrange a contract for the supply of bellows-production equipment and know-how. Schaab told the Iraqis that he could also provide a 'friend' who could assist in providing specific bellows know-how, but Iraq says it did not take Schaab up on this offer. The Iraqis said they judged that they could make the bellows themselves at Rashdiya. Failing in that, they believed Schaab's offer would still be on the table.

Another contradictory piece of information is that Iraq ordered bellows manufacturing equipment from a French firm that would have been suitable for making maraging-steel bellows, although the order was for producing stainless steel bellows. The embargo prevented the delivery of this equipment.

Few believe that Iraq could have successfully operated a 3-metre long machine. However, the assessment during September 1996 remains that Iraq has more to admit about its work on supercritical centrifuges.

Annual output of the centrifuge programme

Iraq has stated that its goal was the operation of 1000-centrifuge cascade in 1994 that would produce 10 kg per year of weapon-grade uranium. Few, however, believe that this number of centrifuges or this annual quantity was the final goal of Iraq's centrifuge programme.

Iraqi centrifuge experts appear genuine in their belief that building and operating 1000 machines would represent a tremendous accomplishment. They have stated that they did not think they could meet this goal by 1994, despite the official declaration. The head of centrifuge programme told an inspector that this goal was unlikely to be reached before about 1997 or 1998.

The Iraqi centrifuge experts have not provided any documents that could substantiate their pessimistic view of their own progress. In addition, other information contradicts their claims. For example, their view is belied by the sheer amount of equipment and know-how that was being supplied by foreign experts and companies just before the war and the ability of the Iraqis to establish new procurement routes through the Far East. In addition, if Iraq succeeded in assembling its centrifuge production facilities on the schedule it has recently declared, it expected to produce 1000 centrifuges per year by the end of 1992. It planned to expand annual centrifuge production several-fold by the end of 1995.

As in the case of EMIS, the centrifuge programme may have faced serious problems as it moved from laboratory testing of gas-centrifuge equipment to industrial production of significant quantities of enriched uranium. Iraq might have encountered difficulties in procuring all the items it would have needed and in achieving the high-quality manufacturing standards necessary for the production of reliable centrifuges. However, inspectors believe the Iraqi centrifuge team was close to the point at which it could build and operate a centrifuge with confidence, although the total economic cost might have been high and some delays might have continued to occur.

A maximal estimate of the Iraqi centrifuge capacity can be derived by using a modification of Iraq's planned manufacturing schedule and assuming adequate procurement of raw materials. This estimate also conservatively assumes that 50–75 per cent of the centrifuges produced will be rejected. The output of each centrifuge is taken as 2.7 SWU per year, which means that Iraq would have optimized its subcritical design prior to manufacturing. Table 11.4 shows that, by the year 2000, Iraq could have an estimated annual capacity of roughly 14 000–28 000 SWU per year. At this rate, and assuming a tails assay of 0.5 per

Table 11.4. Maximal estimated centrifuge production[a]

Year	Annual production	Annual production minus rejects	Cumulative production	Separative output[b] (SWU/y)
1992	500	125–250	125–250	340–675
1993	1 000	250–500	375–750	1 000–2 000
1994	1 000	250–500	625–1 250	1 700–3 400
1995	1 000	250–500	875–1 750	2 400–4 700
1996	2 000	500–1000	1375–2 750	3 700–7 400
1997	3 000	750–1 500	2 125–4 250	5 700–12 000
1998	4 000	1 000–2 000	3 125–6 250	8 400–17 000
1999	4 000	1 000–2 000	4 125–8 250	11 000–22 000
2000	4 000	1 000–2 000	5 125–10 250	14 000–28 000

[a] This estimate assumes that centrifuge production reaches its goal of 1000 machines a year in 1993 and that the expansion in production occurs more slowly than planned. In all cases, a 50–75 per cent rejection rate is assumed. In addition, centrifuge lifetime is not considered, or alternatively, the lifetime is assumed to be greater than 8 years. Actual operational lifetime could be less, even less than five years.

[b] This estimate assumes that each machine has a separative output of 2.7 SWU/y. The actual output could be significantly higher.

cent, Iraq could make about 87–175 kg of WGU per year. At 15 kg per nuclear weapon on average, this is enough HEU for 5–11 weapons a year.

Another upper-bound estimate is based on Iraq's procurement efforts. Iraq has admitted, and the IAEA has verified, that it had obtained enough specialized metals and components to build thousands of machines. In some cases, Iraq acquired enough items for up to 10 000 machines. If it is assumed that Iraq aimed to produce a total of 10 000 machines, and again assuming a 50–75 per cent reject rate, Iraq might have succeeded by 2000 in building 2500–5000 machines although the upper bound would have likely required the procurement of additional materials. Assuming that each machine would have had a capacity of 2.7 SWU per year, the total capacity of all these machines would have been 6750–13 500 SWU per year. Assuming that the tails assay was again high (about 0.5 per cent), Iraq could have produced about 40–80 kg of weapon-grade uranium per year. Assuming 15 kg per weapon, Iraq could build two to five nuclear weapons per year with this amount of HEU.

The available information tends to discount Iraq's ability to make a supercritical centrifuge, at least for several years. As a result, the effect of supercritical machines on these estimates is ignored. However, with time, Iraq could have mastered the production and operation of supercritical machines, if it decided to do so.

If the centrifuge programme would have succeeded, even marginally, Iraq would have had a significant supply of HEU. The supply could have been large enough to reduce dramatically the need for the EMIS programme, while providing for a large number of nuclear weapons (see table 11.5). Because of the flexibility of centrifuges compared to gaseous diffusion, there would have been little value in using the centrifuges to produce LEU for further enrichment to weapon-grade in the EMIS separators. Thus, the centrifuge programme appears to have been Iraq's most successful effort aimed at making fissile material.

In any case, the war and the ensuing inspections crippled Iraq's centrifuge manufacturing programme and ended the programme before HEU was produced. Inspectors now believe that Iraq does not have the capability to make a significant number of centrifuges. However, Iraq has accumulated a large body of information and considerable expertise about Urenco-type centrifuges. If given the opportunity, Iraq could reconstitute its centrifuge programme considerably faster than other states could develop them.

Chemical enrichment

In the spring of 1988, the gaseous-diffusion programme was encountering problems and the gas centrifuge effort was in its infancy. In order to find another way to produce LEU for the EMIS separators, Iraq decided to initiate the development of a chemical enrichment facility.

The Iraqi programme essentially copied the French CHEMEX solvent-extraction process and the Japanese ASAHI ion-exchange technique. These chemical enrichment approaches depend on solvent extraction technologies, that is, reprocessing and ion-exchange technologies. According to the Iraqis, both of these technologies were relatively accessible to Iraq.

It is unclear when Iraq first started research into chemical enrichment or how much design information Iraq obtained from abroad. Iraq has stated that in 1980 France proposed to an Iraqi team that it participate in the development of its CHEMEX enrichment process. Iraq has said it turned down this offer because of its high cost and lack of interest. However, it is unknown whether at that time Iraq somehow obtained design information for the CHEMEX process.

At the time of the Persian Gulf War, Iraqi scientists had accomplished significant laboratory work in both techniques. They had finished designs for pilot plants to produce small amounts of enriched uranium, but they had not implemented construction at Al Tuwaitha.

Iraq procured key equipment, such as distillation units, mixer-settler batteries, pulsed columns and pumps for its laboratory-scale work from France, Sweden and Germany. Much of this equipment was destroyed in the bombing of Al Tuwaitha, but some was salvaged. The Action Team is cataloguing and inspecting this newly declared equipment.

Iraqi success in chemical enrichment would have probably depended on additional, larger-scale foreign procurement. At the time of the Persian Gulf War, however, little of this procurement had occurred, although Iraq has said that it had placed an order with a French company for glass columns, mixer-settlers and other items for its solvent extraction pilot plant. The embargo following Iraq's invasion of Kuwait ended this transaction.

Production-scale plants were designed for the solvent extraction and the ion-exchange techniques and were essentially larger versions of the pilot plants. The design studies called for the production of 4000 kg of 3 per cent enriched uranium (120 kg ^{235}U) per year. In the process of choosing which approach to pursue further, Iraq has said that it conducted paper studies of a hybrid that would have been able to enrich uranium to 8 per cent. In both cases, Iraq has stated that no construction occurred.

There is insufficient information to estimate when or if Iraq could have completed a chemical enrichment plant for uranium isotope separation. Indications are that the programme was encountering serious difficulties as it moved out of the laboratory towards the pilot-stage.

The laser enrichment programme

In May 1994, the IAEA learned from member states that Iraq had pursued uranium enrichment through laser isotope separation (LIS) at the Al Tuwaitha site.[11] According to member states, Iraq had studied both molecular (MLIS) and atomic vapour (AVLIS) technologies.

In 1991, Iraqi officials denied that Iraq had undertaken any laser enrichment activities. During the 26th IAEA inspection in the summer of 1994, after continuing to deny the existence of the programme for four days, Iraqi officials finally admitted that the existing Laser Section 6240 had 'received an objective [in 1981] from the Atomic Energy Commission to work in Laser Isotope Separation.'[12]

Following Hussein Kamel's defection, Iraq provided more information about its laser separation programme. However, the basic story has remained unchanged. In 1987, with little progress to show, the programme was downgraded to a 'watching brief' and a number of key personnel were transferred to other programmes, particularly EMIS. The remaining personnel in the Laser Section worked in basic research and in improving Iraq's scientific expertise.

The IAEA concluded that Iraq's LIS programme was loosely coordinated, largely empirical and had not made much progress. The inspectors found no indications that this programme had reached the point of an integrated experiment that separated any uranium, although a rudimentary laser excitation experiment related to MLIS was carried out on 1 gram of uranium hexafluoride.

According to statements made by Iraqis to the inspectors, export controls and voluntary refusals by several suppliers had prevented the Iraqi programme from importing critical pieces of equipment, such as advanced laser instrumentation and related accessories.

---

[11] Report on the twenty-sixth IAEA on-site inspection in Iraq (note 2).
[12] Report on the twenty-sixth IAEA on-site inspection in Iraq (note 2).

Table 11.5. Projected Iraqi weapon-grade uranium inventories
Weapon-grade uranium figures are in kilograms.

Year	Annual production of weapon-grade uranium			Cumulative[a] total	No. of potential bombs (cumulative)
	EMIS	Centrifuge	Total		
1991	..	..	..	..	..
1992	1	..	1	1	
1993	3	2	5	6	
1994	6	2	8	14	
1995	10	10	20	34	2
1996	10	19	29	63	3
1997	10	45	55	120	7
1998	10	45	55	170	10
1999	10	45	55	230	14
2000	10	45	55	280	17

[a] Cumulative totals are rounded to two significant figures.

V. Projected indigenously produced weapon-grade uranium inventory for Iraq

Although Iraq's nuclear weapon programme has been destroyed, questions remain about when it would have acquired its first nuclear weapon and how many weapons it could have built. Based on the above discussion, an attempt is made below to answer these questions at least speculatively. We will never know the true answers. The figures in table 11.5 are the authors' estimates of the amounts of weapon-grade uranium Iraq could have produced based on the known successes and failures of the programme before the Persian Gulf War. These estimates are based on the following assumptions:

1. Al Tarmiya would have been operated successfully at an average output of 10 kg of weapon-grade uranium (93 per cent) per year starting in 1995. This value assumes that the R120 separators would have sustained currents totalling about 400 milliamps at the collectors, and a machine availability of 55 per cent. Ash Sharqat would not have been finished.

2. Iraq could have operated a 120-machine centrifuge cascade by early 1993, added a 500-machine cascade in early 1995, added another 500 machines by 1996, and reached 2500 operating machines by early 1997. The analysis assumes that after 1997 the number of machines would have remained at 2500 for several years. The separative output of each machine is assumed to be 2.7 SWU per year. The tails assay is taken as 0.5 per cent. Iraq would be expected to achieve a significant expansion in the number of centrifuges or an increase in the separative output per machine soon after the turn of the century.

This increase in separative output could include the construction of supercritical machines.

3. Iraq would not have succeeded in building a LIS plant or chemical enrichment plant able to produce significant quantities of LEU for the EMIS programme; it would not have acquired a significant stock of unsafeguarded LEU or diverted its safeguarded stock.

4. Iraq would have succeeded in designing a nuclear weapon requiring 15 kg of weapon-grade uranium and developing recycling capabilities to maximize its supply of weapon-grade uranium.

As shown in table 11.5, Iraq would have gained enough material for its first nuclear weapon by early 1995, a total of 2 atomic bombs in 1996 and enough for almost 20 bombs by the end of the century had the programme succeeded. These results are uncertain, but they illustrate the probable scale of the planned Iraqi nuclear weapon effort.

VI. The Iraqi plutonium programme

Plutonium separation capabilities

Starting in the early to mid-1970s, Iraq pursued the development of a plutonium separation capability. The work performed during the 1970s, however, was rudimentary in nature.

In 1979, Iraq established a radiochemical laboratory, equipped through a contract with the Italian SNIA-Techint company, suitable for laboratory-scale research on fuel reprocessing. This facility was expanded with additional glove boxes and an analytical cell during the next few years, and from 1983 to 1987 performed 'cold' tests on unirradiated fuel. In 1988 this facility processed a safeguard-exempt spent fuel element from the IRT-5000 reactor and extracted 2.26 grams of plutonium and 920 grams of uranium. Later, in violation of Iraq's safeguards agreement, this facility separated another 2.7 grams of plutonium from small amounts of natural uranium fuel manufactured in Iraq and irradiated in the IRT-5000 reactor. In addition, several small samples of ^{238}Pu were separated in this laboratory from neptunium-237 that was irradiated in the Soviet-supplied reactor. There is no evidence that Iraq tried to enlarge the programme by secretly building a larger plutonium separation facility.

The reactor programme

An underground reactor

Following the Persian Gulf War, persistent reports surfaced that Iraq had a secret underground reactor. During several inspection missions, however, the Action Team found no verifiable information to support the existence of a clandestine reactor, let alone an underground reactor.

Some of the information supporting suspicions of a secret reactor was derived from Iraq's persistent attempts to acquire a power reactor dating back to the early 1970s. According to Iraqi declarations, Iraq first sought a natural uranium reactor from Canada and France, but these efforts failed, partially as a result of tightening export controls following India's detonation of a nuclear device in 1974. Iraq then tried to obtain a light water reactor, but decided to postpone these plans in the late 1970s. By this time, Iraq had negotiated the purchase of two French research reactors.

Following Israel's bombing of the French-supplied Tammuz-1 reactor in 1981, Iraq decided to investigate the possibility of an underground reactor. Iraq approached French, Belgium, Russian, Finnish and Italian companies for the siting and construction of this project. Russia evaluated the siting of a reactor underground or in a fortified site in the mountains but projected huge costs and technical difficulties. A foreign firm also assessed that an Iraqi underground reactor would likely be detected by foreign intelligence agencies through national technical means, further discouraging the Iraqis.

Project 182: a production reactor

In 1985, Iraq initiated Project 182, which was charged with designing and building a 40-MWth reactor. The reactor was modelled on the Canadian NRX reactor. It was moderated by heavy water, fuelled by natural uranium metal fuel and cooled by water. Such a reactor is capable of producing about 8–10 kg of weapon-grade plutonium a year. The Iraqis have declared that this project did not progress beyond the design phase and was halted in 1988.

The reactor design included a containment building able to withstand a direct missile attack. Iraq planned to salvage some equipment, such as heat exchangers, primary circuit pumps and electric generators, from the Tammuz-1 reactor. Some Tammuz-1 designs, such as those for cooling towers, control systems and health physics systems, would also be utilized in the new reactor. Iraq planned to make indigenously the reactor vessel, the primary cooling circuits and the fuel loading and unloading machines.

The site would have been one of those originally characterized for nuclear power reactors. A new plutonium separation plant would have been built and it would have utilized the experience and equipment from the radiochemical laboratories at Al Tuwaitha.

The reactor would have needed about 30–40 tonnes of heavy water. To obtain this amount, a heavy water production plant was planned. Iraq has stated that work in this area continued until January 1991, but no designs of a heavy water production facility were ever finished.

Iraq developed a large uranium metal production programme that continued up to the time of the War. This programme should have been large enough to provide the fuel for this reactor.

Iraq has declared that in 1988, it stopped work on this project, citing a lack of heavy water and competition with the enrichment programmes for manufacturing assistance and expertise. Iraq decided to delay constructing a natural uranium-fuelled reactor and in the meantime to consider building a less costly reactor using LEU fuel. Such a reactor would be smaller in size for a given amount of heat output. This decision helps explain why projects to make heavy water and metallic fuel continued after 1988.

VII. The crash programme

After the sharp international reaction to Iraq's invasion of Kuwait, the Iraqi Government decided to divert its safeguarded HEU fuel in order to make a nuclear weapon. This fuel had originally been supplied by France and Russia along with civil research reactors. The small French-supplied Tammuz-2 reactor had accompanied the sale of the Tammuz-1 reactor and had not been destroyed by the Israeli bombing. It also used 93 per cent enriched uranium. The Russian-supplied IRT-5000 research reactor (5-MWth) used fuel with enrichment below 80 per cent. Table 11.6 lists the quantities of safeguarded enriched uranium the Iraqis have declared they planned to divert. The total amount of HEU (in terms of initial uranium mass) was 39.5 kg with an average enrichment of about 84 per cent contained in 175 fuel elements.

The crash programme called for the HEU to be extracted from the fuel. A portion of this HEU would then be enriched in gas centrifuges up to weapon-grade as necessary for the nuclear explosive programme. The order came from General Kamel, who ordered the nuclear explosive to be finished within six months (e.g., by the end of February 1991).

The Iraqis have said that although the indigenous production of HEU was still far from fruition, the programme to master the steps in building a nuclear explosive was only about one year behind its original schedule of completing a device in 1991. As a result, their plan to use the safeguarded HEU in essence allowed them the opportunity to complete a nuclear explosive on schedule, if they could accelerate the weaponization portion of their work.

Iraq denies it considered diverting safeguarded HEU before its invasion of Kuwait. Nevertheless, strong suspicions remain that this option was included in Iraq's plans from the beginning, despite the lack of documents to substantiate this suspicion.

The Action Team has concluded that Iraq could not have finished a nuclear explosive device within six months. However, several inspectors have estimated that Iraq could have finished a nuclear explosive device by the end of 1991. This estimate is an extrapolation from their understanding of the status of the Iraqi programme during January 1991, when the Allied bombing campaign effectively ended Iraqi efforts. It assumes that Iraq would have succeeded in overcoming the main technical obstacles it still faced in early 1991. In particular, Iraq still needed to finalize its design for high explosive lenses and a neutron initiator and to convert the HEU into bomb components. Because of the difficulty of predicting how fast Iraq would succeed in these tasks, this estimate

A SPECIAL CASE: IRAQ

Table 11.6. Iraq's safeguarded fuel

Reactor	Irradiation level	Initial enrichment (%)	Uranium Mass[a] (gm)	No. of elements
Tammuz-2	Fresh	93	417	1
Tammuz-2	Slight	93	11 874	38
IRT-5000	Fresh	80	13 689	68
IRT-5000	High	80	13 490	68
IRT-5000	High	10	87 760	69

[a] The mass given is the initial mass of the uranium before irradiation, the actual mass of irradiated uranium is less.
Source: IAEA Action Team.

is highly uncertain. Although Iraq is unlikely to have finished earlier than mid-1991, it could have taken many months longer to finish an explosive, in some cases, a year or two longer. In the end, Iraq did not extract any HEU. After the war, the IAEA accounted for the safeguarded material and eventually removed all of it from Iraq.

Iraq is believed to have taken initial steps to reconstitute its entire crash programme right after the war, but it was thwarted by the unexpected intrusiveness and effectiveness of the nuclear inspections. Partial evidence for this plan is that Iraq took extraordinary steps during and after the war to move the fuel to alternative storage locations. In addition, Iraq has stated that it evaluated moving portions of the crash programme to Al Tarmiya after the war (see below).

The major aspects of the crash programme that are related to recovering HEU and turning the fissile material into metal are described below. The weaponization aspects of this programme are not discussed.

Project 601

This project was in charge of designing and erecting a solvent extraction unit at Al Tuwaitha for reprocessing the fuel and recovering the HEU. The process was based on complete dissolution of the fuel elements and multiple extraction and purification stages to recover the HEU. The extraction stage used Swedish-supplied mixer-settlers. Other equipment was made in Iraq.

The staff of Project 601 had prior experience in laboratory-scale reprocessing and were able to finish installing the equipment in the concrete cells of the LAMA building at Al Tuwaitha by mid-January 1991. This French-supplied facility was not intended for reprocessing but rather for handling radioactive material, and Iraq had to remove the old equipment before preparing its concrete 'hot cells' for their new use. However, this building was suitable because it had excellent radiation shielding and was equipped with special ventilation systems, manipulators and decontamination facilities.

After conducting a 'cold test', the facility was ready for 'hot' operation on the eve of 17 January 1991 when the Allied bombing campaign began. Prior to the onset of hostilities, Iraq had cut off the aluminum ends of one fresh 93 per cent element in preparation for reprocessing, but the bombing destroyed most of Al Tuwaitha, including the LAMA facility.

The Iraqis have stated that they planned to process one element a day. They would have started with the easiest material to process, the fresh and lightly irradiated fuel and moved to the most difficult, the highly irradiated material. Under this plan, the HEU would have taken almost six months to recover. According to an IAEA inspector, Project 601 could have sped up the HEU dissolution rate, if desired, although the extraction and purification steps still may have caused delays. Since the weaponization portion of the work was taking longer than was hoped, a speed-up may have been unnecessary.

Project 602

This group was charged with designing and constructing a facility to receive the uranyl nitrate solution from Project 601 and convert it into metallic form. It was also responsible for making natural uranium metal for bomb components. This project had completed the major part of its construction and cold testing activities by the start of the allied bombing campaign.

Building 64 at Al Tuwaitha housed Project 602. Building 73A also contained many of the experimental facilities for the project and a few of the production units involved in removing undesirable impurities from various chemical forms of HEU.

Project 602 used batch operations to convert the HEU in the form of uranyl nitrate first into an oxide, then into uranium tetrafluoride (green cake) and finally into a metal 'button'. The project also included the capability to recycle HEU.

The metallic HEU buttons would have been relatively small, each containing only about 250 gm of HEU. Project 602 personnel have stated that a small batch size reduced criticality concerns and minimized losses in the case that a reduction went drastically wrong. However, the relatively small batch size would have caused higher losses. In addition, melting many small buttons together in a furnace (prior to casting a uranium component) causes much higher losses than if larger buttons were used. The result is that Project 602 would need to supply significantly more HEU than used in the component. Although most of this HEU could be recovered and reused, it meant that each nuclear device would require a great deal more HEU than would end up in components. This approach alone may have prevented the Iraqis from accumulating enough HEU metal for a second device.

Centrifuge programme: Project 521C

The centrifuge group at Rashdiya was given the responsibility of taking the Russian-supplied HEU and further enriching it in a short cascade up to 93 per cent. As described previously, this programme was highly dependent on foreign assistance, and the international embargo prevented it from receiving assistance necessary to complete the plan on schedule. In particular, it was facing serious delays in making sufficient numbers of high-precision centrifuge components.

The plan as of December 1990

The cascade was to be located at Rashdiya in building 9, which was being modified to hold a 50-machine cascade, although the programme was falling behind schedule. In mid-December 1990, the commissioning of the cascade was expected in April 1991 and the introduction of uranium hexafluoride was scheduled for July 1991. The HEU feed material would have been enriched over the next three to six months, depending on feed availability and centrifuge reliability.

The actual enrichment level of this material, half of which was irradiated, varied from 56 to 80 per cent, with an average of 70 per cent (see table 11.6). The cascade was designed to have 49 centrifuges, each with a separative output of 2 SWU per year. The tails assay was 40 per cent and the nominal feed rate was 6.9 kg per month.

The material would have been received as uranium tetrafluoride and it would have been converted into uranium hexafluoride at Rashdiya. As of January 1991, no decision had been taken on the chemical form of the material to be returned to the weapon programme.

The execution

At the time of the bombing, design and construction activities at building 9 were nearing completion. However, the biggest problem encountered in meeting the revised deadline was in manufacturing the individual centrifuges. Although Iraq had imported a significant number of centrifuge components from abroad, it was still missing many of them. For example, as mentioned earlier, Iraq planned to use carbon-fibre rotors, and only about 20 out of a stock of 50 rotors were acceptable.

The embargo not only prevented Iraq from obtaining additional parts overseas, but it also prevented Iraq from receiving all the specialized fixtures for the machine tools already in its possession. Iraq's attempts to make the necessary centrifuge parts indigenously had not succeeded by January.

During the embargo, as discussed above, Iraq was actively trying to obtain a carbon-fibre winding machine to make the rotors. The machine had been purchased in Europe and sent through Singapore to Jordan. Jordan, however, would not send it to Iraq because of the Security Council embargo. Because of the difficulty in getting the carbon-fibre winding machine, the programme was continuing to work on making maraging-steel rotors as an alternative to carbon fibre.

Sufficient components for the cascade, such as piping, valves and frequency converters, were in hand. They had been acquired earlier for the research programme.

Overall, the centrifuge programme is unlikely to have succeeded in making a cascade on the schedule mentioned above. However, the programme is likely to have eventually succeeded.

The EMIS programme

The Iraqis have stated that after reprocessing the safeguarded 10 per cent enriched fuel (after irradiation only 7 per cent enriched), the recovered material was slated to be further enriched in the EMIS programme. They said, however, this plan had not been formulated in detail. They added that material would have been used in the EMIS separators only if it did not pose a radiation hazard to do so.

According to the Action Team, the EMIS programme was de-emphasized after the start of the crash programme in August 1990. Although enriched uranium output increased in the fall at both Al Tuwaitha and Al Tarmiya, as mentioned earlier, the installation of separators was slowed down. The Iraqis have said that this resulted from the decision to transfer many engineers and technicians from Al Tarmiya after September 1990 to Al Tuwaitha to work on designing and installing equipment for projects 601 and 602.

Jaffar said that during the crash programme Iraq would have used its stock of safeguarded LEU at Al Tarmiya, if doing so would have been to some advantage. (This stock contains about 1.7 t of 2.6 per cent enriched material.) He said that in August 1990, they were not ready to use safeguarded LEU at Al Tarmiya. He said that they may not have been prepared to do so by January 1992 and perhaps not until January 1993. These dates would have been too late to meet the crash programme deadline.

A second device

Iraq has stated that it intended to build only one nuclear explosive device from safeguarded HEU. With the losses expected during the preparation of the HEU for this device, this statement appears to be credible. However, it also appears to be incomplete.

One nuclear explosive device could have worked as a deterrent, since most nations would have assumed Iraq had more than one, given the size of the safeguarded HEU stock and the lack of firm information about its indigenous production capabilities; but Iraq is unlikely to have been satisfied with just one.

Iraqi saw the production of the first device as a formidable problem and its declarations have concentrated exclusively on this problem. Even if Iraq did not formulate any formal plans for a second device, it undoubtedly would have done so eventually.

Perhaps Iraq could have successfully made two devices from its safeguarded HEU, but this feat would have been difficult given the expected inefficiencies in its weapon production processes. Therefore, Iraq is likely to have investigated ways to get enough HEU to make a second and perhaps third device.

One of the most likely routes by which to obtain more HEU would entail the successful operation of a gas-centrifuge cascade, Iraq's most developed enrichment process, and involve the use of the safeguarded LEU. Using LEU feed, a 49-machine cascade could make about 1.5 kg per year, and a 120-machine cascade could produce about 3.7 kg per year, assuming in both cases a tails assay of 1 per cent. If the number of centrifuges were increased to roughly 500 machines, Iraq could produce 15 kg of weapon-grade uranium in a year.

VIII. Post-war activities

Although Iraq has denied trying to reconstitute its nuclear weapon programme after the Persian Gulf War, recent information may suggest otherwise. One piece of information is Iraq's own declaration that projects 601 and 602 were ordered to be redesigned towards the end of the war and reports prepared. The resulting Iraqi reports, dated early June 1991, but finished several weeks earlier, called for salvaged and new equipment to be installed at Al Tarmiya. Only in late May did the first inspection team show up at Al Tarmiya, unknowingly halting any Iraqi effort to reconstitute projects 601 and 602.

Perhaps Iraq planned to use Al Tarmiya as a replacement for Al Tuwaitha, which was essentially destroyed by the bombing campaign. Although the buildings were not suited for handling highly radioactive spent fuel, they were adequate for processing fresh or slightly irradiated fuel. Al Tarmiya could have also housed enrichment research activities, such as the chemical enrichment programme.

Suspicions remain that centrifuge activity resumed after the war at Rashdiya for at least some period of time. Little verified information exists for activities at Rashdiya after the war. It was not bombed at all during the war, and it was not inspected until the summer of 1991, and then only in a cursory manner.

Iraq has declared that in March 1991 it started bringing evacuated equipment and materials back to Rashdiya, but it had not finished reconstituting its programme by the time it accepted UN Security Council Resolution 687 in April. Iraq says that it did not resume any centrifuge work at Rashdiya or elsewhere after the war.

Nevertheless, questions remain about why Iraq decided to hide Rashdiya's existence. Although Iraq chose to tell the inspectors about many of the centrifuge programme's accomplishments and the existence of Al Furat, it decided not to reveal Rashdiya or the extent of foreign assistance. Iraq continued to deny the importance of Rashdiya even after defectors had identified the site in

1991. Iraq came clean about Rashdiya and the extent of foreign assistance only after General Kamel's defection in 1995.

Iraq also continues to maintain that all centrifuge programme reports and progress reports were destroyed during the bombing or after the war. This statement is viewed by the Action Team as non-credible because documents from the rest of the Iraqi nuclear programmes continue to surface in abundance.

Determining whether Iraq has conducted any proscribed activities after the Persian Gulf War remains a priority for the Action Team and UNSCOM. Given the nature of the Iraqi regime, few accept that it has given up its nuclear weapon ambitions. However, there is no simple answer to how quickly Iraq could obtain nuclear weapons. Certainly, without sanctions, Iraq would find it far easier to reconstitute its nuclear weapon programme than if sanctions continue indefinitely.

Increasingly, the focus of the Action Team is to create the most effective ongoing monitoring and verification system that can deter proscribed activities or, failing that, promptly detect them. In the end, however, only the vigilance and determination of the Security Council can prevent Iraq from building nuclear weapons.

12. Countries of concern: Iran, Algeria, South Korea and Taiwan

I. Introduction

Several countries are suspected of wanting to acquire nuclear weapons. Iran is currently highest on this list of suspect nations, even though it is a party to the NPT. Debates about the exact status of its programme are highly controversial.

Algeria signed a safeguards agreement for a new research reactor and possible plutonium separation facilities and acceded to the NPT in 1995. It is included in this chapter because it was building these facilities in secret until this activity was exposed by US officials. In addition, political instability in Algeria raises long-term concern about the future of its programme.

South Korea took steps to acquire nuclear weapons in the 1960s and 1970s but, as a result of international pressure, it halted such activities. South Korea has remained a proliferation concern partly because of uncertainties about its response to possible North Korean nuclear activities (see chapter 10).

Despite having entered into agreements on full-scope safeguards, Taiwan is also in this category of countries of concern because its security situation with respect to China remains unsettled. During the 1970s and 1980s, Taiwan twice came under suspicion of having a nuclear weapon programme.

Libya and Cuba have at various times been suspected of developing the capability to make nuclear explosive materials. This chapter does not consider these countries because they do not appear to have made significant progress to date. In addition, Cuba decided to sign the Treaty of Tlatelolco in 1994, removing most fears that it will pursue nuclear weapons

Many other countries have tried or are suspected to have tried to acquire nuclear weapons, but are not seen as concerns now. Perhaps the best known example is Sweden, which had an active programme in the 1950s but decided to abandon its programme under intense domestic opposition. In 1996, Switzerland revealed that it maintained a secret option to build nuclear weapons that could have been exercised if, for example, Germany had acquired nuclear weapons.[1]

Some countries that are NPT states parties in good standing might reverse their commitment under certain conditions. If North Korea obtained a nuclear arsenal, Japan might reverse its long-standing commitment not to develop nuclear weapons. In early 1995, Egyptian officials linked their country's future

---

[1] Stuessi-Lauterburg, J., *Historischer Abriss Frage einer schweizer Nuklearbewaffnung* [Historical summary: question of a Swiss nuclear armament], 1995, provided by the Swiss Embassy in Washington, DC. See also Edwards, R., 'Swiss planned a nuclear bomb', *New Scientist*, 25 May 1996.

in the NPT to Israel taking steps in the direction of nuclear disarmament. Nevertheless, these countries are not covered by this chapter.

None of the countries examined in this chapter is believed to have nuclear weapons or fissile material dedicated to such purposes. Only Iran is believed to be actively engaged in seeking nuclear weapons. Because of this suspicion and the intense controversy about Iran's programme, this chapter focuses on Iran.

This chapter sets out to provide a better understanding, in qualitative terms, of how close these countries are to obtaining indigenously produced stocks of HEU or separated plutonium. Such an exercise is by its very nature highly imprecise. The chapter does not provide a comprehensive compilation of safeguarded stocks of fissile material in each country. Such information can be found in other chapters (see chapters 5 and 8).

II. Iran[2]

Western intelligence agencies have not discovered clandestine Iranian nuclear weapon facilities, nor have they, in fact, developed irrefutable evidence that Iran has a bomb programme. However, they have assembled a substantial body of evidence suggesting that, although Iran signed the NPT, it is secretly pursuing a broad, organized effort to develop nuclear weapons.

Although most reports place the programme under the authority of the Atomic Energy Organization of Iran (AEOI), some implicate the defence ministry in illegal foreign procurement activities and possible nuclear weapon work at military sites. Still other reports indicate that the Revolutionary Guard controls an independent, but fledgling, nuclear weapon effort. If Iraq's pre-1991 approach is taken as a model, an Iranian covert programme could depend on the existing overt, safeguarded programme.

Much of the evidence for an Iranian nuclear weapon programme dates back several years. US and European intelligence agencies have collected considerable information about suspicious procurement efforts in Europe and elsewhere that suggest military intentions. Western intelligence officials have closely followed reports that Iranian agents have travelled throughout the former Soviet Union in search of nuclear materials, know-how and scientists.

In 1992, for example, Iranians visited the Ulba Metallurgical Plant in Kazakhstan, a plant that produces reactor fuel and manufactures specialized metal components for the aerospace, electronics, and other defence industries. The plant also had a surplus inventory of up to about 600 kg of HEU, which the Iranians may have tried to buy. Upon learning of the Iranian interest in Ulba, the US Government asked the Kazakh Government to block any possible transactions with Iran. After deciding that physical security at Ulba was too weak to prevent diversion, the USA bought Ulba's HEU inventory in a secret, complex deal (see chapter 4).

[2] This section is adapted from Albright, D., 'An Iranian bomb', *Bulletin of the Atomic Scientists*, vol. 51, no. 4 (July/Aug. 1995); and Albright, D., 'The Russian–Iranian reactor deal', *Nonproliferation Review*, spring–summer 1995.

Controversy about Iran's possible intentions intensified in January 1995 when Russia signed a contract to complete the construction of a moribund nuclear power plant in Iran. The US Government repeatedly asked Russia to cancel the deal, arguing that Iran could use the reactor deal to bolster its nuclear infrastructure and as a cover to obtain sensitive nuclear technologies and equipment critical for producing separated plutonium or HEU.

Adding to suspicions, Russia's Minister of Atomic Energy, Viktor N. Mikhailov, and Iran's President of the AEOI, Reza Amrollahi, signed a secret protocol on 8 January 1995 declaring their commitment to negotiate additional contracts for research reactors and to develop a uranium mine, train AEOI scientific personnel at Russian academic institutions and build a gas-centrifuge plant for enriching uranium.

The US Government did not learn about the centrifuge plant deal until March or April 1995, at which point it stepped up its efforts to convince Russia to abandon the entire deal. In May 1995 President Yeltsin agreed that Russia would not supply a centrifuge plant. Yeltsin said then that the deal contained components with the 'potential for creating weapons-grade fuel', so 'we have decided to exclude those aspects from the contract'.[3]

Although being implemented slowly, most of the deal remains intact. As a result, this deal continues to be a major issue between the USA and Russia.

On 30 April 1995, President Clinton announced his decision to institute a US trade embargo on Iran. This embargo has attracted limited support from US allies. While they share US concern about Iran's intentions, they reject isolationist policies in this case.[4] In 1996, the USA passed a law requiring the imposition of sanctions on foreign corporations that make substantial investments in Iran. As expected, US allies have sharply criticized this law as counterproductive and driven mainly by domestic political considerations.

The Russian venture

The Russian–Iranian deal highlights the difficulty in determining a country's intentions under the current international regime of limited transparency (see chapter 15). Few dispute that the deal could go a long way towards boosting Iran's nuclear weapon programme. However, determining how, or whether, Iran would actually misuse this deal remains impossible under the present regime.

In addition, Iran has the right under the NPT to a full range of safeguarded fuel-cycle facilities, and Russia is desperate for the hard currency that this deal could provide. These two factors confound further attempts to verify that a legitimate civil activity is not benefiting a covert military one. Because of these considerations, Russian–Iranian nuclear cooperation further exacerbates the

---

[3] Office of the Press Secretary at the White House, 'Remarks by President Clinton and President Yeltsin in a joint press conference', Press Conference Hall, Moscow, 10 May 1995.
[4] Office of the Press Secretary at the White House, 'Remarks by the President at World Jewish Congress Dinner', Waldorf-Astoria Hotel, New York, 30 Apr. 1995.

current international tension over the spread of nuclear technologies into regions of tension.

In a sense, the deal dates back to the 1970s, when FR Germany agreed to build two 1300-MWe light water reactors at Bushehr, about 750 km south of Tehran. Reactor construction began in 1974, but Germany stopped work when Shah Reza Pahlevi's secular government was replaced by a fundamentalist Islamic government in 1979. Further, in 1980, Iran and Iraq became embroiled in an eight-year war, during which Iraqi bombing raids heavily damaged the partially finished reactor site. Germany refused to finish the reactors, partly because of its own suspicions regarding Iranian weapon intentions and partly because the USA exerted pressure.

The Russian–Iranian agreement was reached after several years of negotiation. Russia will receive $800 million to finish one of the two partially completed Bushehr power reactor stations. It also agreed to provide enriched uranium fuel and training in reactor operation for Iranian workers.

Despite the contract, the Russians and Iranians have disagreed about how much of the German-built structures can be used for the 1000 MWe Russian reactor, whose design is different from the German reactor that the buildings were supposed to hold. Russian officials estimated in 1995 that it would take five years to complete the project and supply the reactor, but delays are likely to push back the completion date by several years.[5]

The secret protocol outlined a wide range of nuclear assistance in addition to the building of a gas-centrifuge plant. The protocol pledged each government to instruct the appropriate agencies to prepare and sign—during the first quarter of 1995—contracts for the supply of a 30–50 MWth light water research reactor, 2000 tonnes of natural uranium, and the training of 10–20 AEOI graduates and graduate students annually at Russian academic institutions. The protocol also called for cooperation in building low power research reactors for instructional purposes, and construction of an Iranian desalination plant.

In addition, the protocol instructed both sides to prepare and sign, within six months, a contract for the construction of a shaft for a uranium mine, after which negotiations would be conducted on a contract for the construction of a gas-centrifuge plant. Although Russia has cancelled the centrifuge plant, it still apparently intends to build the mine shaft.

Russia also agreed in September 1995 to supply two VVER-440 LWRs.[6] However, construction is not expected to start in the near future. The projects included in the secret protocol have also been delayed. Any Russian-supplied nuclear facility would be intensely safeguarded by the IAEA, and Russian officials have said that additional steps would be taken to prevent their misuse.

Russian officials have been unclear about whether Russia will require the plutonium-laden spent fuel from power reactors to be returned to Russia. A few

---

[5] See, e.g., Hibbs, M., 'Russia–Iran Bushehr PWR project shows little concrete progress', *Nucleonics Week*, 26 Sep. 1996.

[6] 'Russia signs second contract for reactors in Iran', *Post-Soviet Nuclear and Defense Monitor*, 22 Sep. 1995.

officials have announced Russia's intention to take the fuel back unconditionally. Others have said that the fuel would be left in Iran. Some officials have even said that although the spent fuel could be sent back, Russian environmental law would require its ultimate return to Iran, and, if it were reprocessed, the law would require the return of all waste, recovered enriched uranium and separated plutonium.[7]

A large power reactor—or reactors—would eventually discharge great quantities of plutonium in spent fuel, raising the possibility that Iran could divert some of the plutonium for weapons, if it chose to someday abrogate its NPT commitments. Even if Russia takes back the spent fuel, large quantities will still be stored in cooling ponds at the reactor site at any given time.

Beyond that, Russian–Iranian cooperation will create major connections between their two nuclear programmes. Iran may expand its nuclear infrastructure and build its nuclear expertise by quietly soliciting help from Russian experts or companies, potentially benefiting a secret weapon programme.

Russia has the largest uranium-enrichment centrifuge programme in the world, and Iran's attempt to buy a plant from Russia has alerted many Western governments to the possibility that Iran may covertly seek out centrifuge assistance from Russian experts and companies, many of whom are desperate for business and may be willing to evade weakly enforced national export control laws. Because Russian centrifuges are subcritical models that are relatively unsophisticated compared to the most modern Urenco machines, they may be easier for Iran to develop and manufacture than the European ones. In addition, Russia built millions of them, and many are designed to be interchangeable, raising the possibility that Iran could somehow acquire large numbers of surplus machines (see chapter 4).

Russian officials appear well aware of the threat Iranian nuclear weapons would pose. However, they reject the often-made argument that Iran does not need nuclear power because of its vast oil resources. Iranian officials say that nuclear power will enable Iran to sell more oil on the international market.

In addition, Russian officials often point out that the IAEA has not found Iran in violation of its NPT commitments. As Valery Bogdan, the general manager of Minatom, said in an interview: 'From the strategic point of view, Iran is a close neighbour to Russia and it is very important for us to know what's going on with their nuclear programme. And this is possible only if we conduct joint projects with them.'[8]

None the less, the main motivation for the deal appears to be financial. Minatom needs the money, and Iran is eager to buy, despite its financial limitations. The number of proposed projects has led to speculation in the West that Minatom might be trying to get as much cash as it can. Those who interpret the

---

[7] Hibbs, M., 'Iran may keep Russian spent fuel or take plutonium, REPU and waste', *Nuclear Fuel*, 18 Dec. 1995.
[8] Interview with Valery Bogdan, Minatom, 'On the Russian sale to Iran, Nunn–Lugar, other US–Russian nuclear matters', *Post-Soviet Nuclear and Defence Monitor* (16 May 1995).

deal this way say that many of the items in the protocol are 'sweeteners' designed to keep the Iranians paying as long as possible.

Smoke, but no fire

Several years ago, some members of the Iranian leadership made statements about the country's need for nuclear weapons.[9] In more recent years, however, Iran has consistently denied having nuclear weapon intentions.

On 1 May 1995, Secretary of State Warren Christopher gave the clearest statement of the US position on Iran's nuclear ambitions: 'Based upon a wide variety of data, we know that since the mid-1980s, Iran has had an organized structure dedicated to acquiring and developing nuclear weapons.'[10] He added that, in terms of its 'organization, programmes, procurement, and covert activities, Iran is pursuing the classic route to nuclear weapons which has been followed by almost all states that have recently sought a nuclear capability'. Because Iran's industrial infrastructure cannot support a nuclear weapon effort, it must seek important weapon-related equipment and materials overseas.

Western intelligence officials—burned badly when Iraq's Saddam Hussein mounted an ambitious nuclear weapon programme, much of which escaped Western notice—have been charged by their respective governments with detecting nuclear weapon programmes at an early stage. And, because of their experiences uncovering Iraq's programme following the Persian Gulf War, they have become much better at detecting and interpreting isolated bits of evidence that might once have been overlooked.

US officials say that Iran is attempting to acquire nuclear technologies that are inconsistent with a strictly peaceful programme. Intelligence agencies have detected procurement patterns that point to a weapon programme. Seeking certain specific items or sets of items can be a compelling signature of plutonium separation, uranium enrichment or weaponization activities, but information about attempted purchases gives no reliable picture of what has actually been procured, where the items are located or how the equipment is used.

Although much of the intelligence is secret, according to US officials, it is shared among Western allies. President Clinton reportedly personally gave President Yeltsin a five-page, single-spaced US intelligence document on Iran's programme.[11]

The United States imposed a virtual embargo on the export of nuclear-related equipment to Iran many years ago. Other leading industrialized democracies followed the US lead, ending nuclear trade with Iran and more closely scrutinizing high-tech exports, even if their companies lost business as a result.

Western intelligence officials believe Iran's programme is at an early stage. One IAEA official said any programme to separate plutonium or enrich

[9] See, e.g., Coll, S., 'Tehran ambiguous on its A-arms plans', *Washington Post*, 17 Nov. 1992.
[10] Office of the Spokesman, 'Press briefing by Secretary of State Warren Christopher on the President's Executive Order on Iran', US Department of State, 1 May 1995.
[11] Hoagland, J., 'Briefing Yeltsin on Iran', *Washington Post*, 17 May 1995.

uranium is at a 'pre-research and development stage'. Unlike North Korea, Iran is not believed to be operating secret or unsafeguarded pilot-scale nuclear facilities. News reports about uranium enrichment plants, nuclear weapons plants, or nuclear weapons in Iran are highly speculative.

Although the IAEA has the right to request 'special inspections' of sites in non-nuclear weapon states parties to the NPT, US intelligence officials have repeatedly stated that Iran has no facilities that yet warrant special inspections. To help clarify the situation, however, the IAEA asked for and received permission to visit sites in Iran. IAEA inspectors made two unofficial visits in 1992 and 1993 and reported that activities at all the sites it selected to visit were consistent with peaceful uses. These results cannot be considered conclusive, however, because of the size of the sites and the fact that Iran would not allow the IAEA to deploy its full range of inspection methods, including environmental monitoring. As of September 1996, Iran was still not allowing the IAEA to start environmental monitoring at Iran's declared sites, despite the IAEA's new rights to conduct such inspections under its strengthened safeguards programme (see chapter 15).[12]

It is generally agreed that—as yet—there is no Persian bomb in a Tehran cellar. According to a senior intelligence official, the Iranian nuclear weapon programme suffers from poor management, a paucity of scientifically and technically trained people, and a lack of infrastructure.

On 10 January 1995, then-Director of Central Intelligence James Woolsey told a Senate intelligence committee that the most likely scenario would be for Iran to continue developing its indigenous resources. That might give it the capability to produce a nuclear weapon by early in the next century.[13]

On the same day, then-Israeli Prime Minister Yitzhak Rabin and US Defense Secretary William Perry agreed at a press conference in Israel that it might take Iran—at its present pace—7 to 15 years to develop nuclear weapons.[14] This public estimate seemed to discount an estimate leaked days before to *The New York Times*, in which 'senior' US and Israeli officials apparently said that Iran might be able to build a bomb in five years.[15] Once the full extent of the Russian–Iranian nuclear deal became known, US intelligence officials said that they lowered their estimates to five years or less. Current estimates are not publicly available.

Despite these accounts, little reliable information exists about the status of an Iranian nuclear weapon effort. There are no compelling accounts of whether Iran views a nuclear weapon programme as a major priority or simply part of contingency planning which could include a plan to respond to the resurgence

---

[12] Hibbs, M., 'Iran balking at approval of IAEA environmental monitoring', *Nuclear Fuel*, 23 Sep. 1996.

[13] Director, CIA, James Woolsey, Hearing on *Global Threat Assessment*, Select Committee on Intelligence, US Senate, 10 Jan. 1995.

[14] Haberman, C., 'US and Israel see Iranians "many years" from A-bomb', *New York Times*, 10 Jan. 1995.

[15] Hedges, C., 'Iran may be able to build an atomic bomb in 5 years', *New York Times*, 5 Jan. 1995.

of an Iraqi nuclear weapon programme (see chapter 11). After all, if Iraq had succeeded in building nuclear weapons, Iran may have been an intended target.

What they need

As is pointed out throughout this book, the key to making an atomic bomb is getting the fissile material, and predictions of when Iran might succeed in building an indigenous capability to produce HEU or separated plutonium are fraught with uncertainty. They hinge on estimates of when Iran might overcome organizational difficulties, establish clear priorities, procure key items, and solve technical design problems, all while under scrutiny by intelligence agencies, export control organizations and the IAEA.

Shortly before the full details of the Russian deal became known, a senior US intelligence official told one of the authors that he was comfortable with the minimum estimate of seven years to a bomb mentioned above—if Iran did not get fissile material from abroad and if it was forced to develop production facilities on its own. The right type of foreign assistance or procurement could shorten the time Iran would need to build nuclear weapons, however. Further, Iran has the financial resources to buy black-market fissile material if it becomes available, or to organize a theft from vulnerable facilities in the former USSR.

However, depending solely on fissile material from abroad is risky. Iran could fail to acquire any material or, if successful, it might not obtain enough to build what its leaders would consider a credible arsenal. As a result, Iran's strategy is likely to include plans to develop indigenous facilities to produce fissile materials as well. In addition, it would need other facilities to process the fissile material into bombs and to make other weapon components.

Here, in summary, is what US officials and the public record say about Iran's multifaceted strategy to develop the option to make nuclear weapons:

1. Iran has sought, with limited success, to buy nuclear power and research facilities from many countries, particularly China and Russia. Acquiring such facilities is permitted under the NPT. While they would be under IAEA safeguards, however, these facilities could bring Iran significantly closer to a nuclear weapon capability and provide a cover for secret, illegal procurement activities in supplier countries.

2. Iran has shopped quietly in many countries, particularly in Europe, for a wide range of nuclear-related or 'dual-use' nuclear items that might enable it to put together facilities to enrich uranium, separate plutonium and make nuclear weapons. US officials often point to Iran's attempt to obtain items needed to manufacture Urenco-style gas centrifuges. Although East European countries are mainly mentioned as part of Iran's smuggling routes out of Western Europe, Iran could target them more frequently for sensitive items, if West European countries continue to thwart Iranian purchases. There is little public information

about how effective this clandestine shopping has been or which countries have been contacted.

(A 1993 Russian Foreign Intelligence Service report said that Iran had created a system to purchase dual-use technology that evaded international export controls. It likened the Iranian strategy to the techniques used earlier by Pakistan and Iraq as they pursued their respective weapon programmes.[16])

3. Although US intelligence officials say they have no evidence that Iran has succeeded in buying fissile material abroad, it is not for want of effort. Earlier reports that Iran stole two nuclear weapons from the former USSR have been discredited, but press reports that Iran has tried to obtain fissile material or nuclear weapons in Russia and elsewhere are viewed as credible by intelligence agencies.[17] As a result, intelligence agencies continue to scrutinize behaviours in both the former Soviet republics and Iran, looking for any indication that fissile material is missing and is in Iran.

What they have

Shah Pahlevi had a small nuclear weapon R&D programme until 1979 when he was ousted, but apparently the programme made little progress. Whether or not today's programme is a continuation of the earlier effort is unclear, but most reports date the current programme to the mid-1980s. According to US officials, Iran's current programme may have benefited from the following:

US assistance

During the 1960s, the USA supplied Iran with a 5 MWth research reactor. Located at the Tehran Research Centre, it now runs on low-enriched uranium fuel supplied by Argentina. In addition, the United States supplied Iran with 'hot cells' and training in their use. However, the hot-cell facility is small and is capable of separating only grams of plutonium. It is unclear whether the facility has been expanded recently, but the hot cells are under IAEA safeguards.

The Chinese connection

Since the mid-1980s, China has been Iran's chief supplier of nuclear-related technologies—despite US efforts to stop China from supplying Iran. China has supplied three subcritical and zero-power reactors and EMIS machines. China is also providing a very small, 30 kWth research reactor. None of this hardware is believed to be capable of producing more than minute quantities of nuclear

---

[16] Russian Foreign Intelligence Service Report, *Proliferation of Weapons of Mass Destruction: A New Challenge after the Cold War*, Moscow, 1993 (translated by the US Foreign Broadcast Information Service, Feb. 1993), JPRS Report, JPRS-TND-93-007, 5 Mar. 1993.

[17] See, e.g., Reuter, 'German agency warns of Iran nuclear terrorism report', 20 July 1996. The article quotes a confidential German intelligence (BND) report that there is 'scarcely a doubt remaining that Iran is interested in buying nuclear materials' and that Iran 'is interested with utmost certainty in fissionable material on the black market'.

weapon material, but the small reactors might be useful for training personnel, and the EMIS machines could be reverse-engineered and enlarged.

According to a senior US intelligence official interviewed by one of the authors in 1995, China has also helped Iran create nuclear fuel facilities for uranium mining, fuel fabrication, uranium purification and zirconium tube production. A 1996 media report, based on a US intelligence document, said that China was about to provide a plant near Esfahan that would make uranium products, either natural uranium fuel or uranium hexafluoride.[18] The deal had been under negotiation for several years, and the press report coincided with the visit of a Chinese delegation to Iran slated to start work on the plant.

In 1992 China signed a 'preliminary agreement' to supply Iran with two 300-MWe light water reactors, but it is unlikely that the reactors will ever be supplied. AEOI head Reza Amrollahi told *The New York Times* in May 1995 that Iran made a down payment on the reactors, and China had started to draw up blueprints and engineering reports for a site in southern Iran.[19] However, the two sides did not implement the contract because of economic and technical problems, as well as disagreements as to where to build the reactors. In the fall of 1995, China announced that negotiations for two LWRs had been suspended for the time being. Subsequent Chinese comments appeared less clear, but tended to suggest that China would not sell Iran a LWR.[20]

Despite the extent of its assistance, China is not believed to have helped Iran with weapon design or manufacture. China's aid to Iran is believed to be consistent with the NPT, and the Chinese-built facilities are under IAEA inspection.

Obtaining fissile material

The information available does not indicate how Iran might produce unsafeguarded fissile material, or even if it has chosen one particular method. A senior US intelligence official told one of the authors in 1995 that intelligence information indicates Iran has not yet made a choice.

As noted above, Iran might steal fissile material or buy it on the black market, either in pure or impure forms. Constructing facilities to purify plutonium or to produce HEU and fashion bomb components is probably within Iran's reach, although some foreign procurement might be necessary. Alternatively, or in parallel to attempts to acquire fissile material overseas, Iran may seek to develop the capability to make and separate plutonium or to enrich uranium.

---

[18] Gertz, B., 'Iran gets China's help on nuclear arms', *Washington Times* 17 Apr. 1996; Gertz, B., 'US fears Iran's use of China's know-how', *Washington Times*, 18 Apr. 1996; and Bill Gertz, personal communication, 17 Apr. 1996.
[19] Sciolino, E., 'Iran says it plans 10 nuclear plants but no atom arms', *New York Times*, 14 May 1995.
[20] See, e.g., 'China/Iran: Reactor plans shelved—again?' *Nucleonics Week*, 11 Jan. 1996.

The plutonium path

As yet, Iran does not possess a nuclear reactor that can produce significant quantities of plutonium, despite years of trying to obtain one, nor does it have a plutonium separation plant.

As Warren Christopher said on 1 May 1995: 'For years, Iran has been trying to purchase heavy-water research reactors that are best suited to producing weapon-grade plutonium, not electricity.'[21] Several countries have cancelled the supply of reactors and reactor-related facilities, often under pressure from the USA. Argentina cancelled a contract to supply equipment to convert uranium chemically and manufacture nuclear fuel, probably for a research reactor. India cancelled the supply of a 10-MWth heavy water, natural uranium reactor. China cancelled a 20–30 MWth reactor. The Chinese would have also provided enriched uranium fuel and equipment to allow fuel-cycle research.[22] Russia is reported to have stopped its negotiations to sell Iran a natural uranium reactor in the early 1990s.[23]

Nevertheless, if the current Russian–Iranian deal proceeds, Iran will eventually have at least one large LWR that would routinely discharge in spent fuel as much as a few hundred kilograms of plutonium each year. If the fuel burnup was reduced, perhaps during a national security crisis, the reactor could produce significant quantities of weapon-grade plutonium (see chapter 10).

Meanwhile, Iran's hot cells have given it the capability to learn how to process irradiated fuel. Intelligence agencies have briefed Western governments that Iran has processed irradiated targets in these hot cells outside of safeguards. Although the hot cells are a long way from a facility to separate significant amounts of plutonium, they can give Iran valuable experience in plutonium separation. Iran could pursue a strategy similar to that of Iraq, which used safeguarded reactors secretly to produce small amounts of separated plutonium (see chapter 11).

It is not known whether or not Iran is trying to build an indigenous plutonium production reactor and associated plutonium separation facility. If such facilities were in the design stage, detection would be almost impossible, but if construction were under way, such a site would be vulnerable to detection.

The HEU path

Little information is available about Iran's suspected enrichment programme. Although the West suspects that Iran is pursuing EMIS and laser-enrichment techniques, most experts believe that its enrichment programme concentrates on gas centrifuges. This conclusion was further supported by the revelation that the secret protocol signed by Russia and Iran contained a commitment to supply a gas-centrifuge plant.

---

[21] Christopher (note 10).
[22] Coll, S., 'US halted nuclear bid by Iran', *Washington Post*, 17 Nov. 1991.
[23] Greenhouse, S., 'Russia and China pressed not to sell A-plants to Iran', *New York Times*, 25 Jan. 1995.

EMIS. Iran obtained up to three EMIS machines from China. They were reportedly a part of a deal for the supply of the small reactor at Esfahan. The declared purpose of the EMIS machines is to produce stable isotopes that would be irradiated in the reactor and converted into radioactive materials useful in research and medicine. The IAEA looked at a machine during a 1993 visit to Karaj and confirmed this end-use. Typically, EMIS separators can easily be converted to uranium enrichment, although evidence of such conversion, even if temporary, should remain.

The machine's current is believed to be small, roughly 1 milliampere. Although in theory capable of producing only minuscule amounts of enriched uranium, an operational 1-milliampere EMIS machine could be useful in the development of larger machines (see chapter 11). The small separator could also be reverse-engineered and expanded in size.

However, Iran would first need to put together an industrial infrastructure in order to make the necessary components or find a way to buy them abroad. In an interview with one of the authors, a US official expressed the opinion that Iran is a long way from either prospect.

Laser enrichment. Iran is also periodically reported to be working on a laser uranium enrichment project that was begun under the Shah. Many, however, consider Iran unlikely to make significant progress in this area. Some reports say that Iran recently obtained a copper-vapour laser from China, which suggests an on-going interest in this enrichment technology.

Gas centrifuges. According to a senior US Government official interviewed by one of the authors in 1995, Iran is concentrating on centrifuge designs and looking towards a pilot plant, possibly large enough to produce sufficient HEU for nuclear weapons, with hundreds or thousands of centrifuges connected together in cascades.

Iran currently lacks enough skilled scientists and technicians to manufacture the centrifuges, reflecting its generally low ability to mount advanced science and technology projects. It would have to import both know-how and equipment, and it is finding that getting assistance is difficult. A fraction of its attempts have been detected and are an important part of the evidence that leads the USA to conclude that Iran is pursuing nuclear weapons.

The same official said that Iran must have obtained some early-generation European designs—for G1 and G2 centrifuges. He said that such designs were 'around,' but refused to say more, except that Iran was looking into them.

It is unclear whether Iran has obtained a full set of blueprints, an assembly drawing, or a partial set of component drawings. One indicator that Iran possesses designs is that it approached a company to obtain a sample of a maraging-steel preform, whose dimensions matched those for a centrifuge.

The possession of designs, even if incomplete, and a procurement strategy for key materials, know-how and components could allow Iran to skip many difficult research steps, speeding the ultimate construction and operation of cascades.

It is not likely, this official added, that Iran was operating any centrifuge test stands containing uranium hexafluoride, although he implied that some low-level centrifuge testing is occurring. He was unwilling to specify the type of testing. A report in early 1996 said that US intelligence agencies had accumulated data supporting their allegation that Iran has a programme to 'develop and bench-test' centrifuges.[24] However, the data do not support the claim that Iran has introduced uranium hexafluoride into a test stand. Such a step would also require reporting to the IAEA.

US officials refer to a long list of Iranian procurement attempts in Europe and elsewhere that potentially relate to centrifuges. In addition to the maraging-steel preforms mentioned above, items sought include computer numerically controlled machine tools, high-strength aluminum alloys and uranium hexafluoride. The uranium hexafluoride was placed under IAEA safeguards and involves only kilogram quantities.

Another suspicious procurement attempt, according to a European intelligence agency, was Iran's attempt to procure 'centrifuge components' from an Austrian firm. This company is known for making cobalt-samarium ring magnets, and unwittingly exported many such magnets to Iraq for its centrifuge programme.

Some of the items date to the period of the Shah. In about 1975, Iran obtained a 'horizontal flow-forming machine' suitable for making maraging-steel rotors from a European company.

III. Algeria

In April 1991 the press reported that Algeria was secretly building a Chinese-supplied reactor and research facility at Ain Oussera, a remote site in the Atlas Mountains about 125 km south of Algiers.[25] Soon after the presence of the reactor was revealed, the Algerian Government agreed to place it under IAEA safeguards. A safeguards agreement was signed in late February 1992. The reactor, named Es Salem, was inaugurated in December 1993 and remains subject to IAEA inspections. In January 1995, Algeria became a party to the NPT. It is reported to have allowed the IAEA to start conducting environmental monitoring at the site.

In 1991, Algeria was not under any international obligation to divulge the existence of its nuclear facilities or to put them under IAEA safeguards (unless required to do so by suppliers of its materials and equipment), because it was not then a party to the NPT. Nevertheless, the Government's lack of candour

[24] Hibbs, M., 'IAEA will explore new charges Iran has enrichment programme', *Nucleonics Week*, 22 Feb. 1996.
[25] See, e.g., Gertz, B., 'China helps Algeria develop nuclear weapons', *Washington Times*, 11 Apr. 1991, p. A7; and Gupta, V., 'Algeria's nuclear ambitions', *International Defense Review*, Apr. 1992, pp. 329–31.

about this reactor, even after its existence was exposed by the media, raised concerns about Algeria's intentions. Adding to this concern, Algeria is building a large hot-cell facility that is notable by its connection via a covered canal to the reactor's core, raising suspicion that Algeria intends to reprocess the reactor's spent fuel. Beyond lingering concerns about the current regime, political instabilities in Algeria could lead to a nationalist regime that might not abide by previous governments' commitments.

The Salam reactor

China supplied this reactor under a 1983 agreement. The reason both countries kept the existence of the reactor secret remains unclear.

According to the IAEA, the reactor has a power of 15 MWth, uses LEU fuel, and is moderated and cooled by heavy water. Early press reports, based on analysis of the reactor's cooling towers, said that the power level might go as high as 60 MWth. However, this claim has been discounted.

A 15-MWth reactor of this type is theoretically capable of producing about 2–3 kg of plutonium per year in the LEU fuel.[26] However, this particular type of reactor has many extra tubes that could hold natural uranium targets, boosting production in the targets up to 5 kg of weapon-grade plutonium per year.

Plutonium separation?

The hot cell facility next to the reactor has been under construction since at least 1992. This facility has been declared to the IAEA but, as of August 1996, it has not been inspected because irradiated fuel has not entered it. This facility contains large hot cells, but the public information is not sufficient to determine whether Algeria intends to separate plutonium in it.

According to Western officials, there is another large facility nearby that was not observed by intelligence agencies until its roof was in place, preventing an identification of its purpose via overhead surveillance. Algeria has not declared this facility to the IAEA as a nuclear facility. Nevertheless, some officials believe this facility may have been intended as a large-scale plutonium separation or reprocessing facility. The large building is made out of heavy concrete and may also have hot cells. Next door is a facility containing many large stainless steel tanks with pans below them, raising suspicions that they were built to hold several years' worth of high-level nuclear waste left over after reprocessing.

---

[26] It is assumed that the reactor uses about 3% enriched uranium fuel, operates 70% of the time and produces about 0.6 g of plutonium per MWth-d.

IV. South Korea

South Korea worked to develop nuclear weapons under President Park Chung-hee in the 1970s, but the USA forced South Korea to stop the programme before a device could be finished. Despite this, as long as North Korea is suspected of having nuclear weapons, South Korea will have, at least, a vocal minority that will want South Korea to develop them too.[27]

South Korea is a party to the NPT and has signed an agreement pledging a nuclear weapon-free Korean Peninsula in 1991. This bilateral agreement bans reprocessing and uranium enrichment activities on the Korean Peninsula. Nevertheless, these commitments could be undermined if North Korea reneges on its commitment to dismantle its nuclear weapon programme (see chapter 10).

With a large nuclear power programme and an extensive nuclear R&D infrastructure, including hot cell facilities, South Korea is viewed as capable of rapidly separating plutonium and turning it into nuclear weapons. However, if South Korea decided to renounce its NPT commitments and acquire a nuclear arsenal, it might need to obtain plutonium from its safeguarded power reactors, all of which are restricted to peaceful uses as a condition of their original supply. North Korea could also face a similar situation in the future (see chapter 10).

Because it opposes South Korea stockpiling separated plutonium, the USA has tried to stop any South Korean attempt to develop a reprocessing capability. The US Government has limited the size of a post-irradiation examination (PIE) facility at the Daeduk research site facility, which is responsible for R&D of nuclear fuel-cycle technologies.[28] The PIE hot cells, supplied by France, dissolve US-supplied power reactor spent fuel in order to diagnose fuel problems or improve fuel quality. This type of facility does not normally separate plutonium, although it could be adapted to do so. The USA is evidently worried that if South Korea increases the size of the PIE facility, it could be laying the basis for a domestic plutonium separation capability.

Reports of continued South Korean interest in nuclear weapons surface periodically. In 1994, Suh Sujong, chief policy analyst of the South Korean Democratic Liberal Party and former chief secretary to the head of the Agency for National Security, said that, as recently as 1991, South Korea planned to develop nuclear weapons in response to the North Korean crisis but was forced to stop by the United States.[29] This official did not provide details about the status of the weapons effort during the 1987–93 Administration of President Roh Tae-woo, but he said that nuclear weapon experts employed at the Daeduk research facility were forced to leave under US pressure.

[27] See, e.g., Sanger, D., 'Wary of North Korea, Seoul debates building atomic bomb', *New York Times*, 19 Mar. 1993.
[28] Hibbs, M. and Ryan, M., 'Japan's plutonium plans spotlight Asia's emerging nuclear issues', *Outlook on Asian Nuclear Power*, 30 June 1994, special report published by *Nucleonics Week*.
[29] Shin, P., Associated Press, 'US said to stop South Korea's nuke bomb plans: Official claims project ended in 1991', *Washington Times*, 29 Mar. 1994.

Possibly adding to this concern was South Korea's announcement in the mid-1980s that it was building a 30-MWth research reactor at Daeduk. Although this reactor is heavy water moderated, it uses 19.75 per cent enriched fuel. At this level of enrichment, it will produce insignificant amounts of plutonium. The reactor went critical in early 1995.

V. Taiwan

Although Taiwan is an original NPT signatory and has consistently denied any intention of building nuclear weapons, it is suspected of creating programmes to separate plutonium for nuclear weapons. In addition, its security situation remains uncertain as China continues to subject it to intense political and military pressure. With an extensive civilian nuclear programme, and a self-proclaimed technical capability to build nuclear weapons, Taiwan will remain a proliferation concern until it reaches a political accommodation with China.

Secret efforts

According to a declassified 1974 CIA assessment, Taiwan conducted 'its small nuclear programme with a weapon option clearly in mind'.[30] The 40-MWth Taiwan Research Reactor (TRR), which was supplied by Canada in 1969 and is identical to the Indian Cirus reactor, was widely viewed as the centre-piece of that programme. In 1977 the United States pressured Taiwan to stop construction of a hot cell facility that could have separated plutonium from TRR fuel, possibly for use in nuclear weapons.

In 1988, the US Government leaked that it had learned that Taiwan was secretly building an installation capable of extracting significant amounts of plutonium.[31] As in the earlier case, this plutonium would have come from TRR fuel. In response to US pressure, Taiwan not only stopped work on this plutonium separation facility, but also shut down the TRR.

Spent fuel shipments

Even before the 1988 episode, the USA had decided to act to reduce a growing inventory of high-quality plutonium in TRR spent fuel rods. In 1985 the USA convinced Taiwan to send the spent fuel to the USA, reducing the risk that Taiwan might separate the plutonium for use in nuclear weapons. Because the fuel contains uranium metal, it is difficult to store safely in water for an extended period. US officials worried that Taiwan would insist on reprocessing this fuel, citing health and safety concerns.

---

[30] Central Intelligence Agency, 'Memorandum: Prospects for further proliferation of nuclear weapons', DCI NIO 1945/74, 4 Sep. 1974, sanitized version.
[31] Engelbery, S. and Gordon, M. R., 'Taipei halts work on secret plant to make nuclear bomb ingredient', *New York Times*, 23 Mar. 1988.

Despite the non-proliferation benefits, these fuel shipments conflicted with safety and environmental concerns within the United States about spent fuel transportation. Originally, the fuel rods were to be sent by ship to the west coast of the United States and then by road to the Savannah River Site in South Carolina for reprocessing, but intense local opposition forced the US Government to ship them directly to the east coast and then to the Savannah River Site.[32] Because the uranium that gave rise to the spent fuel was not of US origin, opponents of the shipments worried that this could open the way to the United States becoming a dumping ground for foreign spent fuel. They also objected to the fuel being reprocessed in military facilities, although the US Government said that the Taiwanese plutonium would not be used in weapons.[33]

Most of the spent fuel was shipped to the USA. In 1991, however, a Federal Court blocked the last shipment of 118 fuel rods until officials detail the environmental risk.[34] As of the summer of 1996, it is unclear whether the last batch of TRR spent fuel will be shipped to the USA. Although the parallel legal and political battle over the return of fuel of US origin has been settled, the TRR spent fuel remains outside of this settlement because it is not of US origin (see chapter 8).

Plutonium production

The TRR operated for about 14 years and during this time discharged about 1600 fuel rods, containing plutonium with more than 90 per cent ^{239}Pu.[35] A DOE official told one of the authors in September 1996 that the vast bulk of the plutonium was in fact weapon-grade. The DOE has declassified the information that in total the USA received about 78 kg of plutonium in spent fuel from the TRR, and separated about 63 kg.[36] Since all shipments but the last were received, and this shipment was scheduled to contain 118 rods, the 1480 rods received by the USA contained an average of about 53 grams per rod.

Based on this information, the reactor produced a total of about 85 kg of high-quality plutonium. If 5 kg are assumed per atomic bomb, Taiwan had enough plutonium in TRR spent fuel for about 17 weapons.

Of the 118 rods which remain in Taiwan, about half of them have failed in the sense that they have either visible damage to the cladding or release above a certain amount of radioactive material through small holes or cracks in the clad-

---

[32] Cipriano, R., 'Nuclear fuel rods re-routed to Virginia', *Los Angeles Times*, 11 July 1986.

[33] The US Government actually said that the Taiwanese plutonium would be reprocessed at a military reprocessing site and the plutonium would enter the military stock; however, a different but equivalent amount of plutonium would be committed to peaceful uses.

[34] Roberts, A., 'Judge blocks nuclear waste from port', *Virginian Pilot*, 10 Dec. 1991.

[35] This value is derived from US Department of Energy, 'Environmental assessment on shipment of Taiwanese Research Reactor spent fuel (Phase II)', DOE/EA-0363, June 1988. Table D.1 lists the radioactive inventory of the fuel rods, where a burnup of 1600 MWd/t of heavy metal was assumed. The burnup of many rods was actually considerably less.

[36] US DOE, *Plutonium: The First Fifty Years* (DOE: Washington, DC, Feb. 1996), p. 43.

ding.³⁷Assuming roughly 50 grams of plutonium per fuel rod, these rods contain about 6 kg of plutonium containing over 90 per cent plutonium-239, enough for one nuclear weapon.

Will TRR reopen?

In mid-1996, Taiwan announced that it was considering reopening the TRR.³⁸ According to a knowledgeable US official, Taiwan never promised that it had permanently shut down the reactor. However, Taiwan also sent the USA the TRR's heavy water, without which the natural uranium-fuelled reactor cannot operate. Unless Taiwan has an unknown source of heavy water, the TRR would need to be completely redesigned to use light water and enriched uranium, severely degrading its plutonium production capability and requiring a foreign supply of fuel.

---

³⁷ US Department of Energy, 'Environmental assessment of the risks of the Taiwan Research Reactor spent fuel project', DOE/EA-0515, May 1991.
³⁸ Reuter, 'Taiwan mulls reopening research nuclear reactor', 1 July 1996.

13. Countries backing away from nuclear weapons: Argentina, Brazil and South Africa

I. Introduction

In the 1970s and 1980s, the conventional wisdom was that threshold countries were unlikely to back away from their programmes to produce unsafeguarded fissile materials or nuclear weapons, but several countries have shown that proliferation is reversible.

In the early 1990s, Argentina and Brazil took concrete steps to unveil and roll back their long-secret nuclear programmes and agreed to permit IAEA inspections of all their nuclear facilities. In early 1995, Argentina acceded to the NPT.

In the summer of 1991, South Africa acceded to the NPT and agreed to place all its nuclear materials in the country under IAEA safeguards. South Africa had the largest unsafeguarded enrichment programme outside the nuclear weapon states. In March 1993, President F. W. de Klerk announced that South Africa had had a nuclear weapon arsenal and that his government had decided to dismantle this arsenal of seven nuclear weapons in 1989. This act made South Africa the first country ever to abolish a nuclear weapon arsenal.

II. Argentina and Brazil

In 1991 the parliaments of Argentina and Brazil ratified a bilateral inspection agreement that created the Brazilian–Argentine Agency for Accounting and Control of Nuclear Materials (ABACC). On 13 December of the same year the quadripartite Agreement on the Exclusively Peaceful Utilization of Nuclear Energy between Argentina, Brazil, the IAEA and ABACC was signed, committing Argentina and Brazil to international inspections of all their nuclear facilities. The two countries ratified the Treaty for the Prohibition of Nuclear Weapons in Latin America and the Caribbean (the Treaty of Tlatelolco) in 1994 and, on 10 February 1995, Argentina signed the NPT.

These agreements are the result of several steps the two countries have taken since the mid-1980s to signal their commitment to eliminating any hint of nuclear weapon programmes. Perhaps the most dramatic were taken in September 1990: Brazilian President Fernando Collor de Mellor closed a potential nuclear explosive test site in the Amazon Basin and cancelled a secret 15-year-old atomic bomb project, named 'Solimoes' after a section of the Amazon.

Under the IAEA safeguards agreement, inspections must apply at the uranium enrichment facilities of both countries, a concession that was opposed by some members of their nuclear establishments and the Brazilian military.[1] Safeguarding these particular facilities is important because they are the main ones that many analysts and governments suspected of being built to produce fissile material for nuclear explosives. Brazil also has built a small jet-nozzle enrichment plant supplied by Germany, but it has always been under IAEA safeguards and is not considered further in this chapter.

Both governments have stated as a matter of policy that they will not produce highly enriched uranium in their enrichment plants. ABACC and IAEA safeguards at the enrichment plants are necessary to verify this pledge.

Argentina's secret enrichment facility

Argentina surprised the world with its announcement in November 1983 that for the previous five years it had been secretly building an unsafeguarded gaseous-diffusion uranium enrichment facility in the hamlet of Pilcaniyeu in the Rio Negro province. At the time, the announcement revealed a failure of Western intelligence gathering.

According to official statements at the time of the announcement, the primary purpose of the Pilcaniyeu facility would be to produce up to 20 per cent enriched uranium for use in domestic research reactors and for export as well as to produce slightly enriched uranium for power reactors. It might also be used to make naval reactor fuel if Argentina ever developed a nuclear-powered vessel. Although the civil purpose of this facility is no longer in doubt, strong suspicions remain that the project was initiated in the late 1970s to give Argentina the ability to make nuclear explosives.

When the existence of the plant was announced in 1983, Argentinian nuclear officials expected the plant would soon reach its full capacity of 20 000 SWU per year. This would correspond to the production of 500 kg of 20 per cent enriched uranium. In the longer term, Argentina stated that it planned to expand the capacity of the plant to 100 000 SWU per year.

The plant has been beset with delays because of severe budget cut-backs and serious technical difficulties. It has had problems with short barrier lifetimes, leaking seals and compressor reliability. As a result, it has never functioned well. The Argentinian governments that followed the end of the military dictatorship in 1984 were unwilling to devote the level of resources necessary for overcoming the plant's technical problems and for achieving successful operation of the plant. In the mid-1980s Argentina did manage to finish at least a portion of a cascade. While there is little information about the enrichment levels achieved or the quantities of enriched uranium produced, it is clear that a relatively small amount of enriched material was produced, and only at low levels of enrichment. The cascade was composed of 20 'models', each containing 20 stages, for a total of 400 stages.

[1] Hibbs, M., 'Brazil's military may block safeguards with Argentina', *Nucleonics Week*, 28 Nov. 1991; and Collina, T. Z. and Barros, F. S., 'Transplanting Brazil and Argentina's success', *ISIS Report* (Institute for Science and International Security: Washington, DC, Feb. 1995).

Faced with delays at the plant, Argentina ordered 150 kg of 20 per cent enriched uranium from the Soviet Union. Earlier, it had obtained an unspecified amount of unsafeguarded 20 per cent enriched uranium from China.

Because of problems, the first cascade was closed in 1989. Argentina, however, continued to develop this technology. It opened a renovated pilot or 'mock-up' plant in December 1993, which had about 20 stages with new separation membranes and more advanced equipment. The plant operated for two months and allowed the government, the US DOE and the Argentinian nuclear commission to conduct joint safeguard studies. Because a gaseous-diffusion plant had never been safeguarded before, certain safeguards techniques required development before safeguards could be applied at Pilcaniyeu.

US officials who visited the site in late 1994 said that a refurbishment of the 400-stage cascade was expected to be finished in April 1995.[2] As of May 1996, this cascade was still not operating and did not appear to be nearing readiness. If completed, this cascade would be able to produce a maximum of 5 per cent enriched uranium. Efforts to reach 20 per cent were dropped a couple of years ago. US officials were also told that Argentina might send the 5 per cent material to Brazil for enrichment up to 20 per cent.[3] Argentinian officials also still talk about building a second enrichment building to boost capacity, but few believe this will happen.

Moreover, the reorganization of the Argentinian nuclear establishment has raised doubts that the 400-stage cascade will be renovated. The reorganization is expected to cause a more intense scrutiny of the economics of Pilcaniyeu. Western officials estimate its price tag at hundreds of millions of dollars.

One proposed purpose of this refurbished cascade is to evaluate the costs of producing slightly enriched uranium for the Atucha-1 nuclear power reactor. Argentine officials have said that the use of slightly enriched uranium instead of natural uranium would lead to significant increases in burnup. As a result, the economic advantage gained at these higher burnups would justify operating Pilcaniyeu. US experts, however, have greeted this result with scepticism.

Despite the technical problems, Argentina has produced small quantities of low-enriched uranium at Pilcaniyeu. However, the information available is insufficient to estimate total production or to determine the amount of unsafe-guarded enriched uranium imported by Argentina before inspections started.

Public reports imply that the quantities produced domestically have been significantly smaller than planned. John Redick, a US expert on Latin American nuclear programmes, says that based on interviews he conducted in Argentina in 1994, the enrichment levels at Pilcaniyeu might not have exceeded a few per cent ^{235}U, and only small amounts of enriched uranium were produced. In any case, both ABACC and the IAEA are involved in verifying Argentina's initial

---

[2] Lubenau, J., 'Trip Report, XXII Annual Meeting of the Argentine Association of Nuclear Technology and visits to Argentine nuclear facilities', US Nuclear Regulatory Agency, Washington, DC, 1–9 Nov. 1994.

[3] Lubenau (note 2).

nuclear inventory. The IAEA's initial inventory investigations at Pilcaniyeu began in late February 1995.

Producing highly enriched uranium?

Argentinian nuclear officials involved in building and operating Pilcaniyeu said in an interview with one of the authors in 1988 that the plant was originally designed to produce up to 20 per cent enriched uranium, not highly enriched uranium. Nevertheless, the officials conceded that producing highly enriched uranium in such a plant is possible, although time-consuming. It cannot be ruled out that Argentina might have originally intended to build enough stages to make weapon-grade uranium directly.

Gaseous-diffusion technology is not very flexible, because each enrichment stage increases the fractional amount of ^{235}U only slightly and requires a large volume of uranium feed. Enrichment stages are connected in a long series or cascade. (The higher the enrichment, the more stages needed.) For example, producing 20 per cent enriched uranium from natural uranium would require about 2000 stages, while producing 90 per cent enriched uranium would require almost twice as many stages.

A gaseous-diffusion cascade contains a large amount of uranium gas, which severely complicates another potential route to highly enriched uranium, namely 'batch recycling'. Here, the enriched end-product of the cascade is collected and reintroduced into the cascade at the beginning. In a cascade designed to produce 20 per cent enriched uranium, only one recycle would be necessary to produce 80 per cent enriched uranium, enough for a relatively lightweight crude nuclear explosive. However, the in-process inventory of a 20 000-SWU-per-year plant could easily reach 2000 kg. Producing this much 20 per cent enriched uranium feed in such a plant would require four years of operation at full capacity and at a tails assay of 0.3 per cent.

A practical way to produce highly enriched uranium in a gaseous-diffusion plant is to 'stretch the cascade'. In this method, the enriched uranium product is extracted from the cascade at a greatly reduced rate, causing the enrichment level of the final product to increase dramatically, although at the cost of reduced efficiencies and production rates.[4] In this way, small amounts of uranium enriched to 90 per cent can be produced in a plant configured to produce 20 per cent enriched uranium. Usable amounts of roughly 80–90 per cent enriched uranium can be obtained by withdrawing product at a fraction of the designed value. Estimating the exact amount would require detailed knowledge of the cascade design, which is not available. However, assuming a reduction in

[4] The maximum enrichment level obtainable occurs when no product is removed, where the separative work is zero and the tails concentration reaches its limiting value, which is the feed concentration. In theory, the location of the feed stage moves down the cascade as the product withdrawal rate decreases and the product concentration increases—finally the entire cascade becomes an enricher. The limiting case is called 'total reflux'. The enrichment level of the product can reach 90% ^{235}U in a cascade configured to produce 20% enriched uranium using natural uranium feed. If 20% ^{235}U feed is used instead, the maximum enrichment level will exceed 99% ^{235}U.

product withdrawal rates to 1–2 per cent, a 20 000 SWU per year plant could produce on the order of 5–10 kg a year of 80–90 per cent enriched uranium.

In both batch recycling and stretching, the operator of a gaseous-diffusion plant would have to take special precautions to prevent the possibility of a nuclear criticality accident in the enrichment stages and in uranium withdrawal equipment. In particular, the operator would have to monitor the process very carefully to ensure that high-assay material does not solidify inside the equipment.

Argentina's plutonium separation programme

In 1978, Argentina announced that it would build a small reprocessing facility at the Ezeiza Research Complex near Buenos Aires. The plant suffered many delays and is not expected to operate as a reprocessing plant. At full capacity, the plant could have separated 15 kg of plutonium per year from spent fuel from heavy water reactors. The plutonium was to be recycled into heavy water reactors (see chapter 6).

Because the reprocessing facility could have produced unsafeguarded plutonium, Argentina was suspected in the late 1970s and early 1980s of planning to build an unsafeguarded plutonium production reactor. This suspicion was never substantiated and the Ezeiza facility was gradually accepted as part of a civilian research and development programme aimed at the use of plutonium fuels in breeder reactors and thermal power reactors.

However, after several delays, the Argentinian Government stopped construction of the reprocessing plant in 1990, citing dire economic constraints. As of late 1994, the plant was being modified for testing of spent fuel elements and components.[5]

Brazil's secret gas-centrifuge enrichment programme

The centre-piece of Brazil's autonomous nuclear programme is the Aramar gas-centrifuge enrichment plant, operated by the Navy near Sorocaba in the state of São Paulo. Like Pilcaniyeu, it has been slow to reach planned capacities. The primary reason for the delays is believed to be a shortage of funds.

At the inauguration of the plant in 1988, the then President of Brazil's National Nuclear Energy Commission (CNEN), Rex Nazare Alves, said that the facility would produce low enriched uranium for existing nuclear power, research and submarine reactors. Originally, the centrifuge programme was based at the Institute of Energy and Nuclear Research (IPEN), located on the campus of São Paulo University. It was there, in September 1982, that technicians first succeeded in producing slightly enriched uranium. Two years later, they operated their first cascade, which was composed of nine machines.

[5] Lubenau (note 2).

Although the Brazilian Government has released few details about the capabilities of its centrifuges or its capacity to make them, enough public information is available to draw some conclusions.

Centrifuge type

The design of the centrifuges is unknown but indications are that the original ones with relatively wide rotor tubes and electromagnetic top bearings were subcritical machines. At the inauguration ceremony, Alves stated that the rotors of the centrifuges were made out of maraging steel. Brazil is one of the few countries in the world that produces maraging steel and, according to Brazilian press accounts and US Government sources, it could have produced enough to supply Aramar. The original machine is reported to spin at 60 000 revolutions per minute and to have a lifetime of more than five years.[6] The separative capacity of the machines initially installed at Aramar is 1.8 SWU per year. The latest machines are believed to be supercritical models using carbon-fibre rotors and to have a diameter of about 120 mm. Brazil is not known to produce adequate carbon fibre, but it could have obtained sufficient stocks overseas.

Number of centrifuges

Brazil treats information about its number of centrifuges as secret. Nevertheless, public information suggests that Brazil has not built as many centrifuges as planned.

Brazilian news reports and senior Argentinian nuclear officials who attended the inauguration ceremonies in 1988 stated that the Isotopic Enrichment Facility (later renamed the Isotopic Enrichment Laboratory, LEI) had about 50 operating centrifuges at that time. There are conflicting reports about the number put into operation subsequently. In late 1990, the plant was reported to have 550 new machines and 48 older units in operation.[7] In mid-1991, about 500 machines were reportedly operating.[8] In March 1993 a senior Navy official said that 565 centrifuges had already been installed.[9] The plant's first construction phase, as envisioned in 1994, called for 958 operating centrifuges with a total capacity of about 1700 SWU per year.[10]

In 1990, a working group appointed by then President Collor de Mellor to develop recommendations about the future direction of the Brazilian nuclear

---

[6] 'Ultracentrifuges detailed', *O Estado de Sao Paulo*, 18 Dec. 1990 (in Portuguese).

[7] Godoy, R., 'Ultracentrifuges to raise uranium production', *O Estado de Sao Paulo*, 18 Dec. 1990 (in Portuguese).

[8] Casado, J., 'Reaction to accords with IAEA and Argentina: inspection not intrusive', *Gazeta Mercantil*, 30 July 1991 (in Portuguese).

[9] Casado, J., 'More uranium enrichment capability', *Gazeta Mercantil*, 12 Mar. 1993. English translation in Foreign Broadcast Information Service, *Daily Report–Latin America (FBIS-LAT)*, FBIS-LAT-93-053, 22 Mar. 1993.

[10] Pinguelli Rosa, L. P., Barros, F. de S. and Barreiros, S. R., 'Nuclear technology in Brazil: retrospective, current situation, and outlook on the nuclear power plan and military/nuclear programme', undated (portions translated by US State Department). Luiz Pinguelli Rosa served as a technical adviser to the Brazilian Congressional Commission investigating the autonomous nuclear programme in 1990 and 1991.

programme stated that the second construction phase of the Aramar plant should involve an expansion to 100 000 SWU per year by 1996 at an estimated cost of more than $300 million.[11] This new plant would use improved centrifuges with a capacity of 3 SWU per year, or greater.

The group's report said that machines with a capacity of 3 SWU per year had already been developed, that a 10 SWU per year machine was under development and that a 25 SWU per year machine was being designed. The latter machines would use more advanced carbon-fibre rotors that are capable of much higher speeds than those made with maraging steel. Carbon-fibre rotors have been under development since at least the inauguration of the Aramar facility, according to Alves.[12]

In March 1993, the Brazilian Government announced that it intended to expand the number of machines at the LEI by about 30 per cent over a one-year period, adding 162 centrifuges.[13] The LEI currently has about 725 centrifuges of various separative outputs organized into four cascades. The 'mini' cascade, which contained the original 50 machines, has been removed and replaced by a cascade of more advanced centrifuges using carbon fibre. Three other cascades are in another, larger hall and are reported to use maraging-steel rotors. Assuming 3–5 SWU per year for each machine, the LEI has a total estimated capacity of about 2200–3600 SWU per year. The plant is reported to have a capacity of less than 5000 SWU per year, and this estimate is consistent with this information.

Brazil is building a pilot plant near the LEI that uses a carbon-fibre design similar to the one installed in the mini-cascade. This machine is reported to be about 2 metres high. The facility is slated to hold 3000 centrifuges and to be finished by the year 2000. As of early 1996, one cascade (out of ten) was finished. Assuming this machine has a capacity of 5–7 SWU per year, the pilot plant will have an estimated capacity of 15 000–21 000 SWU per year.

Large-scale expansion has not been approved, although the press periodically publicizes the nuclear industry's hopes for a large centrifuge plant able to produce on the order of 100 000 to 200 000 SWU per year.[14] Lack of funds or actual need have prevented these ambitions from being realized.

Enrichment product

At the time of Aramar's inauguration, the plant was producing a small, but undisclosed, amount of 5 per cent enriched uranium. In mid-1991, it was reported to be producing uranium enriched to a little more than 20 per cent.[15] Later reports raise questions about this claim. The Brazilian press reported in

---

[11] Motta, P., 'Nuclear development recommendations made', *O Globo*, 30 Sep. 1990 (in Portuguese).
[12] See also Fantini, F. and Costa, R., 'History of nuclear parallel programme surveyed', *Istoe*, 13 Apr. 1988 (in Portuguese).
[13] Casado (note 9).
[14] See, e.g., Hibbs, M., 'Brazil mulling new centrifuge plant to serve two reactors at Angra site', *Nuclear Fuel*, 26 Sep. 1994.
[15] Casado (note 8).

July 1994 that, according to Naval Ministry sources, the IAEA had verified that Aramar had produced 10 per cent enriched uranium.[16]

The LEI plant produces 3 per cent enriched uranium, and the pilot plant is also expected to produce 3 per cent enriched material. The LEI may produce some 20 per cent enriched material to establish that it can do so.

The amount of enriched uranium produced so far is not publicly known. In addition, the authors do not have sufficient information to estimate it. Nevertheless, the IAEA and ABACC are verifying Brazil's inventory of enriched uranium as part of their effort to place safeguards on all of Brazil's nuclear material.

Brazil has imported unsafeguarded enriched uranium. About 200 kg of 4.3 per cent enriched uranium were obtained abroad from China for a critical facility started in 1984.[17] This import, and perhaps other ones, have complicated the IAEA's verification of the completeness of Brazil's initial inventory.

Production of WGU

Like a gaseous-diffusion enrichment plant, a centrifuge plant can be used to produce weapon-grade uranium. The centrifuge system is more flexible than the diffusion system, however.

In a gas-centrifuge plant designed to produce low-enriched uranium, batch recycling and connecting cascades in series provide an efficient, relatively quick route to weapon-grade uranium. The small amount of uranium found in the cascade and the modular design of most gas-centrifuge plants make stretching unnecessary. However, by combining a moderate amount of stretching with batch recycling, a plant such as Aramar could produce significant quantities of weapon-grade uranium within a few years.

Laser enrichment

Brazil has pursued laser enrichment for many years although with little success to date. The main effort is at the Institute of Advanced Studies at the Aerospace Technical Center (CTA) in São José dos Campos. This project is investigating both molecular and atomic vapour laser enrichment methods, but it has made little progress so far.

Brazil's secret plutonium programme

While the Navy was pursuing the uranium option to a nuclear weapon capability, the Army was developing the plutonium route. In September 1991, the Army publicly revealed its unsafeguarded nuclear programme, located at the

---

[16] Felicio, C., 'IAEA verifies 10 per cent uranium enrichment in Ipero', *Gazeta Mercantil*, 12 July 1994, p. 5. English translation from *Proliferation Issues*, JPRS Report, Foreign Broadcast Information Service, JPRS-TND-94-015, 22 July 1994, p. 20.

[17] Pinguelli Rosa *et al*. (note 10).

Army Technological Centre (CETEX) in Rio de Janeiro state. It has operated a small subcritical graphite unit containing bars of natural uranium, which can only be operated when an external neutron-emitting source is in the reactor.

The Army had been designing a natural uranium, graphite-moderated, air-cooled reactor, called the Experimental Irradiation Reactor, to be located elsewhere. Originally intended to have a power rating of 20 MWth, large enough to make 4 kg of weapon-grade plutonium a year, the reactor has been scaled back to 2 MWth. According to General Nelson de Almeida Querido of CETEX, the size of the reactor has been reduced to 'make it clear that the Army has no intention of getting plutonium to make nuclear weapons'.[18] At the time, experts stated that the reactor could be finished by the mid-1990s, although the reduction in power has eliminated its plutonium production potential. Few now believe that it will be built.

For several years, Brazil operated an unsafeguarded laboratory-scale plutonium separation facility at IPEN. The head of IPEN said that this process was demonstrated using plutonium simulators, such as thorium, cerium and gadolinium, since all existing spent fuel in Brazil is safeguarded and subject to foreign control.[19] The confidential 1990 nuclear report to President Collor de Mellor recommended that the continuation of the laboratory work on plutonium separation at IPEN be followed eventually by a pilot separation plant. However, the laboratory is reported to have closed in 1989, and plans for a pilot plant have not materialized.

Questions remain about whether uranium targets were irradiated in safeguarded reactors, and processed in the IPEN facility, extracting gram quantities of plutonium. According to a review of the Brazilian nuclear programme prepared by West German intelligence for the Bonn Government, the facility was operated at a throughput of less than 1 gram of plutonium per day.[20] However, a senior Brazilian nuclear official said in an interview in 1988 that the plant would take 'one hundred years to produce a kilogram of plutonium'.

III. South Africa

South Africa made history by being the first threshold state to dismantle a nuclear arsenal. Following a secret decision by former President F. W. de Klerk in 1989, six complete gun-type weapons and a seventh incomplete one were dismantled and the HEU returned to the custody of the Atomic Energy Corporation (AEC). In July 1991, South Africa acceded to the Non-

---

[18] Motta, P., 'Secretary discusses Army nuclear project details', *O Globo*, 16 Sep. 1990 (in Portuguese).
[19] Zygband, F., 'Nuclear waste reprocessing technology mastered', *O Globo*, 16 Sep. 1990 (in Portuguese).
[20] Hibbs, M., 'Germans say Brazil developing two production reactors', *Nucleonics Week*, 27 July 1989.

Proliferation Treaty, and allowed the IAEA to verify its initial declaration of HEU and to inspect all its existing nuclear activities.[21]

On 24 March 1993, President de Klerk publicly revealed that South Africa had had nuclear weapons. His announcement before the Parliament followed growing international and domestic pressure to reveal the programme, which had been widely suspected in any case. De Klerk acknowledged this pressure in his speech, citing allegations in the media and by some countries that South Africa had not fully revealed its stock of HEU. These allegations were harmful to South Africa's effort to commercialize its nuclear infrastructure and build more cooperative relations with other countries.

In his announcement, de Klerk invited the IAEA to verify that all the weapons had been dismantled and the HEU returned to the AEC. The IAEA, utilizing nuclear weapon experts from the declared nuclear weapon states, spent the next five months verifying South Africa's dismantlement statements and the timing and scope of its former nuclear weapon effort.

Meanwhile, the IAEA was in the midst of an unprecedented exercise of verifying South Africa's past production of HEU. It conducted a two-year investigation of the completeness of South Africa's declaration of its HEU inventory. This investigation focused on the Y-Plant, the pilot enrichment plant that South Africa used to make its HEU for weapons. In an initial approach, the IAEA determined that the ^{235}U balance of the highly enriched uranium, low-enriched uranium and depleted uranium produced by the pilot plant was consistent with the uranium feed. This effort faced tremendous difficulties because of large uncertainties regarding the amount of ^{235}U in the tails. Partially because of these difficulties, the IAEA decided to conduct a detailed assessment of the Y-Plant's HEU output. By using detailed daily operating records of the plant and supporting technical data, the IAEA recreated the daily HEU production of the Y-Plant, concluding that the 'amounts of HEU which could have been produced by the plant are consistent with the amounts declared in the initial report.'[22] In practical terms, this conclusion meant that the difference between the IAEA's estimate of HEU production and South Africa's declaration of HEU production was less than a significant quantity, or 25 kg of weapon-grade uranium.

Although South Africa told the IAEA how much HEU it produced, the AEC has refused to release publicly the amounts of HEU produced in South Africa or to provide a credible qualitative explanation of the output of the Y-Plant. This lack of candour on the part of the AEC has sparked considerable controversy, including increasing suspicions both inside and outside Western govern-

---

[21] For a more complete survey of South Africa's former nuclear weapon programme, see Albright, D., 'South Africa's secret nuclear weapons programme', *ISIS Report*, Institute for Science and International Security, Washington, DC, May 1994. For a detailed description of the verification process and a historical survey of the political and security changes that led to the decision to dismantle the arsenal, see Albright, D., *How South Africa Abandoned Nuclear Weapons*, Henry L. Stimson Center, Washington, DC, 1997.

[22] Report of the Director General, IAEA, *The Denuclearization of Africa (GC(XXXVI)/RES/557)*, IAEA document GC(XXXVII)/1075, Vienna, 9 Sep. 1993, Attachment 1, pp. 10–11.

ments, about South Africa's past activities and its future intentions. Although no evidence was found that South Africa's declaration was incomplete, the AEC justification for not publicly revealing its inventory is weak. In fact, its refusal to provide this information or to allow the IAEA to release more information has served to worsen various controversies.

The section below describes the South African enrichment and plutonium production programmes and estimates the output of the Y-Plant. The uncertainty in this estimate remains high.

The enrichment programme

South Africa's Atomic Energy Board (AEB) began secretly developing a uranium enrichment process in the early 1960s.[23] The existence of this programme was known to only a few people within the government until 1970, when Prime Minister John Vorster announced that South African scientists had developed a unique enrichment process based on the aerodynamic technique.

Because of the high energy costs of this process, however, South Africa later developed both gas-centrifuge and molecular laser enrichment processes.[24] In August 1991, the gas-centrifuge programme was terminated for economic reasons, although the laser programme continues.

The pilot enrichment facility

Following Vorster's announcement in 1970, the government formed the state-controlled Uranium Enrichment Corporation (UCOR) to build the Y-Plant at Valindaba, next to the National Nuclear Research Centre at Pelindaba west of Pretoria. The meaning of the name, Valindaba, gives an indication of the secrecy that surrounded this project. Drawn from the indigenous dialects of South Africa, Valindaba literally means 'the council is closed' or, more figuratively, 'no talking about this'.[25] In 1982, UCOR and the AEB were merged into the AEC.

The South African aerodynamic process separates uranium isotopes through centrifugal effects created by the rapid spinning motion of a mixture of uranium hexafluoride and hydrogen carrier gas in a small stationary tube. The gas mixture enters at a high speed through holes in the side of the tube and spirals down the tube. When the mixture reaches the holes at the ends of the tube, its radius of curvature is reduced several-fold, increasing the separation of the uranium isotopes. The heavy fraction, more concentrated in ^{238}U, exits to the side. The light fraction, more concentrated in ^{235}U, exits straight out at the end.

[23] Newby-Fraser, A. R., *Chain Reaction: Twenty Years of Nuclear Research and Development in South Africa* (Atomic Energy Board: Pretoria, 1979), p. 95.
[24] Kemp, D. M. *et al.*, 'Uranium enrichment technologies in South Africa', Atomic Energy Corporation of South Africa, Ltd, Paper presented at the International Symposium on Isotope Separation and Chemical Exchange Uranium Enrichment, 29 Oct.–1 Nov. 1990, Tokyo, Japan.
[25] Newby-Fraser (note 23), p. 104.

According to Anthony Jackson, the leader of the team responsible for the design and commissioning of the Y-Plant, the attainment of an industrial production level required years of trial and error.[26] Because the plant was for a 'strategic' purpose, funding to sort out all the engineering and chemical problems was never an issue.

Construction of the Y-Plant started in late 1970, the first stages of the lower end of the cascade were commissioned by the end of 1974, and the full cascade, designed to produce weapon-grade uranium, started operation in March 1977. Because of the long equilibrium time of the plant, the first and relatively small withdrawal of HEU occurred in January 1978.[27]

The Y-Plant was organized into five consecutive enrichment blocks and one stripper section, which were located in three large buildings, named C, D and E. Natural uranium was fed into block 1 in building C. The enriched product from block 1 went by pipe to blocks 2 and 3 in building D for additional enrichment, and then to blocks 4 and 5 in building E, which discharged the final product. Depleted uranium was discharged at the bottom of the stripper section in building C.

At first, the plant produced only 80 per cent enriched uranium. By the end of 1978, it had produced enough of this material for South Africa's first nuclear explosive. This device was completed in 1979.

An unexpected problem occurred in the Y-Plant in August 1979 when greater than normal chlorine impurities in domestically produced uranium hexafluoride feed caused a massive chemical reaction in the uranium gas and the hydrogen carrier gas. According to Jackson, the result was solid uranium depositing on the inside of the cascade, reducing the output of the top end of the plant to only 30 per cent instead of 80 to 90 per cent enriched uranium. Following 'chaos day', as the incident became known to plant workers, the plant was closed for cleanup and renovation between August 1979 and April 1981. This renovation required the replacement of the separating elements whose holes had become blocked. After restart, the plant did not start producing HEU until July 1981. The chemical reaction involving chlorine continued, but the plant operators learned to control it sufficiently to avoid another catastrophic reaction.

After it was restarted the plant operated successfully, although not without problems, until it ceased HEU production in November 1989. It was officially closed on 1 February 1990. AEC officials have stated that the plant had about 3000 cascade-days of operation during its lifetime.

The Y-Plant employed about 250 people, operating three shifts per day, seven days a week. Waldo Stumpf, Chief Executive Officer of the AEC, has stated that total historical capital costs for the construction of the Y-Plant were about 200 million Rand and annual direct operating costs averaged (historically)

---

[26] Interview in South Africa by one of the authors, Feb. 1994.

[27] Stumpf, W., 'South Africa's nuclear weapons programme', undated. See also Kemp et al. (note 24); and IAEA, *South Africa's Nuclear Capabilities (GC(XXXV)/RES/567)*, IAEA document GC(XXXVI)/1015, Vienna, 4 Sep. 1992.

about 45 million Rand.[28] Stumpf has said that in 1993 exchange rates, the capital costs were about $60 million and the operating costs averaged about $14 million per year. It is uncertain whether Stumpf has included the effect of inflation.

The Y-Plant's original design output was about 10 000 to 15 000 SWU a year, but improvements increased that number to 20 000 SWU per year. UCOR had initiated an 'improvement programme' of the separator elements following the crashing of the cascade in 1979.[29] These improved elements were installed in the mid-1980s. During the installation of the new elements, whole sections of the cascade had to be shut down.

Separative work losses. Losses in the Y-Plant greatly reduced the production of weapon-grade uranium below nominal values. Part of the loss can be attributed to inefficient mechanical processes in the cascade, which led to the remixing of enriched and depleted streams after leaving the separator elements, commonly called 'mixing'. The Y-Plant did not use the more advanced cascade design, called the Helikon technique, which significantly reduced mixing in the semi-commercial plant (see below). It uses a 'Pelsakon backpump cycle' which, according to Jackson, did not work as well as expected.

Other losses resulted from chemical reactions induced by chlorine impurities in the uranium hexafluoride. This type of loss remained a persistent problem, which caused enriched uranium to deposit inside the cascade. The plant operators never understood this problem well enough to prevent it; for example, they never replicated the reaction in a laboratory. During the inventory completeness verification, the IAEA was able to confirm that the problem occurred, but it was also unable to develop a complete understanding of the problem.

This reaction caused significant amounts of enriched uranium to deposit on Teflon filters, or 'candles', which were used throughout the cascade to filter out dust from large compressors. Complicating the situation, the deposition rate was dependent on the radiation level in the cascade, increasing as one went higher in the cascade. Thus, this problem became more serious as the value of the enriched uranium increased.

After the August 1979 crash, the plant operators learned that a similar event could be prevented by carefully recording the operation of the cascade and monitoring the uranium buildup on the filters with a unique, highly collimated gamma-radiation detector. The result was that the operators learned to recognize when a filter was becoming overloaded with uranium and needed to be replaced with a fresh one. In this way, the operators avoided another chaos day. An inadvertent result of this careful record keeping was that the daily operating records were both detailed and maintained over the whole life of the plant.

---

[28] Stumpf (note 27).
[29] Stumpf, W., 'South Africa's limited nuclear deterrent programme and the dismantling thereof prior to South Africa's accession to the NPT', Transcript of talk given at the South African Embassy, Washington, DC, 23 July 1993.

Fortunately for the IAEA, South Africa preserved these records after shutting down the plant.

Over time this plant built up an increasing number of filters containing significant quantities of valuable uranium enriched to many levels. With so much enriched uranium building up in this form, the AEC launched a recovery operation for the HEU and higher quality LEU on the filters, in essence recovering much of the lost separative work. The enriched uranium on the filters was generally in the chemical form of uranium tetrafluoride, which appears typically as green flecks. The recovery process, which involves the 'direct dry conversion process', essentially converts finely divided highly enriched uranium tetrafluoride back into uranium hexafluoride through a fluorination process.[30]

According to plant officials, the recovered HEU was not further enriched up to weapon-grade; instead, it was used to produce fuel for the Safari-1 reactor at Pelindaba and the Koeberg nuclear power reactors near Cape Town. The Safari reactor had been converted to 45 per cent enriched uranium fuel, and recovered HEU was used to make fuel after 1981. Other recovered HEU was blended with LEU to make LEU of the appropriate enrichment level for the Koeberg reactors in the 1980s.

Not all of the enriched uranium deposited on the filters was recovered. In addition, enriched uranium was deposited in oils or on other equipment. According to Stumpf, about 23 000 drums containing over 6000 Teflon filters, powders, sludge oil and other items were sent to a waste site next to Valindaba.[31]

According to AEC officials, the actual output of the plant was on average closer to 10 000 SWU per year than the nominal output of 20 000 SWU per year. Assuming an average output of about 10 000–15 000 SWU per year, natural uranium feed and a tails assay of about 0.46 per cent, the plant could produce about 60–90 kg of 93 per cent enriched uranium a year.[32] This quantity is more than enough for one explosive device a year of the type South Africa built, which was a gun-type device containing about 55 kg of HEU.

The semi-commercial-size enrichment plant

Next door to the Y-Plant, South Africa built a much larger enrichment plant at Valindaba based on the same separation techniques used in the Y-Plant, but it employed the more efficient Helikon approach. Construction of this larger plant began in 1979, and commissioning started in 1984. Because of problems result-

---

[30] For details of this process, see 'The AEC's dry HEU reconversion capability: The "single-step, no effluent" option, as patented by the AEC', Atomic Energy Corporation, Pelindaba, Dec. 1992. This process was also applied to finely divided uranium metal scrap produced by the turning or milling of uranium metal ingots.

[31] Stumpf (note 29).

[32] Tails assays varied greatly from below 0.2% to above 0.6%, with an average of about 0.456% over the life of the plant. Up to about mid-1985, the tails assay fluctuated near 0.5%, and afterwards, it was close to about 0.4%. See notes accompanying Stumpf lecture (note 29).

ing from insufficient prototype experience, production did not begin until 1988.[33] This plant produced 3.25 per cent enriched uranium for the twin Koeberg power reactors, which required about two-thirds of its optimum annual production of 300 000 SWU. Any spare separative capacity was intended to be sold on the world market.[34]

The plant is widely reported not to have been designed to produce highly enriched uranium. However, it could theoretically produce highly enriched uranium through either batch recycling or 'stretching' the cascade. But the use of the plant to produce highly enriched uranium is impractical and could create criticality problems in the plant. The cascades in the plant are capable of achieving an enrichment level of 3.25 per cent only after three successive batching operations, using a 0.3 per cent tails assay.[35]

From 1988 until mid-1993, the semi-commercial plant produced 734 000 SWU, with 95 per cent supplied to the Koeberg reactors and the other 5 per cent supplied to foreign customers.[36] The total output for these years corresponds to the production of about 189 000 kg of 3.25 enriched uranium at a tails assay of 0.3 per cent. The average annual output during each of these five years was about 150 000 SWU per year, or about 38 000 kg per year of 3.25 per cent enriched uranium.

The enrichment process remained highly energy intensive and was not competitive with overseas producers, particularly in an oversupplied world enrichment market. With little prospect of economic viability, the plant ceased operation on 31 March 1995.

Gas-centrifuge and laser enrichment

Because of the high energy costs of the aerodynamic enrichment process, South Africa decided to develop both gas-centrifuge and molecular laser enrichment techniques.

South Africa began centrifuge research in the 1970s.[37] Its centrifuges are based on European-style centrifuges, since South African enrichment experts

---

[33] Kemp et al. (note 24).

[34] Jones, J., 'South Africa enrichment plant now commercial, AEC head says', *Nucleonics Week*, 23 Jan. 1992.

[35] Kemp et al. (note 24).

[36] Ventor, P. J., 'Prospects for the South African frontend nuclear fuel cycle industry', International Conference on Enrichment, sponsored by the US Council for Energy Awareness, Washington, DC, 13–15 June 1993.

[37] Kemp et al. (note 24). South Africa developed both maraging-steel and carbon-fibre/resin centrifuges. Although the latter are more difficult to develop, particularly given South Africa's indigenous capabilities, a plant using carbon-fibre centrifuges would be significantly less expensive than one using maraging-steel machines. The programme aimed to develop a maraging-steel supercritical machine with a rotor 2 m long and 0.145 m in diameter spinning at 450 m/s with a separative capacity of 10 SWU/y. It hoped to develop a carbon-fibre design that would have had a rotor 3 m in length and 0.175 m in diameter, spinning at 550 m/s with a capacity of 30 SWU/y. Kemp et al. write that they had a laboratory with 16 stands, which test various mechanical properties of the centrifuges, and another rig, which used uranium hexafluoride to measure the separative capacity of a machine. Single-rotor carbon-fibre machines were being developed, and cascade development would have begun only after additional single machine tests of bearing lifetimes and of longer rotors. The report does not discuss the maraging-steel centrifuges.

state that a 'large amount of knowledge of these designs was in the public domain'.[38] Because of the difficulty of ever developing a centrifuge plant competitive with existing plants in Europe, and lacking a strategic rationale for the programme, the AEC decided to discontinue the gas-centrifuge programme in late 1991.

South Africa began its efforts in molecular laser isotope separation (MLIS) in 1983.[39] The programme remains in an early stage of development. In early 1996, the AEC announced that it had signed a cooperation agreement with Cogema to develop the MLIS process.[40]

The programme uses uranium hexafluoride in the presence of carrier gases, which is a process the AEC understands well from its experience with its two aerodynamic enrichment plants. Enrichment of uranium hexafluoride in a single step from natural to 4.5 per cent has been demonstrated on a microscopic level involving milligrams of uranium hexafluoride.[41] Similar success was achieved on a macroscopic system that handles grams of uranium hexafluoride.

A 6000 SWU per year demonstration module is being installed in a decommissioned section of the Y-Plant. Its expected completion date is uncertain.

The inventory of highly enriched uranium

South Africa has declared its inventory of highly enriched uranium to the IAEA. However, as mentioned above, South Africa has refused repeated requests to release even summary information about its inventory to the public. Its excessive secrecy about its past production with anybody but the IAEA has caused needless controversy about whether South Africa has declared all its HEU to the IAEA.

Most public reports claiming the existence of significantly more HEU than reported to the IAEA offer little if any substantiation and usually ignore other information that contradicts their findings.[42] The refusal of the AEC to provide credible public information about the operation of the Y-Plant has left these claims unchallenged and has given the impression that the AEC has something to hide.

An attempt is made below to estimate the amount of enriched uranium produced by the Y-Plant. There is not enough public information to recreate the

[38] Kemp *et al.* (note 24).

[39] Kemp *et al.* (note 24).

[40] Loxton, L., 'France, S. Africa agree on nuclear cooperation', Reuter, 29 Feb. 1996; Boyle, B., 'South Africa testing laser uranium process', Reuter, 24 Aug. 1994; and Reuter, 'South Africa in talks to produce enriched uranium', 2 June 1995.

[41] Toit, G. D., Ronander, E. and Birkill, D., 'The Molecular Laser Isotope Separation (MLIS) Program: Status and prospects', International Enrichment Conference, Washington, DC, 13–15 June 1993.

[42] See, e.g., Hounam, P. and McQuillan, S., *The Mini-Nuke Conspiracy* (Faber and Faber: London, 1995). The authors use vague, self-serving statements from South African or Y-Plant officials that the plant worked well throughout its history to try to argue that the Y-Plant operated 2 years longer than declared to the IAEA and ran consistently at its maximum output of 20 000 SWU per year for those 14 years. At this output and over this period, the authors estimate that the plant probably produced over 1500 kg of weapon-grade uranium.

enriched uranium output of the Y-Plant independently or to confirm the IAEA's conclusions about South Africa's declared HEU inventory. These estimates therefore remain highly uncertain, but the available information is sufficient to clarify important questions about the operational history of the Y-Plant and point out where additional information is needed to answer fully the remaining questions about South Africa's HEU output.

The initial declaration

The Y-Plant produced a wide range of enriched uranium during its lifetime. According to Stumpf, South Africa's initial declaration to the IAEA on the amounts and forms of HEU at Pelindaba was one and a half pages long.[43] This material was in many forms, such as metal, uranium hexafluoride and uranium tetrafluoride; and it was at many different enrichments, including a considerable amount above 20 per cent but below 80 per cent. Much of the HEU was enriched to levels between 20 and 45 per cent.

According to media sources, South Africa's initial declaration for the output of the Y-Plant listed about 1500 kg of enriched uranium, containing about 550 kg of uranium-235.[44] These reports do not state how much of this material is HEU or LEU, although the average enrichment of the material is about 37 per cent, implying that most of the material was not weapon-grade.

This inventory includes the HEU taken from the seven dismantled nuclear weapons. It would not have included enriched uranium produced at the Y-Plant and used in the Koeberg reactors.

The total SWU output of the Y-Plant

No public information gives the total output of the Y-Plant. Its total separative work output is therefore estimated below in several different ways.

1. If the Y-Plant had a sustained maximum capacity of 10 000–20 000 SWU per year over slightly more than 10 years of operation, it could have produced between 100 000 and 200 000 SWU. If the entire output of the Y-Plant had gone to produce weapon-grade uranium (93 per cent), it could have produced about 600–1200 kg of weapon-grade uranium, assuming that natural uranium was used as feed and the tails assay averaged about 0.46 per cent. These estimates require a natural uranium feedstock of about 217 750 to 435 510 kg and tails totalling 217 150 to 434 300 kg.

2. The AEC has stated that the plant produced a total of 370 643 kg of depleted uranium with an average tails assay of 0.456 per cent.[45] A source states that the total feed into the plant was about 385 000 kg of natural uranium (0.711 per cent ^{235}U). The total product was therefore 14 357 kg of enriched

---

[43] Presentation by Waldo Stumpf, Carnegie Endowment for International Peace, Washington, DC, 16 Nov. 1993, Seminar on South Africa's nuclear programme organized by ISIS.
[44] Hibbs, M., 'Y Plant MUF "enough for a bomb"'; 350 kg was enriched to 90% U-235', *Nuclear Fuel*, 14 Feb. 1994.
[45] See notes accompanying Stumpf lecture (note 29).

uranium, containing 1047 kg of ^{235}U. This product had an average enrichment of 7.3 per cent, implying again that much of the output of the Y-Plant was not weapon-grade uranium. The production of this amount of product would have required about 135 000 SWU. During about 10 years of operation, the Y-Plant would have produced on average about 13 500 SWU per year.

3. If only 93 per cent enriched uranium was produced and the tails amounted to 370 643 kg, then the product was about 1025 kg. This corresponds to about 170 000 SWU, or 17 000 SWU per year.

The average of these estimates is about 150 000 SWU over about 10 years of actual operation, or 15 000 SWU per year on average. An uncertainty of 20 per cent is assigned to this result. This result is not very sensitive to the specific enrichment level of the product.

Enriched uranium output: material balance approach

The following section attempts to estimate the quantities of enriched uranium produced in the Y-Plant. This estimate is highly uncertain, because the available information is insufficient to understand all the enriched uranium withdrawals that occurred in the plant and many arbitrary assumptions have been made. The attempt is to provide a sensible description of the outputs of the Y-Plant, given the available information. This description, however, is not the only one possible.

HEU from dismantled weapons. The amount of HEU used in the seven weapons dismantled in the early 1990s is relatively well-known. The first device, completed in 1979, utilized only 80 per cent enriched uranium. The rest used weapon-grade uranium. The geometry of each weapon was the same, which means that the first device would have had the same amount of HEU as the rest, but a significantly lower explosive yield than the others. As mentioned above, each device had about 55 kg of HEU, for a total of about 390 kg of HEU. This amount corresponds to roughly 62 000 SWU (see table 13.1).

AEC and South African government officials have stated that this material is the entire stock of HEU enriched above 80 per cent in their possession at the time of the initial declaration. Most of the controversy centres on this claim. Although there is not enough information to independently confirm this statement, no evidence has been found to suggest that it is not true. However, this statement does not exclude the possibility that a significant fraction of weapon-grade uranium was blended down earlier (see below). The available information suggests that roughly 500 kg of material enriched to over 80 per cent was produced in total.

Other enriched uranium withdrawals. Although the Y-Plant was intended to produce weapon-grade uranium for nuclear weapons, the plant was also used to produce enriched uranium for both the Safari research reactor and the Koeberg power reactors.

Table 13.1. Illustrative output of the Y-Plant[a]

Purpose	Enrichment (%)	Uranium (kg)	^{235}U (kg)	Separative work units (SWU)	Feed (kg)	Tails (kg)	Tails Assay (%)
Weapons	80	55	44	7 460	20 821	20 766	0.5
Weapons	93	330	307	55 110	119 764	119 434	0.456
Remainder of initial inventory	21 (av.)	1 090	225	35 980	89 554	88 464	0.46
Sub-total		**1 475**	**576**	**98 550**	**230 139**	**228 664**	
Safari-1	45	70	32	5 350	12 276	12 206	0.46
Koeberg test elements	3.25	1 850	60	5 990	16 990	15 140	0.4
Waste (filters)	1	8 000	80	1 320	17 070	9 070	0.456
Sub-total (cum)		**11 395**	**748**	**111 210**	**276 475**	**265 080**	
Blending stock for imported material	93	100	93	16 700	36 292	36 192	0.456
Blending stock for domestic material	20 (av.)	950	190	29 790	72 810	71 860	0.456
Total		**12 500**	**1 030**	**158 000**	**385 000**	**373 000**	

[a] The numbers in the table are given more precisely than justified by the available information. The reason is that such precision is necessary to maintain an accurate balance in the amounts of feed, product and tails. However, the totals at the bottom of the table are rounded.

South Africa in particular viewed the operation of the Koeberg reactors as a 'strategic' priority. It was therefore prepared to take exceptional steps to ensure their continued operation after overseas supplies of safeguarded fuel were cut off by the USA in response to its apartheid policies and unsafeguarded nuclear programmes. The urgency of supplying fuel for the Koeberg reactors was heightened by the delay in commissioning the semi-commercial plant, which meant that South Africa's indigenous source of LEU was unavailable until 1988. It was under these circumstances that LEU was produced for the Koeberg reactors by blending HEU with a stock of unsafeguarded LEU that had been imported in 1981.

Below are estimates of the major non-weapon uses of enriched uranium from the Y-Plant. This information is summarized in table 13.1. Considerable uncertainties remain in these estimates.

1. Initial declaration. The initial declaration, as reported in *Nuclear Fuel*,[46] indicates that the AEC had on hand a substantial amount of enriched uranium in excess of that taken from nuclear weapons. Possibly, the AEC had over time built up an unused inventory of recovered enriched uranium or had stored enriched uranium removed from the cascade after the plant was shut down in early 1990.

[46] Hibbs (note 44), p. 17.

After subtracting the HEU taken from dismantled weapons, the remainder of the enriched uranium amounts to about 1090 kg of enriched uranium with an average enrichment of 21 per cent, assuming the initial declaration to be true. This amount of enriched uranium would have required roughly 36 000 SWU to produce. The current understanding is that the bulk of this material is enriched below 80 per cent.

2. *Koeberg test elements.* For about six months in 1986, the plant was dedicated to the production of LEU for the first four Koeberg fuel elements for qualification purposes.[47] These elements contained about 1850 kg of 3.25 per cent enriched uranium, and required about 5990 SWU to produce.

3. *Safari-1 fuel.* An unknown quantity of HEU was irradiated in the Safari-1 research reactor. Although designed to operate at 20 MWth, this US-supplied reactor was downgraded to only 5 MWth during the late 1970s and 1980s following the US embargo on additional fuel shipments to South Africa. Designed to use 90 per cent uranium fuel, the reactor was converted to 45 per cent fuel which could be produced in South Africa. The first indigenously manufactured 45 per cent enriched fuel became available in 1981.[48]

The operating record of the Safari research reactor shows that it produced about 11 800 MWth-d of energy from 1981 into 1993.[49] We assume that all this energy was produced by uranium fuel enriched to 45 per cent.[50] In addition, we assume that all this fuel was assigned to the Safari reactor before 30 September 1991 when the AEC submitted its initial inventory of enriched uranium.

About 1.2 grams of ^{235}U are consumed through fission or neutron irradiation per megawatt-day. Therefore, the reactor consumed about 14.16 kg of ^{235}U during these years. Not all the ^{235}U in fuel is consumed. Typically, only about 40 to 50 per cent of it is consumed by the time the fuel is removed from the reactor. If 45 per cent of the ^{235}U was consumed, about 31.5 kg was necessary in the fuel. Since the fuel is 45 per cent enriched, the total estimated amount of HEU irradiated in this reactor is 70 kg, requiring about 5400 SWU.

4. *Waste filters.* Stumpf has stated that 6000 filters containing low and small amounts of highly enriched uranium were sent to a waste site at Valindaba. He has said that in total they contained a 'large quantity of uranium-235', but not enough to warrant recovery. Other AEC officials have stated that there were large uncertainties in the ^{235}U content in 'tens of thousands of waste drums'.[51]

---

[47] Stumpf (note 27).

[48] Mingay, D., Robertson, D. and Niebuhr, H., 'Commercialization of Safari-1 research reactor', Atomic Energy Corporation, IAEA/SR-183/44, undated paper presented to one of the authors by Mingay in Feb. 1993.

[49] Mingay *et al.* (note 48).

[50] Safari may have also used LEU fuel, see figure 2 in Stumpf (note 27).

[51] von Wielligh, N. and Whiting, N., 'Experience of an ex (de facto) nuclear weapon state with the application of post-Iraq safeguards', in IAEA, *International Nuclear Safeguards 1994: Vision for the Future*, vol. 1 (IAEA: Vienna, 1994), pp. 223–30.

Official information is lacking on the amount of enriched and depleted uranium on these filters. Two approaches to understanding this estimate are considered below.

An episode in 1994, in which thieves stole several drums containing contaminated filters, sheds some light on the minimum amount of material in the filters.[52] Stumpf is quoted in an article stating that these 130 drums, containing 260 filters, could contain no more than 5 kg of enriched uranium.[53] Another official was quoted in the same article saying that the enrichment level did not exceed 5 per cent.

This information implies that these filters could each contain up to 20 grams of enriched uranium. Applying this estimate to the 6000 filters mentioned above, about 120 kg of enriched uranium could be in these filters. The average enrichment level of this material is assumed to be 1 per cent and thus would contain about 1.2 kg of ^{235}U. At this average enrichment, this material would have required about 20 SWU.

This estimate provides a lower bound. In fact, considerably more enriched uranium from the Y-Plant is believed to have been sent to the waste site over the lifetime of the plant.

Jackson stated that the loss rate due to the deposition of material inside the cascade was about 0.005 per cent per hour at the lower sections of the cascade and 0.015 per cent per hour at the top of the cascade.[54] Assuming an in-cascade inventory of 500–1000 kg of uranium with an average enrichment of 5 per cent, the total loss over 3000 days of cascade operation was greater than 1800–3600 kg.[55] If an average of 0.01 per cent per hour is applied to the entire cascade, the total loss is estimated at 5400–10 800 kg, or an average of 8100 kg. Most of the material would have been deposited in filters in the lower sections of the cascade, that is, in areas of the cascade where the uranium in the Teflon filters was not recovered. This result implies that a loss estimate of 8000 kg of uranium is more accurate than one of the order of 100 kg.

As a result, the authors assume that the waste included 8000 kg of material with an average enrichment of 1 per cent, requiring 1300 SWU. The waste would have contained about 80 kg of ^{235}U.

5. *Blending stock.* Enriched uranium from the Y-Plant was used to raise the enrichment level of both imported and domestically produced LEU.

(a) Imported LEU. According to AEC officials, enriched uranium from the Y-Plant was used to raise the enrichment level of imported unsafeguarded LEU (other sources give China as the supplier) to the level required for the Koeberg

---

[52] The thieves were actually interested in the large blue barrels holding the filters. After breaking into a locked storage building and taking the drums, they emptied their contents of contaminated filters on the ground near a security fence. See de Lange, F., 'Atomic drums stolen: "no danger"', *The Citizen*, 3 Sep. 1994.

[53] De Lange (note 52).

[54] Interview by one of the authors with South African official, Feb. 1994.

[55] This estimate is based on an interview with plant officials conducted at Pelindaba in Feb. 1994, and assumes that the inventory was constant during the 3000 days of operation.

power reactors, or 3.25 per cent.[56] According to AEC officials, the imported enriched uranium had enrichments of 2.6 per cent, 2.9 per cent and 3.25 per cent.

Nuclear Fuel reported a somewhat different mix of enrichments and it listed quantities imported. It said that a West German middleman arranged for the export from China to South Africa of 30 tonnes of 3 per cent enriched uranium and 30 tonnes of 2.7 per cent enriched uranium in the form of uranium hexafluoride.[57]

An AEC official stated in an interview that about half the necessary blending stock came from the Y-Plant and the rest was produced in the semi-commercial plant. Not all of the imported enriched uranium was used, because some of it contained too high a fraction of uranium-236 for use as power reactor fuel. Whatever remained after South Africa signed the NPT was placed under IAEA safeguards.

Estimating the enrichment level of the blending stock is complicated by the fact that the blending stock may not have been fixed. Officials, however, have said that HEU was used.[58] One official emphasized that at one particular time, a significant amount of weapon-grade uranium was used as blending stock. The actual situation could not be determined unambiguously. Those officials interviewed also did not state when this stock was produced.

In this analysis, the blending stock is arbitrarily selected. The stock is assumed to be 100 kg of weapon-grade uranium, and it was mixed with 30 tonnes of imported LEU with an enrichment of 2.9–3.0 per cent. This accounts for about 17 000 SWU. But this is only one possibility.

(b) Domestic LEU. The IAEA has stated that HEU was also used to upgrade natural uranium to LEU for the Koeberg reactors.[59] On the other hand, AEC officials have stated that HEU was used to upgrade domestically produced LEU for the Koeberg reactors, although the source of this LEU is not given. The implication is that this LEU was from the semi-commercial plant. This process continued after the Y-Plant was closed (see below).

This blending stock is assumed to be about 20 per cent material, and to have resulted from recovery operations. Because Safari-1 was probably slated to receive most of the recovered uranium enriched over 45 per cent, this blending stock would likely have an enrichment level well below 45 per cent on average. In this case, 20 per cent is arbitrarily used and the blending stock is assumed to be 950 kg.

Enriched uranium output: LEU fuel

South Africa could have produced a much larger amount of weapon-grade uranium if it had started with a supply of low-enriched uranium as feedstock in

---

[56] Interviews by one of the authors with AEC officials, Feb. 1994; and figure 2 in Stumpf (note 27).
[57] Hibbs, M., 'British report details heavy water uranium trades by Hempel firms', *Nuclear Fuel*, 25 July 1988.
[58] Stumpf (note 29).
[59] IAEA (note 27).

the Y-Plant. Although the unsafeguarded LEU from China would have been one source, the IAEA said that it found no evidence, through the analysis of samples, that imported LEU had been used as feed for the Y-Plant.[60] According to AEC officials, the imported LEU contained small quantities of uranium-236, and the IAEA found none of this isotope or its daughter products in the Y-Plant.

The AEC has denied that any LEU from the semi-commercial plant was used as feed in the Y-Plant.[61] The IAEA has found no evidence to the contrary.

Current HEU inventory

The amount of HEU currently stored at Pelindaba is not the same as the amount declared to be there in its initial declaration to the IAEA in 1991. More HEU has been added to South Africa's inventory as it was recovered from equipment and components during the decommissioning of the Y-Plant. More significantly, most of the rest of HEU that was not suitable for use in the Safari reactor (< 60 per cent enriched) has been blended down to LEU for use in Koeberg reactor fuel elements.

The Safari reactor has started using fuel made out of weapon-grade uranium. At 20 MWth, Safari cannot use more than about 25 kg of 93 per cent enriched uranium per year. With a stock from dismantled weapons of about 330 kg of weapon-grade uranium, and about 55 kg of 80 per cent enriched, the AEC has enough HEU to fuel the reactor for more than 15 years. US Government efforts to convince South Africa to convert Safari to LEU fuel and blend down all its HEU have so far been resisted by the AEC as too costly (see chapter 8).

Final note about HEU

Although considerable information about the Y-Plant is available, estimating the specific enriched uranium output from the Y-Plant remains a highly uncertain exercise. The above estimates serve to illustrate the likely outputs of the Y-Plant, but more accurate estimates must await more reliable information.

More certain is the total separative work output of the plant. However, the current, preliminary understanding of the Y-Plant is that it produced a wide variety of enrichments and a considerable amount of material. Total weapon-grade uranium output is roughly estimated at about 430 kg, of which about 100 kg was blended down and 330 kg remains in stock. Another 55 kg of 80 per cent enriched material from the first nuclear explosive is also in stock. More or less than 100 kg of weapon-grade uranium could have been produced and blended. Until South Africa decides to release more information, the actual situation will remain unknown.

[60] IAEA (note 27), p. 7.
[61] Stumpf (note 29).

No evidence was found to contradict the IAEA's finding. In particular, this exercise did not provide any information to suggest that HEU was hidden from the IAEA.

Plutonium separation

South Africa was interested in producing and recovering plutonium (and tritium) for nuclear weapons. It was therefore developing the capability to separate plutonium from irradiated fuel, although little information is available about its plutonium recovery activities. South Africa planned to build a 150-MW pressurized water reactor at a site at Gouriqua, near Mosselbay in the Cape Province for both plutonium and tritium production, and a plutonium separation plant may have been planned for this site as well. Little work was accomplished, however, before the reactor was cancelled in 1985.[62]

South Africa began operating a large hot cell facility at Pelindaba in 1987 to examine spent fuel from the Koeberg reactor, although it is not designed to separate plutonium.[63] However, the technologies are similar. Because this facility is indigenously built, the AEC is assumed to have considerable knowledge about plutonium separation.

[62] Albright, ISIS (note 21), p. 12.
[63] South African Embassy, Washington, DC, 24 July 1987; and 'South Africa's aim is "self-sufficiency" as AEC studies casks and reprocessing', *Nuclear Fuel*, 24 Mar. 1986.

Part V
Conclusions

14. Overview of present and future stocks of plutonium and highly enriched uranium

I. Introduction

This chapter provides an overview of plutonium and HEU inventories, drawing upon the findings of chapters 3–13. Aggregate totals for plutonium and HEU are given to the end of 1996. Comprehensive data are provided to the end of 1994, the most recent date for which such data can be compiled. The reader is referred to earlier chapters for a detailed discussion of totals by country or for later dates. The chapter also assesses the amounts of materials placed under international safeguards, and compares the aggregate statistics published by safeguards agencies with the authors' own estimates. Possible trends in plutonium and HEU inventories over the next two decades are also discussed.

It should be emphasized that the figures presented here are not simply updates of those contained in chapter 12 in *World Inventory 1992*. New information and analysis are incorporated here, and there have been some changes in definitions of categories of material. The data in this chapter and in its predecessor are therefore not strictly comparable. As a result, *apparent increases or decreases in estimates may result from definitional or other changes rather than from actual changes in plutonium or HEU inventories.* This chapter should therefore be regarded as *replacing* rather than updating chapter 12 in *World Inventory 1992*.

Throughout this book estimates are presented with approximate error margins. This chapter mainly presents central estimates, without error margins, primarily for reasons of clarity. The reader is encouraged to consult earlier chapters for assessments of relevant levels of uncertainty.

All plutonium figures cited here are quantities of 'total plutonium' (Pu_{tot}), following the convention adopted by international safeguards agencies. Except where otherwise indicated, the unit of measurement used for HEU in this chapter is 'weapon-grade uranium equivalent' (WGU-equivalent).[1]

II. Inventories at the end of 1994

Aggregate totals

The central estimates of the world inventories of plutonium and HEU at the end of 1994, rounded to three significant figures, are:

---

[1] See chapter 2 for an explanation of the differences between Pu_{tot} and Pu_{fiss}. WGU-equivalent refers to the amount of 93% WGU that has the same amount of ^{235}U as the original HEU (see chapter 4).

For plutonium: 1160 tonnes
For HEU: 1770 tonnes

At the end of 1996, the inventory of plutonium had grown to around 1300 tonnes through the continued operation of nuclear power reactors. The inventory of HEU was around 1750 tonnes, the slight reduction coming from the blending down in 1995 and 1996 of a small fraction of Russia's HEU stock to LEU.

Precise error margins cannot be placed on these figures. The authors attach approximate error margins of plus or minus 10 per cent for plutonium, and plus or minus 20 per cent for HEU. These margins are strongly influenced by uncertainties over the sizes of military inventories in the nuclear weapon states, and in Russia in particular. In general, public and private knowledge of HEU inventories is much less developed than knowledge of plutonium inventories. Governments in nuclear weapon states are having particular difficulty characterizing their HEU stocks.

The scope of an inventory is a matter for definition. It is seldom an 'absolute' quantity. The above figures *exclude* (*a*) the amounts of plutonium in operating power reactors, unless it is contained in MOX fuel, and the amounts of ^{241}Pu in spent fuel that have decayed to ^{241}Am (inclusion of these amounts would increase and decrease inventories by approximately 120 tonnes and 40 tonnes respectively); and (*b*) the amounts of HEU in naval fuel cycles, in waste and in production-reactor fuel cycles (except in the USA). The reason for excluding these quantities of HEU is that it is very difficult to assess their magnitude. As chapter 4 demonstrates, the HEU fuel cycles in the NWS have been unusually complex: HEU has been routinely extracted and recycled, and the quantities of material remaining after additional fissioning are hard to judge. In all, perhaps 100–200 tonnes should be added to the stock of weapon-grade uranium equivalent when taking account of these activities.[2]

Civil and military inventories

The gross inventories are broken down in table 14.1 into their civil and military components. Again these are central estimates. This table shows that the great majority of HEU is contained in military inventories, which consist largely but not entirely of weapon-grade uranium. The stock of weapon-grade plutonium is also very large, but it is exceeded by the amounts of fuel- and reactor-grade plutonium existing in the civil sector.

---

[2] This range is based on US figures (see chapter 4). About 90 t of HEU are in the US naval programme or have already been consumed in naval reactors. Another 30 t of HEU are estimated to be in waste. It is not known how much HEU is in Russia's naval and production reactors, but those fuel cycles probably contain less WGU-equivalent than their US equivalents. However, the amount of WGU-equivalent in waste in Russia probably exceeds the US quantities. On balance, it is estimated that the amount dedicated to Russia's naval and production fuel cycles has an upper bound of 100 t of WGU-equivalent. The amounts in the fuel cycles of the other NWS would not affect the aggregate estimates significantly.

Table 14.1. Central estimates for civil and military inventories of plutonium and HEU, 31 December 1994

Figures are in tonnes and do not include material in reactor cores.[a]

	Civil inventory	Military inventory	**Total**
Plutonium	914	249	**1 160**
HEU	20[b]	1 750	**1 770**

[a] Totals do include material recycled in fast and thermal reactor cores.
[b] This includes HEU in civilian programmes in both the NWS and NNWS.

These figures relate to the end of 1994. The inventory of HEU has now probably peaked. No HEU is being produced specifically for civil use. At the time of writing, only China among the nuclear weapon states has not declared that it has halted production of HEU for weapons. France, Russia, the UK and the USA have officially announced the cessation of production. The inventory should begin to fall in the second half of the 1990s as HEU from military stocks is diluted for use in civilian power reactors. The stock of military plutonium is also largely frozen. The Russian Government has announced that the three production reactors still operating in Russia will be converted so that they will no longer produce weapon-grade plutonium, and that any plutonium derived from them will not be used in weapons. In other NWS the supply of plutonium for military purposes has also ceased. However, weapon-grade plutonium is still being produced in India and Israel, and there are concerns that other states (including Pakistan and Iran) might be developing capabilities to make separated plutonium that could be used in nuclear weapons.

In contrast, civil inventories of plutonium continue to grow. The world stock of power reactors is discharging approximately 10 000 tonnes of spent fuel each year, containing some 70 tonnes of plutonium. The inventory of civil plutonium passed 1000 tonnes early in 1996, and will pass 2000 tonnes around the year 2010.

III. Types of inventory

The quantities in table 14.1 can be broken down further, according to the forms and mediums in which they are held. The results are shown in table 14.2. The majority of plutonium in the civil inventory is held in spent-fuel stores. Three-quarters of the plutonium that had been separated by the end of 1994 also remained in store. The 21 tonnes of plutonium in the fast-reactor fuel cycle included material in reactor cores, spent-fuel ponds and being reprocessed. The same applied to the estimated 20 tonnes in the thermal MOX fuel cycle. In neither case does the inventory include plutonium contained in fresh

Table 14.2. Central estimates for inventories of plutonium and HEU by type, 31 December 1994

Figures are in tonnes.

Category	Inventory
Civil plutonium	
In spent reactor fuel	755
Separated in store	118
In fast-reactor fuel cycle	21
In thermal MOX fuel cycle	20
Civil HEU	
In research-reactor fuel cycle	20
Military plutonium	
In operational warheads	70
Weapon-grade outside operational warheads	158
Fuel- and reactor-grade in store	21
Military HEU	
In warheads	450
Outside warheads (not including submarines or waste)	1 300

unirradiated MOX fuel elements.[3] That amount is included in the inventory of separated plutonium.

The inventories of military plutonium and HEU are divided in table 14.2 into two main stocks: those inside and outside operational weapons. For reasons explained in section V, these allocations should be regarded as very rough guides to the actual amounts. They are based on the assumption that around 20 000 operational nuclear weapons existed at the end of 1994. It should be noted that the amounts of weapon-grade plutonium and uranium *outside* operational weapons now substantially exceeds the amounts inside them.[4] As discussed below, the forms in which this material are held are various.

IV. Military inventories in nuclear weapon states

In December 1993, June 1994 and January 1996, the US Government took steps towards declassifying information on its inventories of plutonium and HEU produced for military purposes. As yet, these steps have not been reciprocated by the governments of the other NWS. Their reluctance to follow suit is largely due to entrenched secrecy and to internal difficulties in attaining

---

[3] It therefore excludes the unirradiated MOX core for the Kalkar reactor, and two MOX cores for the Superphénix reactor, both of which were fabricated before 1990 but have not been used, and whose plutonium is counted as belonging to the stock of separated plutonium. This is a departure from the approach adopted in *World Inventory 1992*, where these cores were counted under recycled plutonium. In recategorizing these plutonium inventories, the authors are following the approach adopted by the IAEA and by the Japanese Government in its recent disclosure of Japan's plutonium inventories.

[4] Plutonium and HEU held in warheads that have been withdrawn from service and await dismantlement, or are held in reserve, are counted as being 'outside operational warheads'.

Table 14.3. NWS inventories of highly enriched uranium, after losses and draw-downs, 31 December 1994[a]

Figures are central estimates, in tonnes.

Country	Weapon-grade equivalent	80% ^{235}U-equivalent
FSU	1 050 ± 30%	1 220
USA	645 ± 10%	750[b]
France	24 ± 20%	28
China	20 ± 25%	23
UK	8 ± 25%	9

[a] Losses and draw-downs comprise material consumed in or assigned to naval, civil and production reactors, consumed in nuclear weapon tests, and lost during production and processing.

[b] This figure for the USA is derived by subtracting estimated losses and draw-downs of 245 t from the estimated HEU production of 994 t. The italicized figures for the other NWS are the authors' estimates.

accurate information on their stocks of plutonium and HEU. While the shortcomings in material accounting have become particularly evident in Russia, they may also apply in some degree to China and France where there is no public knowledge of accounting practices in the military sector. In the USA and the UK, there is now claimed to be no difference in standards and methods applied to the accounting of military and civil materials, which are alleged to be the equal of best practices established by the international safeguards agencies. However, these claims cannot be verified, and even in the USA, public and private knowledge of HEU stocks in particular is less than is desirable.

Highly enriched uranium

Little more than 10 tonnes, less than 1 per cent of the world stock, of HEU is located outside the NWS. The central estimate of 1750 tonnes of HEU in the NWS (excluding around 10 t held in research-reactor fuel), broken down by country, is shown in the first column in table 14.3. It shows that over 95 per cent of this stock is located in the former Soviet Union and in the USA. The production of weapon-grade uranium has largely been their preserve. By the same token, they have the greatest responsibility for controlling and reducing HEU stocks.

The estimates in table 14.3 carry large error margins. Alone among the NWS, the United States has published data on HEU, and only then in relation to total historical production and the amounts held in store at 13 identified sites.[5] The other inventories in the NWS have been calculated from knowledge of enrichment plants and their operating histories. Since most of Britain's HEU has

---

[5] See chapter 4. The US Department of Energy has announced that its historical HEU production amounted to about 994 t, and that 259 t are presently stored at 13 sites in the USA (this excludes material located at the Pantex site in Texas where retired weapons are being stored and dismantled).

Table 14.4. NWS inventories of military plutonium, after losses, 31 December 1994[a]
Figures are central estimates, in tonnes.

	Weapon-grade	Fuel- and reactor-grade
FSU	131 ± 25%	..
USA	85 ± 3%	12.9[b]
France	5.0 ± 30%	..
China	4.0 ± 50%	..
UK	3.1 ± 20%	8.4

[a] The losses include consumption in weapon tests and losses incurred during plutonium separation, recycling and warhead fabrication. These losses were not fully accounted for in table 12.5 in *World Inventory 1992*.

[b] This quantity excludes the 1.5 t of plutonium separated from civil reactor fuels in the USA before 1976 and which is now held at sites owned by the US Department of Energy. It is included with civil plutonium inventories recorded in table 14.8.

come from the United States, and since information about this transatlantic trade remains classified, the British inventory is especially difficult to assess.

The second column in table 14.3 presents estimates of the tonnages of HEU that would have been acquired if it had contained 80 per cent of the fissile uranium isotope ^{235}U. The reason for adding this column is that 80 per cent is believed to be the average enrichment level of the 994 tonnes of HEU that the US Government has announced were produced for military purposes (the published amount was not measured in weapon-grade equivalent). In addition to the production of weapon-grade uranium for warhead primaries and for submarine and research-reactor fuels, the United States also produced HEU with lower ^{235}U content, probably for use in the secondaries of thermonuclear weapons. It is not known whether the other NWS followed a similar practice, and whether their HEU stocks contain roughly the same proportions of ^{235}U. The tonnages of HEU actually possessed by the NWS depend on enrichment levels. Inability to specify them may be one of the sources of confusion over the size of Russia's HEU stock in particular.

Plutonium

It is evident from tables 14.1 and 14.2 (and below in table 14.8) that plutonium is more widely distributed than HEU, owing to its production and separation in civilian fuel cycles. However, stocks of weapon-grade plutonium, derived from low burnup fuels discharged from production reactors, are similarly concentrated in the NWS with Russia and the USA holding much the largest stocks. Estimates of inventories in the NWS, broken down by country, are shown in table 14.4. Again, the error margins are high, although lower than in the case of HEU because of the greater knowledge of plutonium production systems and their histories. The US Government alone has released some information on military plutonium stocks.

Only the UK and the USA are known to have produced fuel- and reactor-grade plutonium which remains assigned to their military inventories. The quantities of fuel- and reactor-grade plutonium cited here do not include material that was blended with super-grade plutonium in the USA to provide larger quantities of weapon-grade material. That is counted with the weapon-grade plutonium in table 14.4.

V. Weapon-related inventories and capabilities in countries other than the acknowledged nuclear weapon states

The quantities of weapon-related fissile material produced by the threshold states (including the de facto nuclear weapon states) are relatively small. However, in terms of risk and international concern, these quantities rival, and in some cases outstrip, other larger inventories.

The only countries in which the production of plutonium and HEU for nuclear weapons continues are India, Israel and possibly Pakistan. South Africa has dismantled its nuclear weapon arsenal and is currently storing its stock of HEU from dismantled weapons pending its use in a civilian research reactor. North Korea has frozen its plutonium production. This plutonium is believed to have been intended for weapons, although the North Korean Government denies it.

Iraq is not believed to be producing any fissile material at this time. The exact status of any Iranian production of HEU and separated plutonium is uncertain, but if any such production is occurring, it is considered to be measured in grams or milligrams and not in kilograms.

This section does not cover the small fissile material inventories or capabilities which may have been dedicated to weapon purposes in countries that have subsequently dismantled their weapon programmes. In recent years, these countries mainly include Argentina and Brazil, although South Korea and Taiwan may also belong in this category. The reader is referred to earlier chapters for discussions of these inventories.

Plutonium

De facto nuclear weapon states

Tables 14.5 and 14.6 show that India and Israel are continuing to accumulate weapon-grade plutonium. By the end of 1994, Israel had produced about 440 kg of weapon-grade plutonium, and it could have 510 kg by 2000 (these are central estimates). Through 1994, India had produced about 300 kg of weapon-grade plutonium, and it could have attained 450 kg in 2000, a growth of 50 per cent.

Table 14.5. Central estimates for current and former de facto nuclear weapon states' inventories of plutonium and HEU, produced for nuclear weapon purposes

Figures of mass are central estimates in kilograms on 31 December 1994.

Country	Category	Weapon-grade plutonium	HEU
Israel	De facto	440	–[a]
India	De facto	300	Negligible
Pakistan	De facto	Negligible	210
North Korea	Frozen	25–40	0
South Africa	Dismantled	0	400[b]

[a] Public information provides no indication of whether Israel has an enrichment capability or a stock of HEU.

[b] All HEU enriched above 80%

Table 14.6. Central estimates for de facto nuclear weapon states' inventories of weapon-grade plutonium and WGU, produced for nuclear weapon purposes

Figures of mass are central estimates in kilograms on 31 December 1995 and by 2000.

	31 Dec. 1995		By 2000	
Country	WGPu	WGU	WGPu	WGU
Israel	460	–[a]	510	–[a]
India	330	Negligible	450	?
Pakistan	Negligible	210	Kg quantities	210[b]

[a] Public information provides no indication of whether Israel has an enrichment capability or a stock of HEU.

[b] Pakistan is assumed to continue its moratorium on HEU production.

Currently, few expect India to stop producing and separating weapon-grade plutonium, although it might halt production for nuclear weapons. Israel's Dimona reactor is over 30 years old, and it may need to be shut down soon. Israel may thus end weapon-grade plutonium production unless it develops alternative means of production. It may also judge that it has sufficient material for its weapon requirements.

Pakistan may start producing plutonium in the Khushab reactor that could be used for weapons before the year 2000. However, the exact status of its plutonium programme is unclear.

North Korea

North Korea pledged to freeze its plutonium production in October 1994 under a 'Framework Agreement' with the USA, in exchange for the phased construction of two LWRs. North Korea had previously accumulated enough weapon-grade plutonium for at least 5–8 nuclear weapons. A known quantity of about 25–30 kg of weapon-grade plutonium is contained in spent fuel, currently stored near a small 25-MWth reactor. The North may have produced another 6–9 kg of weapon-grade plutonium in this reactor and separated in a nearby plutonium separation plant in 1989–91.

If North Korea had completed its larger gas-graphite reactors, it would have discharged about 250–300 kg of plutonium by 2000 and 2000–2700 kg of plutonium by 2010. If its LWRs start operating by 2004, as planned, they will discharge about 2800 kg of plutonium by 2010. The two cases differ, however, in important respects. Spent fuel from gas-graphite reactors is hard to store safely and typically requires reprocessing. Furthermore, more than one-quarter (and possible all) of the plutonium produced in these reactors by the end of 2010 would have been weapon-grade, depending mainly on the specific fuelling arrangement of the largest reactor. In the LWR case, the plutonium is almost all reactor-grade and in an oxide fuel that does not require reprocessing. North Korea has also pledged that it will not reprocess this fuel, and that it will allow the supplier of the LWRs to remove the spent fuel from North Korea if misuse is suspected.

Highly enriched uranium

De facto nuclear weapon states

Pakistan is the only state outside the acknowledged NWS that is known to possess a sizeable inventory of unsafeguarded weapon-grade uranium. It has declared unofficially that it stopped making HEU in 1991, but this halt cannot be verified. The central estimate for Pakistan's weapon-grade uranium production is 210 kg.

India, and perhaps Israel, have mounted programmes to develop enrichment capabilities, but there is no public evidence that they have acquired significant stocks of HEU. By 2000, however, India may have accumulated a small stock of HEU.

South Africa

After South Africa abandoned its nuclear weapon programme and signed the NPT, it put its HEU from dismantled weapons and from former military stocks into storage at the Pelindaba research centre, near Pretoria. The announced intention is to use this HEU in the Safari research reactor over the next 15–20 years.

The central estimate of this inventory is about 400 kg of HEU (over 80 per cent enriched). When South Africa signed the NPT, it also possessed an inventory of several hundred kilograms of HEU enriched between 20 and 80 per cent that was not in nuclear weapons. Most of this material has been blended down into LEU.

Table 14.7. Civil stocks of plutonium by NPT status, 31 December 1994 (and 1993)
Figures are central estimates in tonnes of total plutonium.[a]

	NPT/NNWS[b]	Non-NPT	NWS	Total
In spent reactor fuels (not MOX)				755 (703)
By ownership	340 (317)	22 (20)	393 (366)	
By location	246 (227)	22 (20)	487 (456)	
Separated plutonium				118 (107)
By ownership	29 (24)	0.2 (0.2)	89 (83)	
By location	11 (11)	0.2 (0.2)	107 (96)	
Recycled plutonium	19 (18)	0.1 (0.1)	22 (19)	41 (37)
Total (by ownership)	**388 (359)**	**22 (20)**	**504 (468)**	**914 (847)**
Total (by location)	**276 (256)**	**22 (20)**	**616 (571)**	**914 (847)**

[a] Figures in parentheses are for 31 Dec. 1993. They are included here to enable comparison with the IAEA aggregate figures for 1993 (see section VII).

[b] Includes non-Russian states formed after the breakup of the USSR, all of which have now acceded to the NPT. Substantial quantities of plutonium in spent reactor fuel are located in Lithuania and Ukraine in particular. Quantities of weapon-grade plutonium in nuclear weapons located at the ends of 1993 and 1994 in Belarus, Kazakhstan and the Ukraine are counted in the Russian inventory, and are thus included in the third column.

VI. Civil inventories of plutonium and HEU

Around 1 per cent of the world stock of HEU is being used for civil purposes. About half of this is located outside the NWS. Much of this HEU was produced in the USA and is located in western Europe at research reactors and associated fuel fabrication facilities, or is held in fuels discharged from them. The remainder was of Soviet origin and is similarly embedded in research-reactor fuel cycles.

Civil stocks of plutonium are much more extensive. Table 14.7 shows that, excluding material contained in operating reactors, 914 tonnes of civil plutonium had been discharged from power reactors in 31 countries by the end of 1994.[6] However, the quantities discharged are geographically concentrated. Over half of the 914 tonnes had been discharged by the five NWS, and much of the remainder by three NNWS (Canada, Germany and Japan). In addition, around 94 tonnes of plutonium were contained in NNWS spent fuel awaiting treatment at reprocessing plants in the NWS. About 90 per cent of the plutonium separated from civil spent fuels was located in the NWS as a result of the concentration of the reprocessing industry in France, Russia and the UK. This included 13 tonnes of plutonium owned by utilities in NNWS.

[6] Those countries are Argentina, Armenia, Belgium, Brazil, Bulgaria, Canada, China, Czech Republic, Finland, France, Germany, Hungary, India, Italy, Japan, Lithuania, Mexico, the Netherlands, Pakistan, Russia, Slovakia, Slovenia, South Africa, South Korea, Spain, Sweden, Switzerland, Taiwan, Ukraine, the UK and the USA.

Table 14.8. Civil plutonium separation and use, to 31 December 1993
Central estimates in tonnes of total plutonium.

Country	Plutonium separated	Plutonium use		Plutonium balance[a]	
		Fast reactors	Thermal reactors	By ownership	By location
Belgium	2.2	0	0.6	1.6	3.5
France	32.3	8.9	7.8	15.6	27.3
Germany	18.9	2.1	6.9	9.9	2.4
India	0.4	0.1	< 0.01	0.3	0.3
Italy	2.5	1.9	< 0.05	0.6	0
Japan	14.3	3.6	< 0.05	10.7	4.7
Netherlands	1.2	0	0.1	1.1	0
Russia	26.5	0	0	26.5	26.5
Switzerland	1.5	0	1.3	0.2	0
United Kingdom	43.9	4.7	0	39.2	40.9
United States	1.5	0	0	1.5	1.5
Total	**145.2**	**21.3**	**16.8**	**107.1**	**107.1**

[a] 'By ownership' signifies the quantities held in a country that are owned by that country's institutions. 'By location' signifies the quantities held in a country including those owned by foreign institutions (i.e., utilities).

Civil separated plutonium balances (that is, separation less usage) at the end of 1993 for countries which have had plutonium separated in reprocessing are provided in table 14.8. (See chapter 7 and table 7.9 for a detailed explanation of how these figures have been assembled; it is not yet possible to provide detailed figures for the end of 1994.) Establishing accurate balances at a given date is not straightforward, not least because of the lack of precise information on whether unirradiated MOX fuels were at fabrication sites or had been delivered to reactors.[7] Nevertheless, table 14.8 again shows the preponderant roles of reprocessing companies in France, Russia and the UK as plutonium separators and storers. At the end of 1993, these three countries respectively had around 27, 27 and 41 tonnes of civil plutonium from all sources on their territories. As the return of plutonium to countries of origin and its recycling will probably lag significantly behind the rate of its separation, the quantities held in store at La Hague, Sellafield and Chelyabinsk are likely to continue increasing. If MOX recycling plans are not implemented, it is possible that little of this plutonium will leave these sites.

Elsewhere in the world, only Belgium, Germany, Japan and the United States possessed more than 1 tonne of separated civil plutonium, or of plutonium in unirradiated MOX fuel, located on their territories at the end of 1993. Besides its own inventory, Belgium is believed to have had between 2 and 2.5 tonnes of

---

[7] Assessing the plutonium inventory in Belgium, with its large MOX fabrication industry, on 31 Dec. 1993 was a particular problem in this regard.

French plutonium involved in MOX fuel fabrication on its territory. Among non-NPT states, India alone has notable amounts of civil plutonium in store, the quantities coming to a few hundred kilograms.

VII. Material under international safeguards

The primary function of the international safeguards system, as currently organized, is to detect diversions of materials from civil to military use (the IAEA) or from the purpose declared by national authorities (Euratom). Under the NPT, safeguards are mandatory for NNWS but not for NWS parties.[8] The NWS have individually entered into 'voluntary offer' agreements with the IAEA, whereby they inform the Agency of the facilities they are prepared to submit to safeguards.[9] The Agency can in turn choose ('designate') which facilities it will safeguard at a given time. In contrast, Euratom *has* to safeguard declared civil material in all 15 of its member states, including the NWS (France and the UK).

The result is that the IAEA safeguards all known inventories of plutonium and HEU located in the NNWS which are parties to the NPT. However, it inspects only a small proportion of the material held by the NWS, and much of this is material from Japan and European NNWS temporarily or semi-permanently located in France and the UK under reprocessing arrangements. At the end of 1996, only a small fraction of the US and Russian HEU or plutonium stocks was designated for safeguarding by the IAEA. In contrast, all civil inventories of plutonium in France and the UK are routinely inspected by Euratom and a significant proportion were also inspected by the IAEA. Inventories of HEU and weapon-grade plutonium in France and the UK were largely assigned to weapons and are thus outside Euratom safeguards.

It should be noted, however, that the United States has submitted a portion of the HEU and plutonium in its military inventories that is excess to defence requirements to IAEA safeguards under its voluntary offer agreement with the Agency. As a first step, approximately 10 tonnes of 'non-sensitive forms' of HEU, and two tonnes of plutonium, were made available to IAEA safeguards in September 1994 and immediately inspected by the Agency.[10]

The IAEA is also obliged to safeguard certain facilities, equipment and facilities possessed by countries outside the NPT under INFCIRC/66-type agreements accompanying nuclear transfers to those countries. However,

---

[8] The model NPT safeguards agreement is set out in an IAEA information circular (INFCIRC): IAEA, The Structure and Content of Agreements between the International Atomic Energy Agency and States Required in Connection with the Treaty on the Non-Proliferation of Nuclear Weapons (the NPT Model Safeguards Agreement), IAEA document INFCIRC/153, 10 Mar. 1971. Materials in NNWS that are not parties to the NPT are safeguarded under an earlier information circular: IAEA, The IAEA Safeguards System of 1965–68, IAEA document INFCIRC/66/Rev.2, 16 Sep. 1968.

[9] The voluntary offer agreement with France gives it the additional right to designate the materials which may be safeguarded at facilities on the facilities list presented to the IAEA. The INFCIRC numbers of the voluntary offer agreements are 263 (UK), 288 (USA), 290 (France), 327 (USSR) and 369 (China).

[10] McGoldrick, F., 'US fissile material initiatives: implications for the IAEA', *IAEA Bulletin*, vol. 37, no. 1 (Jan. 1995), pp. 49–52.

virtually all of the separated plutonium in India and Israel, and the HEU in Pakistan, remain unsafeguarded.

Each year, the IAEA publishes the aggregate amounts of plutonium and HEU which it safeguards. Euratom publishes such figures every second year, so that figures are available for 1992 but not, at the time of writing, for 1993 or 1994. Both the IAEA and Euratom are barred by rules of confidentiality from disclosing the amounts safeguarded in individual countries. At the end of 1993, the IAEA and Euratom safeguarded around 10 tonnes and 13 tonnes of HEU respectively (the quantities overlap since they both inspect material in the European NNWS).[11] Less than 1 per cent of the world stock of HEU is therefore under IAEA or Euratom safeguards. Furthermore, these figures are not adjusted to take account of burnup and may thus overstate the actual amounts that are safeguarded.

Much larger quantities of plutonium are under international safeguards. Euratom safeguarded 292 tonnes of plutonium in the European Union (EU) at the end of 1992, up from 268 tonnes in 1991, of which 'approximately 72 tonnes' were in separated form.[12] The authors have been informed that the increases in the quantities of plutonium under Euratom safeguards between 1991 and 1992 and between 1992 and 1993 were nearly identical, implying that around 316 tonnes of plutonium were being safeguarded in the EU at the end of 1993. This figure is in close accord with our own estimates of the quantities of civil plutonium produced and separated by EU member countries, or held by them on behalf of countries outside the EU (notably Japan and Switzerland). By those estimates, 275 tonnes of civil plutonium had been discharged from reactors in the EU by the end of 1993, and an additional 40 tonnes of non-EU plutonium were held at the British and French reprocessing sites, giving a total of 315 tonnes.[13] Within this inventory, just under 80 tonnes of separated plutonium were being safeguarded by Euratom according to the authors' estimates.

The plutonium figures published by the IAEA can also be reconciled with the authors' estimates, although in this case comparisons are more complicated. The IAEA's published figures for 1993 are used for the purpose of this comparison (similar figures for 1994 and 1995 are also available in subsequent Annual Reports). Table 14.9 shows the amounts of material under safeguards that the Agency published in *The Annual Report for 1993*. The IAEA includes its own estimate of the quantities of plutonium still held in reactor cores, or still otherwise unreported at the end or 1993, under the heading 'plutonium contained in irradiated fuel'. After this quantity is subtracted, the numbers in the second row in table 14.9 can be compared with our estimates of inventories in the three categories of countries identified in table 14.7.

---

[11] Euratom safeguards a larger amount of HEU because unlike the IAEA it safeguards all civil material in France and the UK. See Euratom, *Report on Operation of Euratom Safeguards*, COM(94) 282 final (Commission of the European Communities: Brussels, 6 July 1994).

[12] Euratom (note 11).

[13] The authors' estimated quantities of discharged plutonium are: Belgium 15.2 t; France 111 t; the FRG 59.3 t; Italy 5.7 t; The Netherlands 2.4 t: Spain 19.4 t; and the UK 62.3 t (see table 5.4.).

Table 14.9. Approximate quantities of plutonium under IAEA safeguards, 31 December 1993

Figures are in tonnes.

Category	NNWS			
	NPT	Non-NPT[a]	NWS	Total
Plutonium contained in irradiated fuel[b]				
Discharged fuel plus fuel in-core	282.7	25.3	104.7	**412.7**
Discharged fuel	218	19.5	104.7	**342.2**
Separated plutonium outside reactors	10.7	–	26.8	**38.5**
Recycled plutonium in fuel elements in reactor cores	2.7	0.4	–	**3.1**

[a] Includes material safeguarded in Taiwan. The People's Republic of China is the only government which has attained the right to represent China in the IAEA. Relations between the Agency and the authorities in Taiwan are therefore 'non-governmental'.

[b] Figures in the lower row are net totals after the IAEA's estimated 70.5 t of plutonium 'which [are] not yet reported to the Agency under the reporting procedures agreed to in reactor cores (the non-reported plutonium is contained in irradiated fuel assemblies to which item accountancy and containment and surveillance measures are applied)' are subtracted on a *pro-rata* basis from the totals for NPT and non-NPT NNWS given in the upper row.

Source: International Atomic Energy Agency, *The Annual Report for 1993* (IAEA: Vienna, 1994), p. 158.

It should be stressed that an identity between the IAEA's reported figures and the authors' estimates is not to be expected. The authors' estimates were based on calculations of the amounts of fuel that had been discharged by the end of 1993, drawing upon knowledge of fuelling arrangements, reactor performance and other factors. Wherever possible, the estimates were cross-checked with utilities and adjusted accordingly. The IAEA figures reflect reports received from governments, their times of receipt depending on the reporting rules agreed with the Agency, and on internal estimates. For instance, six months can legitimately elapse between a discharge of plutonium-containing spent fuel and its report to the Agency.

The following comments should be made about the correspondence between our figures and those of the IAEA:

1. The main point at which IAEA figures can be used to corroborate our estimate is the entry in table 14.9 for plutonium contained in discharged irradiated fuel in NNWS parties to the NPT. The IAEA figure, after adjustment to take account of 'unreported plutonium', is 218 tonnes. However, this does not include plutonium held in irradiated fuel at reactors in Ukraine and Lithuania which are included in table 14.7. The authors estimate that at the end of 1993 about 22 tonnes of plutonium were stored at Ukrainian (17 t) and Lithuanian (5 t) reactors. Taking the figure of 227 tonnes of plutonium in discharged UO_2 fuel in NNWS parties to the NPT from table 14.7, and adding 9 tonnes of

plutonium estimated to be contained in discharged MOX fuel, produces a total of 236 tonnes of plutonium located in NNWS parties to the NPT at the end of 1993. This compares reasonably well with the IAEA figure adjusted to take account of plutonium in Ukraine and Lithuania—240 tonnes (218 t plus 22 t).

2. The 104.7 tonnes of 'plutonium contained in irradiated fuel' and 26.8 tonnes of 'separated plutonium outside reactors' in the NWS column in table 14.9 need to be explained together. These figures relate to plutonium being safeguarded by the IAEA at the La Hague and Sellafield sites in France and the UK. It has to be recalled that no plutonium-bearing facilities in the USA and Russia were designated for safeguarding by the IAEA at the ended of 1993 (inspection of the Novovoronezh facility was withdrawn earlier in the year). Furthermore, neither the RT-1 reprocessing plant where civil spent fuel has hitherto been reprocessed in Russia, nor its spent-fuel ponds and plutonium stores, have been placed on the Russian list of facilities that are open to IAEA safeguarding.

At La Hague, subsidiary arrangements for IAEA safeguarding are in force at the combined UP2 and UP3 spent-fuel ponds. They have not been extended to the reprocessing plants themselves, nor to their plutonium stores. In order to maintain Agency safeguards on foreign plutonium at La Hague during and after reprocessing, particularly on Japanese plutonium which is subject to transfer agreements which require the application of IAEA safeguards, the practice of substitution is invoked.[14] When Japanese spent-fuel is taken from the ponds and fed into the reprocessing plant, an amount of French spent-fuel containing an equivalent quantity of plutonium is brought under IAEA safeguards. Separated plutonium 're-enters' IAEA safeguards on its return to Japan. In table 14.9, it is probable that the 'substituted' amount is entered as 'separated plutonium' even though the plutonium is still embedded in irradiated fuel in ponds which are being inspected.

At Sellafield, the THORP reprocessing plant began to be commissioned in 1994, and the first deliveries of plutonium oxide were expected in late 1995. The spent-fuel ponds attached to THORP, together with a plutonium store, are designated by the IAEA. The B205 reprocessing plant at Sellafield, where Italian and Japanese plutonium has been separated from magnox fuel since the late 1960s, is not safeguarded by the IAEA. There is also no expectation that the THORP process area will be designated for inspection, although it will be subject to full safeguarding by Euratom (as will UP3). In the UK, substitution will therefore only be required in relation to foreign plutonium contained in the process areas of the B205 and THORP reprocessing plants. In contrast to current practice at La Hague, substitution at Sellafield is mainly carried out in the plutonium store rather than in the spent-fuel pond, although Agency safeguards are applied to both facilities.

Not having access to confidential information provided to the IAEA by Britain, France and their foreign customers, it is therefore difficult to assess

[14] Japan has INFCIRC/66-type agreements with both France and the UK.

how the figures recorded for the NWS in table 14.9 have been assembled. As the IAEA has acknowledged, it may not have been possible to avoid some double counting when compiling the figures.[15] The authors have estimated that 90 tonnes of NNWS plutonium were held in unreprocessed spent-fuel at the British and French sites at the end of 1993, all of which has remained under Agency safeguards. Given the total of 104.5 tonnes recorded by the IAEA, this leaves about 15 tonnes of safeguarded plutonium in irradiated fuels located in France and the UK. Presumably, this balance comprised plutonium in British and French spent-fuels that had been brought under safeguards, partly under substitution arrangements, alongside the foreign fuels held in the ponds at La Hague and Sellafield.

It has also been estimated that some 13 tonnes of separated plutonium were held in store at La Hague and Sellafield at the end of 1993 prior to their return to foreign utilities. The reason that this falls so short of the 26.8 tonnes of separated plutonium listed in table 14.9 is, one presumes, that a sizeable quantity of British plutonium in store 9 at Sellafield was also subject to IAEA safeguards in 1993. In this case, as in others, substitution arrangements may have resulted in an overstatement of the amount of foreign plutonium under IAEA safeguards at any one time.

3. The IAEA's figure for plutonium contained in discharged fuel in non-NPT states (and in Taiwan) is reasonably close to the authors' estimate: 19.5 tonnes compared to 21 tonnes.[16] The first of these figures, which may be subject to error, has been arrived at by the authors after subtraction of an estimated quantity held 'in-core' in these countries (see note b, table 14.9).

4. The 10.7 tonnes of 'separated plutonium outside reactor cores' compares well with the 11 tonnes estimated in these pages. The authors may have underestimated the amounts of French plutonium held at Dessel in Belgium on 31 December 1993 in accordance with MOX fabrication contracts. Unused Kalkar fuel elements are also being stored in Belgium.

Overall, table 14.9 suggests that at the end of 1993 the IAEA safeguarded 384 (342.2 plus 38.5 plus 3.1) tonnes of plutonium discharged from power reactors, assuming that the figures reported by the IAEA have involved no double counting. This represented around 35 per cent of the world inventory of plutonium, and around 45 per cent of the world inventory of civil plutonium. Nearly all of this plutonium was reactor-grade. *Almost all weapon-grade plutonium and uranium existing in the world remains outside international safeguards.* The situation has changed little in all these respects since 1993.

---

[15] As the IAEA expressed to the authors, 'the question is complicated because of the interrelationship of transfer agreements and a voluntary offer agreement. Under a voluntary offer, a storage pond containing irradiated fuel may be under safeguards; some of that fuel may be subject to substitution under a transfer agreement but remain under safeguards under a voluntary offer. In preparing the table, efforts are made to count material covered by more than one safeguards agreement only once (but that is not as easy as it sounds)'.

[16] This 21 t is made up from an estimated 1 t of safeguarded plutonium in India, 5.9 t in Argentina, 0.5 t in Brazil, 0.4 t in Pakistan, 2.3 t in South Africa and 11 t in Taiwan.

VIII. Possible future trends in plutonium and HEU inventories

The production of plutonium and HEU for weapon purposes in the acknowledged NWS appears to be approaching its end. The production of weapon-grade materials in de facto nuclear weapon states continues with little apparent prospect of ending soon. Following the adoption of a unanimous resolution by the UN General Assembly in December 1993, efforts are currently being made to formalize the cessation of production in a universal international treaty. The proposed Fissile Material Cut-Off Treaty is discussed in chapter 15, section V.

The major changes to the sizes of inventories will occur in two contexts. One involves the growth of plutonium inventories because of the irradiation of uranium fuels in nuclear power stations which are generating electricity, and the growth of inventories of *separated* plutonium through the reprocessing of spent reactor fuels (the net changes depending on the rate at which the plutonium is recycled). The other involves the extraction of plutonium and HEU from dismantled nuclear weapons, and the withdrawal of excess materials from military usage, as the nuclear weapon states implement arms reduction agreements and reorganize the infrastructures supporting their military programmes. The amounts of material that may come from these two sources are considered below.

Plutonium produced and separated in the civilian fuel cycle

Barring catastrophic accidents, the amount of plutonium that will be produced in operating power reactors over the next two decades is reasonably predictable. The capacity of reactors coming into operation (e.g., in Japan) may be roughly offset by the capacity of reactors being shut down (e.g., in the FSU and the UK). The major changes will occur after 2010 when the many reactors which came into operation in the 1970s and early 1980s will approach the end of their life-times. Generating capacity thereafter will depend on whether new investment programmes can be launched in the late 1990s and in the early part of the next century.

In relation to plutonium arisings, there are some uncertainties attached to the burnups achieved in uranium fuels (see chapters 2 and 5). Table 14.10 gives some projections of spent-fuel and plutonium arisings, based on relatively conservative assumptions about nuclear capacity and about the increases in burnup discussed in chapter 5. Spent fuel discharges are shown to remain steady for the period until 2010 at 10 500–11 000 tonnes per year, with a gradual decline thereafter as fuel burnups increase and reactors are decommissioned. Arisings of plutonium in discharged fuel are also set to remain at around 70 tonnes per year until 2010.

The picture for plutonium separation is different. Owing to the commissioning of new reprocessing plants in the early 1990s, a rapid acceleration in the rate of plutonium separation is now taking place. During the period 1994–2000

Table 14.10. Projection of cumulative spent-fuel discharges, plutonium separation and unrecycled stocks, 31 December 1993, 2000 and 2010

Figures are in tonnes of heavy metal and total plutonium.

Category	31 Dec. 1993	31 Dec. 2000	31 Dec. 2010
Spent-fuel	145 000	220 000	325 000
Plutonium discharged	847	1 400	2 100
Plutonium separation	144	277	437
Plutonium stocks			
MOX fabrication scenario	107	152	79
Utility policy scenario	107	198	245

some 133 tonnes of total plutonium are forecast to be separated, compared with 144 tonnes over the previous 30 years. If reprocessing policies remained unchanged, projections of plutonium separation up to 2002 would be quite reliable since they are based on the working out of binding contracts between utilities and reprocessors. The main uncertainty relates to the operational performance of the Mayak plant in Russia and the THORP plant in the UK.

It is unlikely that these rates of reprocessing can be sustained much beyond the turn of the century since fuel-management policies in all but four countries—France, Japan, Russia and the UK—are coming to rely increasingly on long-term interim storage and direct disposal. Even in those four countries, complete reliance on reprocessing is gradually being abandoned. Despite the anticipated construction of the Rokkasho-mura reprocessing plant in Japan, annual rates of plutonium separation are estimated to decline from an average of about 19 tonnes in 1994–2000 to about 16 tonnes in 2001–2010.

A proportion of this material is expected to be fabricated into mixed-oxide fuel for thermal and fast reactors. Thermal recycling will represent the main route for plutonium use owing to the near collapse of fast-reactor programmes in Europe and Russia in the early 1990s. Between 1994 and 2000, only some 5–6 tonnes of plutonium could at most be inserted into fast reactors, providing France and Japan implement their declared policies. Beyond that, plutonium consumption in fast reactors will depend on whether and how the surviving R&D programmes are maintained.

There are two possible approaches to estimating how much civil plutonium may be recycled in thermal reactors. Under the first, projections reflect the scale of existing and planned MOX fuel fabrication capacity (this is the 'MOX fabrication scenario' in table 14.10). Using this approach, it can be estimated that about 88 tonnes of total plutonium could be disposed of in reactors in the period 1994–2000 while a maximum of about 230 tonnes of plutonium could be fabricated into fuel in 2001–2010.[17] Under the second 'utility-policy' approach, projections are made on the basis of utilities' declared MOX recycl-

---

[17] The discrepancy between the figure of 88 t derived here and the 71 t derived in tables 7.4 and 7.7 results from the use of a slightly different time period (1993–99) for making the latter balance estimate.

ing policies. According to this approach, plutonium consumption in thermal MOX fuel would be about 35 tonnes in 1994–2000. It should be noted that only about 17 tonnes of plutonium had been recycled in thermal reactors by the end of 1993, so that even this amount would involve a significant expansion of recycling activity. Only in France has MOX recycling been developing strongly in the mid-1990s. Most utility policies do not go beyond 2005, so that consumption by 2010 is harder to estimate using this approach. However, assuming that utilities' plans are carried out, about 113 tonnes of plutonium could be recycled in LWRs and fast reactors between 2001 and 2010.

The discrepancy between these two projections reflects utilities' uncertainties over the future attractions of MOX recycling, and their tardiness in preparing for it. It has significant implications for the modelling of national and world civil plutonium stockpiles and surpluses. Under the 'MOX fabrication' approach, the world plutonium surplus, which stood at 118 tonnes at the end of 1994, would grow to about 150 tonnes in 1998/99 before declining back to 79 tonnes by 2010. However, the greater proportion of this will be plutonium belonging to utilities in Russia and the UK. The analysis in chapter 7 showed that according to this approach, plutonium stockpiles in the rest of the world (the BFGJS scenario in figure 7.7) could not be completely consumed by 2010. Under the 'utility-policy' approach, plutonium stockpiles would continue to grow for the entire period being modelled: world stockpiles would increase to about 200 tonnes by 2000. No equilibrium is achieved between rates of separation and rates of consumption in this scenario, so that world stockpiles would grow at a rate of about 4 tonnes per year up to and beyond 2010. This would lead to a world surplus, assuming reprocessing rates were maintained at currently committed levels, of about 245 tonnes in 2010.

In summary, plutonium arisings in discharged spent-fuel are forecast to grow steadily over the next 15 years or so. Plutonium separation in reprocessing is set to increase rapidly during the 1990s, but is likely then to face a decline. The rate at which separated plutonium is consumed in recycled nuclear fuel is difficult to predict. The evolution of world plutonium stockpiles is directly linked to the relative rates at which plutonium is separated and consumed as fuel. As a result, there are several possible outcomes. According to the authors' illustrative estimates, the world surplus of civil plutonium, which is predominantly reactor-grade, could be in the range 79–245 tonnes in the year 2010.

These assessments do not cover the possibility that priority will be given to the disposition of excess weapon-grade plutonium, or to the disposition of existing stocks of reactor-grade plutonium, in years ahead. This possibility is discussed in chapter 15, section IX.

Plutonium and HEU released from dismantled warheads

Before the breakup of the USSR, the US and Soviet governments had already begun to instigate substantial arms reductions. The INF Treaty had contained agreement on the elimination of stocks of intermediate-range nuclear weapons,

Table 14.11. Illustrative inventories of weapon-grade plutonium and HEU inside and outside operational nuclear weapons[a]

Central estimates in tonnes, 31 December 1994.

Category	USA	FSU	France	China	UK	Total
Inside weapons						
Plutonium	31	35	2	2	1	**70**
HEU[b]	202	225	11	10	4.5	**450**[c]
Outside weapons						
Plutonium	54	96	3	2	2	**160**
HEU[b]	443	825	13	10	3.5	**1 300**[c]

[a] The plutonium and HEU figures presented here are based on the following rough estimates of operational warhead numbers: USA 9000; FSU 10 000; France 490; China 450; and UK 200. Average quantities of plutonium and HEU per warhead are assumed to be 3.5 kg and 22.5 kg, respectively.

[b] Calculated in terms of weapon-grade uranium equivalent.

[c] These figures are rounded.

and the START I and II Treaties looked forward to deep reductions in long-range strategic weapons. Once the START II Treaty is ratified, the US and Russian governments will be committed to cutting their operational nuclear arsenals to around 3500 strategic warheads apiece, compared with the tens of thousands that they formerly deployed. They are also taking steps to eliminate large numbers of tactical weapons. The arms reductions include the elimination of the nuclear weapons located on the territories of Belarus, Kazakhstan and Ukraine which have joined the NPT as non-nuclear weapon states since the Soviet Union's demise.

The French and British governments have also announced that they are cutting back their nuclear arsenals, although the numbers of warheads are much smaller. The reductions mainly involve the retirement of old strategic and substrategic weapons, and their replacement by fewer, but more accurate, strategic systems; and the elimination of classes of tactical weapon.

The decommissioning of nuclear weapons involves several steps, none of which is straightforward. They include the gathering together and storage of warheads, their dismantlement, the safe storage of warhead components, and the processing of their fissile content, bringing the plutonium and HEU to a form that is suitable for storage, recycling or disposal. Over the past three years, a total of around 4000 warheads are believed to have been dismantled annually in Russia and the USA. At this rate, the START II dismantlement programme will be completed early in the next century, at least where the warheads are concerned. There are as yet no firm plans to disassemble the warhead components (the 'pits') containing the plutonium. It is expected that the HEU will gradually be separated from warhead components. Under current plans, much of the US weapon-grade uranium will be stockpiled for use in naval reactors. In all NWS, a large fraction of excess material is expected to be diluted eventually so that it can be used as reactor fuel. The elimination of the warheads and their components is expected to take well over 20 years to complete.

Table 14.11 illustrates the amounts of material that may have been located inside and outside operational weapons (i.e., weapons that are still in service) at the end of 1994. The figures here are based on the presumption that an average nuclear warhead contains 3–4 kg of plutonium and 15–30 kg of HEU.

The following points pertaining to table 14.11 deserve emphasis:

1. The table rests upon a distinction drawn between material 'inside and outside operational warheads'. In practice, the forms in which material 'outside operational warheads' are held are extremely diverse. They extend from warheads that are still intact but have been removed from active service (or may be held in reserve), to materials in weapon components, to HEU in submarine reactor fuels, to surplus stocks of separated plutonium and HEU, to scraps and wastes produced in production processes.

2. The quantities of weapon-grade plutonium and HEU existing outside weapons are already very large. In total, there could be as much as 160 tonnes of plutonium and 1300 tonnes of HEU being stored in various forms. As warhead arsenals are reduced in size, the amounts held outside weapons, and outside the weapon production system, are bound to rise. The quantities that may be counted as excess to military requirement are discussed in chapter 15, section VI.

3. The bulk of these inventories is contained in just two countries—Russia and the USA. The amounts of plutonium that could be derived from dismantled British, Chinese and French warheads would be two orders of magnitude less. However, it should be recalled that rather similar quantities of plutonium (approaching 200 t over the next decade) are expected to arise from the expansion of civil reprocessing in Britain and France, although that plutonium will be largely reactor-grade.

This evaluation, which focuses on the trends in large quantities of fissile material, does not imply that relatively small stocks of material, such as those found in India, Israel and Pakistan, are less important. Needless to say, the risks posed by those small stocks are well appreciated by the authors.

The stocks of fissile material dedicated to nuclear weapons in these de facto weapon states will probably continue to increase. Whether other nations will try to acquire stocks of fissile material dedicated to weapons purposes is impossible to predict, but there is justified concern that the security situation in a number of unstable regions may encourage the acquisition of unsafeguarded materials.

Issues relating to the control and disposition of stocks of fissile material discussed above, and to the detection of undeclared production activities, are addressed in chapter 15.

15. The control and disposition of fissile materials: the new policy agenda

I. Introduction

In the nearly four years that have elapsed since *World Inventory 1992* was published, much has been achieved in lessening the threat of nuclear war, reducing the scale of nuclear armament and preventing the spread of nuclear weapon capabilities. One can point in particular to the progress in implementing arms reduction agreements; the successful absorption into the non-proliferation regime of the non-Russian states emerging from the breakup of the Soviet Union; South Africa's voluntary disarmament and the establishment of an African nuclear weapon-free zone; the destruction of Iraq's nuclear weapon capability and the containment of North Korea's nuclear programme; the NPT's expanding membership; the agreement in May 1995 to extend the NPT indefinitely with a more stringent review process; and the conclusion of the Comprehensive Nuclear Test Ban Treaty (the CTBT).

Although many tough problems remain to be solved, a favourable wind seems to have been blowing. This is evident in the perceptible strengthening of international norms in recent years. 'You shall not acquire, use or threaten to use nuclear weapons' is becoming the first rule in the book of good international behaviour. Despite the Chinese and French nuclear tests and the concerns that India might follow suit, despite the delay in ratification of START II and despite the periodically strained relations between Russia and the USA, and China and the USA, the salience of nuclear weapons as instruments of military strategy and great power politics appears to have diminished. Whether this marks an irreversible de-nuclearization of warfare and of international politics, or a hiatus brought about by the end of a distinctive period of hegemonic rivalry, is a matter for conjecture. Either way, there is now a clear opportunity to drain nuclear weapons of their remaining political and military significance, and to erect tougher barriers against their future development and deployment.

Any scheme or set of schemes for achieving these ends must encompass the technical foundations of nuclear weaponry: in particular, the fissile materials used in warheads and the technologies for producing them. The scale and complexity of the regulatory task is evident from this book. Over 3000 tonnes of plutonium and highly enriched uranium have been produced since the birth of nuclear technology. The majority is still outside international safeguards, and a significant proportion may be inadequately protected. Furthermore, events in the 1990s—the revelations about the Iraqi and North Korean clandestine weapon programmes, the huge arms reduction programmes mounted by the USA and the FSU, the threats posed by nuclear smuggling, and the emergence

CONTROL AND DISPOSITION OF FISSILE MATERIALS

of a large overhang of surplus weapon materials—have shown up inadequacies in the present regime. It is acknowledged that they cannot be overcome solely through incremental adjustment: they are systemic in origin and demand major reforms.

A new phase of intensive regime building has therefore begun. This chapter reviews its progress and considers the policy challenges that the international community still has to face in regard to fissile materials. While many of the required innovations have been identified by governments, and are being intensively worked upon in some cases, a clearer strategy is needed to guide the extension and deepening of the regulatory system that has developed around fissile materials. There is a parallel need to develop more coherent strategies for eliminating the large surplus stocks of these materials. These strategies are the other subject of this chapter.

Proposed here is the establishment of a coherent and non-discriminatory regime, embracing countries with and without nuclear weapon programmes, that would provide confidence that fissile materials and associated technologies were being effectively controlled in the new circumstances. Building upon existing institutions, measures and initiatives, the regime's ultimate goal would be to provide a framework for fissile material control capable of supporting complete nuclear disarmament. It would have five main pillars:[1]

1. The deepening of bilateral cooperative measures between Russia, the USA, other countries and international organizations aimed at raising standards of material protection control and accountancy (MPC&A), and at managing nuclear weapon dismantlement and the safe storage and disposition of weapon materials.

2. The ending of production of fissile materials for nuclear weapons through the conclusion of the Fissile Material Cut-Off Treaty; and the negotiation and implementation of a new international treaty or agreements concerning the transparency, verification and management of excess stocks of fissile material.

3. The strengthening of the IAEA safeguards system, including measures to enhance the detection of undeclared nuclear weapon activities, and the extension of IAEA safeguards to all non-military facilities and materials in the nuclear weapon states and non-NPT threshold countries.

4. A disposition strategy that would largely eliminate excess stocks of HEU and plutonium over the next 20–30 years, and that would curtail the unnecessary production of new stocks of HEU and separated plutonium.

5. An effective and universal system of trade regulation consistent with the further development of nuclear power production. (This issue is beyond the scope of this book, but see chapters 11 and 12.)[2]

---

[1] These proposals are consistent with, although not identical to, those contained in the National Academy of Sciences's 1994 report. See National Academy of Sciences, Committee on International Security and Arms Control, *Management and Disposition of Excess Weapons Plutonium* (National Academy Press: Washington, DC, 1994), p. 9.

[2] The other issue that is discussed only in passing in this chapter is the IAEA's budget, which is assumed here, perhaps unrealistically, to be ultimately responsive to need.

In this chapter, emphasis is placed on the development of multilateral measures. This reflects the importance attached to the collective pursuit of nuclear disarmament and strengthening of the non-proliferation regime. The emphasis is also justified because effective multilateral measures are ultimately the best way of serving national interests, especially where nuclear materials are concerned. The sustained support that states have given to the international safeguards system, even at times when its efficacy has been open to question, is testament to this reality.

Sections II–VI examine the political and regulatory steps involved in establishing a stronger and more universal framework of control. Sections II and III consider the main structural defects in the existing framework, not least regarding the NPT's 'dual structure' which has resulted in different obligations being placed on NWS and NNWS. Section IV covers the initiatives being taken to detect and deter clandestine weapon programmes. Sections V and VI are concerned with the situation in the NWS and the non-NPT threshold states: the main bilateral initiatives launched by the US and Russian governments; and the need for a stronger multilateral framework for regulating fissile materials in countries with nuclear weapons, encompassing the Fissile Material Cut-Off Treaty and a new proposal for a verification and transparency regime covering excess weapon materials.

Sections VII to X are concerned with the disposition of excess HEU and plutonium. Technical aspects of disposition, and the 'sizing' of excess stocks, are considered in section VII. Disposition scenarios for HEU and plutonium, and their political and commercial plausibility, are assessed in sections VIII and IX. Section X discusses the conditions for an effective disposition strategy.

The chapter's main conclusions are presented in section X.

II. Two industrial and regulatory systems

The history of fissile materials over the past half century has been largely the tale of two distinctive industrial and regulatory systems. The first of these systems, which was most highly developed in the USA and the FSU, supported the manufacture of nuclear weapons. Its objective was to provide the materials that would give the most effective and predictable explosive yields, at least cost, with production being linked to weapon 'requirements'. Smaller quantities were also provided to fuel nuclear-powered submarines and aircraft-carriers. The second production system, mainly involving industrial nations in North America, Europe, the FSU and East Asia (with uranium also coming from Africa and Australia) was oriented towards electricity supply. It involved the provision of nuclear fuel and its irradiation in power reactors, and the management of radioactive by-products, all in the interest of safely extracting heat and thus electricity from the energetic reactions associated with nuclear fission.

While technological developments in the military sector provided the early foundations for the establishment of civil nuclear production, these two industrial systems developed separately, largely according to their own internal logics and dynamics. In the main, the markets for materials within the two systems were kept apart. Uranium enrichment was the one significant area of overlap, especially in the USA and the USSR where the same facilities were used for enriching materials for civil and military purposes. There was also some overlap in plutonium separation, a legacy of which is today's concentration of civil reprocessing in three nuclear weapon states—Britain, France and Russia.[3]

The dualism of nuclear industrial activity was matched by dualism in the political and institutional settings in which it occurred. Production for military purposes tended to be carried out in autonomous industrial zones. Whether in capitalist or communist countries, or in established or embryonic nuclear weapon states, production was organized with few exceptions on strictly national lines, within enclosed and highly secretive organizations, according to the rules of command economies, subject to separate safety regulations, and largely out of sight of democratic political institutions. All nuclear weapon programmes have been inherently secret, especially in their early stages.

In contrast, production for civil purposes strove (not always successfully) to be a commercial activity, driven by rational economic choices, albeit in a tightly regulated market. The tight regulation arose from the need to guard against the diversion of weapon-usable materials and technologies possessed by countries without nuclear weapons into military programmes or into the hands of sub-national groups, and from the need to protect societies and environments from exposure to harmful nuclear radiation. Markets were also regulated because states everywhere played a large part in the development of civil nuclear industries, partly for reasons of energy security and high R&D costs.

The non-proliferation regime that developed in the period of the cold war largely incorporated this dualism, which came to be expressed in the set of legal presumptions underpinning the NPT:

1. The nation state is the supreme institution under international law, and has jurisdiction over all nuclear materials, facilities and activities on its territory.

2. There are two categories of nation-state: nuclear weapon states and non-nuclear weapon states.

3. There are two categories of nuclear activity: those devoted to peaceful and to weapon purposes.[4]

---

[3] The reprocessing plants at Chelyabinsk, Sellafield and Marcoule in the 1950s and 1960s handled fuel from both military and civilian reactors. Although the new plants constructed at La Hague and Sellafield in the 1980s were dedicated to civilian markets, their political acceptance was partly made possible by their location on sites that had been acquired earlier for military purposes. In all NPT NNWS, the licensing of new sites for the construction of reprocessing plants proved increasingly difficult.

[4] However, the NPT does not prohibit the uses by NNWS parties of nuclear materials and technologies for non-explosive military purposes, e.g., in the powering of submarines. The safeguards implications are addressed in article 14 of the NPT Model Safeguards Agreement. IAEA, The Structure and Content of Agreements between the International Atomic Energy Agency and States Required in Connection with the Treaty on the Non-Proliferation of Nuclear Weapons, IAEA document INFCIRC/153 Rev.2, Vienna, 1983. The provisions in article 14 have never been exercised as no NPT NNWS has acquired a nuclear-powered submarine.

4. NWS parties to the NPT can engage in both peaceful and weapon activities without mandatory international verification (safeguards). NNWS parties can only engage in peaceful activities and must submit them to international safeguards.

The IAEA was appointed the NPT's safeguards agency. It is managed by a Secretariat which answers to a Board of Governors appointed by member states, which in turn report to an annual conference. It was granted no formal supranational powers.[5] The application of NPT safeguards is built upon cooperation between the Agency and the state. This entails *inter alia* the maintenance by NNWS of national systems of material accounting and control, using standard approaches; the requirement for states to provide information to the IAEA, under agreed rules of confidentiality, on nuclear-related materials and facilities on their territories; the definition of circumstances under which inspections may be carried out; and the appointment, duties and rights of IAEA inspectors. These cooperative arrangements, together with the definitions, rules, methods and procedures under which safeguards can be applied, are described in the NPT safeguards document INFCIRC/153, agreed in March 1971. In effect, INFCIRC/153 defines the *limits of intrusiveness and transparency* attached to the Treaty's verification, limits which would hold—or whose interpretations would hold—without major revision for the next 20 years.

In contrast, the 1957 Euratom Treaty, which pre-dated the NPT, avoided the NPT's legal distinctions between states. Instead, the European Community was pre-eminent, and all nation states within it, including any NWS, had the same legal rights and responsibilities. Special fissile materials within its boundaries were the Community's legal property, and all member states had to submit their civil (but only civil) materials to Euratom safeguards.[6] However, the Euratom Treaty remained a regional treaty, and its supranational ambitions were not fully realized. It was regarded as an inappropriate model for the global non-proliferation regime whose construction began in earnest in the 1960s.

Under the NPT, the NWS parties therefore maintained *comprehensive* sovereignty over their internal nuclear activities, whether civil or military. While giving up their rights to undertake certain external transactions, notably the transfer of nuclear weapons to other countries and the export of unsafeguarded materials and equipment to non-nuclear weapon states, they maintained freedom to conduct military and civilian programmes without intrusion and without any requirement for transparency. The 'voluntary offer' safeguards agreements with the IAEA, whereby the Agency could apply safeguards to a list of facilities provided by the NWS in question, were their only concessions. In contrast, the NNWS parties accepted *partial* sovereignty. By acceding to the Treaty, they

---

[5] The IAEA Secretariat was empowered, however, to take certain initiatives, such as informing the United Nations of breaches in safeguards undertakings, without consulting member states. One could argue that this gave it an element of supranational authority.

[6] In the Euratom Treaty, the key distinction is between the 'intended uses' to which nuclear materials are put. Safeguards are applied to civil materials but not to 'materials intended to meet defence requirements'.

accepted that the peaceful nature of their nuclear activities would be subjected to international verification, and to the intrusion and transparency that entailed. Countries that would not accept these conditions, whether for political or military reasons, stayed outside the Treaty.

The NPT's stratification, therefore, came to be reflected in the unequal application of IAEA safeguards noted in chapter 14. In practice, little safeguarding was carried out under the 'voluntary offer' agreements, partly because the diversion of fissile materials in countries that already possessed nuclear weapons was assumed to be irrelevant to non-proliferation policy, and partly because the IAEA was denied the necessary resources (not just by the NWS). Safeguards coverage in the states that had refused to accede to the NPT was also limited. The bulk of safeguarding took place in NNWS parties to the NPT, with power reactors and spent fuel stores accounting for most inspection time, except in Japan where reprocessing and other plutonium-related facilities were in operation.

Why were so many countries prepared to accede to the NPT and thereby accept its dualism? First, countries were prepared to accept a discriminatory NPT in order that an effective non-proliferation regime could be established. They were primarily motivated by their own security interests in preventing the spread of nuclear weapons, particularly in their regions. Second, the numerous NNWS that were members of NATO, the Warsaw Treaty Organization and other military alliances were protected by nuclear umbrellas and therefore had interests in the continuing vitality of the US and Soviet nuclear weapon programmes. Third, the hierarchy implicit in the NPT (and in the permanent membership of the NWS in the Security Council) reflected the international power structure that emerged after the end of World War II. Relatively few countries had the ability, or found it in their interest, to resist their authority.

III. Contemporary pressures to achieve universality and transparency

The nuclear industry's dual structure, in both physical and politico-regulatory terms, has been rendered increasingly anachronistic by the end of the cold war. The nuclear weapon states' insistence on complete autonomy within the NPT was rooted in the psychology and practice of nuclear deterrence, with its premium on secrecy and deception, and on the desire to protect weapon programmes from domestic and foreign scrutiny. As international security has become less reliant on threats of mutual destruction, and as transparency has become an increasingly important strategy for achieving security in a less ordered world, the exemptions granted to the NWS and their nuclear institutions have become harder to justify. Their movement towards the non-nuclear mainstream, involving regulatory 'intrusions' as well as arms reductions, seems unavoidable if a more universal non-proliferation regime is to be achieved. The Chemical Weapons Convention, signed in early 1993, also provides a living example of a universal accord dedicated to the total elimination of a category of weapons of mass destruction.

Much of the industrial system that supported the weapon programmes is also being shut down, taken apart or reoriented towards dismantling its former products or providing other services. The production of fissile materials for weapon purposes has all but ended, and three-quarters or more of the weapon-grade material amassed during the nuclear arms race is no longer needed for military purposes. 'De-militarizing' the residues of the huge nuclear weapon programmes mounted in the cold war will entail bringing institutional practices increasingly into line with the dominant standards applied in civil commerce. The disposition of excess materials will, in particular, rely heavily on their transfer into the civil fuel cycle, possibly including the fuel cycle in countries outside the NWS (see sections VII–IX). This provides a further incentive to 'harmonize' the civil and former military domains within the NPT.

However, more is involved than adjustment to the end of an era. The dual structure now acts directly against the security interests of states, including the nuclear weapon states themselves. Until recently, it was implicitly assumed that the NWS were stable entities whose governments exercised absolute control over nuclear activities in their territories, which had rights to set their own standards of safety, accountancy and physical security, and which were uniquely well qualified to look after the nuclear weapon materials and technologies in their possession. The breakup of the Soviet Union has shown the fallacy of this assumption. A major nuclear power has fractured into a number of independent states. Moreover, a whole political and economic system has crumbled, a system which largely determined how the nuclear industry functioned and was governed, and how physical security and material accounting were implemented. As a stable replacement for this system has yet to be fully established in the FSU, the management of its nuclear legacy has become a source of deep international concern.

The other NWS are not immune to such changes, although it is hard to imagine them experiencing such dramatic upheavals. In various respects, the states that constitute China, France, the UK and the USA seem less stable internally than they were 20 or 30 years ago.

The predicament is therefore that the NWS, and Russia in particular, are potentially the largest and most accessible sources of the materials, technologies and skills required by weapon proliferators, while being largely exempt under international law from the safeguards and other verification measures applied in countries with lesser capabilities, and from the good practices associated with them. This is not to suggest that they have no material control systems in place, but that their systems have not always been as comprehensive and stringent as those established elsewhere, and that they are in some contexts inappropriate to today's circumstances. In Russia and China, in particular, they were designed to keep 'the outside out' rather than 'the inside in' and they relied heavily upon totalitarian instruments of social control. Furthermore, the material control and accounting systems in the NWS have not been transparent.

If China, France, the UK and USA were now to claim that their systems were equal or superior to those applied elsewhere, the truth of their assertions could not be verified.[7]

The gravity with which this problem is now viewed reflects another realization—that nuclear weapons have become easier to design and manufacture. As several chapters in this book attest, various techniques have been developed for producing and separating the necessary fissile materials, some of which involve relatively low technology (Iraq was making steady progress with three enrichment technologies, one of which was last used in the 1940s), and the main design features of fission and thermonuclear weapons are well known, even if the manufacture of effective nuclear explosive devices remains difficult. The pool of dual-use technology has broadened and deepened, while becoming liable to wider diffusion as a result of the progressive internationalization of economic activity and failure to establish universal export controls. The Iraqi and North Korean programmes have shown that nuclear weapons may be within the reach of countries with limited scientific and technical resources.

Furthermore, there is now the possibility that maverick states, or the paramilitary organizations, renegade armies, religious cults or terrorist groups that seem to have multiplied in recent times, could take a short cut to nuclear explosives by acquiring weapon-grade plutonium or uranium directly from ill-protected and/or unsafeguarded installations in an NWS or threshold state. Despite doubts about their abilities to put together explosive devices, their possession of weapon-grade materials would have to be regarded as a grave threat. By avoiding the need to construct the reactors and reprocessing plants which have traditionally been required to produce weapon-grade plutonium, or the enrichment plants required to produce HEU, the time and resources required to build nuclear weapons would be greatly reduced. The detection of clandestine warhead development and manufacturing programmes would also be much more difficult and might occur only after a nuclear explosive was completed.

The need to tighten controls on fissile materials and on the weapon technologies associated with them, and to ensure that controls are applied *everywhere* to the highest standards, has therefore increased. The implication is that nothing short of a universal, non-discriminatory framework of control will be adequate in future. At the same time, the limits of intrusiveness and transparency that were set in the 1960s and 1970s have been called into question, particularly following the revelations about Iraq's undeclared programme. In both regards, sovereignty is at stake. Although engraved in the NPT, the comprehensive sovereignty granted to the NWS no longer seems appropriate in the new circumstances; and the partial sovereignty accepted by the NNWS is having to be redefined.

---

[7] The material accounting and verification procedures applied in the UK and USA may be close to international standards. The authors have been informed that the standards applied by the UK to military facilities and materials are now identical to those required by Euratom in the safeguarding of civil materials; and the material accounting and control system which has been applied by the US Department of Energy since the mid-1960s bears a close resemblance to the IAEA system, in whose design members of the US (and British) weapon laboratories played a major part.

IV. Measures against undeclared activities in NNWS parties to the NPT

In reaction to these concerns, policy makers have focused their efforts in the early and mid-1990s on two tasks. The first has been the detection and thus prevention of clandestine, or undeclared, nuclear weapon programmes among the NNWS parties to the NPT. The second has been the safe management of the materials and technologies released by the end of the cold war and the collapse of the Soviet political and economic systems. The two are linked because political and economic disarray, or plain carelessness, in a major NWS could provide the materials and technologies for clandestine programmes. These tasks have supplemented the traditional non-proliferation objectives of persuading non-NPT threshold states (now India, Israel and Pakistan) to abandon their ambiguous nuclear activities and bring them under full international safeguards.

The first set of policy initiatives gained its impetus from the discoveries that followed the 1991 Persian Gulf War. For several years, Iraq had been running an extensive nuclear weapon programme which had gone largely undetected (see chapter 11), and which had been largely assembled using technologies and procurement techniques which evaded the current trade controls. This had occurred despite Iraq's membership of the NPT and its apparent compliance with safeguards obligations. Along with the revelations about nuclear activities by North Korea and South Africa, events in the early 1990s provoked a thorough re-examination of the whole approach to the detection and prevention of hidden programmes.

The Iraqi episode led to four significant innovations. The first was the direct and continuous engagement of the United Nations in the pursuit of non-proliferation objectives. In particular, the UN Security Council gave legitimacy and leant its authority to intrusive actions, such as the destruction of Iraq's nuclear weapon facilities, which would probably have remained out of bounds without the backing of UN Security Council resolutions. The UN Security Council was entitled to act against Iraq and North Korea because their clear violations of NPT safeguards undertakings threatened international peace and security. The second innovation was the intensification of intelligence gathering and sharing by states, and the use of that intelligence to guide diplomacy and verification (as in North Korea and Iran). The third was the strengthening of technology controls, especially through the incorporation of dual-use technologies in the trigger lists attached to the Nuclear Suppliers Guidelines. This was accompanied by efforts to establish effective export control machineries in countries which had lacked them and in the states formed after the breakup of the Soviet empire.[8]

---

[8] A full description of the steps taken to apply export controls in these new states (including the Russian Federation) is provided in *Nuclear Successor States of the Soviet Union: Nuclear Weapon and Sensitive Export Status Report*, No. 4 (The Carnegie Endowment for International Peace: Washington, DC and The Monterey Institute of International Studies, Monterey, Calif., May 1996).

The fourth innovation was the launching of 'Programme 93+2' by the IAEA, which built upon initiatives taken in 1991–92. Its aim was to enhance the effectiveness of NPT safeguards by increasing the availability of information about the nuclear materials, equipment and facilities possessed by NNWS parties to the Treaty, by broadening access to their sites, and by improving detection techniques. As seen above, the IAEA's actions had hitherto been determined by the rules and procedures laid down in INFCIRC/153, and by the interpretations that had come to be placed upon them.

The measures developed within Programme 93+2 that have been tested in field trials involve the following innovations or extensions of existing safeguards practices:

(*i*) broad access to information involving expanded declarations from States that include a description of their nuclear programme in addition to nuclear material holdings, information on the import-export of certain equipment and material and information from national technical means;

(*ii*) broad physical access to declared locations and managed access to other locations;

(*iii*) the conduct of no-notice inspections where the State is not given advance notification regarding the timing, location and activities associated with an inspection;

(*iv*) the testing of new technical measures such as environmental monitoring for the detection of undeclared activities; and

(*v*) new administrative procedures that include universal inspector designations and the issuance of multi-entry visas.[9]

The proposed measures would increase the IAEA's abilities to detect warhead development and manufacturing programmes (especially important if proliferators could acquire quantities of weapon-grade material in black markets), as well as to detect the production and separation of fissile materials. Where possible, cost increases would be offset by efficiency gains, for instance by utilizing new techniques to lessen the frequency with which reactors and other less sensitive facilities in reliable countries need to be inspected. The intention was to lessen the costs incurred in safeguarding large nuclear power programmes in countries such as Canada, Germany and Japan as well as to increase the ability to detect undeclared activities. In addition, cooperation with regional safeguards systems (notably Euratom) would be deepened in order to spread the burden and increase the regional stake in verification.

Although these measures have won international support, it remains to be seen how successfully they can be implemented. Some can be applied under the current INFCIRC/153 arrangements, while others are requiring that fresh consent ('complementary legal authority') be granted by NPT members. A draft

---

[9] Hooper, R., '"Programme 93+2"—IAEA development programme for strengthened and more cost-effective safeguards', Paper presented at the 36th Annual Meeting of the International Nuclear Materials Management (INMM), Palm Springs, 9–12 July 1995. The title '93+2' was adopted because the programme was officially launched in 1993 and its architects were given 2 years to come up with recommendations. For a useful summary, see also Lewis, P. M., 'Strengthening safeguards', Verification Matters Briefing Paper 95/2 (VERTIC: London, Mar. 1995), p. 4.

protocol to comprehensive safeguards agreements has been under intense discussion since the June 1996 meeting of the IAEA Board of Governors.

Programme 93+2 has significant implications for the non-proliferation regime. It implies a shift of emphasis in international safeguarding away from the confirmation of good behaviour towards the detection of bad behaviour. Implicitly, NPT states parties would grant the IAEA the right to focus attention on fellow members if there were well-founded cause for suspicion. It also implies a further derogation of sovereignty by the NPT's NNWS parties, and an increased burden on industrial operators in those states, while simultaneously increasing the IAEA's reliance on their cooperation in providing information, access and support. For this to be accepted, states will have to perceive that their industrial secrets are protected and that they are being treated fairly and equitably, particularly in comparison with their neighbours and political rivals. Above all, it implies a step-jump in intrusiveness and transparency.

It should be noted that Programme 93+2 has not been designed principally with the non-NPT countries and the NWS in mind. Some of the measures are likely to be incorporated into safeguards practices in those countries, but how, when and where remains undecided. However, it is unlikely that the innovations entailed by the Programme could be confined to one set of countries in the NPT. The whole undertaking could be jeopardized if the 'intrusiveness and transparency gap' between the NPT NNWS on the one hand, and the NWS and non-NPT threshold states on the other hand, was perceived to be widening. The burden carried by Programme 93+2, and by the IAEA, in checking weapon proliferation will also depend significantly on the parallel strengthening of measures to stem leakages of materials and technologies from the NWS and threshold states and to tighten export controls everywhere. There may thus be both political and practical connections between the success of Programme 93+2 and the extension of international controls in all states.

However, it should also be noted that the costs of extending international controls in the NWS and non-NPT threshold states, in the ways discussed in the sections below, would be much higher than the costs of applying Programme 93+2 in the NNWS. The aim is to implement the Programme without significant increases in the IAEA safeguards budget. All assessments of safeguards requirements in the NWS point to a substantial expansion of the budget. This cost differential is one of the reasons for the higher priority being given to the detection and discouragement of clandestine programmes than to the extension of multilateral controls over fissile materials and production facilities in Russia and the USA. These higher costs also provide a pretext for governments in the NWS to resist international encroachment. The risk is that support by the NNWS for the programme will weaken if they alone have to carry the weight of increased multilateral controls (Germany and Japan have already expressed concerns about the increased burdens that they will face), particularly if this discrimination is perceived as acting against their security interests in tightening controls on nuclear materials produced in former weapon programmes. A balanced approach therefore seems essential, even if it involves higher safeguarding costs.

V. Bilateral initiatives to strengthen controls in the FSU

Over the past five years, there has been an immense, multifaceted effort to implement the arms reduction agreements that accompanied the end of the cold war, and to minimize the security and public health risks attending them. With the possible exception of China, the nuclear weapon arsenals and industries in the NWS have been scaled down, to the extent that they are already only a fraction of their former sizes in the United States and former Soviet Union. In all cases, this has involved difficult organizational changes as well as the launching of programmes for managing the remnants of the 40-year nuclear arms race.

Nowhere has this been so problematic as in the FSU. The task has been compounded by a long history of regulatory neglect, and by the difficulties of conducting relations with and within a huge country experiencing political and economic upheaval. The achievements have nevertheless been impressive. The non-Russian states emerging out of the old Soviet Union have been successfully brought into the NPT as non-nuclear weapon states, and their transfer of nuclear warheads and delivery vehicles to Russia appears to be proceeding with only the occasional hitch.[10] In Russia itself, surplus weapons are being transported to central storage sites, deployed weapons are being held at fewer locations, and the dismantlement programme seems to be running according to plan.

Many countries have provided technical and financial support for these developments. Substantial assistance has been given by Japan and the member states of Euratom, whose cooperation with Minatom and Gosatomnadzor (GAN) to improve Russia's nuclear material accounting and control (NMAC) system has become an important programme.[11] However, the main lead has been taken by the United States. The Cooperative Threat Reduction programme announced by the US Congress in 1992 commits resources to a wide range of joint initiatives. Over $600 million of 'Nunn-Lugar' funds had already been obligated in June 1995, with the expectation that the amounts would soon exceed $1 billion. MPC&A has been an important part of this programme. Actions are occurring under four headings:[12]

---

[10] As this chapter is being written, Belarus has suspended its transfer of nuclear weapons to Russia, ostensibly to bargain for additional financial rewards.

[11] Other countries' individual and collective initiatives are described in *Nuclear Successor States of the Soviet Union* (note 8). On Euratom's involvement, see, e.g., Martynov, V. *et al.*, 'Euratom-Russian cooperation: The CISNER project' and Terentiev, V. and Van der Eecken, D., 'MINS project: centralized NMAC system for Minatom of the Russian Federation'. Both papers were presented to the Joint Euratom–Russian Federation Seminar on Nuclear Materials Accounting and Control, Novosibirsk, 3–7 June 1996.

[12] For a summary, see Gibbons, J. H., 'Managing nuclear materials in the post-cold-war era', Keynote Address to the Second International Policy Forum on the Management and Disposition of Nuclear Weapon Materials, Washington, DC, 22 Mar. 1995, p. 7. In financial year 1996, $85 million for MPC&A

1. *Securing nuclear materials.* Beginning with the Kurchatov Laboratory in Moscow, a step-by-step extension of best practices in providing security for nuclear materials is being attempted across the FSU's nuclear infrastructure. This includes the design and construction of storage facilities, together with measures involving customs, police and intelligence services to combat nuclear smuggling. Hitherto, over 90 per cent of expenditures on the US material management programme has been spent under this heading, indicating that it has had the highest priority and the greatest success. Linked initiatives are under way to improve material accounting practices at nuclear facilities.

2. *Building confidence through openness.* At their summit meeting in May 1995, Presidents Clinton and Yeltsin issued a 'Joint Statement on the Transparency and Irreversibility of the Process of Reducing Nuclear Weapons'. This would encompass *inter alia* agreements to exchange information on warhead stocks and fissile material inventories, and to open nuclear storage facilities to reciprocal monitoring. An Agreement for Cooperation between the United States and Russia is being negotiated to enable the exchanges of sensitive information required for this regime to be fully effective.

3. *Halting accumulation of excess stocks.* In June 1994, agreement was reached on the halting of plutonium production for nuclear weapons. Originally under this accord, the three production reactors that still operate in Russia would be shut down by the year 2000, and the plutonium separated from them in the intervening period would not be used in weapons. In mid-1995, the agreement had yet to come into force owing to the failure to agree upon methods of financing the power stations which would provide replacement heat and power to local communities. Attention is now being given to the technical feasibility of converting the reactors' cores so that the reactors can operate more efficiently as electricity producers while no longer producing weapon-grade plutonium.

4. *Disposition of fissile materials.* The US Government, through the US Enrichment Corporation, has already agreed to the purchase of 500 tonnes of HEU from dismantled nuclear warheads (see below). A 'Joint Study Initiative' has been proposed to investigate alternative ways of disposing of excess plutonium, but it is not expected to provide early results. One of the key problems to be overcome is the financing of plutonium disposition in Russia.

What is striking about these international initiatives is that they are being largely pursued through *bilateral* rather than *multilateral* processes. Through bilateral cooperation, the US and other governments have, with assistance from

will be assigned to the programme on 'securing nuclear materials', up to $15 million on 'building confidence through openness', up to $5–10 m. on 'halting accumulation of excess stocks' and up to several hundred thousand dollars on 'disposition' (of plutonium). Tens of millions of dollars are also being spent on the construction of a storage facility, with additional millions being spent on the thwarting of nuclear smuggling. As this book goes to press, we note the publication of Allison, G. T. *et al.*, *Avoiding Nuclear Anarchy: Containing the Threat of Loose Russian Nuclear Weapons and Fissile Materials* (MIT Press: Cambridge, Mass., 1996). It gives a notably pessimistic account of the situation in Russia and of the achievements in improving the security of fissile materials in the Russian military complex.

industries, laboratories and non-government organizations, encouraged the development of more effective administrative, legal and policy-making practices in Russia and the other states formed after the breakup of the USSR. Bilateral processes are being used to induce *domestic* processes, without which little can be achieved. Except in regard to the important task of bringing the former Soviet republics into the NPT, there has been comparatively little recourse to multilateralism, whether in terms of policy processes or instruments. The IAEA and the United Nations have provided valuable support (including helping to coordinate bilateral efforts), but have remained largely in the background, in notable contrast to their pre-eminent roles in developing the measures against undeclared programmes in NPT states parties which were discussed above.

There have been a number of reasons for this emphasis on bilateralism. The first has been the desire for immediate effectiveness. In the early years at least, it was accepted that the greatest progress could probably be achieved through a myriad of bilateral initiatives, rather than through grand multilateral interventions. Until new institutional and legal mechanisms had been put in place, trust had been established, and the dimensions of the task had been defined, multilateral initiatives were less likely to be successful. Furthermore, reaching agreement between and within governments was complicated enough without having to worry about international consensus; and the necessary funds could most quickly and reliably be assembled and channelled through national budgetary processes.

Second, there existed as yet no well-developed multilateral framework, involving rules, instruments and procedures, for strengthening controls in weapon states. The NPT provides no guidance on how disarmament should be managed (this deficiency became clearer when the elimination of South Africa's nuclear weapons was being addressed), and the NWS were exempted from mandatory safeguards and material accounting (excepting France and the UK under the Euratom Treaty). This meant that the only formal obligations, besides the loose commitments made under Article VI of the Treaty and the voluntary offer safeguards agreements, were those emanating from the bilateral arms reduction agreements concluded by the United States and FSU. As such, the steps taken and procedures to be followed in the NWS remained optional and negotiable.

Third, several of the tasks could only be carried out by agencies from the NWS since sensitive information about nuclear weapons was involved. There was also a natural affinity between individuals and institutions that had been involved in nuclear weapon programmes. These linkages proved valuable in bringing about an enhancement of physical security, which is a national responsibility and has never involved multilateral verification.[13]

---

[13] However, many states are parties to the Convention on the Physical Protection of Nuclear Material which entered into force in Feb. 1987, and accept the IAEA's advisory guidelines in IAEA, The Physical Protection of Nuclear Material, INFCIRC/225, Rev. 2, Dec. 1989. The NAS study (note 1) recommended that a 'stored weapons standard' should be applied to the protection of HEU and plutonium, involving a

VI. Extending the multilateral framework for material controls

To date, therefore, problems relating to fissile materials in the NWS have been tackled mainly through domestic and bilateral initiatives, and for good reasons. It is important that these initiatives should continue. However, they are unlikely to remain sufficient. Bringing fissile materials and their production facilities into the multilateral framework of control, albeit in stages, is desirable in view of the need to strengthen and make irreversible disarmament measures, increase the non-proliferation regime's scope, manage the transfer of de-militarized materials and technologies into civilian commerce, and prevent leakage to weapon programmes. It is also desirable because it is essential that the highest standards are attained, consistent with the application of IAEA safeguards, not least in regard to material accountancy. In addition, bilateral arrangements are vulnerable to breakdowns in relations between their parties, and tend to be less stable and weighty than commitments made under international law to the community of states. Multilateral obligations are the best guarantee that the domestic arrangements and processes, upon which effective controls depend, remain stable and efficient. Moreover, a multilateral framework is essential, as Argentina and Brazil found, if the international community is to have full confidence in bilateral arrangements.

This still leaves the class of states (India, Israel and Pakistan) that lie outside the NPT and are believed to have active nuclear weapon programmes. How should controls be exerted over their fissile materials and related capabilities? Traditionally, these countries have been addressed as recalcitrant NNWS: the main policy objective has been to persuade them to join the mainstream and accept the obligations placed on NNWS signatories to the NPT. However, the increasing maturity of their nuclear weapon programmes has meant that a proliferation problem has developed into an arms control and a disarmament problem. As such, the issues now being faced are similar to those affecting the NWS. This has raised the possibility that the situations in non-NPT threshold countries and the declared nuclear weapon states can, in some respects, be tackled through the same multilateral instruments. If this were possible, it would increase the returns on investment in those instruments. The risk, however, is that the threshold states' inclusion in the negotiations would so complicate matters that progress would be slow and unreliable. The obstacles to using multilateral instruments to curtail the nuclear weapon activities of de facto nuclear weapon states have also been well demonstrated by India's recent rejection of the CTBT.

---

substantial increase in physical protection worldwide, and entailing review by the IAEA of member countries' physical protection practices to ensure that the necessary standards were being met.

The Fissile Material Cut-Off Treaty

The first multilateral arrangement to have been proposed is the Fissile Material Cut-Off Treaty (FMCT), formerly referred to as the Cut-Off Convention, which would prohibit the production of highly enriched uranium and plutonium for nuclear explosives.[14] The current initiative has its origins in a resolution passed in the UN General Assembly in December 1993. In the Principles and Objectives agreed at the 1995 NPT Review and Extension Conference, the FMCT was identified as part of the programme of action that would be undertaken in support of the 'full realization and effective implementation of Article VI of the Treaty'. The Conference looked forward to: 'The immediate commencement and early conclusion of negotiations on a non-discriminatory and universally applicable convention banning the production of fissile material for nuclear weapons or other nuclear explosive devices.'[15]

The FMCT's main advocate has been the United States. It has seen three benefits in pursuing the treaty: it would establish a new international norm that would inhibit future weapon programmes; it would help to 'cap' existing weapon programmes in non-NPT threshold countries, while opening a diplomatic process through which the status of their unsafeguarded stocks and facilities could be debated and eventually negotiated; and it would assist disarmament and accelerate the introduction of effective verification measures in the NWS.

In summer 1995, it became apparent that negotiations in the Conference on Disarmament (CD) were unlikely to begin before early 1997, despite the call for haste at the NPT Review and Extension Conference.[16] The FMCT may now be firmly established on the international agenda, but no one expects its conclusion to be quick or easy. Indeed, voices are increasingly being heard in some governments belittling its importance and questioning whether it is practicable or deserves priority. Among other things, they point to the de facto cut-offs already in place in the NWS, the high cost and complexity of verifying a cut-off and the risks that the FMCT might only be attainable at the price of

---

[14] For discussions of the Cut-Off Treaty see Berkhout, F., *et al.*, 'A cutoff in the production of fissile material', *International Security*, vol. 19, no. 3 (winter 1994/95), pp. 167–202; McGoldrick, F., 'US fissile material initiatives: implications for the IAEA', *IAEA Bulletin*, vol. 1 (1995), pp. 49–52; and 'A Cut-Off Treaty and associated costs', IAEA Secretariat Working Paper presented at the Workshop on a Cut-Off Treaty, Toronto, 17–18 Jan. 1995.

[15] Principles and Objectives for Nuclear Non-Proliferation and Disarmament, NPT/CONF.1995/32/DEC.2, New York, 1995, para. 4(b). The negotiating mandate agreed in the Conference on Disarmament runs as follows:

'1. The Conference on Disarmament decides to establish an ad hoc committee on a "ban on the production of fissile material for nuclear weapons or other nuclear explosive devices". 2. The Conference directs the ad hoc committee to negotiate a non-discriminatory, multilateral and internationally and effectively verifiable treaty banning the production of fissile material for nuclear weapons or other explosive devices. 3. The ad hoc committee will report to the Conference on Disarmament on the progress of its work before the conclusion of the 1995 session.'

[16] Some non-aligned countries in the CD, with India in the lead, have not been willing to give their consent to the convening of the Ad Hoc Committee on the Cut-Off Treaty unless the CD also agrees to establish an ad hoc committee on nuclear disarmament. The NWS have been unwilling to accept this precondition.

legitimizing the possession of unsafeguarded stocks of plutonium and HEU in India, Israel and Pakistan.

Behind these arguments lie concerns that the FMCT would impose costs on some participating countries that would outweigh the benefits to their security. In particular, the smaller NWS are yet to be convinced that a treaty will serve their security interests. They appear especially concerned that their weapon programmes would be more constrained by the FMCT than the Russian and US programmes, and that they might face disproportionately high verification costs since fewer of their processing facilities are scheduled for closure.[17] The de facto nuclear weapon states have also regarded the proposed FMCT with some trepidation. A cut-off in fissile material production would have a more direct effect on their nuclear weapon production capacities than on the NWS's capacities, which are already well-established and can be largely sustained, at marginally reduced levels, from existing stocks of material. Disagreements have also surfaced between India and Pakistan over whether stocks should be covered by the Treaty (Pakistan being in favour and India against).

While there are undoubted problems with the FMCT, it is in danger of being seriously undervalued. The Treaty would de-legitimize fissile material production for weapons, and provide an avenue for addressing the problems presented by the de facto nuclear weapon states (in parallel with the search for political settlements). However, the FMCT's greatest significance lies in the part it would play in redefining the political and regulatory status of nuclear production capacities and of their future products in countries with nuclear weapon programmes. Despite the difficulties in negotiating the FMCT, it would be an important policy driver, requiring governments to address fundamental issues about the scope, intensity and implementation of future regulations, including:

1. *Mandatory safeguards*. Current IAEA safeguarding in the NWS is voluntary, in a double sense: the NWS are not required to submit facilities and materials to safeguards (except within Euratom), and the IAEA is not required to 'designate' those that are made available for safeguarding. IAEA inspection of facilities and materials covered by the FMCT would presumably become obligatory.

2. *Ending the right to withdraw materials from safeguards*. Safeguards would be applied under the FMCT to guard against the production of materials for, or their diversion into, nuclear weapon programmes. The NWS' voluntary offer agreements with the IAEA allow them, upon giving notice, to withdraw facilities from the lists provided to the Agency, and to withdraw materials from safeguards once they are applied. These rights would probably have to be foregone for any materials covered by the FMCT, with due allowance made for the withdrawal of materials for use in naval propulsion.[18]

---

[17] Given their extensive submission to Euratom safeguards, France and the UK are also reluctant to accept a treaty that would in practice result in verification measures being applied on their territories long before they were effectively applied in Russia and the USA.

[18] See note 4.

3. *Extending the scope of safeguards*. While it is assumed that safeguards would be applied to reprocessing and enrichment plants, their subsequent reach would have to be negotiated. A number of options have been proposed, including the safeguarding of materials produced in these plants up to their irradiation in reactors, or the extension of safeguards over all materials produced after the FMCT had entered into force.[19]

4. *Safeguarding the 'unsafeguardable'*. Several of the enrichment and reprocessing plants that would be covered by the FMCT have not been designed to be safeguarded at all, let alone in accordance with the IAEA's quantity and timeliness criteria. As retrofitting will probably be impracticable in most cases, second-best solutions might have to be applied, with methods tailored to individual facilities. This would potentially breach the requirement for a non-discriminatory treaty. The only viable approach might be to install verification methods in stages, gradually building up the level of confidence that was achieved, with the aim of approaching if not ultimately meeting the IAEA criteria applied in normal circumstances.[20] In practice, however, few of the facilities operated during the cold war are likely to stay open.

5. *Financial costs*. Whichever approach was adopted, considerable increases in the IAEA's safeguards budget would be involved. Estimates vary from an increase of one-half for the 'minimum' verification, to a trebling if comprehensive safeguards were applied.[21]

However, the FMCT itself can provide no panacea: one of the reasons why its importance is being underplayed is that its full value will only be realized if it is accompanied by other initiatives. The Treaty would cap the stocks of fissile material that could be used in nuclear weapons. The generation of new stocks, and their transfer back into weapon programmes, would be inhibited by the application of safeguards to enrichment and reprocessing plants and their products, or by closing the plants under IAEA seals. However, the FMCT in its proposed form is not retrospective.[22] It would not cover the vast residual stocks of plutonium and HEU, some derived from dismantled weapons, some held in store, whose production and separation took place in the past. As discussed below, this is its principal limitation.

[19] The IAEA has assessed the feasibility and implications of verifying four alternatives. The first would involve comprehensive safeguarding of fuel-cycles, akin to the safeguards applied in NNWS parties to the NPT. The other three options are more limited, entailing 'limited verification of separated fissile material', 'full verification of separated fissile matereial and facilities capable of producing such material', and 'full verification of separated and irradiated fissile material'. See 'A Cut-Off Treaty and associated costs' (note 14).

[20] This is the approach that has been adopted by Euratom in applying safeguards to the older reprocessing and enrichment plants in France and the UK. It is doubtful that any of them could meet the IAEA criteria.

[21] See 'A Cut-Off Treaty and associated costs' (note 14), p. 17.

[22] In the CD, a group of non-aligned countries tried to broaden the mandate to include negotiations of reductions so that 'unsafeguarded stocks are equalized at the lowest possible level'. A compromise was reached which allowed for the discussion of existing stocks while preserving the focus of the negotiations on a production cut-off. Britain, France and India have made clear that they will not enter negotiations on stockpile reductions. Fetter, S. and von Hippel, F., 'A step-by-step approach to a global fissile materials cut-off', *Arms Control Today*, vol. 25, no. 8 (Oct. 1995), pp. 3–8.

Excess stocks of weapon-grade material: a transparency and verification regime

The quantities of fissile material that will come under IAEA safeguards as a result of the successful conclusion of the FMCT would be a small fraction of those previously amassed by the NWS. It is noted in chapter 14 that current military inventories contain an estimated 1750 tonnes of HEU and 250 tonnes of plutonium. These amounts were produced in support of nuclear arsenals many times the sizes of those that will be deployed when the current arms reduction agreements have been implemented. The excess stocks possessed by Russia and the United States will be especially large.

The inventories of plutonium and HEU that in future could be assigned to weapons by the NWS will be comprised of three stocks: one that is contained in deployed weapons; another that is tied up in assembly and disassembly; and another that is held in reserves or wastes. This last inventory is currently held as retired warheads and warhead components, as 'raw' plutonium and HEU, and in scraps and wastes of various kinds. If the reserves held in the military sector are not down-sized in line with the nuclear arsenal, this inventory of plutonium and HEU would become a fissile material reservoir providing a potential 'surge capacity'. It would be a surrogate production system, and one that could provide large quantities of weapon material at very short notice.

It is therefore desirable that a large part of this reserve inventory, belonging mainly to Russia and the USA, is declared to be excess to weapon requirements, placed under effective physical security, and made available to international verification. Besides the arms control benefits, this would provide confidence that the world's main inventories of weapon-grade plutonium and uranium were being held under effective controls prior to their disposition. Without this parallel initiative, the chances of negotiating the FMCT are likely to be seriously weakened. If the Russian and US reservoirs of fissile materials were not also being submitted to international controls, other countries might justifiably question the FMCT's value and fairness.

So far, the NWS governments have exercised their sovereign rights to determine, without consultation among themselves or with other governments, the scale of their excess stocks, and whether and how they will be declared. They are not required by any international treaty to divulge such information, nor to enter discussions on the materials' eventual safeguarding. To date, only the US Government has made any announcement, when President Clinton declared in March 1995 that around 200 tonnes of fissile material would no longer be used for nuclear weapons. The US Government has so far submitted 10 tonnes of excess HEU and 2 tonnes of excess plutonium to IAEA safeguards, as a first step towards wider international verification of its material stocks.[23] The other NWS have been silent on this issue.

---

[23] The 200 t mainly comprised 38 t of weapon-grade plutonium and 164 t of HEU. The US Government has also indicated that the DOE's stock of fuel- and reactor-grade plutonium (14 t) is surplus, together

These arrangements are unsatisfactory. They create uncertainty over intentions and capacities in the NWS, and over the safety and security of their stocks of fissile materials. They are a reminder of the special privileges granted to them under international law, and make it harder to justify greater transparency elsewhere in the regime. They make it easier for conservative groups in government, the armed forces and weapon laboratories to argue in favour of the retention of large reserve stocks and to justify new missions, on the grounds that other countries might be doing the same. Furthermore, they undercut the value of reductions as surge capacity could allow the arsenal quickly to be reconstituted.

In establishing the dimensions of the stock of fissile materials that is excess to requirements, rather than announcing an arbitrary figure, two other stocks have to be defined. The first is the total inventory of weapon-grade plutonium and HEU held, in whichever form, by the country in question. The second is the inventory of plutonium and HEU that is required to sustain a nuclear arsenal at the level to which it is being reduced unilaterally or through bilateral agreement (between 4000 and 5000 warheads apiece, including tactical warheads, in the cases of Russia and the USA if reserve weapons are excluded). The excess stock is the difference between them.

The Russian and US governments have already agreed to exchange information on the first of these inventories, although Russia has yet to comply. The second quantity is politically the most sensitive. Given public knowledge of the overall numbers of warheads being deployed, it should be possible to arrive at an approximate formula linking those numbers to the quantities of fissile material required to sustain them without revealing information about nuclear weapon designs beyond that already published. Arriving at this formula would involve discussion of the acceptable size of reserves, including reserves of intact warheads, which could be beneficial to arms control. In the discussion of disposition in section VI below, it is suggested that a 'significant quantity' (8 kg of plutonium and 25 kg of HEU) per warhead might provide an upper bound. If a precise formula were judged unattainable or undesirable on security grounds, the NWS could declare the quantities that they are assigning to weapons, and defend their figures before the international community.

The task of declaring and verifying excess stocks of weapon material might best begin with a formal agreement, or treaty, between the Russian and US governments, negotiated in consultation with other governments and with the IAEA. As the owners of much the largest stocks of excess fissile materials, and the largest nuclear arsenals, it seems beholden on Russia and the USA to make the first moves. The new accord would expand on aspects of the bilateral agreements already reached under the US–Russian 'Transparency and Irreversibility' initiative. It could then be opened for other countries' accession, including the accession of NNWS that hold or may in future hold stocks of safeguarded

---

with 10 t of HEU which have been placed under IAEA safeguards. The declared surplus therefore now totals 226 t.

weapon-grade material. An advantage of limiting the agreement to Russia and the USA in the first instance is that it would be less vulnerable to dilution or obstruction by the other NWS, or by non-NPT threshold states whose stocks of nuclear weapons and/or of weapon-usable material remain undeclared.

In short, the agreement or treaty would require parties to provide, over a specified time-frame:

1. Declarations of the best available estimates of total inventories of weapon-grade plutonium and HEU, and of fuel- and reactor-grade plutonium, and of the steps that are being taken to establish confidence in those estimates. As knowledge improves, the estimates would be progressively refined.

2. Declarations of the quantities of plutonium and HEU that are being assigned to weapon purposes (after implementation of agreed arms reductions where applicable), and of the quantities of HEU assigned to naval propulsion, and the grounds upon which those quantities have been chosen. These quantities would be reduced upon the announcement of further arms reduction and dismantlement programmes.

3. Declarations of the resulting excess of plutonium and HEU over military requirements, and of the forms in which it is held and the manner in which it is stored.

4. Declarations of the steps that will be taken to submit excess plutonium and HEU to international verification.

Ideally, the accord would be negotiated in parallel with, but without formal linkage to, the FMCT. It should also be noted that the accord would supplement the 'transparency regime' that is currently being negotiated in regard to stocks of civil plutonium. Since December 1992, discussions have been held, at the IAEA and elsewhere, on how to improve confidence in civil plutonium and HEU programmes.[24] Proposals that they should be subject to international management have been discarded in favour of a comparatively limited transparency agreement, whereby countries would declare their inventories in line with agreed definitions and taxonomies. Taken together with the declarations outlined above, they could provide the basis for the comprehensive International Register of Fissile Materials that the authors advocated in *World Inventory 1992* and that has more recently been proposed in a report by the Carnegie Commission on Preventing Deadly Conflict.[25]

---

[24] Rioux, J.-F., 'Options for the management of highly-fissionable civilian materials', *Non-proliferation Review* (Monterey Institute of International Studies: Monterey, Calif.), vol. 2, no. 3 (spring–summer 1995), pp. 52–57.

[25] Steinbruner, J. et al., 'Comprehensive disclosure of fissionable materials: a suggested initiative', Discussion Paper prepared for the Carnegie Commission on Preventing Deadly Conflict, Carnegie Corporation of New York, Apr. 1995, p. 6. For an excellent overview of these and other measures towards HEU and plutonium, see Shea, T. E., 'Confidence-building measures for plutonium and highly enriched uranium', International Atomic Energy Agency, IAEA-SM-333/216, Vienna, 1994. A fissile material register would also be consistent with recent suggestions to establish a nuclear weapons register. See Müller, H., 'Transparency in nuclear arms: toward a nuclear weapons register', *Arms Control Today*, vol. 24, no. 8 (Oct. 1994), pp. 3–8.

In September 1996, the IAEA, Russia and the USA jointly announced at the IAEA General Conference that steps would be taken to facilitate the application of safeguards to fissile materials of weapon-origin. The steps would include discussion of technical methods of protecting sensitive weapon information; the formation of a joint group to address technical, legal and financial issues associated with IAEA verification of the pertinent materials; and trilateral visits to relevant sites. This is an important announcement which is consonant with the above proposals.

Controls on weapon assembly, disassembly and support systems

Finally, the proliferation risks posed by the weapon assembly, disassembly and support industries should not be overlooked. As elsewhere, effective physical security and material accounts need to be maintained, and procedures followed so that the chances of theft or diversion are kept to a minimum. These issues cannot be addressed in a multilateral framework. However, this does not exonerate the NWS or non-NPT threshold states from providing assurances, backed where possible by evidence, that the relevant activities are well managed and secured. Several measures are already being developed under bilateral arrangements between Russia and the USA. They include mutual inspection of weapon dismantlement and access to sites. For instance, 'under [the US Government's] proposal, all of the sites in both the United States and Russia where plutonium or HEU are stored would be opened to reciprocal visits, excluding only nuclear weapons that will remain in the enduring stockpile, and naval reactor fuel.'[26]

The Russian and US governments should be encouraged to develop these measures further, and to seek their extension to other countries with nuclear weapon programmes.

VII. The disposition of excess plutonium and HEU

Come what may, the international community will have to cope with large excess stocks of weapon plutonium and HEU in years ahead. One approach is to hold them in long-term storage, under effective safeguards and security arrangements. Another is to render them inaccessible, via 'disposition', by altering their physical forms and recoverability. Both approaches seem essential, but while storage is already receiving attention, disposition has until recently tended to take a back seat, particularly where plutonium is concerned. A common view has been that the disposition of these materials is a desirable objective, but that there is no urgency in establishing the programmes for achieving it. Indeed, delay may be desirable.

Four main arguments are put forward for taking a very gradual approach to disposition, with stocks possibly being eliminated over a 30–50 year time-

[26] Gibbons (note 12), p. 4.

frame. The first is that time is required to develop and demonstrate new techniques and to make necessary investments in facilities. The second is that the acts of moving, processing and disposing of the fissile materials could exacerbate the control problems. The third reason, which is linked to the second, is that the political and economic situation in the FSU is too fragile to sustain a large disposition programme without endangering public health and international security. The fourth is that whereas HEU disposition would result in cost savings, plutonium disposition would involve considerable expense, including the expense of implementing safeguards on the various facilities engaged in the activity.

There is, however, another reason for the interest in delay which is less frequently expressed and which may be the most influential. Large-scale programmes to dispose of weapon-grade material could have substantial consequences for the commercial fuel cycle. In the case of HEU, they could displace enrichment capacity and depress prices of natural uranium and LEU in world markets. In the case of plutonium, they might come into conflict with civil reprocessing by competing for plutonium recycling capacities (in the forms of MOX fabrication facilities and reactors licensed to take MOX fuels), and by drawing further political attention to the validity of reprocessing policies when there is already an abundance of plutonium.[27] In addition, plutonium disposition raises various health and environmental questions which are awkward for governments.

However, there are also telling arguments in favour of early and large-scale disposition of surplus fissile materials. If a pessimistic view is taken of the internal situation in the FSU, there is a case for acting quickly to diminish the stocks of materials over which control has to be exercised. The disposition of materials would also bring income and activity to sections of the nuclear industry, while requiring that effort is put into establishing more effective MPC&A procedures. Leaving large stocks of fissile material in store with no clear programme for eliminating them would also set a bad example. The argument that such materials may legitimately and safely be stockpiled in NWS, but not in NNWS, seems less persuasive today. Finally, the stock of power reactors that is technically and politically suitable for plutonium recycling may be smaller in 20 years' time than it is today, because of obsolescence, resulting in a narrowing of recycling options.

Whichever conclusion is reached, it is important that there should be a thorough debate over the prospects for reducing or even eliminating stocks of excess weapon material. This debate is still in its infancy, and is only beginning to be joined by governments at an international level. The purpose of sections VII–IX is to assess the scale of the task, and how it might be carried out and over which time scales depending on the priority that is attached to it.

---

[27] Paradoxically, some opponents of civil reprocessing also argue against early disposition through recycling, on the grounds that it might increase the political legitimacy of recycling and thereby *strengthen* the reprocessing industry. Neither side seems able to predict the outcome, and tends to fear for the worst.

CONTROL AND DISPOSITION OF FISSILE MATERIALS 439

This section begins with brief discussions of technical issues and the quantities that can be defined as excess to requirements and thus available for disposition.

Technical aspects of HEU and plutonium disposition

Disposing of HEU is technically straightforward. HEU is uranium containing a high concentration of the fissile isotope ^{235}U (see chapter 2 for a full explanation). As a result of the successful programme to convert research reactors to low-enriched uranium fuels, there is today little civil demand for HEU (see chapter 8).[28] HEU can be rendered useless to weapon designers by diluting it with the common non-fissile isotope ^{238}U, achieved through the blending of HEU with depleted, natural or slightly enriched uranium. The result is the LEU which is commonly used to fuel power reactors.[29] The critical mass required for a nuclear explosion cannot be attained with LEU, and the ^{235}U contained in it could only be retrieved for weapon purposes by re-enrichment. This would be expensive and time-consuming. Such an activity would also violate international agreements if a verified FMCT were in place or if the LEU were otherwise under safeguards that barred re-enrichment.

The situation regarding plutonium is more problematic. Isotopic blending provides no solution as critical masses can be formed from all relevant isotopic mixtures of plutonium; and plutonium can be readily separated by chemical means from MOX fuel (i.e., from plutonium that has been blended with ^{238}U), causing it to be much more accessible than the ^{235}U in LEU. Furthermore, MOX fuel is usually more expensive than LEU fuel because of its higher fabrication costs, so that LEU fuel is likely to remain utilities' first choice for the foreseeable future. In consequence, the consumption of plutonium has to be 'forced', whether through subsidy or through a politically motivated requirement to eliminate stocks or avoid stockpiling. Using the analogy of magnetism, the market has an attraction for diluted HEU and repulsion for plutonium.

In considering how to deal with excess plutonium, the US National Academy of Sciences has argued persuasively in favour of a 'spent fuel standard':

Options for the long-term disposition of weapons plutonium should seek to meet a 'spent fuel standard'—that is, to make this plutonium roughly as inaccessible for weapons use as the much larger and growing stock of plutonium in civilian spent fuel. Options that left the weapons plutonium more accessible would mean that this material would continue to pose a unique safeguards problem indefinitely. Conversely, the

---

[28] This RERTR programme is discussed in chapter 8. It is regrettable that the German Government continues to give its consent to the construction of an HEU-fuelled reactor at Munich (see chapter 8). See Schaper, A., 'The first new HEU-fuelled research reactor after a moratorium of 15 years and its impact on non-proliferation policy', Paper presented at the Hearing in the Bundestag, Bonn, 1 Nov. 1995 (Peace Research Institute Frankfurt: Frankfurt, 1995). The aim must be to phase out the usage of HEU in all civilian reactors, and even in naval reactors over the longer term.

[29] The Russian HEU being sold to the US Enrichment Corporation (see below) is being diluted with slightly enriched uranium containing around 1.5% ^{235}U. By increasing the volume of LEU in the final deliveries, this allows the concentration of the troublesome isotope ^{236}U (which is relatively abundant in HEU) to be kept within the bounds acceptable to the operators of nuclear power stations.

costs, complexities, risks, and delays of going beyond the spent fuel standard to eliminate the excess weapons plutonium completely or nearly so would not offer substantial additional security benefits unless society were prepared to take the same approach with the global stock of civilian plutonium.[30]

Two main options for achieving this standard have been proposed, both having the effect of re-embedding the plutonium in radioactive fission and transuranic products. One is to burn the plutonium in MOX fuel in power reactors.[31] The other is to blend it with high-level wastes that have already been separated through reprocessing. In effect, this would restore the isotopic mixture contained in the spent fuel discharged from power reactors, the resulting mixture being vitrified prior to long-term storage or disposal.

These two technical approaches have been assessed in detail in a study by the US National Academy of Sciences (NAS).[32] While the first approach is reasonably well tested and extracts useful energy from the plutonium, the second sacrifices the energy content and requires further development and testing. The study concluded that there was little between them in terms of their overall balances of advantage and disadvantage and its authors recommended that steps should be taken to establish both options. A third approach, involving the disposal of plutonium in deep boreholes, is also being investigated, although it is not expected to be pursued and is left aside here.

There have also been proposals that advanced reactors and other technologies such as particle accelerators should be designed specifically to burn plutonium. These proposals are not considered here because of their high cost and uncertainty, and the long lead-times that would be required to demonstrate and build the reactors. It is assumed that the disposition of plutonium through recycling would mostly be carried out in operating light water reactors.

Plutonium disposition also entails greater regulatory costs and complexities than HEU disposition. Increased safeguarding costs would have to be borne, especially because of the need to submit MOX plants to continuous inspection. The NWS would also have to ensure that certain technical steps were taken prior to disposition, notably in relation to the conversion of the plutonium metal contained in weapon pits to safer chemical forms, such as plutonium oxide. The

---

[30] See NAS (note 1), p. 34.

[31] Fast reactors are no longer widely regarded as capable of playing a substantial role, although France and Russia are exploring the possibilities of converting fast reactors to plutonium 'burners'. In Japan, it has been pointed out that the Monju and other R&D reactors could absorb about 2 t of weapon plutonium per year. Suzuki, A., 'Japan's civil use of foreign military plutonium', *Proceedings of the International Conference on Evaluation of Emerging Nuclear Fuel Cycle Systems*, Versailles, 11–14 Sep. 1995, pp. 661–68. However, this would substitute for the consumption of plutonium separated from civil spent fuels. The future of the Monju reactor has, in any case, been thrown into doubt by the large sodium leak which has recently forced its shutdown.

[32] National Academy of Sciences, Committee on International Security and Arms Control, *Management and Disposition of Excess Weapons Plutonium: Reactor-Related Options* (National Academy Press: Washington, DC, 1995). Plutonium disposition is also discussed in Berkhout, F. *et al.*, 'Disposition of separated plutonium', *Science and Global Security*, vol. 3 (1993).

industrial facilities for carrying all the necessary steps are not yet in place on the required scale.[33]

Gauging the quantities of 'excess' plutonium and HEU

Excess stocks of weapon-grade material

The amounts of weapon-grade plutonium and uranium that should be declared excess to future military requirements have not been decided, as noted in section VI. In the analysis below, the 'significant quantities' (SQ) recognized by the IAEA for safeguards purposes (8 kg of plutonium and 25 kg of HEU) are adopted for simplicity as the average quantities per warhead that will be required to sustain the strategic nuclear arsenals in the NWS in future. These quantities may be twice the average amounts used in modern nuclear warheads, but allowance has to be made for reserves and for materials involved in assembly and disassembly operations. None the less, *this is a very generous allocation and should be regarded as an upper bound.*

Under this assumption, the future US and Russian arsenals anticipated in the START II Treaty (approximately 3500 warheads each) would be sustained, at maximum, by 28 tonnes of plutonium and 87 tonnes of HEU in either country. A further inventory of, say, 1000 tactical warheads apiece would raise these quantities to 36 tonnes and 112 tonnes, respectively. Taking the central estimates presented in chapter 14, tables 14.3 and 14.4, of their total inventories of weapon-grade material, it follows that 49 tonnes of plutonium and 533 tonnes of HEU are excess to weapon requirements in the United States, and 95 tonnes of plutonium and 938 tonnes of HEU are excess to weapon requirements in Russia according to these definitions. With a less generous allocation per warhead, the excess quantities would be higher.

An amount of HEU has to be subtracted from these figures to allow for the fuelling of the US and Russian submarine fleets. The assumption here is that a reasonable upper limit is 50 tonnes. Each country probably consumes less than 1 tonne per year, and a cut-off treaty would not bar the future production of HEU for this purpose if stocks were depleted. Furthermore, HEU requirements for naval reactors could be substantially reduced over the next decades through a programme, mounted for security and/or economic reasons, to substitute lower enriched uranium or even LEU for fuels which used uranium that was at or near weapon-grade.[34] Taking these naval requirements into account, it is assumed below that 480 and 890 tonnes of HEU (rounded to two significant

---

[33] Possible US approaches, and the technical issues, are discussed in Cremers T. L. *et al.*, 'A system for automated dismantlement of plutonium weapon components', Proceedings of the Fifth International Conference on Radioactive Waste Management and Environmental Remediation, 3–7 Sep. 1995, Berlin, pp. 553–56.

[34] In Russia, economic pressures are leading to substantial reductions in the numbers of operational submarines. There are even suggestions that Russia may be forced to abandon its submarine-based deterrent and concentrate on land-based forces. See Fyodorov, Y., 'Prospects and conflicts of Russian nuclear deterrence', *Yaderny Kontrol No. 1* (Center for Policy Studies in Russia), spring 1996, pp. 12–15.

figures) are excess to requirements in the USA and Russia respectively, or 1370 tonnes in total. Including the much smaller excess quantities in the other NWS, and recalling that a generous allocation is being made per warhead, the excess can be rounded up to 1400 tonnes of HEU.

What percentages of the excess stocks of plutonium and HEU are retrievable? How much is contained in low-quality scraps, wastes and other forms that are virtually irretrievable? There are currently few answers to these questions, particularly where Russia is concerned. It is assumed here that all material will ultimately be available for disposition, whether that entails utilization as fuel or disposal as waste. The reader should nevertheless be aware that a proportion (one hopes a small proportion) may never be retrieved, let alone accounted for.

Excess stocks of reactor-grade plutonium

Large stocks of separated reactor-grade plutonium are also accumulating in the civil fuel cycle as a result of the reprocessing programmes discussed in chapter 6.[35] Table 15.1 shows that Russia and the USA are not alone in having to deal with large stocks of excess material: over the coming decade, six countries (including Germany and Japan) will have over 40 tonnes of plutonium that will require disposition if civil reprocessing proceeds as planned. (China is excluded from tables 15.1 and 15.3 because of the great uncertainties over the sizes of its plutonium and warhead inventories.) The inventories of the reactor-grade plutonium available for disposition can be divided into those that are commercially 'active' and 'inactive'. Inactive stocks, which mainly comprise stocks in Russia, the UK and the USA, are those for which there are no current or plausible recycling plans. Active stocks are those that have been separated, or will in future be separated under firm contracts between utilities and reprocessors, and for which utilities have recycling plans. The largest of these stocks belong to France, Germany and Japan.

It is worth noting that the 'inactive' stocks of fuel- and reactor-grade plutonium in Russia (around 30 t) and the USA (13 t) are currently unsafeguarded. An estimated 8 tonnes of the British stock are outside international safeguards.

Whereas the inactive stocks are clearly excess to market requirements, the status of the active stocks is more ambiguous. If plutonium fuels were competitive with uranium fuels in price terms, no stocks of separated plutonium would be 'inactive' for long—they would be cleared by the market. In reality, reprocessing is not being carried out today in response to an economic demand for plutonium (see chapter 7). Large-scale reprocessing continues for two main reasons. First, many utilities in Europe and Japan, expecting that plutonium fuels would be needed in the 1990s and beyond, signed binding contracts with the British and French reprocessing companies in the 1970s and 1980s from which there is no easy escape (see chapter 6). Second, capacities for storing and disposing of spent fuel are today the scarce resource for which utilities are

---

[35] These stocks contain a small proportion of fuel-grade plutonium. For simplicity, 'reactor-grade' plutonium is assumed here to comprise both fuel- and reactor-grade material.

Table 15.1. Illustrative inventories of plutonium available for disposition
Figures are in tonnes.

Country	Excess weapon-grade plutonium	Fuel- and reactor-grade plutonium stocks (31 Dec. 1994)		Future arisings from civil reprocessing[b]	Total
		Inactive[a]	Active[a]		
USA	49	15[c]	..	–	64
Russia	95	30	..	10[d]	135
UK	1	50[e]	..	24[d]	75
France	2	..	22	48	72
Germany	–	..	12	31	43
Japan	–	..	11	39	50
Other NPT NNWS	–	..	2	10	12
India	?	..	0.2	1	1
Israel	?	..	..	..	..
Total	**147**	**95**	**47**	**163**	**452**

[a] 'Active' and 'inactive' plutonium are materials held in store for which there are and are not commercial recycling plans respectively. Figures in the 'active' column are updates of the figures for Dec. 1993 contained in table 7.9.

[b] Plutonium due for separation under existing baseload contracts which cover the first 10 years of operation of UP3, UP2-800 and Thorp, and from RT-1 and Tokai-mura over the next decade. Additional plutonium which will be separated if post-baseload contracts are concluded and honoured is not counted here. Nor is the plutonium that would be separated if the Rokkasho-mura reprocessing plant were completed and came into operation.

[c] Includes 13.7 t of military and 1.5 t of civil plutonium.

[d] Future Russian and British stocks are assumed to remain inactive.

[e] Includes 8 t of military and 42 t of civil plutonium.

prepared to pay a premium. If reprocessing contracts were cancelled, utilities might have difficulty finding alternative storage space for their spent fuels. Once reprocessed, the responsibility for the radioactive wastes contained in the spent fuel also tends to shift from utilities to regional and central governments. These factors have placed the reprocessors in a strong bargaining position, at least where the next decade's 'baseload' contracts are concerned.

In most cases, utilities with reprocessing contracts are required to accept the return of their plutonium once it has been separated. In NNWS parties to the NPT, it is widely perceived that there is a de facto obligation to minimize the stockpiling of plutonium.[36] Utilities in Germany, Japan and other NNWS with reprocessing contracts have therefore tried to establish, with considerable difficulty, recycling programmes that would absorb all of their plutonium arisings.

---

[36] For instance, the Japanese Government has stated that 'it is a national principle that Japan will not possess plutonium beyond the amount required to implement its nuclear fuel recycling programmes'. See 'Nuclear fuel recycling in Japan', Advisory Committee on Nuclear Fuel Recycling (Japan Atomic Energy Commission: Tokyo, Aug. 1991).

As Britain, France and Russia are NWS, stockpiling has not carried the same political sensitivity. For them, recycling has been optional, with France and Russia opting in favour (the latter in principle rather than in practice), and Britain opting against partly because its gas-cooled reactors are less suited to burning plutonium fuels. Wherever plutonium recycling is occurring, whether in NWS or NNWS, it has largely become a disposition strategy aimed at maintaining the legitimacy of reprocessing policies. The alternative disposition strategy would be to leave the plutonium undisturbed in spent fuel: by definition, this would satisfy the spent-fuel standard, at a considerably lower cost.

'Active' plutonium can therefore also strictly be regarded as excess to market requirements, justifying its inclusion with the other stocks. Furthermore, these stocks of separated plutonium are 'avoidable' whereas the other stocks are now 'unavoidable'. Enriched uranium fuels could easily substitute for plutonium fuels (they are utilities' preferred choice), and plutonium separation could be averted if consent was granted for the long-term storage of spent fuel that is already held at reprocessing sites.

Four stocks of 'excess' material

In summary, disposition policies will have to address four stocks of fissile material over the next decade (the figures are rounded to two significant figures, and are current estimates):

1. The stock of excess HEU (an estimated 1400 t of WGU equivalent).
2. The stock of excess weapon-grade plutonium (an estimated 150 t).
3. The 'inactive' but 'unavoidable' stock of reactor-grade plutonium (an estimated 130 t).
4. The 'active' but 'avoidable' stock of reactor-grade plutonium arising from reprocessing contracts for which recycling plans already exist (an estimated 180 t).[37]

VIII. HEU disposition

The world stock of power reactors is loaded annually with LEU requiring just under 30 million separative work units of enrichment. The 500 tonnes of Russian HEU that have been purchased by the US Enrichment Corporation (USEC) acting as executive agency for the US Government are roughly equivalent to an expenditure of 90 million SWU, or closer to 100 million SWU after taking account of the 1.5 per cent enriched uranium used by Russia to dilute the HEU. This amount of separative work is sufficient to fuel the world stock of reactors for around three years. If the total estimated stock of excess HEU (1400 t) were made available for electricity generation, it would be equivalent to around 250 million SWU or just over eight years' supply of LEU fuels for

---

[37] The 176 t (180 t when rounded up) comprise 163 t of future arisings less 34 t of 'inactive' plutonium in Russia and the UK, plus 47 t of 'active' plutonium that have already been separated.

Table 15.2. HEU disposition scenarios[a]

	Disposition rate (t/y)	Displaced HEU (m. SWU/y)	Share of current world LEU requirement (%)
Low-disposition (500 t)	25	4.5	15
High-disposition (1400 t)	70	12.6	42

[a] Assumes a 20-year disposition period.

reactors. If it were released into world markets over a 15-year period, a world enrichment capacity of less than 15 million SWU/y would be sufficient to meet the balance of demand; if released over a 20-year period, a capacity of around 18 million SWU/y would suffice. This compares with today's underutilized capacity of around 54 million SWU/y, so that enrichment capacity could be more than halved in either case without creating supply scarcities.[38]

In a 'perfect market', utilities would be free to exploit this windfall of low-cost enriched uranium and to pass its benefits on to electricity consumers through lower prices, forcing production capacity to be mothballed or scrapped to the extent required to balance supply and demand. The market for nuclear fuels seldom functions in this way. As nuclear reactors take many years to construct and require stable fuelling arrangements, demand is heavily influenced by decisions taken 10 or 20 years previously, and most nuclear fuel supplies are subject to long-term contracts. There is also an expectation that preference will be given to national or regional suppliers of fuel-cycle services. Nuclear fuel markets are 'sticky' and do not adjust readily.

Decisions on which capacities would be scrapped if large-scale HEU disposition were carried out could also be politically complex. The release of large quantities of HEU from military stocks would force large cuts in US and Russian enrichment capacities unless most of the diluted HEU could be sold in other markets. That would simply shift the burden on to the European enrichment industry which has shown itself in the past to be adept at gaining protection from states and utilities. However, the European Union will have to play its part in HEU disposition, and thus accept some adjustment in European enrichment capacities. A high-disposition strategy would therefore probably have to entail formal or informal agreements among the countries supplying enrichment services on how the costs and benefits of the windfall would be distributed.

The 'high-disposition scenario' in table 15.2, involving all excess HEU, would inflict considerable pain on the enrichment industry. Under the Russian deal with USEC, however, less than 2 million SWU/y of production capacity will be displaced annually in the first 5 years, rising to just over 5 million SWU/y (or 9 per cent of current world capacity) in the remaining 15 years. If these 500 tonnes were alone destined for disposition (the 'low-disposition

---

[38] This capacity mainly comprises 20 million SWU/y in Russia, 14 million SWU/y in Europe and 19.6 million SWU/y in the USA.

scenario' in table 15.2), industries could probably adjust without difficulty. The Russian Government has recently announced that the rate of conversion into LEU will rise from 12 tonnes of HEU in 1996 to 30 tonnes per year over an as yet unspecified period.[39]

The interests of enrichment suppliers in maintaining their activities and market shares have therefore to be weighed against the security interests in reducing stocks of HEU and the economic interests of utilities and consumers in minimizing the prices paid for nuclear fuel. How much market power are governments and utilities prepared to concede to the enrichment industry? The rate of disposition of excess HEU will depend, above all else, on the answer given to this question.

Whether or not the product is sold commercially in the short and medium terms, the rapid blending down of excess HEU to LEU deserves a high priority on grounds of security. Current HEU inventories are much larger, and in some contexts less well protected, than plutonium inventories. HEU is also easier to fashion into a crude but reliable nuclear weapon. If governments decide that enrichment industries must be protected, the creation of LEU stockpiles out of diluted HEU is a price that should be paid.

IX. Plutonium disposition

The quantities of fissile material contained in inventories of excess plutonium are much smaller than those in inventories of HEU. The combined military and civilian surpluses might equate to half a year's supply of fissile material to the world's power reactors. Plutonium disposition would therefore have little effect on markets for uranium and for enrichment services. Its effect on fuel-cycle activities would be mainly felt through the ability to absorb plutonium from civil reprocessing, thereby affecting the viability of reprocessing itself.

Over the next 10–15 years, between 400 and 500 tonnes of plutonium are likely to become available for disposition (assuming that civil reprocessing continues as planned). The 452 tonnes recorded in tables 15.1 and 15.3 comprise a proportion of the inventory of weapon-grade plutonium that could conservatively be regarded as excess to military requirements; the fuel- and reactor-grade plutonium that was held in stock at the end of 1993; the balance of plutonium that remains to be separated under baseload contracts, mainly with British and French reprocessors; and an additional 10 tonnes that might be separated in Russia over the same period. These quantities exclude the relatively small quantities that might be separated under the post-baseload contracts that have been signed to date, and the plutonium that would be separated from

---

[39] See Statement by V. N. Mikhailov, Head of the Russian Federation Delegation, to the 40th Session of the IAEA General Conference, Vienna, 16 Sep. 1996. A lengthy analysis of the US–Russian HEU agreement is contained in Falkenrath, R. A., 'The HEU deal', *Avoiding Nuclear Anarchy* (note 12), appendix C, pp. 229–92. Falkenrath criticizes the institutional arrangements in the USA which depress the price offered to Russia for the blended-down HEU, reducing Russia's incentives to implement the deal.

Table 15.3. Capacities for disposition through plutonium recycling in power reactors

Country	Anticipated Pu inventories for disposition (t Pu_{tot})	Anticipated Pu inventories for disposition (t Pu_{fiss})[a]	Annual MOX fabrication capacity (t Pu_{fiss}/y)[b]	Potential absorptive capacity in power reactors (t Pu_{fiss}/y)[c]
USA	64	56	zero	*25*
Russia	135	118	zero	*2?*[d]
UK	75	56	(3.8)	*0.25*
France	72	54	7.5	*14*(5)
Germany	43	32	(4)	*6*(3)
Japan	50	37	–	*9*
Canada	–	–	–	*3*
Other NPT NNWS	12	9	1.5	(1)
India	1	1	0.1?	0.1?
Total	**452**	**363**	**16.9**	***64*(9)**

[a] Assumes that the fissile contents of reactor- and weapon-grade plutonium are 75% and 93% respectively.

[b] Assumes that MOX fuel is enriched with 3.8% Pu_{fiss} on average, and assumes that one-third of a reactor core is loaded with MOX fuel in every case. Figures in brackets indicate that plants are not yet completed (UK) or licensed for operation (Germany and UK).

[c] Figures in italics indicate reactor capacity that is technically appropriate for recycling. Roman numerals in brackets indicate reactor capacity that has been licensed (1996) for recycling. It is assumed that one-third core is loaded with MOX fuel, and that 20 t of fuel are replaced on average per reactor. Approximately 250 kg of fissile plutonium would be loaded per gigawatt of generating capacity under these assumptions.

[d] It is unclear how many reactors might be suitable for MOX recycling in Russia. The capacity cited here involves the 7 operating VVER-1000 reactors. There is a common assumption outside Russia that substantial investment in new safety measures would be required to enable large-scale thermal MOX recycling

Japan's Rokkasho-mura reprocessing plant if it began operating before 2010 (this seems increasingly improbable).

The figures in the first column in table 15.3 are quantities of 'total plutonium'. Their fissile content varies: in reactor-grade inventories, the fissile isotopes ^{239}Pu and ^{241}Pu may average 75 per cent of the total isotopic mixture, whereas the proportion for weapon-grade inventories will be slightly over 93 per cent. It is customary to measure plutonium in terms of its fissile content when MOX fuel fabrication and plutonium recycling are the issues. The second column therefore records the estimated quantities of 'fissile plutonium' that are available for disposition.

Table 15.3 shows that technically there is sufficient reactor capacity around the world to reduce 360 tonnes of fissile plutonium to the 'spent-fuel standard' in just under 11 years, assuming that it is loaded sequentially and has five years' residence in core (time would also have to be allowed for fuel fabrication).⁴⁰ By the same token, the excess weapon-grade plutonium could be consumed in less than seven years. However, there are two immediate constraints on the rate at which plutonium could be recycled and thus returned to the 'spent-fuel standard'.

The first is that thermal MOX fabrication capacity has been largely sized to absorb plutonium from the British and French commercial reprocessing operations, and is located in its entirety in western Europe. It is no longer expected that the plant that has been constructed in Germany will receive its operating licence, certainly for the fabrication of MOX fuels from civil plutonium. Current fabrication capacity could therefore only handle a small fraction of the excess plutonium unless recycling were carried out over several decades with no further plutonium separation. Gaining political consent to build new fabrication plants is never easy, as experience in Belgium and Germany attests.

The second problem is that fewer than 30 reactors, with the combined capacity to accept around 9 tonnes of plutonium each year, have so far been licensed to take MOX fuel. There are no reactors licensed, and little recent experience of using MOX fuels in power reactors, in three of the countries with the largest inventories—Russia, the UK and the USA. Although each of these countries has substantial nuclear-generating capacity, MOX recycling in Russia may be unwise in all but a handful of reactors for safety reasons, only one reactor in Britain is capable of using MOX fuel, while major political obstacles would need to be overcome for plutonium disposition in reactors to be adopted in the USA.⁴¹ Even in countries with large numbers of reactors that could be made available for plutonium recycling (notably Canada, Germany and Japan), the licensing of reactors for this purpose tends to be politically controversial. Only in France has reactor licensing seemed unproblematic so far.

Table 15.3 indicates that there are two basic options for disposing of excess weapon-grade plutonium, assuming that MOX recycling will figure large in any disposition strategy. The first option is to use established MOX recycling capacities (i.e., fabrication plants and licensed reactors), and the accumulated expertise associated with them, to reduce the plutonium to the 'spent fuel standard'. The second is to construct new capacities where they do not currently exist, notably in Russia and the USA. Let us briefly consider these two approaches.

1. *Using existing MOX recycling capacities.* In 1995, it was proposed that the German MOX fabrication plant at Hanau, which had been denied an operating licence to manufacture MOX fuel using civil plutonium, should instead be used to help eliminate Russia's excess stock of weapon-grade material.⁴² Plutonium from Russia would be fabricated into MOX fuel and irradiated in reactors in Germany, the spent-fuel discharges being returned to Russia for storage and

---

⁴⁰ It is assumed here that plutonium could only be recycled in light water and CANDU reactors.

⁴¹ Russia's proposals to recycle plutonium in fast reactors seem unrealistic and are discounted here.

⁴² For a discussion, see Schaper, A., 'A potential German contribution to the disposition of plutonium from dismantled Russian warheads', Paper presented at the 45th Pugwash Conference on Science and World Affairs, Hiroshima, Japan, 23–29 July 1995.

eventual disposal. The German facility had the capacity to absorb between 4 and 5 tonnes of fissile plutonium per year, so that 40–50 tonnes of weapon-grade plutonium could be neutralized in this way over a 15-year period (10 years plus 5 years in reactors).

The proposal was considered at a high level by the German, Russian, US and other governments. It has since foundered, partly for German domestic reasons, partly as a result of Russia's apparent reluctance to send its plutonium abroad and partly because the political grounds for creating international confidence in the arrangements had not been sufficiently established.

A proposal from the European Union, involving cooperation between the member countries which have MOX recycling capacities, might carry greater political weight.[43] If put into effect, it could also eliminate a larger proportion of Russia's *and* the USA's excess stock over a shorter time than the German proposal. The 16.9 tonne per year MOX fabrication capacity identified in table 15.3 could lead to the elimination of the entire excess stock of weapon-grade plutonium in around 14 years (9 years plus 5 years in reactors). This would involve the fabrication plants in Belgium, France, Germany and the UK being dedicated to the disposition of weapon-grade material, and the burning of the resultant MOX fuels in various reactors across the European Union.

This proposal would have two great advantages. The first is that the disposition of weapon plutonium would be carried out in the countries that have the greatest experience of plutonium recycling, and that have invested heavily in the necessary techniques and productive capacities over many years. Once the arrangements had been made, this would probably provide the safest and quickest route. The second advantage is that the safeguarding and physical protection of MOX recycling is well established in Europe. In particular, the whole exercise, from the arrival of the separated plutonium to its departure in spent fuel, could be conducted under Euratom safeguards. This might save IAEA resources and avoid any difficulties that might arise over IAEA safeguarding of the relevant materials and facilities in Russia and the USA.

However, the proposal would have two equally great disadvantages. One is that it would require substantial movements of plutonium, in many shipments, from Russia and the USA to fabrication sites in Europe, and from those sites to power reactors within Europe. These movements would carry safety and security risks, would require intensive physical protection and might encounter serious political opposition. The return of spent fuels containing the irradiated plutonium to Russia and the USA would also be controversial. The second disadvantage, at least in the eyes of the European fuel-cycle industry, is that the recycling of weapon plutonium in Europe could probably only be carried out at the expense of civil plutonium recycling, and thus at the expense of civil reprocessing. It seems unlikely that political consent could be gained for doubling recycling capacities so that the recycling of both civil and military plutonium

---

[43] It is assumed here that the blending of US and Russian weapon plutonium with high-level wastes in Europe would be a political non-starter.

might be achieved. Although the curtailment of civil reprocessing would be welcomed in many quarters, including parts of the electricity supply industry, the reprocessing and MOX fabrication capacities in Europe are bound by contracts between utilities and fuel-cycle companies that would be difficult to unravel.

A decision to give priority to the elimination of Russian and US weapon plutonium through recycling would therefore entail a complex series of negotiations accompanied by a willingness to make significant changes in reprocessing policies. On the other hand, MOX capacity could still become available in Europe if the recycling of separated civilian plutonium is blocked or delayed (especially in Germany and Japan), or the rate of reprocessing is less than anticipated (especially in the UK).

In addition, it should be recalled that the stocks of weapon-grade plutonium are not the only excess stocks requiring disposition. Over 140 tonnes of 'active' and 'inactive' fuel- and reactor-grade plutonium are also held in store in separated form, mostly in Europe. The priority in Europe should arguably be given to the disposition of these other stocks, and to the establishment of a new balance between civilian supply and demand that is consistent with market forces. So long as the rate of civil reprocessing continues to exceed the rate of civil plutonium recycling (see chapters 7 and 14), current European and Japanese fuel-cycle policies will result in net *accumulations* rather than dispositions of plutonium.

The unusual situation regarding Britain's stock of over 40 tonnes of 'inactive' reactor-grade plutonium is worth noting. It would take more than a century to burn this plutonium in Britain's only light water reactor, which has a nominal life-time of 30 years. Given the British utilities' recent announcement that no more nuclear reactors will be constructed for the foreseeable future, Britain's abandonment of its fast reactor programme, and the absence of foreign demand for Britain's plutonium, there are few prospects for eliminating this stock through plutonium recycling. Vitrification of a mixture of plutonium and high-level waste might be the only viable means of achieving the 'spent-fuel standard' in this instance. This is not just a local matter, as Britain's stance on this issue has implications for the disposition of the other large stock of inactive reactor-grade plutonium—the 30 tonnes held in Russia.[44]

2. *Building new disposition capacities in Russia and the USA.* The alternative approach is to establish new capacities, and/or re-engineer existing capacities, in Russia and the USA, the countries possessing much the largest stocks of weapon-grade plutonium awaiting disposition. Its attractions, which may prove decisive, are that the Russian and US industrial stakes in disposition would be maximized, international transportation would be kept to a minimum, Russian

---

[44] As yet, the British Government has formulated no policy towards the disposition of excess plutonium. Nor, to the authors' knowledge, is R&D being carried out in the UK on the mixing and subsequent vitrification of plutonium and high-level waste. As reactor-grade plutonium contains isotopes that are strong neutron emitters and give rise to significant heat output, the technical problems may be more substantial than with weapon-grade plutonium.

and US qualms about allowing other countries to handle such large quantities of weapon-grade plutonium would be overcome and a cooperative strategy could be developed within the framework of the bilateral arrangements discussed in section V. Set against this, the Russian and US capacities are limited at present. While both countries have long experience of plutonium-related technologies, they have not in recent years invested heavily in the techniques required for plutonium disposition. Establishing these capacities would take time. Nor can there be complete confidence that either country can muster enough political support for programmes that would be expensive, give rise to environmental risks and stir up old controversies in the United States over the security risks attached to plutonium recycling.

Nevertheless, the 1995 NAS study concludes that the USA could mount a substantial disposition programme within a few years of a decision to launch it. There are many reactors that could be used for MOX recycling, including four reactors (three operating and one under construction) that were designed to be fuelled with full MOX cores.[45] If licences could be attained, sufficient reactor capacity exists to enable the disposition of the entire US stock of excess weapon-grade plutonium in a few years. There is, however, no MOX fabrication capacity at present in the United States. Construction of a plant was begun in the late 1970s to produce fuel for the Fast Flux Test Facility, but it was not operated. The completion and updating of this facility is a possibility, but it seems likely that a new plant would have to be built, possibly to a Franco-Belgian or German design.

There is also the option in the USA and elsewhere for eliminating plutonium stocks by mixing them with high-level wastes prior to immobilization, especially through vitrification, and underground disposal. Vitrification plants are already being constructed at Savannah River and West Valley which could be adapted to incorporate plutonium in the radioactive wastes. The NAS study estimates that the Savannah River facility could alone eliminate 50 tonnes of plutonium in this way over an eight-year period. Once established, these techniques could be equally well applied in the other NWS, each of which possesses sizeable stocks of high-level wastes which are currently held in store (France and the UK already have extensive experience of waste vitrification). However, significant uncertainties remain over the technical performance of such facilities, and the NAS panel questions the wisdom of relying on one large facility. Again, it would probably be necessary to establish new capacities if a large-scale plutonium disposition programme which relies partly or wholly on vitrification is to be successfully mounted.

The lack of the necessary experience and technological capacities in Russia is still more problematic. Only a handful of thermal reactors may be judged suitable for plutonium recycling, MOX experience has been gained only from pilot-scale facilities, and there are doubts about the viability of the vitrification

---

[45] See *Management and Disposition of Excess Weapons Plutonium: Reactor-Related Options* (note 32), pp. 117–28. The 3 operating System-80 reactors are at the Palo Verde Nuclear Generating Station, Arizona, and the System-80 reactor under construction belongs to the Washington Power Supply System.

techniques developed hitherto in Russia (reliance has been placed on phosphate glasses rather than on the borosilicate glasses that have become standard elsewhere). Even if a substantial programme were launched to utilize plutonium fuels in Russian power reactors, the elimination of the approximately 100 tonnes of excess Russian weapon-grade plutonium would take many decades. The Russian Government and industry are insisting that all plutonium must be recycled, on the grounds that it is 'national treasure'. This policy, which has little technical or economic logic behind it, is to be regretted, not least because Russia's outright rejection of alternatives has encouraged other countries (including the USA) to downplay vitrification.

Come what may, a serious disposition programme involving recycling would have to entail the absorption by other countries of a sizeable proportion of the Russian inventory of excess plutonium. One option is to undertake MOX fuel fabrication in Russia and transfer the MOX assemblies to Europe, Canada or elsewhere for insertion in power reactors. The Canadian option has recently received considerable attention, since the CANDU reactors appear capable of being loaded with a full MOX core without the need for significant adjustment, although in mid-1996 it seems to be falling out of favour.[46] However, as the debate has largely been based on paper studies, full-blown trials would be required to verify their findings, the economics are questionable and it is uncertain whether political consent could be gained in Canada for the large-scale usage of plutonium in power reactors. Transportation issues would also have to be addressed.

Neither of the two recycling approaches discussed above—the first largely entailing the exploitation of existing capacities in western Europe, the second the establishment of new capacities in Russia and the United States—is therefore straightforward. Both deserve to be given serious consideration. While the politics of the US–Russian approach may be the more amenable to solutions, the European option should not be rejected out of hand. Whichever choices were made, European technical assistance would probably be required, and a reassessment of Europe's (and Japan's) own disposition policies regarding its stocks of reactor-grade plutonium is overdue.

Nor should the vitrification option be dismissed, however, as has tended to happen since the NAS study was published. Given the political and economic problems that have bedevilled plutonium recycling in recent years, vitrification may turn out to be the quickest and most reliable technique for reducing unwanted stocks of separated plutonium. It should therefore be pursued with equal determination. This is the implicit conclusion of a study of plutonium disposition, containing the most detailed technical and economic assessment so

---

[46] Hink, A. D., et al., 'Plutonium disposition in CANDU reactors', Paper presented at the Fifth International Conference on Radioactive Waste Management and Environmental Remediation (note 33), pp. 475–77.

far attempted, that was published by the US Department of Energy in July 1996.[47]

X. Obstacles to an international disposition strategy

Unless a deliberate strategy is developed for disposing of excess fissile materials, probably little will happen. A portion of the HEU would be removed from military stocks and used to fuel power reactors, but it would remain a small part of the total inventory. The military plutonium is likely to go nowhere.

Some would argue that this is the best outcome. The materials should be placed in safe storage, and held there until better times. But what guarantee is there that those times will turn out better? There should at least be a thorough international investigation of the paths that could lead to all excess plutonium and HEU being brought into line with the 'spent-fuel standard'. While it would probably be unwise to set a target date for the task's completion, it would be unsatisfactory if the majority of these materials were still unable to meet this standard 20 years from now.

This discussion suggests that the large-scale disposition of excess fissile material would only take place if six conditions were met. First, states would have to be prepared to invest resources in order, *inter alia*, to meet the higher costs that utilities and other industrial players might incur when pursuing development and demonstration programmes, pay for additional facilities and their operation, and allow standards of safety, security and environmental protection to be fully met.[48] Alternatively, they would have to commit themselves to using the windfall gains from HEU disposition to cross-subsidize plutonium disposition.

Second, extensive international cooperation would be required. Broad agreement would have to be reached on how uranium and enrichment capacities were to be adjusted to allow utilities full access to the HEU windfall; on the implications of disposition strategies for civil reprocessing and recycling and for the disposal of high-level wastes if the vitrification route were followed; on technology transfer; and on the international control measures and administrative arrangements that would have to be applied.

Third, there would have to be international reciprocity. One cannot expect Russia to consent to the elimination of its excess stocks of fissile material if US

---

[47] Office of Fissile Materials Disposition, *Technical Summary Report for Surplus Weapons-Usable Plutonium Disposition* (US Department of Energy: Washington, DC, 17 July 1996).

[48] It has been estimated that the extra cost of recycling Russia's stock of excess weapon-grade plutonium would be in the range $1–2 billion. Chow, B. G., 'Assessing alternative strategies for the disposition of weapons-grade uranium and plutonium', Proceedings of the Fifth International Conference on Radioactive Waste Management and Environmental Remediation (note 33), pp. 527–29. The NAS study estimates that the disposition cost for the US weapon-grade plutonium would lie in the range $0.5–2 billion for the vitrification option, and $0.5–5 billion for the MOX recycling option, with $1–2 billion again possibly being the most likely outcome. See *Management and Disposition of Excess Weapons Plutonium: Reactor-Related Options* (note 32), pp. 11–12.

stocks remained untouched, or *vice-versa*. As such, cooperation would be required to ensure that disposition programmes developed in tandem, irrespective of the approach that was taken.

Fourth, there would have to be public participation in the debates in order to establish confidence that environmental and other objectives were being met. This would be especially important in regard to plutonium. Public support for its recycling in power reactors or for waste disposal might be more forthcoming if the cause of nuclear disarmament was being clearly served by the initiatives being proposed.

Fifth, whatever is done, the activities should not increase the chances that nuclear proliferation will occur. In particular, great care will have to be taken to render secure the transport and storage of weapon material, to apply effective safeguards and to ensure that disposition does not become a pretext for pursuing industrial activities that will diminish rather than enhance international security. Security interests should always be given precedence over institutional and commercial interests.

Last but not least, there would have to be flexibility in all quarters. An effective disposition strategy would probably require the United States and Russia to soften their respective positions on plutonium recycling and vitrification, require the European fuel-cycle companies to accept adjustments to their enrichment and reprocessing programmes, require the Russian nuclear community to end its insistence on regarding plutonium as 'national treasure', and require environmental lobbies to accept industrial activities that have hitherto been regarded as anathema by some of their members.

The obstacles to plutonium and HEU disposition are therefore not technical. Nor, ultimately, are they economic, even if the financial constraints are significant. They are institutional and political.

XI. Conclusions: moving towards the framework of control required for complete nuclear disarmament

In the Principles and Objectives for Nuclear Non-Proliferation and Disarmament agreed at the 1995 NPT Review and Extension Conference[49] (henceforth referred to as 'the Principles and Objectives'), the Treaty's parties reaffirmed 'the ultimate goal of the complete elimination of nuclear weapons'. Let us imagine that, at some time in the next century, this goal is about to be realized and that the NWS and threshold countries which possess nuclear weapons have decided to destroy their nuclear arsenals and join the throng of NNWS.

This momentous step would entail placing *all* fissile materials and associated production facilities in *all* countries under IAEA safeguards. For disarmament to be stable and acceptable, an intensification of verification and confidence-building activities would be required in order to minimize the chances of break-

[49] See note 15.

outs.⁵⁰ In addition, full-scope safeguards would provide one of the few convincing means by which the five NWS, which have eternal rights to possess nuclear weapons under the NPT, could demonstrate their renunciation of nuclear arms and have it verified by the international community.⁵¹

The prospect of total nuclear disarmament would also increase pressures to eliminate stockpiles of nuclear weapon material, and to halt further accumulations. As so much knowledge of weapon design and manufacture would still exist, such stockpiles would provide their holders with an essential and visible part of the capability needed to acquire or re-acquire nuclear weapons. The elimination of unnecessary stocks would in these circumstances become an important confidence-building measure. The option of placing stocks of separated plutonium and HEU under multinational supervision, as provided for instance by Article IX of the IAEA Statute, might also have to be revisited.

Complete nuclear disarmament may seem a distant prospect, although it is today being addressed with a seriousness that has hitherto been absent.⁵² Today, neither the NWS nor the non-NPT threshold states show any desire to abandon their nuclear weaponry. However, the essential point here is that today's *need for stringent control over nuclear materials and facilities is little different from that which would be required in conditions of total disarmament*. When the current nuclear arms reduction programmes are completed, more than 80 per cent of the nuclear weapons that were deployed at the height of the cold war will have been dismantled. This means that much of the material associated with them is also having to be de-militarized, requiring strict oversight to ensure that it is not diverted back into weapon programmes or into the hands of rogue entities. Furthermore, the breakup of the Soviet Union, and the evident shortcomings in material accountancy and control in nuclear weapon programmes, have shown the fallacy of assuming that sensitive materials and technologies are uniquely safe in the hands of the NWS. The risks and consequences of diversion or theft may today be graver in the NWS and non-NPT threshold states than in the NNWS, not least because the former states possess the great majority of weapon-grade plutonium and uranium.

Despite this reality, the steps that are now being taken to exercise control in these countries fall well short of those required by total disarmament. Notwithstanding the impressive efforts described in section IV to strengthen the security of nuclear materials through national and bilateral programmes, the inven-

---

⁵⁰ An interesting discussion of the challenges of achieving nuclear disarmament, especially the 'virtual' disarmament which envisages the retention of residual capabilities which would deter break-outs, is provided by Mazarr, M., 'Virtual nuclear arsenals', *Survival*, vol. 37, no. 3 (autumn 1995), pp. 7–26.

⁵¹ Even if they had discarded their weapons, full-scope safeguards would not be mandatory for the NWS under Article III of the NPT. They would have to volunteer to make them mandatory, thereby accepting the obligations placed on the Treaty's NNWS. For a discussion of the steps required so that an NWS that is a member of the NPT could become an NNWS under international law, see Dombey, N., Fischer, D. and Walker, W., 'Becoming a non-nuclear weapon state: Britain, the NPT and safeguards', *International Affairs*, vol. 63, no. 2 (spring 1987), pp. 191–205.

⁵² This is illustrated by the recent work of the Canberra Commission on the Elimination of Nuclear Weapons which was appointed by the Australian Government and comprised distinguished persons from many countries, including the NWS.

tories of fissile material held by the NWS and threshold states remain ill-defined, a tiny proportion of the weapon-grade plutonium and uranium in former military stocks is being opened to IAEA safeguards, and plans to eliminate excess materials are in their infancy.

Given the precipitous changes that have occurred since the late 1980s, and the political stresses and bureaucratic workload associated with them, it is perhaps not surprising that achievements have so far been rather limited in this field. Besides the task of managing arms reduction programmes, securing the NPT's extension and negotiating the CTBT, the international community has had to cope with the Iraqi and North Korean crises, the dismantlement of South Africa's nuclear weapon programme, and the absorption of Belarus, Kazakhstan, the Ukraine and other states into the NPT regime. Among other things, the IAEA's financial and administrative resources have been stretched to the limit by these developments. The situation in the NWS may, in any case, have been best handled through bilateral measures at this stage. In Russia, for instance, more was probably achievable through cooperative programmes with the USA and other countries, and through the sticks and carrots that states can exercise, than could have occurred as a result of multilateral initiatives.

There is a preference in some quarters for keeping the management of the nuclear weapon states' nuclear legacy outside the multilateral framework. Remedial measures should, according to this view, be built up gradually through domestic and bilateral processes, partly to prevent the IAEA's resources becoming overstretched. However, there is a more powerful case for taking the achievements of the past few years and building them into a set of multilateral measures consistent with those applied in the NNWS. For both practical and political reasons, the regulatory situation in all countries, including the NWS, should be approached *as if the world is preparing for total nuclear disarmament*, whether or not that is a desirable or realistic prospect. This would provide a clear sense of purpose; ensure that the highest standards of management and verification are being applied, and that disarmament undertakings are as near to being 'irreversible' as is possible; satisfy concerns that the 'intrusiveness and transparency gap' between NWS and NNWS is being narrowed; and provide confidence that the NWS are respecting the goals set out in the NPT Principles and Objectives, and are not simply intent on bringing about the 'recrystallization' of their nuclear armaments and privileges. The final step towards nuclear disarmament would also be less daunting, and hence more achievable, if there had already been substantial movement towards the adoption of the measures required to bring it into effect.

It should be noted in this context that the international community has learned a great deal from the recent management and verification of disarmament in Iraq and South Africa, and from the steps taken to eliminate surplus weapons and capabilities in the NWS. The techniques of disarmament are now broadly understood.

However, no one should underestimate the magnitude of the task involved in submitting the nuclear weapon states' nuclear materials and facilities to effective international verification (although the task is comparatively straightforward in France and the UK due to the extensive application of Euratom safeguards on their territories). The quantities are very large, materials are held in many forms, accurate historical records are often lacking and production facilities were seldom designed with NPT-type safeguards in mind. Even if it were accepted that all non-military materials and facilities in the NWS should be submitted to routine inspection by the IAEA, and that additional resources should be made available for the purpose, the task would take many years to complete. The recent experience in bringing South Africa's much more limited nuclear infrastructure and materials under IAEA safeguards has shown how meticulous the preparation has to be.

Furthermore, there is general recognition, brought about by the revelation of Iraq's clandestine weapon programme discussed in chapter 11, that the IAEA safeguards system requires considerable strengthening. At present, the innovations being tested in Programme 93+2 are mainly being addressed to situations in NNWS parties to the NPT where full-scope safeguards are already applied and disarmament is, or is supposed to be, a permanent fact of life. However, the measures aimed at detecting undeclared programmes have equal relevance to the verification of disarmament in the NWS and non-NPT threshold states. They are fundamental if the ultimate project of eliminating all nuclear weapons is to have any chance of realization.

The substantial deepening and extension of international verification is therefore bound to take years to achieve. This is implicitly recognized in the Principles and Objectives, which state that 'nuclear fissile material transferred from military use to peaceful nuclear activities should, as soon as practicable, be placed under IAEA safeguards in the framework of the voluntary safeguards agreements in place with the nuclear-weapon States'. Safeguards cannot become instantly practicable in these situations, nor can fresh resources be committed by the stroke of a pen.

However, these difficulties must not become an excuse for endless procrastination. The needle of the policy compass should be set, pointing firmly towards the arrangements that would have to be applied in conditions of total nuclear disarmament. This direction is consistent with the NPT's goals as elaborated in the Principles and Objectives. It also serves the vital practical purpose of placing the world's huge inventories of fissile material under the most effective control.

In this light, the initiatives that have been reviewed in this chapter—the strengthening of international safeguards and physical protection, the Fissile Material Cut-Off Treaty, the newly proposed treaty on excess fissile material, and the international programme on fissile material disposition—are not optional. Together, they form a package of policy measures that serve the common interest in preventing nuclear proliferation, and in bringing the international community closer towards the effective management of partial or complete nuclear disarmament. Although their detailed design will require complex and lengthy negotiation, the four guiding principles can be simply stated:

1. *Universality*. All stocks of fissile material (military and non-military), and all facilities capable of producing them, should be subject in all countries to the most exacting standards of accountancy, verification and physical protection.

2. *Transparency*. All states should regularly publish summaries of inventories of fissile materials (military and non-military) held on their territories, and held on their behalf by other states. Detailed inventories of fissile materials should be assembled by all states, and details of non-military stocks should be routinely made available to safeguards agencies so that they can be submitted to international verification. Capabilities should also be transparent.

3. *Minimization*. Production of fissile material for weapon purposes should be ended. The production and separation of fissile material for other purposes should only occur where there are well-founded commercial or other requirements. As far as is practicable, steps should be taken to eliminate all stocks of excess fissile material: excess plutonium should be made to conform with the spent-fuel standards and excess HEU should be diluted to LEU.

4. *Access*. International inspection agencies should have greater access, under agreed conditions, to facilities and information relevant to nuclear or nuclear-related activities in all states.

A final remark is in place. This book is concerned with plutonium and HEU, the fissile materials used in nuclear weapons. Their control and disposition can only be one facet of nuclear non-proliferation and disarmament policy. The focus here on nuclear materials and capabilities is not intended to imply any prioritization. Equal importance has to be attached to the other goals described in the NPT Principles and Objectives.

Nor can reliance be placed only on multilateral measures and international regimes and undertakings. Achieving the goals of nuclear disarmament and non-proliferation will also depend, as always, on the reduction of tensions in several parts of the world and on the judicious exercise of power and diplomacy.

In every respect, the essential task is to create the conditions in which nuclear weapons can be stripped of their remaining political and military potency. Ultimately, they will only be defeated if people and states cease to value them, and if international society denies them legitimacy.

Appendices

Appendix A. Weapon-grade plutonium and highly enriched uranium production

The purpose of this appendix is to provide some technical background information on the production of weapon-grade plutonium and highly enriched uranium. This appendix is not intended to be comprehensive but to provide information that is needed to supplement the main text. Appendix B treats the production of reactor-grade plutonium in power reactors.

I. Weapon-grade plutonium production

As described in chapter 2, weapon-grade plutonium is made through neutron irradiation of uranium. Almost all weapon-grade plutonium has been made in reactors specifically designed for such production, although it can be made in any type of reactor.

However, it is simplest to make weapon-grade plutonium in production reactors and certain types of research reactor. They share certain characteristics that both maximize plutonium production and minimize operational problems. Characteristics that have been favoured include natural uranium metal fuel, easy fuel reloading, simple fuel cladding, and graphite or heavy water moderators.

This book uses two main approaches to estimating the amount of weapon-grade plutonium produced in production reactors. The first approach involves a simple approximation of average annual weapon-grade plutonium production. It can be represented by the following equation:

$$Pu_{\text{yearly total}} = P_{\text{thermal power}} \times C \times (365 \text{ days}) \times F$$

where P is the nominal thermal power of the reactor and C, often called the capacity or 'innage' factor, represents the ratio of the total annual heat output to the annual heat output based on continual full-power operation. This ratio is often stated to be the fraction of the year the reactor operates at full power. The last factor in the equation, F, represents a plutonium conversion factor which gives the amount of plutonium produced per megawatt-day of thermal output (MWth-d) from the reactor. For weapon-grade plutonium produced in a graphite-moderated, natural uranium reactor, the factor is about 0.9 g weapon-grade plutonium per MWth-d (g/MWth-d).[1] Because weapon-grade plutonium includes all plutonium with more than 93 per cent ^{239}Pu, this conversion factor can vary by up to 10 per cent (ignoring the uncertainty of the estimate itself). This variation reflects the range of burnup for irradiated fuel containing weapon-grade plutonium. Table A.1 contains typical conversion factors for several types of reactor.

---

[1] Turner, S. E., et al., *Criticality Studies of Graphite-Moderated Production Reactors*, Report prepared for the US Arms Control and Disarmament Agency, SSA-125, Southern Sciences Applications, Washington, DC, Jan. 1980.

Table A.1. Representative conversion factors for reactors producing weapon-grade plutonium

Reactor type	Conversion factor (g/MWth-d)	Burnup (MWd/t)	Percentage ^{239}Pu
Gas-cooled, graphite-moderated[a]	0.9	500	96.3
Air-cooled, graphite-moderated[a]	0.9	500	96.6
Heavy water moderated[b]	0.9	1 000	93.8
LWR[b]	0.5	1 000	98.5
RBMK[b]	0.5	1 000	96.2

[a] Turner, S. E., et al., *Criticality Studies of Graphite-Moderated Production Reactors*, Report prepared for the US Arms Control and Disarmament Agency, SSA-125, Southern Sciences Applications, Washington, DC, Jan. 1980.

[b] Libby, D., 'Nuclear fuel cycle and nuclear reactors', Pacific Northwest Laboratory, view-graphs presented at the Department of Energy Nuclear Nonproliferation Workshop, Washington, DC, Dec. 1993.

The second approach used in this book relies on more detailed information about the amount and isotopic composition of plutonium in irradiated fuel as a function of burnup. This section of the appendix focuses on two types of graphite-moderated reactor that differ in their cooling medium. The first type is the gas-graphite reactor, which is cooled by carbon-dioxide, and the second type is air-cooled. In these cases, the fuel is assumed to be natural uranium metal. Graphite-moderated, water-cooled reactors and heavy water reactors are not covered here, but they have similar conversion factors.

The amount of weapon-grade plutonium produced in gas-graphite reactors depends on several factors that involve reactor design decisions. Table A.2 lists some standard estimates for plutonium production in this type of reactor, based on information about the British Calder Hall reactors and the French G2 and G3 reactors (see chapter 3). Given a specific burnup, the uncertainty in these numbers is taken as about 10 per cent.

Table A.3 lists standard production estimates for an air-cooled gas-graphite reactor such as the French G1 reactor. This table lists values to 1000 MWd/t because the fuel still contains weapon-grade plutonium. However, this level of burnup is not usually achieved if the purpose of the reactor is the production of plutonium for a military programme. Since the power of this type of reactor is usually small, typical burnups would usually be about 100–200 MWd/t. Otherwise, the fuel would need to remain in the reactor for an extended period. For example, for the fuel to reach 500 MWd/t in the G1 reactor, it would need to remain in the reactor for 4 years, assuming an 80 per cent capacity factor and a typical initial fuel loading of 100 tonnes.

For comparison, table A.4 lists the plutonium in irradiated fuel at typical discharge burnups for several types of power reactor, as calculated by a US nuclear expert.

Table A.2. Conversion factors for gas-graphite reactor with natural uranium fuel

Burnup (MWd/t)	Percentage ^{240}Pu, ^{241}Pu and ^{242}Pu	Concentration in spent fuel (kg/t)	Conversion factor (g/MWth-d)
100	0.75	0.1	1
200	1.5	0.19	0.95
300	2.3	0.28	0.933
400	3.1	0.36	0.925
500	3.7	0.45	0.9
600	4.4	0.535	0.892
700	5.1	0.62	0.886
800	5.7	0.7	0.875
900	6.3	0.78	0.867
1 000	6.9	0.86	0.86
1 100	~7.5	~0.94	~0.855
1 200	~8.1	~1.02	~0.85

Source: Turner, S. E., *et al.*, *Criticality Studies of Graphite-Moderated Production Reactors*, Report prepared for the US Arms Control and Disarmament Agency, SSA-125, Southern Sciences Applications, Washington, DC, Jan. 1980, pp. 68–70. The values in this table are for a 20.3 cm pitch, which is the distance between adjacent fuel elements. For the reactor specifications evaluated in Turner *et al.*, this pitch is not optimal, although Turner *et al.* say that this pitch is more accurate for the Calder Hall and Marcoule reactors. At the optimal pitch, the plutonium production values are a few per cent lower. The values 1100 MWd/t and 1200 MWd/t are extrapolated from data in Turner *et al.*

Table A.3. Gas-graphite, air-cooled reactor with natural uranium fuel

Burnup (MWd/t)	Percentage ^{240}Pu, ^{241}Pu and ^{242}Pu	Concentration in spent fuel (kg/t)	Conversion factor (g/MWth-d)
100	0.7	0.1	1
200	1.4	0.195	0.975
300	2.1	0.285	0.95
400	2.8	0.37	0.925
500	3.4	0.45	0.9
600	4.1	0.53	0.883
700	4.7	0.61	0.871
800	5.2	0.685	0.856
900	5.8	0.76	0.844
1 000	6.4	0.83	0.83

Source: Turner, S. E., *et al.*, *Criticality Studies of Graphite-Moderated Production Reactors*, Report prepared for the US Arms Control and Disarmament Agency, SSA-125, Southern Sciences Applications, Washington, DC, Jan. 1980, pp. 50–52. These values are for the optimal pitch of 21.2 cm.

II. Highly enriched uranium production

Highly enriched uranium is made in uranium isotope separation facilities. These facilities enrich the fraction of ^{235}U relative to ^{238}U, and are therefore called enrichment plants. Currently, virtually the entire capacity of enrichment plants worldwide is dedicated to the production of LEU. Historically, the opposite has been true, but HEU production for weapons has ceased in all the nuclear weapon states. Nevertheless, almost all enrichment plants could be modified to make HEU, if such an output were desired.

Since the isotopes of an element have very similar chemical and physical properties, their separation is especially difficult. In an effort to reduce costs and best use available resources, countries have developed several different types of enrichment technology. Table A.4 lists the most common types. Figure A.1 shows the basic elements of a gaseous diffusion plant, figures A.2 and A.3 show the components of a gas centrifuge, and figure A.4 shows an EMIS separator.

For most enrichment processes, the basic measure of the amount of isotope separation achieved is the separative work unit (SWU). The technical definition of this concept is beyond the scope of this appendix, but several treatments are available.[2] For the purposes of this study, knowledge of the separative output of an enrichment plant allows simple estimates of the rate of LEU or HEU production from that facility. For example, if a facility has a capacity of 100 000 SWU/y, it can produce about 500 kg of 90 per cent enriched uranium, if natural uranium is used as feed into the plant and depleted uranium waste containing 0.3 per cent enriched ^{235}U (a 0.3% tails assay) is discarded. Similarly, 100 000 SWU will produce enough LEU to fuel a 1000 MWe power reactor for one year. Table A.5 lists several additional examples of enriched uranium production. In general, the more separative work that is used to produce a given quantity of enriched uranium, the less natural uranium required as feed.

Table A.6 shows the amount of weapon-grade uranium, taken here as uranium containing 93 per cent of the isotope ^{235}U, that can be produced given different assumptions about total SWU, the enrichment of uranium feed and the proportion of ^{235}U left in the tails. The separative capacity of large plants may be expressed in units of tonnes of separative work. One tonne of separative work equals 1000 SWU.

Enrichment plants are located throughout the world. Table A.7 lists the current activities on uranium isotope separation. Table A.8 lists the main commercial enrichment plants.

---

[2] For a definition and a wide selection of technical references, see Krass, A. S., *et al.*, SIPRI, *Uranium Enrichment and Nuclear Weapon Proliferation* (Taylor & Francis: London, 1983).

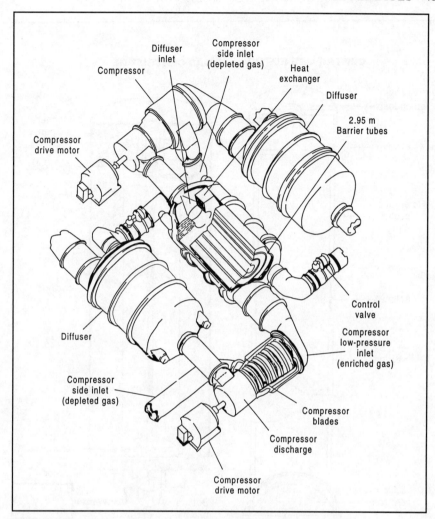

Figure A.1. The basic elements of a gaseous-diffusion plant

Source: Redrawn for Kokoski, R., SIPRI, *Technology and the Proliferation of Nuclear Weapons* (Oxford University Press: Oxford, 1995), p. 24, after Tait, J. H., 'Uranium enrichment', ed. W. Marshall, *Nuclear Power Technology. Vol. 2, Fuel Cycle* (Clarendon Press: Oxford, 1983), p. 120.

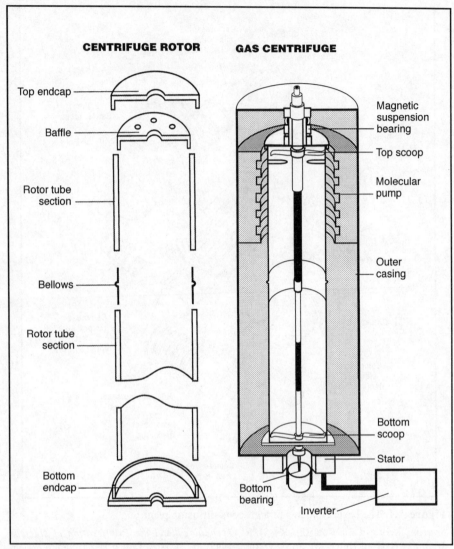

Figure A.2. A gas centrifuge and a centrifuge rotor.
Source: Albright, D. and Hibbs, M., 'Iraq's shop-till-you-drop nuclear program', *Bulletin of the Atomic Scientists*, vol. 48, no. 3 (Apr. 1992), pp. 32 and 33.

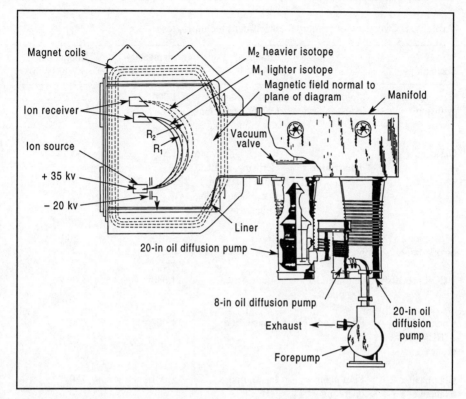

Figure A.3. EMIS configuration
Source: Redrawn for Kokoski, R., SIPRI, *Technology and the Proliferation of Nuclear Weapons* (Oxford University Press: Oxford, 1995), p. 24, with permission from Love, L. O., 'Electromagnetic separation of isotopes at Oak Ridge', *Science*, vol. 182, no. 4110 (26 Oct. 1973), p. 344. © 1973 American Association for the Advancement of Science.

Table A.4. Overview of principal enrichment technologies[a]

Technology	Status[b]	Countries involved[c]	Elementary separation factor[d]	Energy use (kWh/SWU)[e]
Gaseous diffusion	Major industrial use	USA, Russia, UK, France, China, Argentina, Iraq	1.004	2 500
Gas centrifuge	Major industrial use	Russia, UK, France, Germany, Holland, Japan, Pakistan, Brazil, Iraq, India, China, Iran	1.5	135
Separation nozzle	Large pilot plant built	Brazil	1.015	3 600
UCOR process	Commercial-size plant	South Africa	1.030	4 000
Chemical (CHEMEX process)	Pilot plant built	France, Iraq	1.002	600
Chemical exchange (ASAHI process)	Pilot plant operating	Japan, Iraq	1.001	150
Atomic Vapor Laser Isotope Separation (AVLIS)	Engineering development research only	USA, France, UK, Japan, Israel, Brazil, Iraq	High	c. 100–200
Molecular laser (MLIS)	Development	Japan, Germany, South Africa, France, Iraq	High	c. 235
Plasma separation (PSP)	Development	France	High	c. 330

[a] The primary source of this table is *Fiscal Year 1989 Arms Control Impact Statements,* Committees on Foreign Affairs and Foreign Relations of the House of Representatives and Senate, 100th Congress, 2nd session (US Government Printing Office: Washington, DC, Apr. 1988), p. 142. This table includes cancelled programmes.

[b] This column on the status of the technology refers only to the largest programmes.

[c] Russia and China have R&D programmes in various enrichment technologies that are not listed in this table. Iraq's programmes have ended.

[d] The elementary separation factor is defined as the ratio of the relative amount of ^{235}U in the product stream to the amount in the waste stream.

[e] The specific energy requirement, usually given in units of kilowatt-hour per separative work unit (kWh/SWU), is given for comparative purposes only. Actual values can vary.

Table A.5. Common examples of enriched uranium output

Amount of separative work (SWU)	Product Amount (kg)	Product Percentage enrichment	Feed Amount (kg)	Feed Percentage enrichment	Tails assay (%)
200	1	93	226	0.711	0.3
160	1	93	440	0.711	0.5
50	1	93	20	5.0	0.5
40	1	93	23	5.0	1.0
5 000	25	93	5050	0.711	0.3
3.9	1	3.25	7.2	0.711	0.3
7.2	1	5.0	11.5	0.711	0.3

Table A.6. Weapon-grade uranium production (93% enriched)

Capacity (SWU/y)	Uranium feed (% ^{235}U)	Uranium[a] tails (% ^{235}U)	WGU[b] (kg/y)	No. of days to produce 25 kg
5 000	0.71	0.2	21	435
	0.71	0.3	24	380
	0.71	0.5	31	295
5 000	5.0	1.0	122	75
	5.0	2.0	156	60
	5.0	4.0	208	45
15 000	0.71	0.2	63	145
	0.71	0.3	72	125
	5.0	1.0	366	25
70 000	0.71	0.2	294	31
	5.0	1.0	1 700	5

[a] As the tails assay is raised, the amount of uranium feed required rises dramatically. Assuming that the uranium feed is natural uranium (0.71% ^{235}U), the production of 25 kg of WGU requires 4600 kg of natural uranium at a 0.2% tails assay, and about 11 000 kg at a 0.5% tails assay.

[b] The following equations relating separative work units (D), feed (F), tails (T), product (P), feed assay (N_F), tails assay (N_T) and product enrichment (N_P) are from Krass, A. S., et al., SIPRI, *Uranium Enrichment and Nuclear Weapon Proliferation* (Taylor & Francis: London, 1983), pp. 97, 99 and 100.

$$F/P = (N_P - N_T)/(N_F - N_T)$$

$$T/P = (N_P - N_F)/(N_F - N_T)$$

$$D/P = V(N_P) + V(N_T)[T/P] - V(N_F)[F/P]$$

$$V(N) = (2N-1) \ln[N/(1-N)]$$

Table A.7. Current activities on uranium isotope separation and the level of such activities, by country[a]

The level of uranium isotope separation activities is given in parentheses.[b]

Enrichment	NWS	De facto NWS[c]	NNWS
Gaseous diffusion	China (4) France (5) USA (5)	–	Argentina (3)
Gas centrifuge	UK (5) Russia (5) China (2)	Pakistan (3) India (3) Israel?	Brazil (3) Germany (5) Netherlands (5) Japan (4–5) Iran?
Laser	France (3) USA (3–4) UK (2) China (1–2) Russia?	India (1–2) Israel (2)	Japan (2–3) Brazil (1–2) South Africa (2–3)
Ion/Chemical exchange	China (1–2)		Japan (3) North Korea? (1)

[a] This table includes publicly available information. Cancelled programmes and Iraq's pre-Persian Gulf War programme are excluded.

[b] Level of activity is defined as: (1) research, (2) R&D, (3) building or operating small pilot plant, (4) building or operating demonstration plant or medium-sized plant and (5) building or operating large-scale plant.

[c] The de facto nuclear weapon states are Israel, India and Pakistan (see chapter 9).

Source: Starr, D., 'Enrichment', Martin Marietta Energy Systems, viewgraphs presented at the Department of Energy Nuclear Nonproliferation Workshop, Washington, DC, Dec. 1993; and authors' estimates.

Table A.8. Commercial enrichment plants

Nation/Group	Process	Date of Completion	Nominal capacity (SWU/y (x 1000))
China			
Lanchou	Gaseous diffusion	1960s	200
Heping[a]	Gaseous diffusion	1970s	> 200
Eurodif			
Tricastin, France	Gaseous diffusion	1982	10 800
Japan			
Rokkashomura	Centrifuge	–2000	1 500
TBD[b]	Centrifuge, laser, or ion exchange	–2005	1 500
Urenco			
Alemelo, The Netherlands	Centrifuge	1980– Continuing	1 250
Capenhurst, UK	Centrifuge	1980– Continuing	750
Gronau Germany	Centrifuge	1985– Continuing	560
United States			
Paducah and Portsmouth	Gaseous diffusion	1950s; Updated 1970s	19 600
Russia			
Ekaterinburg, Tomsk, Krasnoyarsk, Angarsk (all in Siberia)	Centrifuge	1950s– Continuing	–20 000

[a] See chapter 4.
[b] Site to be determined.

Source: Starr, D., 'Enrichment', Martin Marietta Energy Systems, viewgraphs presented at the Department of Energy Nuclear Nonproliferation Workshop, Washington, DC, Dec. 1993.

Appendix B. Calculation of plutonium production in power reactors

The method used to derive estimates of plutonium production in power reactors in chapter 5 follows closely the method used by the Systems Studies Section in the Department of Safeguards of the International Atomic Energy Agency in studies published during the 1980s.[1] This method models the discharge of spent fuel from nuclear reactors, together with an estimate of their burnup. Knowledge of mean fuel burnup allows estimation of plutonium production in the fuel using standard conversion factors. These factors are shown as graphs in chapter 5—figures 5.3 and 5.4. In this appendix, only the method of calculation is explained. In the interests of brevity, the spreadsheets detailing electricity, fuel and plutonium production at each of the world's operating, decommissioned and planned reactors are not included. All the main results are discussed in chapter 5.

For each operating reactor, an estimate of heat output was calculated using published electricity generation data over six time periods: to the end of 1970; to the end of 1980; to the end of 1990; to the end of 1993; to the end of 2000; and to the end of 2010. Using published data or assumptions about reactor fuelling, mean fuel burnup was estimated by reactor for each of these time periods. Plutonium production was estimated by employing conversion functions which relate mean burnup in nuclear fuel to the generation of plutonium.

A simpler approximation for estimating plutonium in spent fuel was used by one of the authors in earlier work.[2] In his previous work Albright used a conversion factor which matched electrical output with plutonium production by reactor *type*. For example, for LWRs, 330 kg of plutonium are estimated to be produced per GWe (net)-year. In general, each reactor of a particular type was assumed to have equivalent characteristics. The values in table B.1 closely match those used in the earlier Albright study.

There are important differences in the approach used in this book. In particular, an attempt is made to use all available information on fuelling, achieved fuel burnups and actual thermal efficiencies of reactors. The effect is to lower estimates of plutonium discharges, and to improve confidence levels in estimates.

---

[1] For example, Bilyk, A., *Forecast of Amounts of Plutonium at Power Reactors subject to Safeguards (1981–1990)*, STR-125 (IAEA Department of Safeguards: Vienna, June 1982); and Mal'ko, M., *Estimation of Plutonium Production in Light Water Reactors*, STR-226 (IAEA Department of Safeguards: Vienna, Nov. 1986).

[2] Albright, D., *World Inventories of Plutonium*, PU/CEES Report no. 195 (Center for Energy and Environmental Studies: Princeton University, Princeton, N.J., June 1987).

Table B.1. Plutonium discharge rate by reactor type
Discharge is measured at 100% capacity factor.

Reactor type	Pu discharged[a] kg Pu per GWe (net)	Pu discharged[a] kg Pu per GWe (gross)
PWR	325	315
BWR	345	330
CANDU	630	570
GCR	815	735

[a] We assume mean burnups for PWR, 30.4 GWd/t; BWR, 28 GWd/t; CANDU, 7.5 GWd/t; and GCR, 4 GWd/t. Gross and net thermal efficiencies are typical values for these types of reactor (PWR [net: 0.320; gross: 0.332]; BWR [net: 0.312; gross: 0.328]; CANDU [net: 0.29; gross: 0.32]; GCR [net: 0.28; gross: 0.30]). The quantity of fuel discharged by each reactor type could then be calculated as follows:

Spent fuel = (365 * 1/n)/ burnup,

where 365 is the electricity production in GWd from a 1-GW reactor operating at 100% capacity for one year, n is the thermal efficiency, and burnup is expressed in GWd/t.

Finally, this fuel quantity was multiplied by the appropriate plutonium concentration by weight of fuel shown in figures 5.3 and 5.4.

There are a number of advantages to the more detailed approach used in chapter 5. First, for all reactors, except those in the former Soviet Union, weight and average burnup characteristics of fuel discharged by individual reactors are included in the models. This method does not assume that all reactors of a certain type always operate in the same way. Mean fuel burnups do fluctuate significantly from reactor to reactor, and this will have an effect on plutonium production, from reactor to reactor and within the lifetime of any single reactor. For instance, we estimate that mean burnups at British Magnox reactors in the decade 1981–90 varied between 3710 MWd/t (Hunterston 'A', total plutonium produced, 2.3 kg/t fuel) to 5080 MWd/t (Hinkley Point 'A', total plutonium produced, 2.86 kg/t fuel). A difference between the fuel burnup of 27% represents a 20% difference in plutonium production per kilogram of fuel.[3] An attempt to encompass the non-linear relation between burnup and plutonium production is central to our analysis.

Second, our approach takes account of the lower inventories of plutonium found in fuel discharged during the first few years of a reactor's operating life. For instance, at a PWR with a one-third core refuelling scheme, the first discharge of fuel after one year of operation will have a mean burnup of one-third of equilibrium burnup. In the second year, mean burnup will be two-thirds of equilibrium, and only at the third discharge will the mean burnup reach around 33 000 MWd/t. The effect on plutonium production is significant in the first decade in which a reactor starts up or is decommissioned. Again, the main effect is to bring plutonium estimates down.[4]

[3] The earlier method of plutonium calculation for these two reactors gives totals of 3.27 t Pu_{tot} for Hinkley Point 'A' (4.45 GWy(e) (gross) generated in 1981–90) and 2.06 t Pu_{tot} for Hunterston 'A' (2.8 GWy(e) (gross)). By comparison, the current method gives plutonium totals of 3.23 t Pu_{tot} for Hinkley Point 'A' and 2.09 t Pu_{tot} for Hunterston 'A'. Using values from table B.1, therefore, gives results which in this case agree closely with those in the present method.

[4] In the case of a PWR, the earlier method will overestimate cumulative plutonium discharges for the first decade by about 7–10%, assuming similar burnups to those in table B.1. For the first 5 years of

Third, our approach allows a more sophisticated perspective on how different fuel cycles work, and in particular generates spent fuel figures. These are interesting in themselves, but have the advantage that they can be corroborated with spent fuel figures published in the UK, the USA, Germany, Japan and France. This provides another means of corroborating plutonium estimates.

Fourth, our approach allows the generation of forecasts for plutonium production which can take account of trends in fuel management—higher fuel burnups, for instance. This will have the overall effect of bringing down fuel and plutonium discharges from reactors for a given electrical output.

This approach still contains many averaging assumptions. The initial enrichment of fuel and the neutron flux within a reactor core are not uniform. Fuel burnup within the core, and within each individual fuel rod, therefore varies from place to place. Within any batch of fuel the burnup of different fuel pellets may differ by as much as 20%, and will depend on the way the reactor has been operated and fuel within it has been moved around. With the data currently available in the public domain it would be impossible to embark on this type of analysis, and indeed the benefits of so doing are likely to be marginal.

Outline of method

A variety of approaches were used to calculate plutonium inventories, depending on the information available. In general, an assumption has been made that reactor operators have tried to achieve design burnups in order to reduce the costs of fresh fuel and the inconvenience of managing spent fuel. For further information on the different strategies used to derive plutonium figures, see chapter 5, section IV.

For non-nuclear weapon states, we have used as a benchmark design equilibrium burnup assumptions for each reactor published by the IAEA[5] to make estimates of the amount of fuel discharged using the cumulative gross electricity outputs published by *Nucleonics Week*. For those countries where utilities have increased fuel burnups beyond design levels, a set of standard assumptions have been made.[6] For reactors not under IAEA safeguards, such as those in the United Kingdom, France, Russia, the United States and India, we have estimated mean burnups using electricity outputs and assumptions about fuel discharges derived from published sources, or from fuel core characteristics published in the *World Nuclear Industry Handbook 1995*.[7] In many cases it has been possible to normalize these fuel discharge estimates by referring to the SPRU Spent Fuel and Plutonium Survey 1994 and to published totals.

operation, the previous method overestimates plutonium discharges by 15–25%. The reason is that burnups during the initial years are significantly less, lowering plutonium values sharply. During the second and third decades of operation, the two methods give similar results. For 20 to 30 years of cumulative discharges, the estimates differ by only a few per cent, again assuming similar burnups.

[5] Mal'ko, M., *Plutonium Production in Power Reactors subject to IAEA Safeguards (1985–1994)*, STR-228 (IAEA Department of Safeguards: Vienna, Dec. 1986), Annex: 'Input data for plutonium production forecasting'.

[6] Burnup increases are included in our analysis only for reactors which operate throughout a given decade (i.e., 2000–10). Germany, Japan, Sweden, Switzerland and Belgium are assumed to raise mean PWR fuel burnups to 40 GWd/t in 1991–2000 and to 45 GWd/t in 2000–10. Mean BWR burnups are assumed to rise from a mean of 30 GWd/t in 1991–2000 to 35 GWd/t in 2001–10. Mean LWR burnups in South Korea, Taiwan and Spain are taken to come into line after 2000.

[7] Nuclear Engineering International, *World Nuclear Industry Handbook 1995* (Reed International: London, 1994).

Electricity output figures began being published for the former Soviet republics (Russia, Ukraine, Lithuania) in 1990. For the period before 1990, we have made capacity factor assumptions to calculate electricity production and derived plutonium arisings directly using the factors in table B.1.

The basic modelling steps for each of these three categories are as follows.

Power reactors in NNWS safeguarded by the IAEA and in France, Russia (since 1990) and the United States

Step 1: Calculate the amount of fuel discharged by the reactor in a given time period. The total weight of irradiated fuel discharged is calculated using the function:

$$F_t = E_t / (24B [n/100]) - C \qquad (1)$$

where F_t is the total amount of fuel (tonnes) discharged in period t, E_t is gross electrical output of the reactor in time period t (GWh), B is mean equilibrium design fuel burnup (GWd/t), n is the gross thermal efficiency of the reactor (%), and C is a half-core of fuel (t) added in the first and last time period in which a reactor operates (that is, if a reactor with a 30-year life began operating in 1978, C would be added in the time periods 'to 1980' and '2001–10').

The factor C is introduced to take account of the heat produced by partially irradiated fuel still held in the core (in the first time period) and by partially irradiated fuel discharged after the reactor has been decommissioned. Without this factor, estimates of fuel discharges would be over-estimates.

Example:

The Atucha 1 heavy water reactor in Argentina has a core of 38 t, a nominal design fuel burnup of 5.6 GWd/t and a gross thermal efficiency of 30.3%. By the end of 1980 it had generated 15 689 GWh of electricity. We estimate that the amount of fuel discharged by Atucha 1 up to the end of 1980 is:

$$(15\,689 / [24 \times 5.6 \times 0.303]) - 19 = 366\,t$$

For nuclear weapon states such as France and the United States, whose reactors are not included in previous IAEA analyses, mean fuel burnups were estimated *by reactor type*, based on published figures for aggregate spent fuel discharges.

For example, US PWR discharges were estimated for each of the six time cohorts using a simple assumption of a one-third core reload per year. These estimates were compared with US Department of Energy, Energy Information Administration published data on fuel discharges by US reactors.[8] The fuel curve was adjusted iteratively by changing the fuel burnup in equation 1 associated with each reactor until a 'best fit' was achieved. In the process, new figures were derived for the amount of fuel discharged by each reactor.

---

[8] Energy Information Administration, *Spent Nuclear Fuel Discharges from US Reactors 1993*, SR/CNEAF/95-01, Washington, DC, Feb. 1995.

Although this method does not provide much confidence in the accuracy of fuel and plutonium figures for individual reactors, it does allow the normalization of aggregate estimates of plutonium discharges data with available fuel discharge information.[9]

For United Kingdom reactors

Step 1: Calculate the mean burnup of discharged spent fuel.

For the UK there exists a uniquely good base of information on reactor fuel discharges.[10] The mean fuel burnup is estimated using the function:

$$B = E_t / (24n [F_t + C]).$$

Example:

The Oldbury Magnox station with a thermal efficiency of 28.8% generated 33 170 GWh of electricity in the period 1981 to 1990. During this period it discharged about 960 tonnes of fuel. The mean burnup of this fuel was therefore:

$$33\ 170 / (24 \times 0.288 \times 960) = 4.99\ \text{GWd/t of fuel}.$$

For all power reactors (including Russia, Ukraine and Lithuania since 1990)

Step 2: Calculate the mean plutonium content in the fuel.

Mean plutonium content in the fuel is derived using published or estimated mean fuel burnup using functions derived from the work of Bilyk and Mal'ko at the IAEA.[11] In deriving figures for plutonium production in natural uranium fuel at low burnups (less than 2000 MWd/t) we have used data produced by Turner.[12]

Apart from burnup, the main determinant of the plutonium in irradiated fuel is the initial uranium-235 enrichment of the fuel. Five categories of enrichment are used: natural uranium (0.7% ^{235}U); advanced gas-cooled reactors (1.5%); boiling water reactors (2.7–2.8%); pressurized water reactors (3.2–3.3%); and VVERs (3.6%). The functions used to calculate plutonium inventories in these fuel types are as follows:

---

[9] The effect on plutonium estimates when compared with the simpler approximation in table B.1 varies from reactor to reactor. For example, in 1981–90 the Fort Calhoun 1 PWR generated 3.45 GWy(e), while the Peach Bottom 3 BWR generated 4.63 GWy(e). According to table B.1 these reactors would have discharged 1.09 t Pu$_{tot}$ and 1.53 t Pu$_{tot}$ respectively, compared with 1.0 t Pu$_{tot}$ and 1.6 t Pu$_{tot}$ in the authors' estimates.

[10] This figure is derived from Barnham, K. W. J. *et al.*, 'The production and destination of British civil plutonium', *Nature*, vol. 317 (19 Sep. 1985), pp. 213–17, and from official declarations in *Parliamentary Debates, House of Commons Official Report [Hansard]*, 1 Apr. 1982, 27 July 1983, 25 Jan. 1985, 23 July 1985 and 21 July 1986, and in British Department of Energy Press Releases, 16 Dec. 1987, 13 Oct. 1988, 5 Dec. 1989, 18 Oct. 1990, 17 Oct. 1991, 4 Feb. 1993, 1 Mar. 1994, 19 July 1994, 13 July 1995 and 18 July 1996.

[11] Bilyk (note 1) and Mal'ko (note 1).

[12] Turner, S. E., *et al.*, *Criticality Studies of Graphite-Moderated Production Reactors*, SSA-125, Southern Sciences Applications, prepared for the US Arms Control and Disarmament Agency, Jan. 1980.

Natural uranium fuel (Magnox and CANDU):

$$Pu_{tot} = 0.9235 \, B^{0.6946},$$

where Pu_{tot} is the mean plutonium content (kg/t of fuel).

This overestimates plutonium production at low burnups. The effect of using the Turner figures is to bring down plutonium estimates below 2000 MWd/t by up to 30%.

This function is shown graphically in figure 5.3.

Enriched uranium fuel

BWR:

$$Pu_{tot} = (1.138 \times 10^{-4}) \, B^3 - 0.011 \, B^2 + 0.508 \, B + 0.144$$

PWR:

$$Pu_{tot} = (0.978 \times 10^{-4}) \, B^3 - 0.011 \, B^2 + 0.523 \, B + 0.193$$

These two curves are shown in figure 5.4.

VVER:

$$Pu_{tot} = (0.551 \times 10^{-4}) \, B^3 - 0.008 \, B^2 + 0.480 \, B + 0.193$$

The curve for plutonium generation in AGR fuel shown in figure 5.4 was derived primarily from plutonium and fuel discharge data published by the British Department of Energy annual plutonium figures.[13]

For all power reactors (including Russia, Ukraine and Lithuania since 1990)

Step 3: Calculate amount of plutonium in discharged spent fuel.

The total amount of plutonium discharged from each reactor is estimated by multiplying mean plutonium content (Pu_{tot}) by the amount of fuel discharged (F_t).

Example:

The Biblis B PWR generated 76 846 GWh (gross) of electricity between 1981 and 1990. We estimate that it discharged 283 t of fuel at a mean discharge of 32.5 GWd/t. Mean plutonium content at this burnup is 8.93 kg/t of fuel. The reactor therefore discharged a total of 2.53 t Pu_{tot}.[14]

---

[13] Department of Energy (note 9).
[14] By comparison, using the earlier factors yields a result of 2.76 t Pu_{tot}, an overestimate of about 8%.

For power reactors in Russia, Ukraine and Lithuania up to 1990

Steps 1 to 3: Estimate electricity production at each reactor and estimate plutonium production.

For reactors in Russia, Lithuania, the Ukraine and Armenia we have been forced to use a simpler estimation approach. Plutonium production for these reactors is based on standard electricity–plutonium conversion factors for VVER and RBMK reactors. For VVERs we assume 330 kg Pu per GWe (net)-year, for RBMKs we assume 300 kg plutonium per GWe (net)-year. We have estimated electricity generation by using standard capacity factors. We assume capacity factors of 75% for VVER 440s, and 65% for RBMKs and VVER 1000s.

Example:

The Beloyarsk 1 RBMK began operating in 1964 with a design rating of 102 MWe (net). Using the above assumptions we can estimate that by the end of 1980 the reactor had discharged,

$$(1980-1964) \times (0.102 \times 0.65) \times 300 = 320 \text{ kg of plutonium.}$$

Separate estimates of fuel discharge weights were made using nominal values for annual discharges published in the design reload values given in the *World Nuclear Industry Handbook* (note 6).

Appendix C. Separation of plutonium from power reactor fuel at reprocessing plants

As with estimates of plutonium production in reactors, estimates of the amount of plutonium separated are composed of data derived from a wide variety of sources. In most cases these sources are given in footnotes in the main text. In this appendix historical data compiled by the authors are reproduced for six reprocessing plants: B205, UP1, UP2, UP3, WAK and Tokai-mura.

The main parameters in estimating plutonium separation are the throughput of fuel at a reprocessing plant, the fuel's burnup at discharge, the length of time that the spent fuel was stored prior to reprocessing and the efficiency of the separation process. For many of the plants reasonably good fuel throughput figures are published or can be estimated from what is known about reactor fuel discharges. Burnup information is more scarce, being available only for plants in France, Germany and the United States. In other cases we have made assumptions about burnups using a variety of sources, some published (as with the Tokai plant for which mean burnups over periods of years have been published) and some based on estimates made in our own calculations of plutonium production in power reactors. Details about the amount of plutonium lost in waste streams at reprocessing plants (process losses) are much harder to come by. We have assumed fixed values of 5% for older plants processing metallic fuel (B205 and UP1) and 1% for all other plants.

Table C.1. Plutonium separation from British Magnox power reactor fuel at B205[a]

Year	Fuel throughput (t)	Burnup (MWd/t)	Plutonium separated (kg)	Cumulative Pu separated (kg)
1965	173	913	141	141
1966	370	1 436	417	558
1967	483	1 500	561	1 119
1968	688	1 909	944	2 063
1969	938	2 000	1 331	3 394
1970	1 150	2 391	1 847	5 241
1971	1 125	2 500	1 865	7 106
1972	425	2 500	705	7 811
1973	500	2 950	930	8 741
1974	713	3 000	1 341	10 082
1975	1 013	3 407	2 117	12 199
1976	725	3 500	1 550	13 749
1977	900	3 917	2 081	15 830
1978	925	4 000	2 170	18 000
1979	825	4 145	1 984	19 984
1980	866	4 200	2 102	22 086

(Table C.1 continued)

Year	Fuel throughput (t)	Burnup (MWd/t)	Plutonium separated (kg)	Cumulative Pu separated (kg)
1981	953	4 353	2 372	24 458
1982	1 012	4 324	2 506	26 964
1983	1 036	4 225	2 525	29 489
1984	846	3 438	1 782	31 272
1985	833	3 942	1 947	33 219
1986	858	3 598	1 889	35 107
1987	794	4 275	1 968	37 076
1988	783	4 751	2 092	39 168
1989	938	4 879	2 552	41 720
1990	763	4 968	2 102	43 822
1991	651	4 559	1 688	45 510
1992	667	4 400	1 690	47 200
1993	1 032	4 233	2 544	49 744
1994	978	3 988	2 314	52 058
1995	905	3 975	2 136	54 195

[a] The assumptions used in making these estimates are described in chapter 6, section V. The burnup figures used here are taken from mean burnup calculations for British reactors. See appendix B for the methodology. We assume process losses of 5%.

Table C.2. Plutonium separated from foreign magnox fuel at B205[a]

Year	Fuel throughput (t)	Burnup (MWd/t)	Plutonium separated (kg)	Cumulative Pu separated (kg)
1965	0	0	0	0
1966	45	1 000	54	54
1967	45	1 200	84	138
1968	85	1 500	115	253
1969	85	2 000	136	389
1970	85	2 500	155	544
1971	85	3 000	160	704
1972	85	3 000	160	864
1973	85	3 000	160	1 024
1974	85	3 000	160	1 184
1975	85	3 000	162	1 347
1976	85	3 000	163	1 510
1977	85	3 000	163	1 673
1978	85	3 000	177	1 850
1979	85	3 500	182	2 032
1980	85	3 500	182	2 214
1981	85	3 500	182	2 396
1982	85	3 500	182	2 577
1983	85	3 500	182	2 759
1984	85	3 500	206	2 965
1985	100	3 500	216	3 181

Year	Fuel throughput (t)	Burnup (MWd/t)	Plutonium separated (kg)	Cumulative Pu separated (kg)
1986	100	3 500	250	3 431
1987	121	3 500	117	3 548
1988	32	3 500	145	3 693
1989	88	3 000	248	3 941
1990	141	3 000	143	4 084
1991	51	3 000	288	4 372
1992	181	3 000	162	4 534
1993	51	3 000	105	4 639
1994	55	3 000	177	4 816
1995	117	2 500	139	4 954

[a] Estimates of throughputs of Japanese and Italian fuel at B205 before 1987 are based on fuel discharge estimates made in chapter 5. From 1987, the Department of Energy has published fuel and plutonium data (Department of Energy Press Releases, 16 Dec. 1987, 13 Oct. 1988, 5 Dec. 1989, 18 Oct. 1990, 17 Oct. 1991, 4 Feb. 1993, 1 Mar. 1994, 19 July 1994 and 13 July 1995). Burnup assumptions are also based on estimates made in chapter 5.

Table C.3. Plutonium separated from Magnox power reactor fuel at UP1 (France)

Year	Fuel throughput (t)	Burnup (MWd/t)	Plutonium separated (kg)	Cumulative Pu separated (kg)
1965	40	500	20	20
1966	49	500	20	40
1967–70	0	..	..	..
1971–72	3	2 000	5	45
1973	0	..	..	..
1974	111	2 000	160	205
1975	8	2 000	10	215
1976	21	2 000	30	245
1977	120	2 000	170	415
1978	245	2 000	350	765
1979	280	2 000	400	1 165
1980	267	2 000	380	1 545
1981	208	2 500	345	1 890
1982	300	2 500	495	2 385
1983	125	3 000	235	2 620
1984	315	3 000	595	3 215
1985	436	3 500	915	4 130
1986	387	3 500	810	4 940
1987	350	3 500	735	5 675
1988	327	3 500	685	6 360
1989	330	3 500	690	7 126
1990	350	3 500	805	7 931
1991	350	3 500	805	8 736
1992	390	3 500	780	9 516

(Table C.3 continued)

Year	Fuel throughput (t)	Burnup (MWd/t)	Plutonium separated (kg)	Cumulative Pu separated (kg)
1993	390	3 500	780	10 296
1994	400	3 500	800	11 096
1995	404	3 500	606	11 702

Sources: Syndicat CFDT de l'Energie atomique, *Le dossier électronucléaire* (Éditions du Seuil: Paris, 1981), pp. 186–91; Hirsch, H. and Schneider, M., *Wiederaufarbeitung in Europa: Wackersdorf ist tot—es lebe La Hague?*, Rest-Risiko, nr. 6, Greenpeace, Hamburg, Apr. 1990; Commissariat à l'Energie Atomique (CEA), 'Le retraitement des combustibles irradiés', *Industrie Nucléaire Française*, Paris, 1982, pp. 154–64; Couture, J., 'Status of the French reprocessing industry', Paper presented at the American Nuclear Society Conference, *Fuel Processing and Waste Management*, 26–29 Aug. 1984, Jackson, Wyo.; Delange, M., 'Operating Experience with Reprocessing Plants', *Atomwirtschaft*, Jan. 1985, pp. 24–28; Delange, M., 'LWR spent fuel reprocessing at La Hague: Ten years on', *RECOD 87* conference, Paris, 1987, pp. 187–93; 'Reprocessing and waste management, country: France, pt 1', *NUKEM Market Report*, no 3 (1988), pp. 15–18; Lewiner, C. and Gloaguen, A., 'The French reprocessing programme', *Atomwirtschaft*, May 1988, pp. 227–29; CEA, *Cycle du combustible nucléaire: retraitement*, Paris, Mar. 1989; 'Reprocessing and waste management: review 1989', *NUKEM Market Report*, no. 2 (1990), pp. 14–23; EdF, 'Retraitement recyclage', Paper by Service des Combustibles, Paris, 6 Mar. 1990; and Cogema, 'Reprocessing–recycling: The industrial stakes', Paper presented at Konrad-Adenauer Stiftung, Bonn, 9 May 1995.

Table C.4. Plutonium separated from Magnox power reactor fuel at UP2 (France)[a]

Year	Fuel throughput (t)	Burnup (MWd/t)	Plutonium separated (kg)	Cumulative Pu separated (kg)
1966	52	500	25	25
1967	98	800	65	90
1968	189	1166	180	270
1969	157	986	130	400
1970	245	1 079	215	615
1971	126	2 287	205	820
1972	250	2 164	385	1 205
1973	213	2 385	355	1 560
1974	635	2 331	1 035	2 595
1975	443	3 038	870	3 465
1976	218	2 783	400	3 865
1977	355	2 947	680	4 545
1978	371	3 345	775	5 320
1979	240	3 590	530	5 850
1980	252	3 317	525	6 375
1981	250	3 672	560	6 935
1982	226	3 720	510	7 445
1983	117	3 272	240	7 685
1984	185	3 865	430	8 115
1985	120	3 900	280	8 395

Year	Fuel throughput (t)	Burnup (MWd/t)	Plutonium separated (kg)	Cumulative Pu separated (kg)
1986	76	3 900	175	8 570
1987	77	3 900	180	8 750

[a] Magnox fuel was reprocessed at UP2 from 1966 to 1987. For sources, see table C.3.

Table C.5. Plutonium separated from oxide fuel at UP2 (France)[a]

Year	Fuel throughput (t)	Burnup (MWd/t)	Plutonium separated (kg)	Cumulative Pu separated (kg)
1976	15	15 800	90	90
1977	18	28 000	150	240
1978	38	27 300	310	550
1979	79	20 400	560	1 110
1980	105	21 000	750	1 860
1981	101	25 400	800	2 660
1982	154	21 100	1 105	3 765
1983	221	23 200	1 670	5 435
1984	255	23 200	1 925	7 360
1985	343	24 000	2 635	9 995
1986	333	21 000	2 385	12 380
1987	425	23 500	3 250	15 630
1988	355	21 000	2 540	18 170
1989	430	21 000	3 080	21 250
1990	331	22 000	2 163	28 983
1991	311	23 000	2 032	31 015
1992	220	24 000	1 459	32 474
1993	354	26 000	2 348	34 822
1994	575	29 000	3 928	38 750
1995	814	29 000	5 560	44 311

[a] Throughput and fuel figures for 1989, 1990 and 1991 are estimated.

Table C.6. Plutonium separated from oxide fuel at UP3 (France)

Year	Fuel throughput (t)	Burnup (MWd/t)	Plutonium separated (kg)	Cumulative Pu separated (kg)
1989	30	25 000	220	220
1990	195	27 000	1 350	1 570
1991	351	29 000	2 430	4 000
1992	448	31 000	3 100	7 100
1993	600	33 000	4 400	11 500
1994	702	33 000	5 140	16 640
1995	839	33 000	6 400	23 040

Table C.7. Plutonium separated from oxide fuel at WAK (Germany)[a]

Year	Fuel throughput (t)	Plutonium separated (kg)	Cumulative Pu separated (kg)
1971	3.5	14	14
1972	13.6	53	67
1973	7.3	21	88
1974	5.9	1	89
1975	11.8	53	142
1976	13.7	105	247
1977	15.8	127	374
1978	13.7	55	429
1979	19.2	81	510
1980	9.2	37	547
1981	..	..	547
1982	6.6	27	574
1983	16.8	110	684
1984	11.8	51	735
1985	11.7	105	840
1986	14.1	115	955
1987	14.3	103	1 058
1988	5.4	54	1 112
1989	3.0	25	1 137
1990	9.5	43	1 180

[a] These figures are actual outturns. Personal communication from Dr Lausch and Dr Zabel, WAK, Leopoldshaven, Karlsruhe, 20 Jan. 1992.

Table C.8. Plutonium separated from oxide fuel at Tokai-mura (Japan)

Year	Fuel throughput (t)	Burnup (MWd/t)	Plutonium separated (kg)	Cumulative Pu separated (kg)
1976	0	0	0	0
1977	8	10 000	35	35
1978	11	15 000	65	100
1979	12	17 000	77	177
1980	55	17 000	351	528
1981	53	18 000	351	879
1982	33	18 000	218	1 097
1983	2	19 000	14	1 110
1984	5	19 000	34	1 145
1985	74	21 000	535	1 680
1986	69	21 000	499	2 179
1987	51	23 000	387	2 566
1988	19	23 000	144	2 710
1989	45	23 000	342	3 052
1990	86	25 000	681	3 733
1991	82	25 000	650	4 383
1992	71	25 000	562	4 945

Year	Fuel throughput (t)	Burnup (MWd/t)	Plutonium separated (kg)	Cumulative Pu separated (kg)
1993	37	27 000	304	5 249
1994	71	27 000	584	5 833
1995	89	27 000	732	6 564

Sources: 'Tokai marks reprocessing of 500 tonnes of fuel', *PNC Review*, no. 17 (spring 1991), p. 6; *PNC Annual Report 1993*, Tokyo, 1993; and *PNC Annual Report 1994*, Tokyo, 1994.

Table C.9. World annual separation of civil plutonium, 1965–2000

Year	Total plutonium (kg)
1965	171
1966	481
1967	701
1968	1 237
1969	1 595
1970	2 338
1971	2 363
1972	1 427
1973	1 464
1974	2 695
1975	3 212
1976	3 069
1977	4 836
1978	5 333
1979	5 236
1980	5 858
1981	6 135
1982	6 575
1983	6 505
1984	6 649
1985	8 260
1986	7 755
1987	8 448
1988	7 392
1989	8 970
1990	9 220
1991	9 459
1992	9 013
1993	11 657
1994	14 483
1995	17 158
1996	18 342
1997	19 733
1998	20 412
1999	21 146
2000	21 831

Appendix D. Research reactors (>1 MWth) using HEU fuel

These tables list the civil research and test reactors with a power rating greater than 1 MWth that use HEU fuel. The listings for the former Soviet Union and China are derived from information provided by these countries to the IAEA and member states. Where information is available, they indicate the current position regarding the conversion of research reactors to use low-enriched fuels. Reactors with a power of less than 1 MWth are not included because they typically require only one fuel loading during their lifetime.

Table D.1. US operating research and test reactors with power >1 MWth using HEU fuel (as of mid-1995)[a]

Reactor[b]	Power (MWth)	Enrichment (%)	Fresh ^{235}U/y (kg)	Conversion
DOE reactors				
ATR	250	93	~175	Not planned
HFIR	100	93	~150	Not planned
HFBR	60	93	~64	Not planned
BMRR	3	90	~0.2	Not planned
Total	**415**		**~390**	
NRC-licensed reactors				
NBSR	20	93	8.7	Not planned
MURR	10	93	19	Not planned
MITR	5	93	6.7	Not planned
GTRR	5	93	2	In process
RINSC	2	93	..	Yes
UVAR	2	93	..	Yes
ULR	1	93	1	In process
4 TRIGA reactors	4	70	0.6	In process
Total	**49**		**38**	

[a] IAEA, *Nuclear Research Reactors in the World* (IAEA: Vienna, Dec. 1995); and Matos, J. E., *Estimated Uranium Densities with Reduced Enrichment for DOE Research and Test Reactors*, Argonne National Laboratory, Argonne, Ill., 21 Oct. 1986.

[b] Fast reactors are excluded, as are pulse-type reactors that use HEU fuel, in particular the EBR-2 which uses 67% enriched uranium fuel.

Table D.2. US-supplied operating research and test reactors with power >1 MWth using HEU (>90%) fuel (as of mid-1995)[a]

Reactor	Power (MWth)	Enrichment (%)	Fresh ^{235}U/y (kg)	Conversion[b]
RA-3 (Argentina)	2.8	20	..	Yes
HIFAR (Australia)	10	75	7.5	
ASTRA (Austria)	8	20	..	Yes
BR-2 (Belgium)	80	93	27	Not agreed
IEA-R1 (Brazil)	2	20–93	1	
NRU (Canada)	125	20	..	Yes
MNR (Canada)	5	93	1.9	
LO AGUIRRE (Chile)	10[c]	90	0	
LA REINA (Chile)	5	45–80	1	
DR-3 (Denmark)	10	20	..	Yes
OSIRIS (France)	70	7	..	Yes
RHF (France)	57	93	51	Not agreed
SILOE (France)	35	93	23	
SCARABEE (France)	20	?	0	
ORPHEE (France)	14	93	14.7	Not agreed
FRJ-2 (Germany)	23	80	18	
FRG-2 (Germany)	15	93	11	
BER-2 (Germany)	10	92	4.8	
FRG-1 (Germany)	5	20	..	Yes
FMRB (Germany)	1	93	1	
GRR-1 (Greece)	5	93	2.7	
NRCRR (Iran)	5	20	..	Yes
IRR-1 (Israel)	5	93[d]	0	?
JMTR (Japan)	50	20	..	Yes
JRR-2 (Japan)	10	45	10	Shut in 1995?
KUR (Japan)	5	93	2.1	
JRR-4 (Japan)	3.5	93	0.9	
TRIGA (Mexico)	1	70	0.7	
HFR Petten (Netherlands)	45	90	36	Not agreed, but can convert
HOR (Netherlands)	2	93	1.7	
PARR (Pakistan)	5	20	..	Yes
RPI (Portugal)	1	93	0.9	
PRR (Philippines)	1	20	..	Yes
SSR (Romania)	14	20–70	11	
SAFARI (S. Africa)	20	93	15	Not agreed, but can convert
TRIGA (S. Korea)	2	20–70	1	
R-2 (Sweden)	50	20	..	Yes
THOR (Taiwan)	1	20	..	Yes
TR-2 (Turkey)	5	93	1.5	
HERALD (UK)	5	70	3	
Total	**410**[e]		**248**[e]	

[a] Travelli, A., 'The RERTR Program: Status and progress', Paper presented at the 1995 International Meeting on the Reduced Enrichment for Research and Test Reactors', 18–21 Sep. 1995; Letter from Armando Travelli, RERTR Program Manager, to Dr Leonard Weiss, Staff

Director, US Senate Committee on Governmental Affairs, 19 Apr. 1988; Travelli, A., 'The RERTR Program: a status report', Paper presented at the 1991 International Meeting on Reduced Enrichment for Research and Test Reactors, 4–7 Nov. 1991, Jakarta, Indonesia; and IAEA, *Nuclear Research Reactors in the World* (IAEA: Vienna, Dec. 1995).

[b] The absence of a comment in this column means that the reactor is in the process of conversion.

[c] Operates sporadically.

[d] Out of HEU fuel, might not be operating.

[e] Total includes only reactors that have not been converted.

Table D.3. Russian operating research and test reactors with power >1 MWth using HEU fuel (as of mid-1995)[a]

Reactor[b]	Power (MWth)	Enrichment (%)	Fresh ^{235}U/y (kg)
IRT-A MEPI	2.5	90	1.0
IRT-8	8	90	2.0
MR[c]	40	90	35
IRV[d]	2	90	< 0.5?
WWR-M	18	90	13
IRT-T	6	90	5.0
IVV-2M	15	90	8.0
MIR-M1	100	90	56
SM-3	100	90	100
WWR-TS	15	36	8.0
RBT-6	6	63	0[e]
RBT-10/1	10	63	0[e]
RBT-10/2	10	63	0[e]
Total	**332.5**		**228**

[a] IAEA, *Nuclear Research Reactors in the World* (IAEA: Vienna, Dec. 1995).

[b] Fast reactors are excluded, as are pulse-type reactors that use HEU fuel. The latter reactors typically require only one fuel loading during their lifetime.

[c] Not operating as of 1995, but may restart.

[d] Reactor is being rebuilt.

[e] These reactors use HEU recycled from the SM-3 reactor.

Table D.4. Russian-supplied operating research and test reactors with power >1 MWth using HEU fuel (as of mid-1995)[a]

Reactor[b]	Power (MWth)	Enrichment (%)	Fresh ^{235}U/y (kg)
Former Soviet republics			
IRT-7 (Latvia)[c]	5	90	1.5
WWR-CM (Uzbekistan)	10	90	7.0
WWR-K (Kazakhstan)[d]	10	36	4.0
WWR-M (Ukraine)	10	36	5.0
Subtotal	**35**		**17.5**
Eastern Europe			
LWR-15 (Czech Republic)	15	80	5.0
R-A (Yugoslavia)[e]	6.5	80	3.0

Reactor[b]	Power (MWth)	Enrichment (%)	Fresh ^{235}U/y (kg)
MARIA (Poland)	30	80–36	27
IRT-2000 (Bulgaria)	2	36	< 1.0?
WWR-SZM (Hungary)	10	36	4.0
WWR-S (Romania)	2	36	< 1.0?
Subtotal	**65.5**		**41**
Other states			
IRT-DPRK (North Korea)[f]	8	36	?
IVV-7 (Libya)[f]	10	80	?
DRR (Vietnam)	0.5	36	0.2
Subtotal	**18.5**		**0.2?**
Total	**119**		**59**

[a] IAEA, *Nuclear Research Reactors in the World* (IAEA: Vienna, Dec. 1995)

[b] Fast reactors are excluded, as are pulse-type reactors that use HEU fuel. The latter reactors typically require only one fuel loading during their lifetime.

[c] Scheduled for shut-down in 1997.

[d] Not operating because of concerns of seismic activity.

[e] Reactor being renovated, future uncertain.

[f] The exact operating status of these reactors is not known.

Table D.5. Chinese and Chinese-supplied operating research and test reactors using HEU fuel (as of mid-1995)[a]

Reactor[b]	Power (MWth)	Enrichment (%)	Fresh ^{235}U/y (kg)
Chinese reactors			
HFETR	125	90	75
MJTR	5	90	3
HFETR Critical	0	90	0
Zero Power Fast Critical	0	90	0
MNSR-IAE (Beijing)[c]	0.027	90	0
MNSR-SZ (Shenzhen University)	0.027	90	0
MNSR-SD (Shan Dong)	0.027	90	0
Subtotal	**130**		**78**
Chinese-supplied reactors			
MNSR (Pakistan)	0.027	90	0
MNSR (Iran)	0.027	90	0
MNSR (Ghana)	0.027	90	0
Subtotal	**0.08**		**0**
Total	**130**		**78**

[a] IAEA, *Nuclear Research Reactors in the World* (IAEA: Vienna, Dec. 1995). Reactors with power less than 1 MWth are included in this table because so little is publicly available about Chinese research reactors.

[b] Pulse-type reactors that use HEU fuel are excluded. These reactors typically require only one fuel loading during their lifetime.

[c] The Miniature Neutron Source Reactor, which has a critical mass of about 900 gm of weapon-grade uranium and needs only one core loading.

Index

For information on Russia and the Union of Soviet Socialist Republics readers should consult both headings.

Afghanistan 37
Africa, nuclear weapon-free zone 416
Agreement on the Exclusively Peaceful
  Utilization of Nuclear Energy (1991) 369
Algeria:
  Ain Oussera 363
  Es Salem 363, 364
  IAEA and 363, 364
  NPT and 351, 364
  plutonium separation 364
Al Hussein missile 311
Almeida Querido, General Nelson de 377
Alves, Rex Nazare 373, 374, 375
America *see* United States of America
americium-241 18, 43, 211, 396
Amrollahi, Reza 353, 360
Ardenne, Manfred von 95–96
Argentina:
  Atucha 1 reactor 183, 371
  CNEA 183
  IAEA and 369, 371
  Iran and 361
  NPT and 369
  nuclear programme ceased 369
  Pilcaniyeu plant 371, 372
  plutonium:
    reprocessing 183
    separation 373
  uranium enrichment 369–73
  uranium purchase 371
  see also Brazilian–Argentine Agency for
    Accounting and Control of Nuclear
    Materials
Armenia 189
Aspin, Les 296
Australia 247
Austria 335
AVLIS project, United States of America 16

Barwich, Heinz 97
Beams, Jesse 99, 329
Becker jet nozzle 16

Beg, General Mirza Aslam 271–72
Belarus:
  NPT and 50
  nuclear weapons and 115
  uranium and 115
Belgium:
  Belgonucléaire 204, 215
  Dessel plant 204, 214, 215, 219, 223, 225
  Fragema 215
  Mol plant 150, 156, 215
  MOX fuel 210, 215, 223, 224, 227
  plutonium:
    inventory 230, 232, 233, 235, 405
    reprocessing 179, 184, 215–16, 227
  Synatom 169
Benedict, Manson 98
Beriya, L. P. 98
Bhutto, Benazir 274, 279, 281
Blix, Hans 283, 284, 288, 289, 303
Bogdan, Valery 355
Brazil:
  Aramar plant 373, 374, 375
  CETEX 377
  CNEN 373
  CTA 376
  Germany and 370
  IAEA and 369, 376
  IPEN 373, 377
  Isotopic Enrichment Facility/Laboratory
    374, 375, 376
  nuclear programme ceased 369
  plutonium programme 376–77
  uranium:
    enrichment 16, 370, 373–76
    inventory 376
    see also following entry
Brazilian–Argentine Agency for Accounting
  and Control of Nuclear Materials (ABACC)
  369, 370, 371, 376
Bukharin, Oleg 111
Bulgaria 53, 175, 189
Bush, President George 52

Busse, Walter 330

calutrons 16, 318
Canada:
    CANDU reactor 21, 137
    India and 269
    plutonium, reprocessing 448
    plutonium disposition 447
    reactors 144, 145
Carnegie Commission on Preventing Deadly Conflict 436
Carter, President Jimmy 64, 118, 152, 180, 210, 289
CD (Conference on Disarmament) 431
ceramic oxide fuels 135
Chemical Weapons Convention (1993) 421–22
CHEMIX solvent extraction process 339
China:
    fissile material control 422, 423
    Guangyuan plant 32, 76
    Heping 127
    IAEA and 78
    Iran and 358, 359–60
    Jiuquan plant 32, 76
    Korea, North and 289, 304
    Lanzhou 126
    nuclear accountability in 78
    nuclear exports 359
    nuclear warheads, plutonium in 77
    nuclear weapons 77
    nuclear weapon tests 126, 129
    plutonium:
        inventory 20, 76, 400, 414
        losses 77
        production 76, 77
    reactors 32, 36, 129, 359
    South Africa and 389, 390, 391
    Taiwan and 366
    USA and 244
    uranium:
        enrichment 15, 77, 126–30
        exports 240, 276
        inventory 128–30, 253, 399, 414
        losses 129
        non-weapon uses 128–29
    USSR, nuclear assistance from 80
Chirac, President Jacques 121
Christopher, Warren 356, 361
Clinton, President Bill 93, 239, 289, 290, 353, 356, 434

Collor de Mellor, President Fernando 360, 374, 377
Comecon 153, 175
Council for Mutual Economic Assistance *see* Comecon
Cuba, nuclear weapons and 351
Czechoslovakia 153, 189
Czech Republic 244

Davis, Mary 260
de Klerk, President F. W. 369, 377, 378
deuterium 271
deuterium oxide 56
Diakov, Anatoliy 55

Egypt:
    Israel and 351–52
    NPT and 351–52
Emelyanov, V. S. 101
EU:
    MOX fuel and 449
    plutonium safeguarded by 407
    uranium disposition and 445
Euratom:
    amounts safeguarded by 407
    confidentiality and 6, 407
    MOX recycling and 449
    research reactors 251
    safeguards 406, 407, 432, 457
    uranium imports 248, 250–51
Euratom Treaty (1957) 420
Eurochemic 149, 150, 179
Eurodif 123
European Community 420

Fangataufa 122
Finland 151, 189
Fissile Material Cut-Off Treaty 38, 431–33, 441
fissile materials:
    amounts of, accuracy and 8–9
    amounts of, ignorance of 7, 8
    black markets in 6
    civil–military 418–21
    control of:
        bilateral measures 417, 427, 428–29, 456
        defects in present regime 418–23
        multilateral measures 26, 418, 426, 429, 430–37
        nuclear disarmament and 454–58
    dual nature of 418–21

electricity and 418
end of production for nuclear weapons 422
excess stocks 434–37
International Register of 436
material inventories necessary 5, 7–8
policy on 417
significant quantities 8
surpluses 417
theft of 6
transparency 6–7, 421–23
see also plutonium; uranium, highly enriched
France:
  Bugey plant 67, 68, 71, 72, 166
  Cadarache plant 73, 195, 198, 200
  CAPRA project 171
  CEA 67, 71, 122, 150, 170, 198
  Célestin plant 67, 68, 74, 124–25
  CERCA 251
  CHEMIX process 339
  Chinon plant 32, 67, 68, 70, 71–72, 165
  Cogema 67, 68, 154, 167, 169, 170, 191, 215, 384
  Côtes du Rhône plant 32
  Curien Report 172, 201
  EdF 71, 165, 168, 200, 216–17
  fast reactors 67, 73, 170, 171, 194, 199–201, 208, 209
  fissile material control 423
  Fontenay-aux-Roses plant 170
  IAEA and 406
  Iraq, nuclear reactor sold to 313
  Israel, nuclear exports to 258, 260
  La Hague plant 66, 67, 72–73, 151, 152, 156, 165, 166, 167, 168, 169, 170, 183, 187, 188, 191, 199, 215, 235, 248, 405, 409
  Le Bouchet plant 165
  Magnox reactor 21
  Marcoule plant 32, 66, 67, 68–70, 73, 74, 156, 165, 166, 168, 170, 171, 185, 199, 211, 223, 260
  Melox plant 214, 219, 223, 228–29
  MOX fuel 168, 169, 171, 195, 199, 200, 210, 216–17, 223, 224, 225, 227, 413
  nuclear accounting in 75
  nuclear weapon tests 75, 122, 125
  Pakistan and 281
  Phénix plant 67, 68, 73, 170, 171, 200, 208
  Pierrelatte plant 74, 121, 122–23
  plutonium:
    inventory 68, 69, 70, 72–75, 201, 230, 232, 233, 235, 400, 405, 414
    losses 75
    production 66–75
    reprocessing 5, 25, 73, 150–51, 165–73, 184, 185, 186, 187, 191, 192, 208, 209, 216–18, 228
    safeguarded 409, 410
    separation 66, 73
    surplus 443
  plutonium disposition 447
  reactor exported to Israel 258
  reactors 32, 36, 124, 125, 144, 145, 240
  research reactors 240
  St Laurent plant 67, 68, 70, 71, 72, 217
  SPIN project 172, 200
  Superphénix 171, 194, 195, 200–201, 208
  Tricastin plant 15
  uranium:
    enrichment 15, 80, 121–26
    exports 240
    imports 240
    inventory 126, 399, 414
    losses 125
    reprocessing 248
FSU *see* Union of Soviet Socialist Republics, former

Georgia 115
German Democratic Republic 189
Germany:
  fast reactors 203–204
  H&H Metalform 328, 330, 331
  Hanau plant 204, 211, 213, 214, 218, 221
  INTERATOM 331, 333
  Iran and 354
  Kalkar reactor 203
  Karlsruhe plant 156, 178, 203
  KNK plants 203, 204, 251
  MAN Technologie 330, 336
  MOX fuel 210, 213, 214, 223, 224, 225, 448–49
  nuclear weapons and 351
  NUKEM 242, 251
  plutonium:
    inventory 230, 232, 233, 235, 405
    reprocessing 147, 150–51, 178–79, 184, 189, 213, 227, 448
    surplus 443
  plutonium disposition 447

reactors 14
research reactor 241–42, 244, 251
ROSCH 334, 335
Siemens 98, 204, 213, 215, 218, 334
Social Democrat–Green coalition 213
spent fuel management 161–62, 169–70
uranium enrichment 15, 16
Wackersdorf plant 178
WAK plant 178–79
Gorbachev, President Mikhail 52
Great Britain *see* United Kingdom

Hersh, Seymour 257
Hiroshima 79
Hungary 189
Hussein, President Saddam 238

IAEA (International Atomic Energy Agency):
  Actinide Database 140
  amounts of material safeguarded 407, 408
  confidentiality and 6, 407
  figures of 408–409, 410
  management 420
  Programme 93+2 425–26, 457
  resources stretched 456
  safeguards 8, 406, 407, 409:
    'voluntary offer' 421, 432
  for relations with individual countries see under names of countries
India:
  Atomic Energy Agency 265
  Bhabha Atomic Research Centre (Trombay) 180, 206, 222, 265, 267, 270
  Canada and 269
  CANDU reactors 180, 181, 265, 266–67
  Cirus plant 265–66, 267, 268, 269
  CTBT and 430
  Department of Atomic Energy 267, 270
  Dhruva plant 265–66, 267, 268, 269
  fast reactors 180, 182, 206–207, 209
  IAEA and 181
  Iran and 361
  Kalpakkam plant 156, 182, 183, 222
  KAPS plant 266, 268
  Madras plant 181
  MAPS plant 266, 267, 268
  MOX fuel 182, 206, 207, 222, 223
  NAPS plant 266, 268
  nuclear explosion 267, 268
  nuclear programme 264–71
  nuclear weapons 265

  plutonium:
    inventory 182, 183, 230, 232, 268, 269, 401, 402, 405
    production 265–67
    reprocessing 5, 151, 180–83, 184, 185, 186, 191, 206–207, 221–22
    separation 265, 267–69
    surplus 237, 443
    unsafeguarded 407
  plutonium disposition 447
  PREFRE facility 180, 181, 182, 267, 268
  Purnima plant 206, 268
  Rajasthan plant 181
  reactors 31, 265, 268
  research reactors 180, 271
  RMP plant 270
  submarine 271
  Tarapur plant 156, 180, 181, 182, 185, 222
  uranium enrichment 265, 269–71, 403
INF Treaty (1987) 413–14
International Plutonium Management group 233
International Register of Fissile Materials 436
inventory differences 45–47
Iran:
  Atomic Energy Organization of Iran 352, 360
  China and 358, 359–60
  EMIS machines 359–60, 361, 362
  fissile material:
    attempts to obtain 359, 360–63
    need for 358
  Germany and 354
  IAEA and 356–57, 362
  India and 361
  Kazakhstan and 352
  NPT and 351, 355, 424
  nuclear weapons and 352–63
  oil 355
  plutonium:
    acquiring 355
    making 360, 361
  research facilities, attempts to buy 358
  Russia and 353–56, 357, 361
  Tehran Research Centre 359
  Ulba Metallurgical Plant and 352
  uranium enrichment 360, 361–63
  USA and 352, 353, 354, 356, 357, 358–59, 360
Iraq:
  Al Furat 333, 334, 335, 336

Al Jesira 319
Allied bombing campaign 311, 317, 319, 344, 345
Al Taji 335

Al Tarmiya 314, 315, 318, 319, 321, 322, 323, 324, 325, 326, 327, 341, 345, 348, 349
Al Tuwaitha Nuclear Research Centre 315, 317, 319, 320, 321, 326, 327, 339, 343, 345, 348, 349
Ash Sharqat 315, 321, 323–24, 326
carbon-fibre manufacturing 334–335
crash programme 344–49
economy 312
Engineering Design Centre 315
fuel diversion by 238
heavy water plant 343
IAEA and 309, 310, 311, 317, 319, 320, 322, 338, 340, 342, 344, 350
Kuwait, invasion of 309
MIMI 309
missiles 311
nuclear know-how 311
nuclear testing site 311
nuclear weapon programme 311–13
oil sales 312
plutonium, separation 342
post-war activities 349–50
Project 521C 347–48
Project 601 345–46, 348
Project 602 346, 348
Rashdiya 315, 327, 328, 332, 334, 335, 336, 347, 349–50
reactors 313, 342–44
spent fuel sent to Russia 247
UN resolutions on 309, 349
UN sanctions on 312, 313, 333, 334, 347
UNSCOM 309, 350
uranium acquisitions 324, 344
uranium diversion 344
uranium enrichment:
    chemical enrichment 315, 339–41, 342
    EMIS 314–15, 317, 318, 338, 339, 348
    gas-centrifuge 315–16, 328–39, 349, 350
    gaseous diffusion 315, 317, 327–28
    laser 317–18, 340, 342
weapon capabilities destroyed 309
weapon-grade inventory 341–42
isotopes 12

Israel:
  Dimona plant 257–63, 402
  Iraq's reactor bombed by 313
  Mochon 2 260, 263
  Negev Nuclear Research Centre 257
  nuclear weapons 257, 263
  plutonium:
    inventory 261–62, 401, 402
    production 257–63
    separation 258, 260–61
    surplus 443
    unsafeguarded 407
  reactors 31, 257–58
  uranium enrichment 264, 403
  yellowcake imports 259
Italy:
  ENEL 200
  Latina plant 158, 188
  plutonium:
    inventory 230, 232, 405
    reprocessing 151, 188, 189, 409
  SNIA-Techint 342
  uranium supplied to Iraq 214
Iyengar, P. K. 265, 270

Jackson, Anthony 380, 381, 389
Jaffar, Jaffar D. 314, 323, 325, 348
Japan:
  ASAHI technique 339
  Atomic Energy Commission 219
  Atomic Energy Research Institute 202
  European reprocessing and 153, 162, 176, 188, 189, 220, 225, 409
  fast reactors 195, 201–203, 208, 209
  Fugen Advanced Thermal Reactor 202, 203
  IAEA and 406
  JFNS 177
  Kansai plant 220
  Korea, North and 289, 290, 304
  MITI 219–20
  Monju reactor 202–203
  MOX fuel 195, 201, 202, 210, 220, 223, 224, 228
  nuclear weapons and 351
  plutonium:
    inventory 230, 232, 233, 235, 405
    reprocessing 151, 176–78, 184, 191, 192, 194, 201–203, 210, 219–21, 228, 448
    surplus 443
  plutonium disposition 447
  Plutonium Fuel Production Facility 202

plutonium stocks, declaration of 232
reactors 145, 219–21
Rokkasho-mura plant 5, 156, 177, 228, 229
Tepco plant 220
Tokai-mura plant 156, 158, 176, 177, 178, 195, 202
uranium enrichment 15
Jordan 335, 347

Kamel, General Hussein 309, 310, 312, 327, 335, 336, 340, 344, 350
Kazakhstan:
   Aktau site 50, 204
   fast reactors 204–206
   MOX fuel 204
   NPT and 50
   plutonium reprocessing 204–206
   Ulba Metallurgical plant 115, 116, 352
   uranium and 115
Khan, A. Q. 274
Khan, Shahryar 271, 274
Kim Il Sung 289, 290
Kim Jong Il 290
Kissinger, Henry 273
Korea, North:
   Agreed Framework with USA 282, 290–94, 307, 402
   China and 289, 304
   IAEA and 283–90, 291, 294, 298, 305, 307
   Japan and 289, 290, 304
   NPT and 283, 287–88
   nuclear freeze 289–90, 292
   nuclear weapons 307–308
   plutonium:
      inventory 306–307, 402, 403
      production 295, 296–99, 300–303
      reprocessing 306
      separation 283, 303–306
   Radiochemical Laboratory 283, 284, 286, 287, 290, 297, 304, 305
   reactor defuelling 288
   reactors 31, 282–83, 288, 290, 291, 292, 293, 294, 295–303
   Russia and 283, 304
   spent fuel storage 294
   Taechon 284, 300
   uranium enrichment 307
   uranium imports 244
   USA and 282, 283, 288, 289, 290–94, 296, 300, 303, 308
   waste sites, suspect 286–87
   Yongbyon plant 283, 284, 286, 287, 290, 297, 304, 305
Korea, South:
   Daeduk facility 365, 366
   NPT and 365
   nuclear weapons and 65–66, 351, 365–66
   plutonium and 354, 402
Korean Peninsula, North–South Joint Declaration on the Denuclearization of (1991) 306
Korean Peninsula Energy Development Organization 292, 293
krypton-85 33, 34–35, 56–59

Latvia 115
Libya, nuclear weapons and 351
lithium 36, 56
Lithuania, reactors and 408–409

Mikerin, Evgeni 97, 103, 107, 108
Mikhailov, Viktor N. 51, 94
Millon, Charles 121
Mitterrand, President François 66
MOX fuels:
   committed capacity 234
   constituents 195
   costs 24, 195, 211
   fabrication plants, list of 197
   over-capacity 226–27
   plutonium, summary of use in 223–29
   production process 210–12
   reprocessing rate and 212
   stock reduction and 447–50
   see also under names of countries
Mururoa 122

neptunium-239 18
Netherlands:
   plutonium:
      inventory 230, 232, 405
      reprocessing 147, 189
   uranium enrichment 15
nitric acid 22
NPT (Non-Proliferation Treaty, 1968):
   dualism in 419–20
   NNWS parties' undeclared activities 424–27
   NWS–NNWS dichotomy 420
   Principles and Objectives 454, 456, 457, 458
   safeguards document 420
   sovereignty and 420

NPT Review and Extension Conference (1995) 431, 454
nuclear power reactors, fuels discharged from 5
Nuclear Suppliers Group 293, 424
nuclear weapons:
  design easier 423
  dismantlement of 5, 17, 25, 414, 437
  fizzling 19
  manufacture easier 423
  plutonium, amounts in 8, 413–415
  plutonium recycling from 25
  reduction talks 52
  role diminished 416

OECD (Organisation for Economic Co-operation and Development) 179, 210, 211
O'Leary, Hazel 79, 83

Pacific Rim countries 143, 146
Pahlevi, Shah Reza 354, 359
Pakistan:
  Chasma plant 281
  China and 276, 281
  France and 281
  German design plans stolen 274
  Golra plant 273
  Kahuta plant 271, 272, 273, 274–75, 276, 277, 278
  Khushab reactor 279–81
  nuclear weapons 271–72, 276
  Pinstech plant 281
  plutonium:
    inventory 402
    production 279–81
    separation 271, 281
  uranium:
    enrichment 272–79, 403
    imports 276
    inventory 276–77
  USA and 271, 272, 273, 275, 279, 281
Park Chung-hee 365
Péan, Pierre 258
Perry, William 288, 300, 357
plutonium:
  characteristics 18–25
  civil balances 233
  derivation of 4, 12, 18
  disposition of 437–44, 446–54
  enrichment 20–21
  fast reactors, use in 193–209

fissile isotopes 18
fission cross-sections 194
fissioning 13, 19
fuel, use as 193–237
grades of 19
half-life 13, 18, 20
handling 195
inventories:
  difficulties of 29–30
  estimating methods 33–37
  future trends 411–15
  new information on 3
  see also under names of countries
inventory differences, United States of America 45–47
isotopes 18, 211
lost 31
national balances 231–37
overview of 395–415
power reactors and 133–47
production 20–23
production process 30–33
production reactors 17, 30, 35–37
protection needed from 13, 18
R&D use 229–37
reactor fuel 23–25
recycling 18–19, 21, 23–25, 141, 193–94:
  economic disincentives 25
reprocessing 20, 22, 141:
  amounts 184, 190, 229, 230
  commercial aspects 152–53, 157–83
  concentration of 185–86, 419
  continuance, reasons for 442
  definition 4, 148
  fast reactors and 151, 193, 195–209
  future of 411–13
  future projections 190–92, 225–29
  international 150, 187, 220
  justification for 148
  losses during 149
  method of 149
  MOX fabrication rate and 212
  ownership and 187–89
  proliferation and 152
  radioactive releases 152
  reactor supply agreements 150
  significance 190
  summary 183–90
  thermal reactors 193–94, 209–29, 412–13
scraps 31
separation 20

separation plants 31
significant quantity 46, 435, 441
stocks, international register of 7
storage problems 25
surpluses 236, 236–37, 413, 434–44
vitrification 451–52, 453, 454
warheads, recycling from 37, 413–15
weapon grade, definition of 30
see also fissile materials; MOX fuels
Poland, uranium imports 244
Prasad, A. N. 182
Pressler, Larry 271, 272
Purex process 22, 37

Rabin, Prime MInister Yitzak 357
reactors:
  advanced 20
  breeder 14
  fast:
    core size 195
    expectations of 37, 193
    fuel cycles 194–95
    list of 196
    prospects for 24
  high-temperature 14
  Magnox 135, 136, 137, 150
  naval 14
  power 14:
    burnup 136–38
    estimating methods 138–41
    fuel cycle 134, 135–36
    fuelling strategy 136–38
    fuel types 136
    on-load refuelling 135
    plutonium produced in 133–47
    power densities 136
    reprocessing 135, 155–57
  research 14, 238
  thermal 13, 14, 21
  see also under names of countries
Reagan, President Ronald 273
Redick, John 371
Rickover, Admiral Hyman G. 87
Roh Tae-woo 365
Russia:
  Chelyabinsk 205, 223
  Dimitrovgrad 204, 205
  economic situation 438
  fast reactors 204–206, 209, 222, 223
  fissile material control 422
  Foreign Intelligence Service 279, 359
  IAEA and 93, 406
  Iran and 353–56, 357, 361
  Iraqi spent fuel sent to 247
  Korea, North and 283, 304
  MOX fuel 204, 205, 206, 448–49, 452
  MOX recycling 451–52
  nuclear exports 114, 353–54
  nuclear warhead dismantlement 414
  plutonium:
    disposition 447, 451–52
    inventory 230, 232, 233, 400, 405
    production 241, 397
    recycling 25
    reprocessing 5, 173–76, 184, 185, 186, 188–89, 204–206, 210, 222–23
    surplus 237, 443
  reactors 36, 144, 145
  reactors supplied by 238, 245, 344
  RERTR programme 244–45
  research reactor 242, 244
  spent fuel take-back 355
  uranium:
    enrichment 15, 94–116, 355
    excess 442
    exports 17, 240, 244
    form of 114–16
    inventory 94
    location 114–16
  USA buys fuel from 17, 444
  USA, cooperation with about fissile materials 417
  VVER reactors 354–55
  see also Union of Soviet Socialist Republics, former

SBK 200
Schaab, Karl Heinz 331, 332, 334, 335, 336
Scheffel, Rudolph 98, 101
Sharif, Nawaz 272
Sinev, N. M. 101–102
Singapore 335, 347
Slovakia 189
SNEAK fuel 252
sources 10–11
South Africa:
  Atomic Energy Board 379
  Atomic Energy Corporation 244, 377, 378, 379, 382, 384, 385, 387, 389, 390, 392
  China 389, 390, 391
  Gouriqua 392
  IAEA and 254, 378, 382, 384, 457
  Koeberg plant 382, 383, 386, 387, 388, 389, 391, 392

NPT and 369, 377
nuclear arsenal dismantled 377
nuclear weapons 378
plutonium separation 392
Safari reactor 244, 382, 386, 388, 390, 391
UCOR 379
uranium:
   enrichment 14, 16, 378, 379–84, 403–404
   inventory 254, 378, 384
   USA and 387
   Valindaba plant 379, 382, 384, 403
   yellowcake to Israel 259
Spain, plutonium reprocessing 189
spent fuel:
   dangers posed by 147
   discharges:
      estimates of 138–46
      geography of 144–46
   management 5
   storage 146, 153
   take-back 245–48
   see also plutonium: reprocessing
spent fuel standard 439–40, 448
Srinivasan, M. R. 271
START I Treaty 52, 414
START II Treaty 52, 90, 414, 416, 441
steel cladding 195
steel, maraging 332, 334, 336, 337
Steenbeck, Max 98, 99, 101–102
Stemmler, Bruno 330, 331, 336
Stumpf, Waldo 380–81, 382, 384, 389
Suh Sujong 365
Sweden 147, 351
Switzerland:
   ALWO 334, 335
   MOX fuel 210, 218, 224
   nuclear weapons and 351
   plutonium:
      inventory 230, 232, 405
      reprocessing 189, 218–19, 227
   USA and 218

Taiwan:
   China and 366
   NPT and 366
   nuclear weapons and 351, 366–68
   plutonium and 367
   TTR 366
   USA and 366
terrorism 6, 238, 423

thermonuclear weapons 13
Thiessen, Peter Adolf 96
Threshold Test Ban Treaty (1974) 89
Tlatelolco Treaty (1967) 351, 369
Transparency and Irreversibility initiative 435–36
tributyl phosphate 22
Trident submarines 60, 62
TRIGA fuel 246, 251
tritium:
   France 31, 67
   India 271
   Israel 263
   South Africa 392
   UK 33, 43, 60, 118
   USA 31, 33, 36, 37, 43, 84, 85
   USSR 31, 54, 111–12

Ukraine:
   NPT and 50
   nuclear weapons and 15
   plutonium, reprocessing 189
   reactors 175, 408–409
   uranium and 115
Union of Soviet Socialist Republics, former:
   Angarsk 97
   Chelyabinsk 32, 53, 54, 56, 58, 151, 153, 154, 156, 173, 174, 175, 185, 187, 188, 189, 205
   Chernobyl plant 38
   Dodonovo 32, 53
   fast reactors 152, 173, 175
   fissile material, information on 50–51
   fissile material controls 427–29
   GAN 427
   Germans aid nuclear programme 95–96, 98–99, 101
   Iraqi spent fuel and 247
   Kirov plant 99
   Krasnoyarsk 53, 54, 56, 97, 156, 173, 174, 175, 176, 191
   Kurchatov Institute 32
   Kyshtym 32, 52
   MAPI 150, 151
   Minatom 153, 154, 189, 427
   nuclear accidents 52, 56
   nuclear aid to China 80
   nuclear exports 111, 151–52
   nuclear smuggling from 3
   nuclear weapons:
      dismantlement 427
      numbers of 51–52

plutonium:
  estimates uncertain 58
  inventory 58, 59, 400, 414
  losses 56
  production 33, 34, 36, 52–59
  recycling 17
  reprocessing 56, 152, 173–76, 192
  separation 56
RBMK reactors 17, 152, 173
reactors 31, 32, 52, 58, 112 *see also under names of*
reactors supplied by 244–45
spent fuel management 153
Sukhumi 95, 98, 99, 100
Tomsk 32, 53, 54, 56, 58, 97
uranium:
  enrichment 80, 94–116
  inventory 113–14, 399, 414
  non-weapon uses 111
  reprocessing 247, 253
  sale to USA 428
  secrecy about 94
Verkh Neyvinsk plant 96, 102, 103
VVER reactor 53, 151, 173, 174, 175, 176
warheads, plutonium in 51–52
see also Armenia; Kazakhstan; Russia; Ukraine; Uzbekistan
United Kingdom:
  AEA Technology 160, 164
  Atomic Energy Authority 62–63, 150, 160, 198, 221
  Berkeley plant 157
  BNFL 148, 154, 163, 191, 221, 228
  British Energy 160
  Butex plant 157
  Calder Hall plant 43, 59, 60–61, 62, 63, 64, 65–66, 157
  Capenhurst 116, 117–18
  Chapelcross plant 43, 60–61, 63, 64, 65–66, 157
  Chevaline programme 60, 61
  Dounreay reactor 120, 156, 160, 163–64, 198, 247
  Euratom and 66
  fast reactors 120, 194, 195, 198–99
  fissile material control 423
  Harwell plant 164
  IAEA and 406
  ITSC 333
  Magnox Electric 159
  Magnox reactors 21, 31, 60, 62, 63, 133, 145, 157–60

MOX fuel 198–99, 221, 223, 225, 227
non-weapon needs 120–21
nuclear accounting 66
Nuclear Electric 163
nuclear exports 61, 63–65, 116, 119
nuclear imports 116, 118–19, 120
nuclear weapons, numbers 121
nuclear weapon tests 121
plutonium:
  civil–military distinction 63
  excess 66, 442, 443
  inventory 61, 62, 63, 65, 230, 232, 232–33, 233, 400, 405, 414
  production 33, 59–63, 159
  reprocessing 5, 25, 150, 157–65, 184, 185, 186, 187, 191, 192, 198–99, 220
  safeguarded 409, 410
  separation 63, 65
  storage 221
  surpluses 237
plutonium disposition 447, 450
privatization 159
reactors 32, 36, 60–61, 144, 145 *see also under names of*
research reactors 160, 164
Scottish Nuclear 163
Sellafield plant 32, 59, 63, 149, 151, 152–53, 156, 157, 159, 160, 161, 163, 179, 183, 187, 188, 198–99, 211, 219, 221, 235, 409
Sizewell B 221
THORP plant 160, 161, 163, 185, 187, 191, 221, 409–10
uranium:
  enrichment 15, 80, 116, 117–21
  inventory 117, 118, 120–21, 399, 414
  losses 121
  reprocessing 247
USA, nuclear trade with 33, 42, 61, 63–65
Windscale *see* Sellafield plant
United Nations:
  Fissile Material Cut-Off Treaty and 431
  non-proliferation activities 424
United States of America:
  AEC 150
  Argonne National Laboratory 44, 45, 46, 48
  AVLIS project 16
  Barnwell plant 179, 180
  Brookhaven National Laboratory 82, 92
  China and 244
  CIA:
    China and 126

INDEX   501

France and 70, 71
India and 265, 270
Iran and 357
Iraq and 312
Korea, North and 282, 307, 308
Pakistan and 271, 272
Taiwan and 366
Cooperative Threat Reduction programme 427
Defence Intelligence Agency 127
DOE, spent fuel and 246–47
Dresden plant 10
Enrichment Corporation 444, 445
Fast Flux Test Facility 451
fast reactors 207–208
fissile material control 423
Fissile Material Cut-Off Treaty and 38, 431
fissile materials cut-off initiative 38, 93
Hanford plant 32, 36, 37, 38, 41, 44, 45, 46, 47, 48, 54–55, 82, 92, 180
IAEA and 93, 406
Idaho National Laboratory 17, 44, 45, 46, 47, 48, 82, 85, 86, 89, 92, 207, 245
Iran and 352, 353, 354, 357, 358–59, 360
Kerr-McGee Corporation 207
Knolls Atomic Power Laboratory 82, 89
Korea, North and 282, 283, 288, 289, 290–94, 296, 300, 303, 308
Korea, South and 365
Lawrence Livermore National Laboratory 44, 45, 46, 48, 82
Los Alamos National Laboratory 19, 44, 45, 46, 47, 48, 82, 92, 207
Manhattan Project 79, 81, 99
Morris plant 179, 180
National Academy of Sciences 439, 440, 451, 452
Natural Resources Defense Council 85, 88
Nevada test site 19, 48
nuclear exports 43, 89, 116, 118, 238
nuclear imports 114, 115, 116, 117, 119
Nuclear Non-Proliferation Act 218
Nuclear Regulatory Commission 243, 246, 248–51
nuclear warhead dismantlement 414
nuclear weapons, dismantlement 91
nuclear weapon tests 89
Oak Ridge National Laboratory 15, 32, 48, 81, 82, 83, 92, 93, 254
Paducah plant 81
Pakistan and 271, 272, 273, 275, 279, 281
Pantex plant 44, 45, 47, 90, 91, 92

plutonium:
    acquisition of 41–43, 61, 118
    disposition 451
    excess 43, 434, 443
    information on released 29, 36, 38
    inventory 6, 40, 41–50, 230, 232, 233, 235, 400, 405, 414
    inventory differences 45–47
    production 30, 33, 36, 37–40
    recycling 25
    reprocessing 37, 151, 152, 179–80, 184
    wastes 47–49
Plutonium Credit Activity 41
plutonium disposition 447
Portsmouth plant 81, 82, 83, 92
Princeton University 83
reactor fuel supplied by 238
reactors 14, 31, 32, 81, 86–89, 144, 145
RERTR programme 243
research reactors 243–44, 245
Rocky Flats 44, 45, 46, 47, 48, 82, 91, 92, 93
Russia, cooperation with about fissile materials 417
Russian HEU bought by 17, 444
Sandia National Laboratories 82, 92
Savannah River plant 17, 32, 37, 38, 41, 44, 45, 46, 47, 48, 82, 84–85, 86, 88, 89, 92, 179, 245, 451
ships, nuclear powered 86–89
spent fuel management 162
spent fuel returns to 252
spent fuel take-back policy 246–47, 252
Switzerland and 218
Taiwan and 366
TRIGA fuel 246
UK, nuclear trade with 33, 42, 61, 63–65, 89
uranium:
    amount consumed 252–53
    blending down 94
    declassification of information on 83
    enrichment 15, 39, 79–80, 81–90
    enrichment, cessation 81
    exports 239, 248–51
    foreign reactor fuel 238
    HEU, excess 434
    HEU use discouraged by 241, 246
    inventory 83–84, 90–94, 93, 399, 414
    LEU, conversion to 243–44
    losses 90
    recycling 85, 245

reprocessing 245–47, 253
Russia, purchase from 428
surplus 93–94, 442
withdrawn from weapon use 93, 94, 253–54
warheads, plutonium in 49–50
warheads, recycling from 37
West Valley plant 156, 179, 180, 451
uranium:
bunrup 21
critical mass 14
depleted 13
enrichment:
aerodynamic 14, 16
centrifuge 14, 15
chemical 14
definition 13
electromagnetic 14, 16, 80
gaseous diffusion 14, 15, 80
laser 14, 16
plasma separation 16–17
programmes overview 79–80
separative work units 15
tails assay 14–15
techniques 14–17
fissioning 13
grades of 13
isotopes 12–14
mining 18
price 17, 23, 24, 438
weapon grade 13, 79
see also following entries and **MOX** fuels
uranium, highly enriched:
blending down 439, 446
characteristics of 12–18, 79
civil reactors using 241–42, 253
definition 13
derivation of 4
disposition of 437–44
IAEA safeguards and 238
inventories 238–54, 395–415
inventories, growth in 238
new information on 3
production of 12
recycling 17–18
reprocessing 238, 245–48
role of diminished 238, 241
significant quantity 435, 441
stocks, international register of 7
surplus stocks 434–37, 437–44
see also fissile materials
uranium, low-enriched:

conversion to 242–45, 246
chain reaction and 14
definition of 13
uranium hexafluoride gas 15, 16, 328, 335, 360, 363, 384, 385
Urenco 15, 118, 274, 331, 336, 339, 355, 358, 374
Uzbekistan 115

Vanunu, Mordecai 258, 259, 260
Vorster, John 379

Woolsey, James 357

Yeltsin, President Boris 52, 53, 175, 176, 353, 356

Zippe, Gernot 95, 97, 98, 99, 101, 102, 105